The NASA STI Program…in Profile

Since its founding, NASA has been dedicated to the advancement of aeronautics and space science. The NASA Scientific and Technical Information (STI) Program Office plays a key part in helping NASA maintain this important role.

The NASA STI Program Office is operated by Langley Research Center, the lead center for NASA's scientific and technical information. The NASA STI Program Office provides access to the NASA STI Database, the largest collection of aeronautical and space science STI in the world. The Program Office is also NASA's institutional mechanism for disseminating the results of its research and development activities. These results are published by NASA in the NASA STI Report Series, which includes the following report types:

- TECHNICAL PUBLICATION. Reports of completed research or a major significant phase of research that present the results of NASA programs and include extensive data or theoretical analysis. Includes compilations of significant scientific and technical data and information deemed to be of continuing reference value. NASA's counterpart of peer-reviewed formal professional papers but has less stringent limitations on manuscript length and extent of graphic presentations.

- TECHNICAL MEMORANDUM. Scientific and technical findings that are preliminary or of specialized interest, e.g., quick release reports, working papers, and bibliographies that contain minimal annotation. Does not contain extensive analysis.

- CONTRACTOR REPORT. Scientific and technical findings by NASA-sponsored contractors and grantees.

- CONFERENCE PUBLICATION. Collected papers from scientific and technical conferences, symposia, seminars, or other meetings sponsored or cosponsored by NASA.

- SPECIAL PUBLICATION. Scientific, technical, or historical information from NASA programs, projects, and mission, often concerned with subjects having substantial public interest.

- TECHNICAL TRANSLATION. English-language translations of foreign scientific and technical material pertinent to NASA's mission.

Specialized services that complement the STI Program Office's diverse offerings include creating custom thesauri, building customized databases, organizing and publishing research results…even providing videos.

For more information about the NASA STI Program Office, see the following:

- Access the NASA STI program home page at <http://www.sti.nasa.gov>

- E-mail your question via the Internet to <help@sti.nasa.gov>

- Fax your question to the NASA STI Help Desk at 443–757–5803

- Phone the NASA STI Help Desk at 443–757–5802

- Write to:
  NASA STI Help Desk
  NASA Center for AeroSpace Information
  7115 Standard Drive
  Hanover, MD 21076–1320

NASA/CP—2011–216469

Meteoroids: The Smallest Solar System Bodies

W.J. Cooke, Sponsor
Marshall Space Flight Center, Huntsville, Alabama

D.E. Moser and B.F. Hardin, Compilers
Dynetics Technical Services, Huntsville, Alabama

D. Janches, Compiler
Goddard Space Flight Center, Greenbelt, Maryland

Proceedings of the Meteroids Conference held in Breckenridge, Colorado, USA, May 24–28, 2010. Conference sponsored by the NASA Meteorid Environment Office, NASA Orbital Debris Program Office, National Science Foundation, Office of Naval Research, Los Alamos National Laboratory, and the NorthWest Research Associates, CORA Division,

National Aeronautics and
Space Administration

Marshall Space Flight Center • Huntsville, Alabama 35812

July 2011

Available from:

NASA Center for AeroSpace Information
7115 Standard Drive
Hanover, MD 21076–1320
443–757–5802

PREFACE

The technical report embodied in this volume is a compilation of articles reflecting the current state of knowledge on the physics, chemistry, astronomy, and aeronomy of small bodies in the Solar System. The articles reported here represent the most recent scientific results in meteor, meteoroid, and related research fields and were presented at the Meteoroids 2010 Conference. Meteoroids 2010 was the seventh conference in a series of meetings on meteoroids and related topics, which have been held approximately every 3 years since the first one celebrated in 1992 in Smolenice Castle, Slovakia. The 2010 edition was the first time the conference was held in the U.S.; the last three meetings were held in Barcelona, Spain (Meteoroids 2007), London, Ontario, Canada (University of Western Ontario, Meteoroids 2004), and Kiruna, Sweden (Swedish Institute for Space Physics, Meteoroids 2001). The 2010 meeting took place at the Beaver Run Resort in Breckenridge, CO, USA on May 24–28, 2010, surrounded by the spectacular scenery offered by the Continental Divide in the Rocky Mountains. Researchers and students representing more than 20 countries participated at this international conference where 145 presentations were delivered in oral and poster forms. Sadly, for the 2010 Conference, the meteor community lost two of their giants. Prof. Zdenek Ceplecha of the Ondrejov Astronomical Observatory passed away at age 81 in Prague on December 4, 2009. And, shockingly, only a few weeks before the meeting on May 2, Dr. Douglas ReVelle of Los Alamos National Laboratory passed away in Los Alamos, New Mexico at age 65. Two special lectures were given remembering the unique scientific and personal contributions that Zednek and Doug gave throughout the years and the legacy they have left behind.

The conference gave a comprehensive overview on meteoroid and meteor science organized in several broad themes. The first themes to be covered were related to the astronomical aspects of the field. The scientific sessions during the first 2 days discussed the relation of comets and meteor showers—in particular, their activity and forecasting. Other topics addressed were the case of the Geminids Shower as a prime example of asteroids as meteor shower parents and asteroids as a source of meteorites and the need for awareness and alert programs for large body impacts. An always present and exciting topic is the study of the Sporadic Meteor Complex (SMC). New results were presented addressing the nature and characteristics of the SMC sources and their relation to comet and asteroid populations as well as the origin of interstellar meteoroids. Special attention was given to satellite impact hazard, both mechanical as well as electromagnetic, and due to the upcoming Hayabusa sample return capsule, a session was dedicated to artificial meteors. Almost 2 days were focused on the physics and chemistry of the meteor phenomenon and their effects on Earth's atmosphere as well as other terrestrial planets. In particular, there were sessions devoted to the physical properties of meteoroids and meteorites, physical and chemical processes resulting from the meteoroid interactions with Earth's atmosphere, and the physical conditions in meteors, bolides, and impacts. The last portion of the meeting concentrated on the ever-evolving observational techniques utilized for the study of meteors, current detection programs, and the future developments and upgrades of the various detection schemes.

Technological advances in meteor and meteoroid detection, the ever-increasing sophistication of computer modeling, and the proliferation of autonomous monitoring stations continue to create new niches for exciting research in this field, allowing the compilation of long-term databases which provide a much needed statistical view of the nature and effects of these small Solar System bodies. This progress is fundamental in providing the insight required to understand their origins and distributions and accurately assess their impact on human life.

In particular, the choice of members for the scientific organizing committee (listed below) was key for the success of the conference. Their broad expertise and vision is reflected in the meeting agenda, which successfully covers long-term research directions and objectives while also exploiting opportunities and testing new directions and interactions. This was also reflected by the large presence of student presentations showing that new generations of scientists are continuously joining this area of research. These goals were achieved by judicious choices of invited, regular and poster presentations and are reflected in the compilation of articles presented in this book. The meeting also included an invited public lecture by Prof. Iwan Williams from Queen Mary College, celebrating his 70th birthday and more than 40 years of service to the community. The lecture was entitled, "The Origin and Evolution of Meteor Showers and Meteoroid Streams" now published in Astronomy and Geophysics (April 2011, Vol. 52, pages 2.2–2.26). We would like to take this opportunity to acknowledge and thank the members of the local organizing committee (LOC, listed also below). Their dedicated work as well as the support received from the staff of the Beaver Run Resort resulted in a flawless meeting. We look forward to the next Meteoroids conference, which will be held in the Poznan, Poland in 2013 and wish the best of luck to their organizers.

Finally, we would like to acknowledge the sponsors for this conference, including the NASA Meteoroid Environment Office (MEO), the NASA Orbital Debris Program Office, the Office of Naval Research (ONR), Los Alamos National Laboratory (LANL), the National Science Foundation (NSF), and NorthWest Research Associates. Their financial contributions made it possible to have a successful and exciting scientific meeting.

Sincerely,

Diego Janches
William J. Cooke
Danielle Moser

Scientific Organizing Committee

Dr. Diego Janches, Chair (NorthWest Research Associates, now at NASA Goddard Space Flight Center, MD, USA)
Dr. William Cooke (NASA Marshall Space Flight Center, AL, USA)
Prof. Peter Brown (University of Western Ontario, Canada)
Dr. Pavel Spurny (Ondrejov Observatory, Czech Republic)
Prof. Iwan Williams (Queen Mary College, U.K.)
Prof. Jun-Ichi Watanabe (National Astronomical Observatory of Japan, Japan)
Dr. Lars Dyrud (Applied Research Lab, John Hopkins University, MD, USA)
Prof. John Plane (University of Leeds, U.K.)
Dr. Sigrid Close (Los Alamos National Lab, NM, USA, now at Stanford University)
Dr. Olga Popova (Institute for Dynamics of Geospheres, Moscow, Russia)
Dr. Josep M. Trigo-Rodríguez (Institute of Space Sciences, CSIC-IEEC, Barcelona, Spain)
Prof. Frans Rietmeijer (University of New Mexico, NM, USA)
Dr. Douglas ReVelle (Los Alamos National Laboratory, NM, USA)
Dr. William Bottke (South West Research Institute, Boulder, CO, USA)
Dr. Peter Jenniskens (SETI Institute, CA, USA)

Local Organizing Committee

Dr. Diego Janches (Chair)
Dr. Jonathan Fentzke (NWRA)
Janet Biggs (NWRA)
Andrew Frahm (NWRA)

TABLE OF CONTENTS

TABLE OF CONTENTS (Continued)

TABLE OF CONTENTS (Continued)

TABLE OF CONTENTS (Continued)

CONFERENCE PUBLICATION

METEOROIDS: THE SMALLEST SOLAR SYSTEM BODIES

CHAPTER 1:

COMETS AND METEOR SHOWERS: ACTIVITY AND FORECASTING

Dynamical Evolution of Meteoroid Streams, Developments Over the Last 30 Years

I. P. Williams

Abstract As soon as reliable methods for observationally determining the heliocentric orbits of meteoroids and hence the mean orbit of a meteoroid stream in the 1950s and 60s, astronomers strived to investigate the evolution of the orbit under the effects of gravitational perturbations from the planets. At first, the limitations in the capabilities of computers, both in terms of speed and memory, placed severe restrictions on what was possible to do. As a consequence, secular perturbation methods, where the perturbations are averaged over one orbit became the norm. The most popular of these is the Halphen-Goryachev method which was used extensively until the early 1980s. The main disadvantage of these methods lies in the fact that close encounter can be missed, however they remain useful for performing very long-term integrations.

Direct integration methods determine the effects of the perturbing forces at many points on an orbit. This give a better picture of the orbital evolution of an individual meteoroid, but many meteoroids have to be integrated in order to obtain a realistic picture of the evolution of a meteoroid stream. The notion of generating a family of hypothetical meteoroids to represent a stream and directly integrate the motion of each was probably first used by Williams Murray & Hughes (1979), to investigate the Quadrantids. Because of computing limitations, only 10 test meteoroids were used. Only two years later, Hughes et. al. (1981) had increased the number of particles 20-fold to 200 while after a further year, Fox Williams and Hughes used 500 000 test meteoroids to model the Geminid stream. With such a number of meteoroids it was possible for the first time to produce a realistic cross-section of the stream on the ecliptic.

From that point on there has been a continued increase in the number of meteoroids, the length of time over which integration is carried out and the frequency with which results can be plotted so that it is now possible to produce moving images of the stream. As a consequence, over recent years, emphasis has moved to considering stream formation and the role fragmentation plays in this.

Keywords meteors · numerical integration · modeling

1 Introduction

Understanding the basic physics involved in meteoroid stream evolution is relatively easy. First, some model for the ejection of material from the parent body, that is time (location), speed and direction is needed. From this the initial orbit of each meteoroid can be calculated. Some means of calculating the effects of gravity from the Sun and Planets on the orbits of these meteoroids is then required which should also incorporate the effects of Solar Radiation (Pressure and the Poynting-Robertson effect). Hence the orbit of each meteoroid can be calculated at any desired time after the initial formation. Finally if the meteoroid position coincides with that of the Earth, there is a need to understand the

I. P. Williams (✉)
Queen Mary University of London, Mile End Rd, E1 4NS, UK. E-mail: i.p.williams@qmul.ac.uk

interaction between the meteoroids and the atmosphere so that the observed meteor shower can be tied in with the meteoroid stream.

Walker (1843) drew attention to the similarity, in terms of eccentricity, between meteor and comet orbits, but it was left to Kirkwood (1861) to propose that shower meteors were debris of ancient comets. At that time, the standard model for comets was essentially the flying sandbank model, so that initially the velocity of the meteoroids were essentially the same as that of the comet, there was no need for an ejection model. LeVerrier (1867) correctly pointed out that, given sufficient time, planetary perturbations would spread the meteoroids all around the orbits. Newton (1864 a, b) showed that the node of the Leonid orbits advanced relative to a fixed point in space at 52.4 arc seconds per year and Adams (1867) showed that a 33.25 year period was the only period that was consistent with the observed nodal advancement. Thus, early workers were incorporating the principles laid down above into their thoughts but computers were human assistants rather than machines and of necessity rather slow.

2 New Techniques and Thoughts

Nagaoka (1929) had suggested that meteors could affect the propagation of radio waves, a suggestion also made by Skellet (1931, 1932), but little was done. Hey realized that radar could be used as a tool to investigate meteors and at the end of the war ensured that military radar equipment became available for civil use allowing astronomers to start meteor work. There was a strong storm of Draconid meteors in 1946. This resulted in several papers being published on radar observations of the Draconids (Clegg et. al. 1947, Hey et. al. 1947, Lovell et. al. 1947). Radar can detect smaller meteoroids (down to sub-millimetre size) and so detected many more meteors. Radar also had the advantage of working in the day as well as by night, thus doubling the coverage and discovering many new streams (Ellyett 1949) and orbits of thousands of meteors were obtained.

Whipple (1950) proposed a new model for a comet, replacing the flying sandbank model. According to this model, a comet had an icy nucleus with dust grains embedded within it, the dirty snowball model. As a comet approaches the sun, solar heating causes the ices to sublimate and the resulting gas outflow carries away small dust grains with it, the larger ones becoming meteoroids and the very small ones forming the dust tail. Whipple, (1951) modelled this and produced an expression for the ejection velocity, V of the meteoroids relative to the cometary nucleus at a heliocentric distance r as

$$V^2 = 4.3 \times 10^5 R_c \left(\frac{1}{b\sigma r^{2.25}} - 0.013 R_c\right)$$

where σ is the bulk density of the meteoroid and r the heliocentric distance in astronomical units. R_c is the nucleus radius in kilometers and all other quantities are in cgs units. Others (e.g. Gustafson 1989, Crifo 1995, Ma et al, 2002), have modified this model, but the general result is the same, namely that the outflow speed of the meteoroids is much less than the orbital speed of the comet. Thus there is little change in the specific energy and momentum of these meteoroids and so they move on similar orbits to that of the comet, in other words, they form a stream. If the ejection velocity is known relative to the nucleus, then the heliocentric velocity can be calculated and from this, the initial orbit. The mathematics involved in this and the relevant equations are given in detail in Williams (2002).

Initially, computing capabilities were too limited to allow direct integration of a significant set of meteoroids and so secular perturbations were commonly used, generally based on an algorithm by

Brouwer (1947) that could be applied to orbits with high eccentricity, all previous methods relied on using a series expansion that was valid only for low values of e. This mathematical development allowed Whipple & Hamid (1950) to follow the evolution of the mean Taurid stream over an interval of 4700 years. Secular perturbation methods were the prime method of investigation and became quite sophisticated, the most popular being the Halphen-Goryachev method described in Hagihara (1972). This was used by Galibina & Terentjeva (1980) to determine the effect of gravitational perturbations on the stability of a number of meteoroid streams over a time interval of tens of thousands of years. Babadzhanov & Obrubov (1980, 1983) also used the Halphen-Goryachev method to investigate the evolution of both the Geminid and the Quadrantid streams. The major draw-back of any secular perturbation method is that it deals with the evolution of orbits rather than determining the position of individual meteoroids (that is, no account is taken of true anomaly). Hence, the method may show that the orbits of meteoroids intersect the Earth's orbit, but unless meteoroids are present at that location at that time, no meteors will be seen. This consideration is particularly important for showers like the Leonids as was discussed by Wu & Williams (1996), Asher et. al. (1999).

3 Direct Integration Methods

Direct integration methods integrate the path of each individual meteoroid and this was done by Hamid & Youssef (1963) for the six meteoroids then known to belong to the Quadrantid stream. The difficulty is that as there are at least 10^{16} meteoroids in a typical stream so that the six observed meteors are almost certainly not a representative sample of the whole stream. However, a smaller sample has to be taken to represent the stream, in reality a set of test particles have to be generated to represent the stream. This was done 30 years ago by Williams et. al. (1979), who represented the Quadrantid stream by 10 test particles, spread in uniformly in true anomaly around the orbit and integrated over an interval of 200 years using the self adjusting step-length Runge-Kutta 4th order method.

Four years later, Fox et al. (1983) were using 500 000 meteoroids and were able to produce a theoretical cross section on the ecliptic for the Geminid stream which gives vital information about the properties of the resulting shower. Jones (1985) used similar methods to produce a stream cross section. In four years computer technology had advanced from allowing only a handful of meteoroids to be integrated to the situation where numbers to be used did not present a problem.

By the mid eighties, complex dynamical evolution was being investigated, Froeschlé and Scholl (1986), Wu & Williams (1992) were showing that the Quadrantid stream, experiencing close encounters with Jupiter, was behaving chaotically. A new peak in the activity profile of the Perseids also caused interest with models being generated by Wu & Williams (1993) for example. Williams & Wu (1994) were able to show how the cross-section of the Perseid shower should vary from year to year. Babadzhanov et al. (1991) looked at the possibility that the break-up of comet 3D/Biela was caused when it passed through the most heavily populated part of the Leonid stream.

By now calculating from models the likely cross-section at any given time has become routine (Jenniskens & Vaubaillon 2008, 2010).

4 A Problem Emerges

The Quadrantid shower is a prolific and regular shower seen at Northern latitudes around the beginning of January. It is arguably the only major meteor shower that does not have a body that is generally accepted as being its parent. Part of the problem of identifying the parent undoubtedly lies in the fact that orbits in this region of the Solar System evolve very rapidly so that claims can be made based on a similarity of orbits at some epoch in the past. Equally, a similarity of orbits at the current time alone is not a proof of parenthood. The history of the Quadrantid meteoroid stream, including a discussion of most of the suggested parent bodies can be found in Williams & Collander-Brown (1998).

One of the suggestions for the parent of the Quadrantids is comet C/1490 Y1 (Hasegawa, 1979), the claim being based on orbital similarity around 1490 AD. In the Quadrantid shower there is both a strong narrow peak and a broad background showing the existence of both an old stream and a new one (Jenniskens et. al. 1997). There is an asteroid, 2003 EH1 with an orbit that is currently almost identical to the mean orbit of the Quadrantids and it has been argued that this asteroid may be a surviving remnant of the comet of 1491, following its catastrophic break-up (Jenniskens 2004, Williams et. al. 2004). We now know that comet break-up is fairly common and so one might expect meteor streams with such an origin to be also common. The Taurid complex is also generally considered to consist of comet 2P/Encke, a significant number of asteroids and of course the Taurid meteor streams, suggesting a past fragmentation (Babadzhanov et. al. 2008, Napier 2010).

5 Conclusions

In the last 30 years, the field appears to have gone full circle. In the beginning it was generally agreed that we knew how meteor streams formed, but were struggling to follow the effects of perturbations on the orbits. Now we are confident that we can follow the evolution of any given set of orbits but are struggling to model the stream formation process when partial or total disintegration takes place.

References

Adams J.C. On the orbit of the November meteors, *MNRAS*, 27:247-252, 1867

Asher D.J. Bailey M. E. Emel'Yanenko V.V. Resonant meteors from comet Tempel-Tuttle in 1333: the cause of the unexpected Leonid outburst in 1998, *MNRAS*, 304:L53-57, 1999

Babadzhanov P.B. Obrubov Y.Y. Evolution of orbits and intersection conditions with the Earth of Geminid and Quadrantid meteor streams, in *Solid particles in the Solar System*, Eds Halliday I. McIntosh B.A., D.Reidel, Dordrecht, 157-162, 1980

Babadzhanov P.B. Obrubov Y.Y. Some features of evolution of meteor streams, in *Highlights in Astronomy*, Ed West R.M., D. Reidel Dordrecht, 411-419, 1983

Babadzhanov P.B. Williams I.P. Kokhirova G. I. Near-Earth Objects in the Taurid complex, *MNRAS*, 386:1436-1442 2008

Babadzhanov P.B. Wu Z. Williams I.P. Hughes D.W. The Leonids, Comet Biela and Biela's associated Meteoroid Stream, *MNRAS* 253:69-74, 1991

Brouwer D. Secular variations of the elements of Enckes comet, *AJ*, 52:190-198, 1947

Clegg J.A. Hughes V.A. Lovell A.C.B. The Daylight Meteor Streams of 1947 May-August, *MNRAS*, 107:369-378, 1947

Crifo J.F. A general physiochemical model of the inner coma of active comets I. Implications of spatially distributed gas and dust production, *Ap.J* 445:470-488, 1995

Ellyett C.D. The daytime meteor streams of 1949: measurement of velocities, *MNRAS*, 109:359-364, 1949

Froeschlé C. Scholl H. Gravitational splitting of Quadrantid-like meteor streams in resonance with Jupiter, *A&A*, 158:259-265, 1986

Fox K. Williams I.P. Hughes D.W. The rate profile of the Geminid meteor stream, *MNRAS*, 205:1155-1169, 1983

Galibina I. V. Terentjeva A. K. Evolution of meteors over milenia in *Solid particles in the Solar System*, Eds Halliday I. McIntosh B.A., D.Reidel, Dordrecht, 145-148, 1980

Gustafson B.A.S. Comet ejection and dynamics of non-spherical dust particles and meteoroids, *Ap.J*, 337:945-949, 1989

Hagihara Y. *Celestial Mechanics*, MIT Cambridge Mass 1972

Hamid S. E. Youssef M. N. A short note on the origin and age of the Quadrantids *Smithson. Contr. Astrophys.* 7:309-311, 1963

Hasegawa I. Historical records of meteor showers, in *Meteors and their parent bodies*, Eds Stohl J. Williams I.P., Astronomical Institute, Slovak Academy of Sciences, Bratislava, 209-223, 1983

Hey J.S. Parsons S.J. Stewart G.S. Radio Observations of the Giacobinid Meteor shower, *MNRAS*, 197:176-183, 1947

Hughes D.W. Williams I.P. Fox K. The mass segregation and nodal retrogression of the Quadrantid meteor stream, *MNRAS*, 195:625-637, 1981

Jenniskens P. 2003 EH1 Is the Quadrantid Shower Parent Comet, *AJ*, 127:3018-3022, 2004

Jenniskens P. Vaubailllon J. Minor Planet 2008 ED69 and the Kappa Cygnid Meteor Shower, *AJ*, 136:725-730, 2008

Jenniskens P. Vaubaillion J Minor Planet 2002EX12 (169P/Neat) and the Alpha Capricornid shower, *AJ*, 139:1822-1830, 2010

Jenniskens P. Betlen H. De linge M. Langbroek M. Van Vliet M. Meteor stream activity V. The Quadrantids, a very young stream, *A&A*, 327:1242-1252, 1997

Jones J The structure of the Geminid Meteor Stream: I the effect of planetary perturbations, *MNRAS*, 217:523-532, 1985

Kirkwood D. Cometary astronomy, *Danville Quarterly Review*, 1:614-618, 1861

LeVerrier U.J.J. Sur les etoiles filantes de 13 Novembre et du 10 Aut, *Comptes rendus*, 64:94-99, 1867

Lovell A.C.B.Banwell C.J. Clegg J.A. Radio Echo observations of the Giacobinid Meteors, *MNRAS*, 107:164-175, 1947

Ma Y. Williams I.P. Chen W. On the ejection velocity of meteoroids from comets, *MNRAS*, 337:1081-1086, 2002,

Nagaoka, H. Possibility of the radio Transmission being disturbed by Meteoric showers, *Proc. Imp. Acad. Tokyo*, 5:233-236, 1929

Napier W.M. Palaeolithic extinctions and the Taurid Complex, *MNRAS*, 405:1901-1906, 2010

Newton H.A. The original accounts of the displays in former times of the November star-shower, together with a determination of the length of its cycle, its annual period, and the probable orbit of the group of bodies around the Sun, *American Jl of Science and Arts series 2*, 37:377-389, 1864a

Newton H.A. The original accounts of the displays in former times of the November star-shower, together with a determination of the length of its cycle, its annual period, and the probable orbit of the group of bodies around the Sun, *American Jl of Science and Arts series 2*, 38:53-61, 1864b

Skellett A.M. The effect of Meteors on Radio transmission through the Kennelly-Heavyside Layer, *Phys. Rev.*, 37:1668, 1931

Skellett A.M. The ionizing effect of Meteors in relation to Radio Propagation, *Proc. Inst. Radio Eng.*, 20:1933-1941, 1932

Williams I.P. Wu Z. The Quadrantid meteoroid stream and comet 1491 I, *MNRAS*, 264:659-664, 1993

Walker S. E. Rearches concerning the Periodic Meteors of August and November, *Trans. American Phil. Soc.*, 8:87-140, 1843

Whipple F.L. A comet model I; The acceleration of comet Encke, *Ap.J*, 111:375-394, 1950

Whipple F.L. A comet model II: Physical relations for comets and meteors, *Ap.J*, 113:464-474, 1951

Whipple F.L. Hamid S.E. On the origin of the Taurid meteors, *AJ*, 55:185-186, 1950

Williams I.P., 2002, The evolution of meteoroid streams, in *Meteors in the Earth's Atmosphere*, Eds Murad E Williams I.P., CUP Cambridge, 13-32

Williams, I. P. Collander-Brown, S. J. The parent of the Quadrantid meteoroid stream, *MNRAS*, 294:127-138, 1998

Williams I.P. Murray C.D. Hughes D.W. The long-term orbital evolution of the Quadrantid stream, *MNRAS*, 189:483-492, 1979

Williams I.P. Ryabova G.O. Baturin A.P. Chernitsov A.M. The parent of the Quadrantid meteoroid stream and asteroid 2003 EH1, *MNRAS*, 355:1171-1181, 2004

Williams I.P. Wu Z. The Quadrantid meteor stream and comet 1491 I, *MNRAS*, 264:659-664, 1993

Williams I. P. Wu Z. The current Perseid meteor shower, *MNRAS*, 269:524-528, 1994

Wu Z. Williams I. P. On the Quadrantid meteoroid stream complex, *MNRAS,* 259:617-628

Wu Z. Williams I. P. The Perseid meteor shower at the current time, *MNRAS,* 264:980-990, 1993

Wu Z. Williams I. P. Leonid meteor storms, *MNRAS*, 280:1210-1218, 1996

The Working Group on Meteor Showers Nomenclature: a History, Current Status and a Call for Contributions

T. J. Jopek • P. M. Jenniskens

Abstract During the IAU General Assembly in Rio de Janeiro in 2009, the members of Commission 22 established the Working Group on Meteor Shower Nomenclature, from what was formerly the Task Group on Meteor Shower Nomenclature. The Task Group had completed its mission to propose a first list of established meteor showers that could receive officially names. At the business meeting of Commission 22 the list of 64 established showers was approved and consequently officially accepted by the IAU.

A two-step process is adopted for showers to receive an official name from the IAU: i) before publication, all new showers discussed in the literature are first added to the Working List of Meteor Showers, thereby receiving a unique name, IAU number and three-letter code; ii) all showers which come up to the verification criterion are selected for inclusion in the List of Established Meteor Showers, before being officially named at the next IAU General Assembly. Both lists are accessible on the Web at www.astro.amu.edu.pl/~jopek/MDC2007.

Keywords meteor shower · meteoroid stream · methods: nomenclature

1 Introduction

The naming conventions for celestial objects, and the method of announcement of their discovery, has been the prerogative of the International Astronomical Union (IAU) since years. At its inaugural meeting in Rome in 1922, the IAU standardized the constellation names and abbreviations. More recently the IAU Committee on Small Body Nomenclature has certified the names of asteroids and comets, e.g. see Kilmartin (2003), Ticha et al. (2010) or enter the website *www.ss.astro.umd.edu/IAU/csbn/.*

Until 2009, however, the IAU has never named a meteor shower. The need to settle on official nomenclature rules was widely discussed, but the problem was not settled by the community of meteor astronomers. As a result, there was much confusion in the meteor shower literature. Some well defined showers had multiple names (Draconids, Giacobinids, ...), while many showers were given a different name in each new detection.

This situation changed during the IAU General Assembly in Prague in 2006, when Commission 22 established a *Task Group on Meteor Shower Nomenclature.* The task of this group was to formulate a descriptive list of established meteor showers that could receive official names during the next IAU General Assembly in Rio (Jenniskens 2007; Spurný et al. 2007, 2008). Task Groups are established for

T. J. Jopek (✉)
Institute Astronomical Observatory UAM, Sloneczna 36, 60-286 Poznań, Poland. E-mail: jopek@amu.edu.pl

P. M. Jenniskens
SETI Institute, 515 N. Whisman Road, Mountain View, CA 94043, USA

periods of three years, and serve until the next General Assembly. The members of the first Task Group on Meteor Shower Nomenclature were: Peter Jenniskens (chair), Vladimír Porubčan, Pavel Spurný, William J. Baggaley, Juergen Rendtl, Shinsuke Abe, Robert Hawkes and Tadeusz J. Jopek.

2 Nomenclature Rules and the Working List of Meteor Showers

To make this task possible, the traditional meteor shower nomenclature practices were formalized, and a set of nomenclature rules was adopted:

- a meteor shower should be named after the constellation of stars that contains the radiant, using the possessive Latin form of the constellation and replacing the Latin declension for "id" or "ids",
- if in doubt, the radiant position at the time of the peak of the shower (at the year of discovery) should be chosen,
- to distinguish among showers from the same constellation:
  - the shower may be named after the nearest (brightest) star with a Greek or Roman letter assigned ("η Lyrids", "*c* Andromedids"),
  - the name of the month (months) may be added (May Lyncids, September-October Lyncids),
- for the shower with a radiant elongated less than 32 degrees from the Sun, one should add "Daytime" before the shower name ("Daytime Arietids", "Daytime April Piscids"),
- by adding "South" and "North" one refers to the branches of a single meteoroid stream, both branches are active over about the same period of time. The radiants of these branches are located south and north of the ecliptic plane,
- showers that move through two constellations can be named by giving the two constellations in successive order using a "-" symbol, e.g., Librids-Luppids,
- a composed name of a shower is allowed (Northern Daytime ω Cetids),

In case of confusion, *The Task Group on Meteor Shower Nomenclature* will select among the proposed names a unique name for each shower. For further details related to all above rules see (Jenniskens 2006a, 2008).

The second part of the task – to create a descriptive list of established meteor showers – is a much more complicated issue. As a starting point a Working List of ~ 230 showers was compiled using data collected and published in the book by Jenniskens (2006b). Each shower was given a name, a unique number and a three-letter code to be used in future publications (η Aquariids, 31, ETA). The Working List, and the list of nomenclature rules, was posted on a newly established IAU Meteor Data Center website (Jopek 2007).

During the Meteoroids 2007 meeting in Barcelona, the Task Group worked out the logistics of adding new streams to the Working List, and of adding new information on streams already in the Working List:

- the institute responsible for maintaining the Working List is the IAU Meteor Data Center, managed currently by Vladimír Porubčan of SAS, Slovakia,
- already known and newly discovered streams should be reported in the literature only with a designated IAU name, number and code,

- Tadeusz J. Jopek of the UAM Astronomical Observatory, Poland, is the person currently responsible for:
  - maintaining the shower part of the IAU MDC website,
  - reporting new streams and new data on existing streams,
  - giving out new IAU numbers and codes. To obtain new numbers and codes the author should contact T.J. Jopek directly[3],
- the International Meteor Organization takes a role in coordinating the reporting of newly discovered showers. It facilitates the inclusion of showers that are recognized by amateur astronomers, for example from visual observations.

To inform the scientific community of newly discovered showers, the IAU's Central Bureau for Astronomical Telegrams (CBAT) issues an electronic telegram (CBET) with a brief summary of each new find. Those telegrams are prepared by the Task Group, as a part of the process of reporting new streams, when new showers are added to the working list. Following this CBET, all publications discussing that new shower should use the newly established name, number, and shower code.

During the 2006-2009 triennium, the Working List was updated several times (Kashcheev et al. 1967; Uehara et al. 2006; Brown et al. 2008; Molau and Kac 2009; Molau and Rendtel 2009; SonotaCo 2009; Brown et al. 2010; Jopek et al. 2010). In July of 2009, the Working List of all Meteor Showers consisted of 365 meteor showers.

3 The List of Established Meteor Shower

The Task Group met again at the May 2009 Bolides Meeting in Prague, where the Task Group settled on the list of established meteor showers. Established showers are those meteor showers that have certainly manifested. 64 meteor showers from the Working List were moved to the *List of Established Showers*. As the main grounds for this action, two factors were considered — definite shower activity (for example because of a strong meteor outburst) or confirmation from the detection of a shower in at least two recent meteor orbit surveys. The decision to move a shower into the list of established showers was to some extend subjective and border cases were decided by the democratic process of voting in the Task Group. Goal was to leave out any showers that were not certain to exist. The list was subsequently posted on the Meteor Data Center website for review.

In August of 2009, during the Commission 22 business meeting held in Rio de Janeiro, the content of the List of Established Showers was approved without changes (Watanabe et al. 2010), and this decision was confirmed by the subsequent Division III business meeting, see Bowell et al. (2010). As a result, for the first time in the history of meteor astronomy, meteor showers were officially named by the IAU. All these showers are listed in Table 1.

4 The Working Group for Meteor Shower Nomenclature

To facilitate the future update of the Working List and the List of Established Showers, Commission 22 (C22) has accepted a two step process:

[3] Email: jopek@amu.edu.pl, web:http://www.astro.amu.edu.pl/~jopek/JopekTJ/.)

Table 1. Geocentric data of 64 showers officially named during XXVII IAU General Assembly held in Rio de Janeiro in 2009. For each shower, the solar ecliptic longitude λ_S, the radiant right ascension and declination α_g, δ_g are given for J2000.0.

No	IAU No & code		Stream name	λ_S (deg)	α_g (deg)	δ_g (deg)	V_g (km/s)
1	1	CAP	α Capricornids	127	306.6	-8.2	22.2
2	2	STA	Southern Taurids	224	49.4	13	28
3	3	SIA	Southern ι Aquariids	131.7	339	-15.6	34.8
4	4	GEM	Geminids	262.1	113.2	32.5	34.6
5	5	SDA	Southern δ Aquariids	125.6	342.1	-15.4	40.5
6	6	LYR	April Lyrids	32.4	272	33.3	46.6
7	7	PER	Perseids	140.2	48.3	58	59.4
8	8	ORI	Orionids	208.6	95.4	15.9	66.2
9	9	DRA	October Draconids	195.1	264.1	57.6	20.4
10	10	QUA	Quadrantids	283.3	230	49.5	41.4
11	12	KCG	κ Cygnids	145.2	284	52.7	24
12	13	LEO	Leonids	235.1	154.2	21.6	70.7
13	15	URS	Ursids	271	219.4	75.3	33
14	16	HYD	σ Hydrids	265.5	131.9	0.2	58
15	17	NTA	North. Taurids	224	58.6	21.6	28.3
16	18	AND	Andromedids	232	24.2	32.5	17.2
17	19	MON	December Monocerotids	260.9	101.8	8.1	42
18	20	COM	December Comae Berenicids	274	175.2	22.2	63.7
19	22	LMI	Leonis Minorids	209	159.5	36.7	61.9
20	27	KSE	κ Serpentids	15.7	230.6	17.8	45
21	31	ETA	η Aquariids	46.9	336.9	-1.5	65.9
22	33	NIA	North. ι Aquariids	147.7	328	-4.7	27.6
23	61	TAH	τ Herculids	72	228.5	39.8	15
24	63	COR	Corvids	94.9	192.6	-19.4	9.1
25	102	ACE	α Centaurids	319.4	210.9	-58.2	59.3
26	110	AAN	α Antliids	313.1	140	-10	42.6
27	137	PPU	π Puppids	33.6	110.4	-45.1	15
28	144	APS	Daytime April Piscids	30.3	7.6	3.3	28.9
29	145	ELY	η Lyrids	49.1	292.5	39.7	45.3
30	152	NOC	North. Daytime ω Cetids	46.7	2.3	17.8	33
31	153	OCE	South. Daytime ω Cetids	46.7	22.5	-3.6	36.6
32	156	SMA	South. Daytime May Arietids	55	33.7	9.2	28.9
33	164	NZC	North. June Aquilids	86	298.3	-7.1	36.3
34	165	SZC	South. June Aquilids	80	297.8	-33.9	33.2
35	170	JBO	June Bootids	96.3	222.9	47.9	14.1
36	171	ARI	Daytime Arietids	76.7	40.2	23.8	35.7
37	172	ZPE	Daytime ζ Perseids	78.6	64.5	27.5	25.1
38	173	BTA	Daytime β Taurids	96.7	84.9	23.5	29
39	183	PAU	Piscis Austrinids	123.7	347.9	-23.7	44.1
40	187	PCA	ψ Cassiopeiids	106	389.4	71.5	40.3
41	188	XRI	Daytime ξ Orionids	117.7	94.5	15	44
42	191	ERI	η Eridanids	137.5	45	-12.9	64
43	198	BHY	β Hydrusids	143.8	36.3	-74.5	22.8
44	206	AUR	Aurigids	158.7	89.8	38.7	65.7
45	208	SPE	September ε Perseids	170	50.2	39.4	64.5

Table 1 (continued). Geocentric data of 64 showers officially named during XXVII IAU General Assembly held in Rio de Janeiro in 2009. For each shower, the solar ecliptic longitude λ_S, the radiant right ascension and declination α_g, δ_g are given for J2000.0.

No	IAU No & code		Stream name	λ_S (deg)	α_g (deg)	δ_g (deg)	V_g (km/s)
46	212	KLE	Daytime κ Leonids	181	162.7	15.7	43.6
47	221	DSX	Daytime Sextantids	188.4	154.5	-1.5	31.2
48	233	OCC	October Capricornids	189.7	303	-10	10
49	246	AMO	α Monocerotids	239.3	117.1	0.8	63
50	250	NOO	November Orionids	245	90.6	15.7	43.7
51	254	PHO	Phoenicids	253	15.6	-44.7	11.7
52	281	OCT	October Camelopardalids	193	166	79.1	46.6
53	319	JLE	January Leonids	282.5	148.3	23.9	52.7
54	320	OSE	ω Serpentids	275.5	242.7	0.5	38.9
55	321	TCB	θ Coronae Borealids	296.5	232.3	35.8	38.66
56	322	LBO	λ Bootids	295.5	219.6	43.2	41.75
57	323	XCB	ξ Coronae Borealids	294.5	244.8	31.1	44.25
58	324	EPR	ε Perseids	95.5	58.2	37.9	44.8
59	325	DLT	Daytime λ Taurids	85.5	56.7	11.5	36.4
60	326	EPG	ε Pegasids	105.5	326.3	14.7	29.9
61	327	BEQ	β Equuleids	106.5	321.5	8.7	31.6
62	328	ALA	α Lacertids	105.5	343	49.6	38.9
63	330	SSE	σ Serpentids	275.5	242.8	-0.1	42.67
64	331	AHY	α Hydrids	285.5	127.6	-7.9	43.6

- before being published, each new shower will obtain a unique name, the IAU number and three letter code. After publication, the shower will be added to the Working List of Meteor Showers and the discovery announced,
- all showers which come up to the verification criterion will be included in the List of Established Showers, and after their approving by the C22 business meeting during the next General Assembly, all new established showers will from thereon be known by their official name.

This makes the naming of meteor showers an ongoing effort. During the business meeting in Rio, the present members of Commission 22 agreed that the *Task Group on Meteor Shower Nomenclature* should be transformed into the *Working group on Meteor Shower Nomenclature*. The current members of the Working Group in the 2009-2012 triennium are: Peter Jenniskens (chair), Tadeusz J. Jopek (vice-chair), Vladimír Porubčan, William J. Baggaley, Juergen Rendtl, Shinsuke Abe, Peter Brown and Pavel Koten. The main goal of the Working Group is similar to that in the previous triennium: maintaining and improving the Working List of meteor showers on the IAU Meteor Data Center website; assigning new names, numbers and three letter codes for the showers discovered in new surveys; and decide which new showers can be moved to the List of Established Showers, and thus obtain official names during the next IAU General Assembly in Beijing in 2012.

5 Conclusions and Call for Contributions

In 2009 for the first time in history of the Meteor Astronomy, 64 showers were officially named by the IAU. Their names are given in Table 1, and are posted on the IAU MDC website, see Jopek (2007). The current Working List of Meteor Showers has already 301 candidate showers that could receive official names if their existence can be confirmed.

Meteor astronomers can contribute to minimizing the confusion in the literature by checking the correct name of a shower when minor showers are discussed and by adhering to the newly adopted names (e.g., "δ Aquariids", not "δ Aquarids"). Showers that are not yet in the Working List should be reported before they are mentioned in new (amateur or professional) literature.

Nomenclature is important in astronomy because it regulates the language used by astronomers. In our meteor community we started with this task quite recently. Our first experiences taught us that there is a real need to assign a particular name to a particular shower, but that this task alone is not simple. We needed to check, and check again, that those names were unique and did not lead to confusion. The task to establish if a new shower is a real entity or only ill defined, is even more difficult. To establish a shower is the end of a long process that can take many years. At the beginning of the process, no one can predict all problems that wait for a solution in a given case.

In the near future, the Working Group on Meteor Shower Nomenclature has several tasks to solve. At this moment, we are expanding the information on meteor showers included in the Working List to make the list more descriptive. As a very important next step, we consider developing more objective criteria to be used for verification whether a given shower can be considered an established one. More precise and regular meteor observation can be of invaluable help in this task. In addition, our community needs new theoretical concepts and studies that can make us more confident in recognizing meteor showers among a sporadic meteor background.

Acknowledgements

TJJ work on this paper was partly supported by the MNiSW Project N N203 302335.

References

Bowell, E. L. G., and 16 colleagues: 2010, Division III: Planetary Systems Science. Transactions of the International Astronomical Union, Series B 27, 158-167.

Brown, P., Weryk, R. J., Wong, D. K., Jones, J.: 2008, A meteoroid stream survey using the Canadian Meteor Orbit Radar. I. Methodology and radiant catalogue. Icarus 195, 317-339

Brown, P., Wong, D. K., Weryk, R. J., Wiegert, P.: 2010, A meteoroid stream survey using the Canadian Meteor Orbit Radar. II: Identification of minor showers using a 3D wavelet transform. Icarus 207, 66-81

Jenniskens, P.: 2006, Meteor Showers and their Parent Bodies, Cambridge UP, UK, 790 pp

Jenniskens, P.: 2006, The I.A.U. meteor shower nomenclature rules. WGN, Journal of the International Meteor Organization 34, 127-128

Jenniskens, P.: 2007, Div.III, Comm.22, WG Task Group for Meteor Shower Nomenclature, IAU Information Bulletin 99, January 2007, 60–62

Jenniskens, P.: 2008, The IAU Meteor Shower Nomenclature Rules, Earth, Moon and Planet, 102, 5–9

Jopek T.J.: 2007, www.astro.amu.edu.pl/~jopek/MDC2007, or www.ta3.sk/IAUC22DB/MDC2007

Jopek T. J., Koten P., Pecina P., 2010, MNRAS, 404, 867

Kashcheev, B.L., Lebedinets, V.N., Lagutin, M.F.: 1967, Meteoric Phenomena in the Earth's Atmosphere, No 2, Moscow, Nauka, 1967, pp 168

Kilmartin, P.M.: 2003, Committee on Small Body Nomenclature, Transaction of the International Astronomical Union, 25A, 143-144

Molau, S., Kac, J.: 2009, Results of the IMO Video Meteor Network -March 2009. WGN, Journal of the International Meteor Organization 37, 92-93

Molau, S., Rendtel, J.: 2009, A Comprehensive List of Meteor Showers Obtained from 10 Years of Observations with the IMO Video Meteor Network. WGN, Journal of the International Meteor Organization 37, 98-121

SonotaCo: 2009, A meteor shower catalog based on video observations in 2007-2008. WGN, Journal of the International Meteor Organization 37, 55-62

Spurný, P., and 11 colleagues: 2007, Commission 22: Meteors, Meteorites and Interplanetary Dust. Transactions of the International Astronomical Union, Series B 26, 140-141

Spurný, P., and 11 colleagues: 2008, Commission 22: Meteors, Meteorites and Interplanetary Dust. Transactions of the International Astronomical Union, Series A 27, 174-178

Ticha, J., and 15 colleagues, 2010, Division Iii: Committee on Small Body Nomenclature. Transactions of the International Astronomical Union, Series B 27, 184-185

Uehara, S., and 11 colleagues: 2006, Detection of October Ursa Majorids in 2006. WGN, Journal of the International Meteor Organization 34, 157-162

Watanabe, J., Jenniskens, P., Spurný, P., Borovička, J., Campbell-Brown, M., Consolmagno, G., Jopek, T., Vaubaillon, J., Williams, I. P., Zhu, J.: 2010, Commission 22: Meters, Meteorites and Interplanetary Dust. Transactions of the International Astronomical Union, Series B 27, 177-179.

Large Bodies Associated with Meteoroid Streams

P. B. Babadzhanov • I. P. Williams • G. I. Kokhirova

Abstract It is now accepted that some near-Earth objects (NEOs) may be dormant or dead comets. One strong indicator of cometary nature is the existence of an associated meteoroid stream with its consequently observed meteor showers. The complexes of NEOs which have very similar orbits and a likely common progenitor have been identified. The theoretical parameters for any meteor shower that may be associated with these complexes were calculated. As a result of a search of existing catalogues of meteor showers, activity has been observed corresponding to each of the theoretically predicted showers was found. We conclude that these asteroid-meteoroid complexes of four NEOs moving within the Piscids stream, three NEOs moving within the Iota Aquariids stream, and six new NEOs added to the Taurid complex are the result of a cometary break-up.

Keywords near-Earth object · dormant comet · meteoroid streams · meteor showers · orbital evolution · Piscids stream · Iota Aquariids stream · Taurid complex

1 Introduction

Though there had been some prior speculation that Near Earth Asteroids could be responsible for some minor meteoroid streams, the first definite association was between the Geminid stream and asteroid 3200 Phaethon (Whipple 1983, Fox et al. 1984). A number of Near Earth Asteroids were also found to be moving on orbits within the Taurid complex, though comet 2P/Encke also moves in this complex (Asher et al. 1993). More recently asteroid 2003EH1 was identified as moving on the same orbit as the Quadrantids (Jenniskens 2003, Williams et al. 2004) and the generally accepted hypothesis is that these are the result of the fragmentation of a larger comet so that these 'asteroids' are in reality comet fragments that are dormant or dead. All the associations mentioned above are based on the similarity of the orbits of the NEO and the meteor stream that gives rise to the observed shower at roughly the present time.

2 Orbital Evolution

Gravitational perturbations from the planets change all orbits over a period of time. However in the region of the Solar system that is of interest to us (the Earth-Jupiter region), ω (the argument of

P. B. Badadzhanov
Institute of Astrophysics of the Academy of Sciences of the Republic of Tajikistan

I. P. Williams
Astronomy Unit, Queen Mary University of London, E1 4NS, UK

G. I. Kokhirova (✉)
Institute of Astrophysics of the Academy of Sciences of the Republic of Tajikistan. E-mail: kokhirova2004@mail.ru

perihelion) passes through the range of values from 0 to 2π in a period of several thousand years. We call this one cycle of ω. Though there may be short term variations, the changes in the three orbital elements q (perihelion distance), e (eccentricity) and i (inclination) over one cycle of ω are all essentially sinusoidal. The nodal distances, R_a and R_d also show the same characteristic variation. This variation is shown in Figure 1 for the three NEOs, 2002JS2, 2002PD11 and 2003 MT9. Some time ago (Babadzhanov and Obrubov 1992) pointed out that as the nodal distance will be equal to 1 AU at four different values of ω, during one cycle (clearly seen in Figure 1), four meteor showers originating from a single meteoroid stream can be formed. These four meteor showers consist of a night-time shower with northern and southern branches and of a day-time shower also with northern and southern branches.

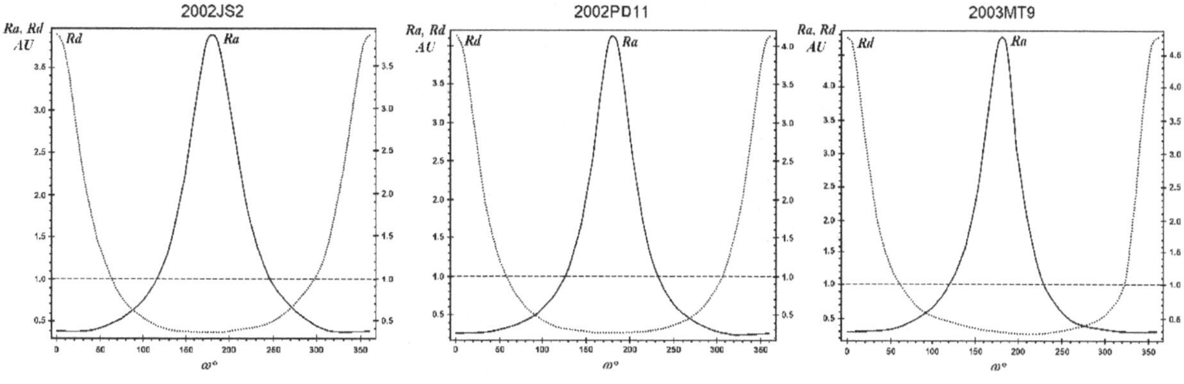

Figure 1. The variation in the nodal distances for three NEOs.

With a large number of NEOs currently being discovered, the probability that one has an orbit that is similar to a meteoroid stream at the present time by chance is high and, in order to establish a relationship with a stream, similarity of orbital evolution must be shown. This was carried out by Porubcan et al. (2004) through numerically integrating both the orbital evolution of the NEO and the meteoroid stream. Integrating the evolution of a meteoroid stream can be expensive due to the large number of particles involved and here we describe an alternative, and computationally cheaper, approach to the problem.

If the break-up of a comet was part of its history, then one might expect several large fragments to be present within the meteoroid stream. Such fragments should show the same evolutionary pattern as the stream. We thus integrate only the orbits of NEOs that might be suspected of being such fragments and calculate the characteristics of a theoretical meteor shower that would be formed at each location where the nodal distance of the NEO is 1 AU, assuming the orbital elements to be those of the NEO. We then have to ascertain whether a known meteor shower has these characteristics.

3 'Asteroids' Associated with Meteor Showers and Meteorite Streams

Babadzhanov et al. (2008a, 2008b, 2009) have used the procedure described above in order to identify NEOs that can be associated with meteor showers that are related to three well know showers, the Piscids, the Taurids and the Iota Aquariids. Such associations indicate that they are likely to be fragments of a comet. The results are summarized in Tables 1, 2 and 3. In Tables 1-3 the values of the

D-criterion which quantifies the similarity between the orbits of a meteor shower and an NEO are also given, calculated using the formula given by Steel et al. (1991) namely

$$D^2 = (q_1 - q_2)^2 + (e_1 - e_2)^2 + \{2\sin[(i_1 - i_2)/2]\}^2.$$

All the determined values of the *D*-criterion satisfy $D < 0.3$ showing that the meteor showers and the NEOs under investigation move on very similar orbits implying that the meteoroid stream also contains large fragments of the parent comets.

Table 1. Orbital elements for NEOs and showers in the Piscid Complex.

Name	q	e	i	λ	α	δ	D
N.Piscids	0.40	0.80	6	174	3	8	-
1997GL3	0.49	0.78	7	178	0	8	0.09
2000PG3	0.34	0.88	12	172	0	10	0.14
2002JC9	0.38	0.85	6	169	357	4	0.05
S.Piscids	0.44	0.82	3	179	7	-1	-
1997GL3	0.45	0.80	6	173	3	-6	0.06
2000PG3	0.37	0.87	14	174	6	-1	0.21
2002JC9	0.38	0.83	6	167	0	-6	0.08
Ass.25	0.34	0.78	6	31	13	10	-
1997GL3	0.45	0.80	6	21	10	11	0.11
2000PG3	0.36	0.87	13	37	19	20	0.15
2002JC9	0.38	0.83	6	30	16	13	0.06
Ass.30	0.27	0.83	11	30	14	3	-
1997GL3	0.49	0.78	7	17	14	-2	0.24
2000PG3	0.35	0.88	13	38	28	1	0.10
2002JC9	0.38	0.85	5	28	19	3	0.15

Table 2. Orbital elements for NEOs and showers in the Taurid Complex.

Name	q	e	i	λ	α	δ	D
N.Taurids	0.36	0.86	2.4	231	59	22	-
16960	0.27	0.88	18.2	200	31	25	0.29
1998VD31	0.49	0.81	6.7	244	65	29	0.16
1999VK12	0.46	0.79	7.3	230	52	27	0.15
1999VR6	0.53	0.76	7.9	231	50	28	0.22
2003UL3	0.41	0.82	3.7	239	65	25	0.07
2003WP21	0.45	0.80	1.5	239	63	23	0.11
2004TG10	0.31	0.86	3.2	224	55	22	0.05
S.Taurids	0.37	0.81	5.2	221	51	14	-
16960	0.30	0.87	19.9	202	41	1.4	0.27
1998VD31	0.52	0.81	9.4	247	70	11	0.17
1999VK12	0.50	0.78	9.4	233	59	9	0.15
1999VR6	0.49	0.78	9.1	228	53	8	0.14
2003UL3	0.44	0.81	5.9	241	68	16	0.07
2003WP21	0.49	0.79	3.8	242	66	17	0.12
2004TG10	0.29	0.87	5.0	221	54	16	0.10
ζ-Perseids	0.34	0.79	0.0	79	62	23	-
16960	0.29	0.87	19.5	85	64	35	0.30

Table 2. (continued) Orbital elements for NEOs and showers in the Taurid Complex.

Name	q	e	i	λ	α	δ	D
1998VD31	0.52	0.81	9.3	76	69	33	0.24
1999VK12	0.50	0.78	9.3	70	61	31	0.23
1999VR6	0.49	0.78	9.1	66	56	30	0.22
2003UL3	0.44	0.81	6.0	91	82	29	0.15
2003WP21	0.48	0.79	3.8	81	75	27	0.15
2004TG10	0.33	0.85	5.4	99	86	28	0.11
β-Taurids	0.33	0.85	6.0	97	87	19	-
16960	0.31	0.86	18.0	83	68	9	0.21
1998VD31	0.49	0.81	6.6	79	74	15	0.17
1999VK12	0.47	0.79	7.1	74	67	14	0.15
1999VR6	0.53	0.76	7.8	62	59	11	0.22
2003UL3	0.41	0.82	3.5	94	85	20	0.10
2003WP21	0.45	0.80	1.4	85	77	21	0.15
2004TG10	0.31	0.86	2.9	101	87	21	0.06

Table 3. Orbital elements for NEOs and showers in the Iota Aquariid complex.

Name	q	e	i	λ	α	δ	D
N.ι-Aquariids	0.26	0.86	8	132	330	-5	-
2002PD11	0.32	0.85	7	150	342	-2	0.06
2002JS2	0.38	0.83	8	150	341	-1	0.12
2003MT9	0.16	0.94	4	116	319	-14	0.15
S. ι-Aquariids	0.26	0.86	8	134	337	-13	-
2002PD11	0.29	0.87	7	146	344	-12	0.04
2002JS2	0.34	0.84	7	147	344	-13	0.08
2003MT9	0.29	0.88	2	133	330	-13	0.11
April Piscids	0.31	0.80	4	29	10	8	-
2002PD11	0.29	0.87	7	29	11	10	0.09
2002JS2	0.33	0.84	7	22	5	8	0.07
2003MT9	0.16	0.94	4	323	10	2	0.21
April Cetids	0.28	0.83	9	29	10	1	-
2002PD11	0.32	0.86	7	26	13	0	0.06
2002JS2	0.36	0.83	8	18	8	-4	0.08
2003MT9	0.30	0.88	2	13	-1	1	0.13

4 Conclusions

In all three cases a number of NEOs were found that could have formed observable meteor showers. We thus conclude that the break up of a comet nucleus, leaving a number of fragments as well as a meteoroid stream, is common, supporting the view of Asher et al. (1993), and Jenniskens and Vaubillion (2008). We also conclude that a number of objects, currently classified as asteroids, are in fact cometary fragments.

References

D.J. Asher, S.V.M. Clube, D.I. Steel, Asteroids in the Taurid Complex. R.A.S. Monthly Notices **264**, 93-105 (1993)

P.B. Babadzhanov, Yu.V. Obrubov, Evolution of short-period meteoroid streams. Celestial Mech. Dyn. Astron. **54**, 111-127 (1992)

P.B. Babadzhanov, I.P. Williams, G.I. Kokhirova, Near-Earth asteroids amongst the Piscids meteoroid stream. Astron. & Astrophys. **479**, 149-255 (2008a)

P.B. Babadzhanov, I.P. Williams, G.I. Kokhirova, Near-Earth asteroids in the Taurid complex. R.A.S. Monthly Notices **386**, 1436-1442 (2008b)

P.B. Babadzhanov, I.P. Williams, G.I. Kokhirova, Near-Earth asteroids amongst the Iota Aquariid meteoroid stream. Astron. & Astrophys. **507**, 1067-1072 (2009)

K. Fox, I.P. Williams, D.W. Hughes, The Geminid asteroid (1983TB) and its orbital evolution. R.A.S. Monthly Notices **208**, 11-15 (1984)

P. Jenniskens, 2003EH1 is the Quadrantid Shower parent comet, Astrophys. J. **127**, 3018-3022 (2003)

P. Jenniskens, J. Vaubaillion, Minor Planet 2008ED69 and the Kappa Cygnid meteor shower. Astrophys. J. **136**, 725-730 (2008)

V. Porubcan, I.P. Williams, L. Kornos, Associations between asteroids and meteoroid streams. Earth Moon Planets **95**, 697-712 (2004)

D.I. Steel, D.J. Asher, S.V.M. Clube, The Structiure and evolution of the Taurid complex. R.A.S. Monthly Notices **251**, 632-648 (1991)

F.L. Whipple, 1983 TB and the Geminid Meteors. IAU Circular **3881**, 1W (1983)

I.P. Williams, G.A. Ryabova, A.P. Baturin, A.M. Chernitsov, The parent of the Quadrantid meteoroid stream and asteroid 2003EH1. R.A.S. Monthly Notices **355**, 1171-1181 (2004)

Stream Lifetimes Against Planetary Encounters

G. B. Valsecchi • E. Lega • Cl. Froeschlé

Abstract We study, both analytically and numerically, the perturbation induced by an encounter with a planet on a meteoroid stream. Our analytical tool is the extension of Öpik's theory of close encounters, that we apply to streams described by geocentric variables. The resulting formulae are used to compute the rate at which a stream is dispersed by planetary encounters into the sporadic background. We have verified the accuracy of the analytical model using a numerical test.

Keywords meteoroid streams · planetary close encounters

1 Introduction

Meteoroids stream orbits can intersect, in specific phases of their evolution, the orbit of the Earth, leading to meteor showers. This causes not only the removal of particles from the stream due to collisions, but also potentially large perturbations of the remaining stream members due to planetary encounters.

We here examine the role of planetary encounters on the dispersion of streams using results from the analytical theory of close encounters. The reason for an analytical approach, which is inevitably affected by some approximations, is to be able to generalize the results to most orbits of interest.

To this purpose, we use the extension of Öpik's theory of planetary close encounters [Öpik 1976] developed in recent years [Valsecchi et al. 2003]. In it, the gravitational model is a restricted, circular, 3-dimensional 3-body problem in which, far from the planet, the small body moves on an unperturbed heliocentric keplerian orbit. The encounter with the planet is modeled as an instantaneous transition from the incoming asymptote of the planetocentric hyperbola to the outgoing one, taking place when the small body crosses the b-plane, the plane centered on the Earth and normal to the incoming asymptote of the planetocentric hyperbola (i.e., normal to the unperturbed geocentric velocity U of the small body). The direction of the latter is defined by two angles, $\theta(U, a)$and $\phi(a, e, i)$ (see Figure 1), such that

$$U = \sqrt{3 - \frac{1}{a} - 2\sqrt{a(1-e^2)}\cos i}$$

and

$$\begin{aligned} U_x &= U \sin\theta \sin\phi \\ &= \pm\sqrt{2 - \frac{1}{a} - a(1-e^2)} \end{aligned}$$

G. B. Valsecchi (✉)
IASF-Roma, INAF, Roma (Italy). E-mail: giovanni@iasf-roma.inaf.it

E. Lega • Cl. Froeschlé
Observatoire de la Côte d'Azur, Nice (France)

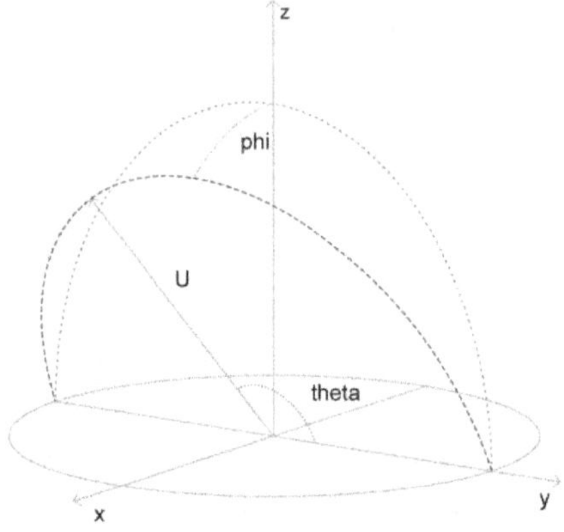

Figure 1. The geometric set up of Öpik's theory: the Earth is at the origin of axes and moves in the direction of the y-axis, while the Sun is on the negative x-axis; the geocentric velocity vector of the small body is $\boldsymbol{U}$, θ is the angle between $\boldsymbol{U}$ and the y-axis, and ϕ is the angle between the plane containing $\boldsymbol{U}$ and the y-axis, and the y-z plane.

$$\begin{aligned} U_y &= U\cos\theta \\ &= \sqrt{a(1-e^2)}\cos i - 1 \\ U_z &= U\sin\theta\cos\phi \\ &= \pm\sqrt{a(1-e^2)}\sin i \end{aligned}$$

and

$$\cos\theta = \frac{1-U^2-\frac{1}{a}}{2U}$$

$$\sin\phi = \pm\frac{\sqrt{2-\frac{1}{a}-a(1-e^2)}}{\sqrt{2-\frac{1}{a}-a(1-e^2)cos^2 i}}$$

$$\cos\phi = \pm\frac{\sqrt{a(1-e^2)}\sin i}{\sqrt{2-\frac{1}{a}-a(1-e^2)cos^2 i}}$$

where the upper sign in the expressions for U_z and *cos* ϕ apply to encounters at the ascending node, and in the expressions for U_x and *sin* ϕ apply to post-perihelion encounters, while a is in AU and U is in units of the orbital velocity of the Earth.

2 Earth Cross-section

As already noted, collisions with the Earth remove meteoroids from a stream. For a given stream, the collisional cross-section of the Earth on the b-plane is $\pi b_\oplus^2$ with

$$b_\oplus = \sqrt{r_\oplus^2 + 2cr_\oplus},$$

where $r_\oplus$ is the radius of the Earth in AU, $c = m/U^2$ and m is the mass of the Earth in solar masses. The values of c and $b_\oplus$ are tabulated for various streams in Table 1; the values of U of the stream orbits are taken from [Jopek et al. 1999].

Table 1. Values of c and $b_{\oplus}$, in Earth radii, for various streams of interest.

Stream	c	$b_{\oplus}$
Leonids	0.013	1.01
Perseids	0.018	1.02
Lyrids	0.029	1.03
Quadrantids	0.038	1.04
Southern δ-Aquariids	0.038	1.04
Geminids	0.052	1.05
Northern Taurids	0.070	1.07
Northern α-Capricornids	0.12	1.11

Starting from [Valsecchi 2006], [Valsecchi et al. 2005] derived an algorithm to pass from *b*-plane coordinates, close to a collision with the planet, to pairs of orbital elements (assuming that all the other elements are kept constant), in the framework of the extended Öpik's theory. The algorithm neglects second and higher order terms in the distance from the origin.

We here apply it to meteoroid streams encountering the Earth, keeping fixed a, e, i, Ω (and thus U, θ, ϕ, λ), and computing in the ω-M plane the area of the ellipse corresponding to a circle centered in the origin of the *b*-plane.

Figure 2 shows the collisional cross-section of the Earth on the *b*-plane for the Northern Taurids and the corresponding ellipse, computed analytically, in the $\delta\omega$-δM plane, where $\delta\omega$ and δM are the displacements in the respective angles relative to a central collision with the Earth.

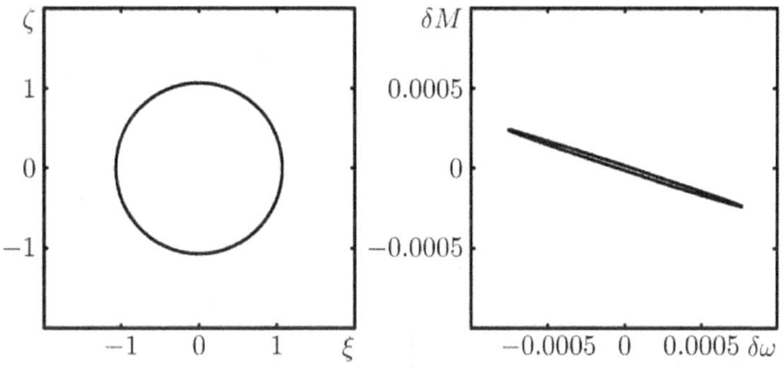

Figure 2. Left: the collisional cross-section of the Earth on the *b*-plane for the Northern Taurids; right: the same cross-section in the $\delta\omega$-δM plane.

An explicit computation, along the lines of [Valsecchi et al. 2005], shows that the area of the ellipse in the $\delta\omega$-δM plane is

$$A(b_{\oplus}) = \frac{\pi {b_{\oplus}}^2}{(a^{3/2} \sin i \sin\theta \, |\sin\phi|)},$$

where we take the values for a, i, θ, ϕ for the stream from [Jopek et al. 1999].

To test the validity of the analytic approach, we have set up the following numerical experiment: in the restricted, circular, 3-dimensional 3-body problem, we start a suitable number of meteor particles at a large distance from the Earth; all the particle orbits have the same a, e, i, Ω, while ω, M are distributed on a regularly spaced grid. We follow the particles through an encounter with the Earth, and check which of them actually collide with it (i.e., those for which the minimum geocentric distance along the perturbed trajectory is less or equal to $r_\oplus$); interpolating in the grid, we can then find the initial values of ω, M for which the minimum geocentric distance is exactly $r_\oplus$.

We have used a fourth order Runge-Kutta integrator on the equations of motion regularized through Kustaanheimo-Stiefel regularization [Kustaanheimo and Stiefel 1965]. The reader can find in [Froeschlé 1970] a detailed derivation of the regularized equations of motions using the Lagrangian formalism and in [Celletti 2002] a review of regularization theory. As recently shown in [Celletti et al. 2010] and in [Lega et al. 2010], when integrating orbits undergoing close encounters or even collisions, the existence of the singularity cannot be canceled, neither by changing the integration scheme, nor through a better precision computation; by singularity we mean that the solution does not behave as a power series about the point, while usual integration schemes are based on the development in power series of the solution.

The results of these computations are compared to those of our analytical approach in Figure 3 and, as the plots show, are definitely satisfactory.

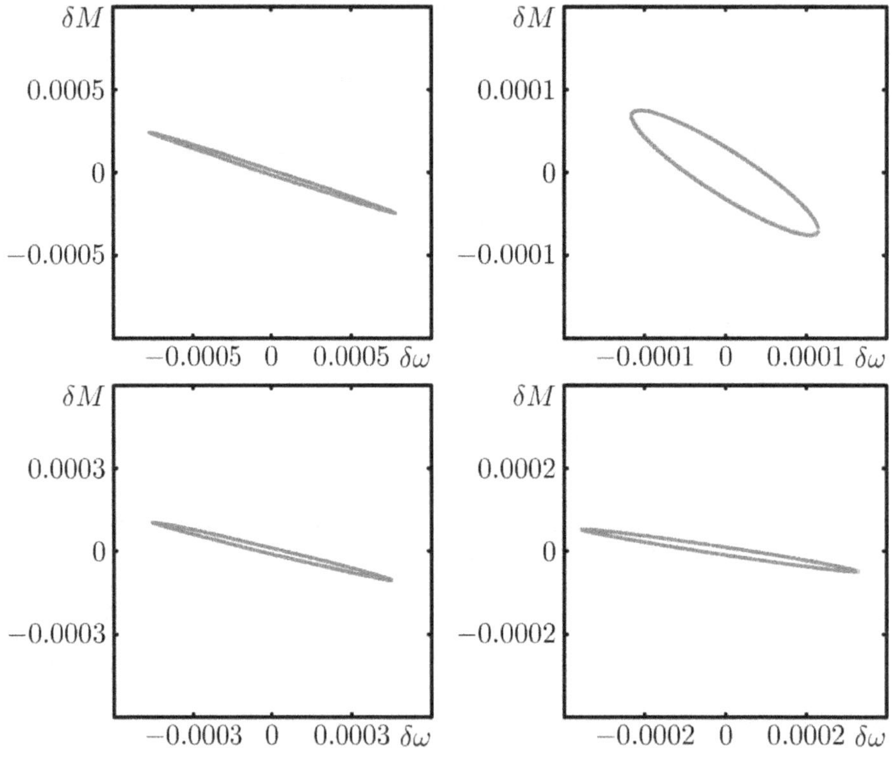

Figure 3. The collisional cross-section of the Earth in the $\delta\omega$-δM plane for the Northern Taurids (top left), the Geminids (top right), the Northern α-Capricornids (bottom left), and the Quadrantids (bottom right); superimposed on the analytical estimates (green lines) are the results of a numerical computation (red dots).

3 Stream Dispersion

The agreement between the analytic computation and the numerical check encourages us in the use of the former in order to study the dispersion induced in a stream by its passage close to the Earth. To get a quantitative description of stream dispersion, we start by recalling the orbital similarity criterion based on U, $\cos\theta$, ϕ and λ, the longitude of the Earth at the time of the meteor fall, introduced by [Valsecchi et al. 1999] to classify meteoroids in streams; it is based on the quantity D_N defined by:

$$D_N^2 = [U_2 - U_1]^2 + [\cos\theta_2 - \cos\theta_1]^2 + \Delta\Xi^2$$

where

$$\Delta\Xi^2 = \min\,[\Delta\phi_{\mathrm{I}}^2 + \Delta\lambda_{\mathrm{I}}^2, \Delta\phi_{\mathrm{II}}^2 + \Delta\lambda_{\mathrm{II}}^2]$$

and

$$\Delta\phi_{\mathrm{I}} = 2\sin\frac{\phi_2 - \phi_1}{2} \qquad \Delta\phi_{\mathrm{II}} = 2\sin\frac{\pi + \phi_2 - \phi_1}{2}$$
$$\Delta\lambda_{\mathrm{I}} = 2\sin\frac{\lambda_2 - \lambda_1}{2} \qquad \Delta\lambda_{\mathrm{II}} = 2\sin\frac{\pi + \lambda_2 - \lambda_1}{2}$$

As discussed in [Valsecchi et al. 1999], this criterion basically uses the geocentric speed, the anti-radiant coordinates (in a frame rotating with the Earth about the Sun) and the date of meteor fall, instead of the usual orbital elements. Of these quantities, encounters with the Earth affect only the anti-radiant coordinates, and therefore θ, ϕ, since U is an invariant; we disregard changes in λ, since they are far smaller than those in θ and ϕ.

In the approximation $c^2 \ll b^2$, valid for all of the streams we examined, we have that an encounter with the Earth at unperturbed distance b rotates the geocentric velocity vector by an angle γ given by:

$$\sin\gamma \approx \frac{2c}{b}.$$

Thus, in order to turn the direction of U (i.e., to change the radiant) by a significant quantity, say by $\gamma \geq 0.1$ rad, we need an encounter with the Earth taking place at:

$$b_{\gamma\geq 0.1} \leq \frac{2c}{0.1}.$$

Table 2 gives the values of $b_{\gamma\geq 0.1}$ for the same streams of Table 1; note that for all the tabulated streams, with the exception of the Northern Taurids and the Northern α-Capricornids, the collisional cross-section is larger than the deflection cross-section.

Table 2. Values of $b_{\gamma\geq0.1}$ and of $b_{\oplus}$, in Earth radii, for various streams of interest.

Stream	c	$b_{\gamma\geq0.1}$	$b_{\oplus}$
Leonids	0.013	0.25	1.01
Perseids	0.018	0.36	1.02
Lyrids	0.029	0.57	1.03
Quadrantids	0.038	0.75	1.04
Southern δ-Aquariids	0.038	0.75	1.04
Geminids	0.052	1.03	1.05
Northern Taurids	0.070	1.4	1.07
Northern α-Capricornids	0.12	2.3	1.11

For a generic, not too large value of b, the area $A(b)$ on the $\delta\omega$-δM plane covered by the ellipse corresponding to a circle of radius b centered in the origin of the b-plane is given, as seen before, by:

$$A(b) = \frac{\pi b^2}{a^{3/2} \sin i \sin \theta \, |\sin \phi|}.$$

Every year, when the Earth crosses a stream, a fraction of the latter will be removed, either by collision or by deflection by an angle larger than a suitable threshold. On the other hand, at the Earth crossing the meteoroids have, in general, $0 \leq \delta M \leq 2\pi$, while the values of $\delta\omega$ are characterized by $(\delta\omega)_{min} \leq \delta\omega \leq (\delta\omega)_{max}$, with values of $(\delta\omega)_{min}$ and $(\delta\omega)_{max}$ different for each stream, as function of, among other things, the age of the stream itself.

Thus, the fraction $f(b)$ removed each year from a stream, either by collision or by

$$f(b) = \frac{A(b)}{2\pi\Delta\omega},$$

where $\Delta\omega = (\delta\omega)_{max} - (\delta\omega)_{min}$.

Table 3 gives the fraction, to be divided by a suitable value of $\Delta\omega$ for each stream, eliminated yearly by collisions and/or close encounter with the Earth; as it is readily seen, the Earth seems not to have a major effect in dispersing streams. Note, however, that some of the tabulated streams intersect the orbit of Jupiter, something that would greatly accelerate their dispersion.

Table 3. Values of $f(b_{\gamma\geq0.1})\cdot\Delta\omega$ and of $f(b_{\oplus})\cdot\Delta\omega$, in Earth radii, for various streams of interest.

Stream	$f(b_{\gamma\geq0.1})\cdot\Delta\omega$	$f(b_{\oplus})\cdot\Delta\omega$
Leonids	–	$1.9\cdot10^{-9}$
Perseids	–	$6.8\cdot10^{-11}$
Lyrids	–	$4.1\cdot10^{-12}$
Quadrantids	–	$3.1\cdot10^{-9}$
Southern δ-Aquariids	–	$5.7\cdot10^{-10}$
Geminids	–	$1.7\cdot10^{-9}$
Northern Taurids	$1.1\cdot10^{-8}$	$6.6\cdot10^{-9}$
Northern α-Capricornids	$1.2\cdot10^{-8}$	$2.5\cdot10^{-9}$

4 Conclusion

We have presented an analytic formula relevant for the rate of dispersion of a meteoroid stream induced by encounters and collisions with the Earth, and have checked its validity by numerical computations in the restricted, circular 3-body problem.

We plan to pursue this work, extending it to encounters with more than one planet, and investigating the coupled effects of planetary close encounters and of secular perturbations.

References

A. Celletti, Singularities, collisions and regularization theory, in Singularities in Gravitational Systems. Applications to Chaotic Transport in the Solar System (D. Benest and C. Froeschlé eds.), Lecture Notes in Physics 590, Springer Verlag, Heidelberg, Germany, pp. 1-24 (2002)

A. Celletti, L. Stefanelli, E. Lega Cl. Froeschlé, Some results on the global dynamics of the regularized restricted three-body problem with dissipation, submitted to Cel. Mech. & Dynam. Astron. (2010)

Cl. Froeschlé, Numerical Study of Dynamical Systems with Three Degrees of Freedom. I. Graphical Displays of Four-Dimensional Sections, Astron. Astrophys. 4, 115-128 (1970)

T. J. Jopek, G. B. Valsecchi and Cl. Froeschlé, Meteoroid stream identification: a new approach. II. Application to 865 photographic meteor orbits, Mon. Not. R. Astron. Soc. 304, 751-758 (1999)

P. Kustaanheimo and E. Stiefel, Perturbation theory of Kepler motion based on spinor regularization, J. Reine Angew. Math. 218, 204-219 (1965) [Lega et al. 2010] E. Lega, M. Guzzo and Cl. Froeschlé, Close encounters and resonances detection in three body problems through Kustaanheimo Stiefel regularization, preprint (2010)

E. J. Öpik Interplanetary Encounters, Elsevier, New York (1976)

G. B. Valsecchi, Geometric Conditions for Quasi-Collisions in Öpik's Theory, in Dynamics of Extended Celestial Bodies And Rings (J. Souchay ed.), Lecture Notes in Physics 682, Springer Verlag, Heidelberg, Germany, pp. 145-158 (2006)

G. B. Valsecchi, T. J. Jopek and Cl. Froeschlé, Meteoroid stream identification: a new approach. I. Theory, Mon. Not. R. Astron. Soc. 304, 743-750 (1999)

G. B. Valsecchi, A. Milani, G. F. Gronchi and S. R. Chesley, Resonant returns to close approaches: analytical theory, Astron. Astrophys. 408, 1179-1196 (2003)

G. B. Valsecchi, A. Rossi, A. Milani and S. R. Chesley, The size of collision solutions in orbital elements space, in Dynamics of Populations of Planetary Systems (Z. Knezević and A. Milani, eds.), I.A.U. Colloquium 197, Cambridge Univ. Press, pp. 249-254 (2005)

Numerical Modeling of Cometary Meteoroid Streams Encountering Mars and Venus

A. A. Christou • J.Vaubaillon

Abstract We have simulated numerically the existence of meteoroid streams that encounter the orbits of Mars and Venus, potentially producing meteor showers at those planets. We find that 17 known comets can produce such showers, the intensity of which can be determined through observations. Six of these streams contain dense dust trails capable of producing meteor outbursts.

Keywords Mars · Venus · meteoroid streams · meteors · meteor showers · meteor outbursts

1 Introduction

Although no meteor showers have yet been observed at Venus and Mars, the undertaking of projects such as the U.S. rover Curiosity and the JAXA orbiter Akatsuki leads one to expect that such observations will be made, serendipitously or otherwise, in the near future. In support of these and follow-on missions with some meteor-detecting capability we carried out numerical simulations of cometary streams identified by previous work as potentially Mars- or Venus-encountering. We hope that these results will be used to guide future meteor surveys but also to interpret observations.

2 Method

Our method is that of Vaubaillon et al (2005a; 2005b). The motion of test particles initially ejected from the comet near perihelion according to the model of Crifo and Rodionov (1997) is propagated forward in time by numerical integration. The software then records all particles that approach the planet to within a few hundredths of an AU within the period 2000-2050. The initial sample of cometary candidates, either Intermediate Long Period Comets (ILPCs; $P > 200$ yr) or Halley Type Comets (HTCs; $P < 200$ yr), is taken from Christou (2010). Cometary orbits from HORIZONS (Giorgini et al, 1996) were back-integrated in time, to simulate past perihelion passages. Relevant physical and orbital characteristics of the comets themselves may be found in Tables 2 and 4 of the work by Christou. As these comets' orbital periods span two orders of magnitude, we have varied the number of perihelion passages considered for particle ejection on a case-by-case basis as shown in Table 1. In some cases, we have considered non-consecutive perihelion passages (eg one out of every five) in order to extend the time period over which the comet's, and hence the stream's, orbital evolution can be investigated.

A. A. Christou (✉)
Armagh Observatory, College Hill, Armagh BT61 9DG, Northern Ireland, UK. E-mail: aac@arm.ac.uk

J. Vaubaillon
IMCCE, Observatoire de Paris, 77 Avenue Denfert-Rochereau, F-75014 Paris, France

Table 1. Characteristics of all ILPCs and HTCs from Christou (2010) that satisfied the "shower" criterion in the numerical simulations. Column 2 gives the orbital period in years. Column 3 identifies the relevant planetary body as Venus (V) or Mars (M). Column 4 gives the number of perihelion passages where test particles where ejected from the comet. In cases where a mixture of consecutive and non-consecutive perihelion passages were considered for particle ejection we provide the number of said passages x and the increment y in the format (y)x in Column 5. Column 6 gives the date of the earliest perihelion passage considered in the simulations.

Comet	Period (yr)	Planet	Number of per. pass. considered	Step	Start Year
13P/Olbers	70	M	11	1	1313
27P/Crommelin	27	V	30	5(10) + 1(20)	326
35P/Herschel-Rigollet	155	V	17	1	−6160
161P/Hartley-IRAS	21	M	21	5(6) + 1(15)	1104
177P/Barnard	119	M	10	1	1038
P/2005 T4 (SWAN)	29	V	11	1	1720
P/2006 HR30 (Siding Spring)	22	M	11	1	1751
C/1769 P1 (Messier)	2100	M	9	1	−12087
C/1857 O1 (Peters)	235	V	8	1	−336
C/1858 L1 (Donati)	2000	V	9	1	−18372
C/1917 F1 (Mellish)	145	V	16	1	−5496
C/1939 B1 (Kozik-Peltier)	1800	V	5	1	−4689
C/1964 L1 (Tomita-Gerber Honda)	1400	V	4	1	−3600
C/1984 U2 (Shoemaker)	270	M	10	1	11
C/1998 U5 (LINEAR)	1000	M	5	1	−1989
C/2007 H2 (Skiff)	348	M	4	1	1016
5335 Damocles	41	M	18	5(8) + 1(10)	6

The simulation of the generation and evolution of these meteoroid streams was run on 5 to 50 parallel processors at CINES (France). Three size bins, equally log-spaced from 0.1 mm to 100 mm were considered. Ten thousand (10^4) particles per size bin and per perihelion passage were simulated. In the analysis reported in this work, we do not discriminate between the different particle sizes.

3 Results

The results of the numerical simulations consist of state vectors of planet-encountering particles as defined in the previous Section. If the distribution of the particle orbit nodes on the planetary orbital plane encompasses the planetary orbit then we can say that a shower is present at that planet. This condition was quantified by highlighting all those test particles (TPs) that approached the planetary orbit to within 0.005 AU and binning them in the direction parallel to the planetary orbit in units of time. Bins of angular width corresponding to one hour of time were used. In the resulting distribution plot, the comet tests positive for a shower if any one of the bins contains more than one particle. 17 comets in our sample satisfied this criterion, which we will hereafter refer to as the "shower" criterion.

We separate those into two groups. The first group consists of those streams which exhibit a smooth distribution of particles on the planetary orbit plane, in other words a smooth "background" flux of meteoroids. An example of such a stream is shown in Figure 1. The left panel shows the spatial distribution of Venus-encountering particles from comet C/1858L1 (Donati). The orbit of Venus, indicated by the black curve, passes well within the distribution of particles in its orbit plane, indicating that this planet samples the core of the Donati stream. The right panel shows the distribution of particles that satisfy the shower criterion along the Venusian orbit as a function of the astronomical solar longitude λ_S .The profile of this shower, and all other showers in Group I, appears to be well-behaved, in the sense that the distribution is fairly symmetric with a gradually varying slope and a single maximum. From this information, basic properties of the shower can be predicted.

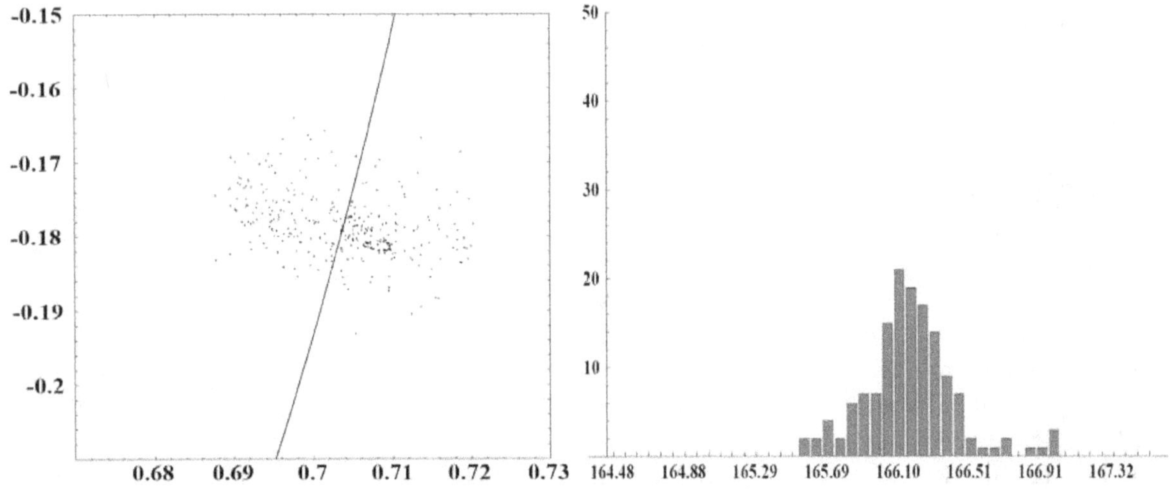

Figure 1. Left panel: Distribution of test particles ejected from comet C/1858 L1(Donati) that encountered Venus between the years 2000 and 2050. The points represent the locations of the particles, in cartesian heliocentric J2000 coordinates and units of AU, as they cross the orbital plane of the planet. The black curve represents the orbit of Venus, with the direction of motion of the planet being from bottom to top. Right panel: Histogram of those particles shown on the left panel that approach the planet's orbit to within 0.005 AU as a function of solar longitude in units of degrees. The size of each bin corresponds to one hour of time.

In Table 2 we provide such properties in the form of the solar longitude of the peak of the histogram (Column4), the shower duration in terms of the solar longitudes at which the first and last bins with more than one TP are encountered (Column5), the peak count of test particles per bin (Column6) and the total number of TPs that satisfied the shower criterion for that comet (Column7).

Group II, also listed in Table 2, consists of those cometary streams, six in total, containing multiple density enhancements orders of magnitude higher than the background value. The fact that these enhancements only appear on certain years lead us to conclude that they correspond to individual dust trails which can yield meteor outbursts at the corresponding planet. An example of such a case, for comet C/2007 H2 (Skiff), is shown in Figure 2. A number of planet-approaching trails are embedded in the background (left panel) resulting in at least two maxima in the corresponding shower density histogram (right panel). For two cases belonging to this group, that of 13P/Olbers and C/1998 U5 (LINEAR), the background component is not well defined as its particle density is too low. These are indicated by a question mark (?).

Table 2. Simulation results for all ILPCs and HTCs from Christou (2010) that satisfied the "shower" criterion. Whether a stream tested positive for membership in Group I or II as defined in the text is indicated in Columns 2 and 3 respectively.

Comet	Back ground	Out bursts	Peak λ_S (°)	Width (°), (hr)	Peak Count	Total Count
13P/Olbers	Y?	Y	256.0	255.8-256.6 (33)	20	175
27P/Crommelin	Y	N	251.7	250.0-253.2 (47)	12	205
35P/Herschel-Rigollet	Y	N	175.1	175.0-175.2 (4)	2	11
161P/Hartley-IRAS	Y	Y	176.9	176.7-177.4 (31)	20	124
177P/Barnard	Y	Y	100.1	100.1-100.4 (16)	20	57
P/2005 T4 (SWAN)	Y	Y	210.5, 213.4	210.1-213.8 (56)	130	869
P/2006 HR30 (Siding Spring)	Y	N	303.1	− (1)	2	30
C/1769 P1 (Messier)	Y	N	175.6	− (1)	2	6
C/1857 O1 (Peters)	Y	N	207.6	206.6-208.0 (22)	28	273
C/1858 L1 (Donati)	Y	N	166.1	165.6-166.9 (21)	25	143
C/1917 F1 (Mellish)	Y	N	271.7	270.2-272.6 (35)	5	30
C/1939 B1 (Kozik-Peltier)	Y	N	289.1	288.8-289.6 (12)	25	133
C/1964 L1 (Tomita-Gerber Honda)	Y	N	139.2	137.8-139.8 (27)	6	43
C/1984 U2 (Shoemaker)	Y	N	214.6	214.3-214.6 (11)	10	51
C/1998 U5 (LINEAR)	Y?	Y	235.7	235.4-235.9 (21)	100	524
C/2007 H2 (Skiff)	Y	Y	32.7	32.2-32.8 (28)	420	1526
5335 Damocles	Y	N	308.7	308.4-308.8 (18)	5	35

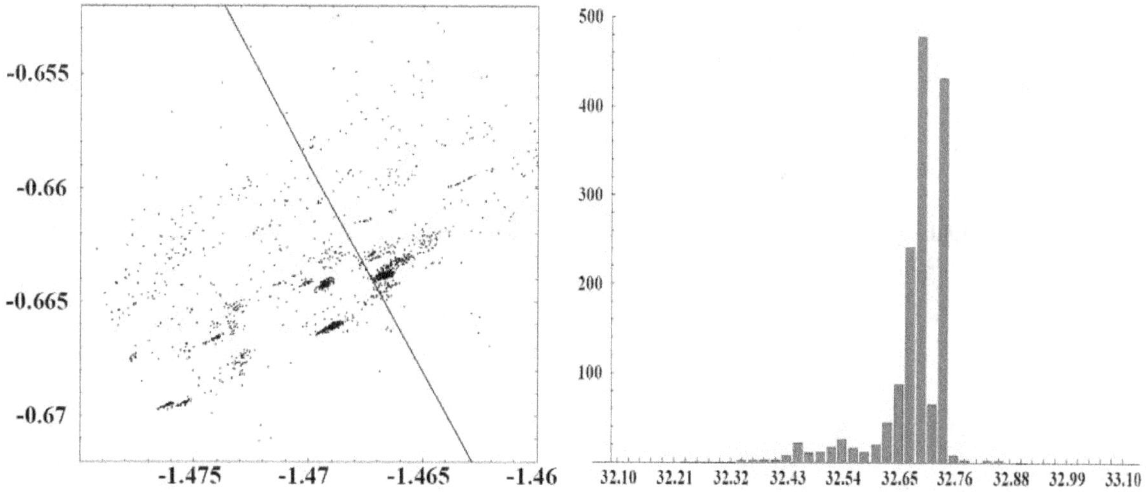

Figure 2. As Figure 1 but for Mars-encountering test particles ejected from comet C/2007 H2 (Skiff). In the left-hand panel, the direction of the planet's motion is from top to bottom. Note the numerous concentrations of particles within the stream's cross-section. These result in multiple maxima well above the background intensity of the shower in the histogram on the right-hand side.

4 Conclusions and Future Work

In this work we have simulated numerically the structure of meteoroid streams that encounter the orbits of Mars and Venus. We have highlighted seventeen of those streams where the planet-encountering density of test particles is sufficiently high to allow estimation of the solar longitude of maximum meteor activity, constrain the duration of said activity and determine whether the stream cross-section as sampled by the planet contains denser trails of particles that could give rise to meteor outbursts.

To convert the density histograms into actual meteor activity profiles would require observations of these showers at Venus and Mars (Vaubaillon et al, 2005b). In the meantime, we intend to use the information in Tables 1 and 2 in combination with available knowledge of the properties of these comets from observations and dynamical studies to calibrate these histograms in the relative sense and conduct intra-sample comparisons.

We also intend to follow up on our discovery of outburst activity from some of these comets by initiating a new series of numerical experiments to model any such outbursts occurring in the near future. These would be prime targets for meteor searches at those planets in coming years.

Acknowledgements

The authors wish to thank the CINES team for the use of the super-computer. Part of the work reported in this paper was carried out during JV's visit to Armagh Observatory in May 2009 funded by Science & Technology Facilities (STFC) Grant PP/E002242/1. Astronomical research at the Armagh Observatory is funded by the Northern Ireland Department of Culture, Arts and Leisure (DCAL).

References

Christou, A. A., 2010. Annual meteor showers at Venus and Mars: lessons from the Earth. MNRAS, 402, 2759–2770.

Crifo, J.F.,Rodionov, A.V., 1997. The dependence of the circumnuclear coma structure on the properties of the nucleus. Icarus 129, 72–93.

Giorgini, J.D., Yeomans, D.K., Chamberlin, A.B., Chodas, P.W., Jacobson, R.A., Keesey, M.S., Lieske, J. H., Ostro, S. J., Standish, E. M., Wimberly, R. N., 1996. JPL's on-line solar system data service. Bull. Am. Astron. Soc. 28, 1158.

Vaubaillon, J., Colas, F., Jorda, L., 2005. A new method to predict meteor showers I. Description of the model. Astron. Astrophys. 439, 751–760.

Vaubaillon, J., Colas, F., Jorda, L., 2005. A new method to predict meteor showers II. Application to the Leonids.. Astron. Astrophys. 439, 761–770.

Meteor Shower Activity Derived from "Meteor Watching Public-Campaign" in Japan

M. Sato • J. Watanabe • NAOJ Campaign Team

Abstract We tried to analyze activities of meteor showers from accumulated data collected by public-campaigns for meteor showers which were performed as outreach programs. The analyzed campaigns are Geminids (in 2007 and 2009), Perseids (in 2008 and 2009), Quadrantids (in 2009) and Orionids (in 2009). Thanks to the huge number of reports, the derived time variations of the activities of meteor showers is very similar to those obtained by skilled visual observers. The values of hourly rates are about one-fifth (Geminids 2007) or about one-fourth (Perseids 2008) compared with the data of skilled observers, mainly due to poor observational sites such as large cities and urban areas, together with the immature skill of participants in the campaign. It was shown to be highly possible to estimate time variation in the meteor shower activity from our campaign.

Keywords meteor showers · Geminids · Perseids · Orionids · public-campaign

1 Introduction

The public-campaign is one of the outreach programs which we perform in Japan, such as "Watch a comet", "Watch planets", "Watch a meteor shower" and "Watch an eclipse". This is widely announced to the public by the National Astronomical Observatory of Japan, and we received more than a few thousands of reports every time. The main purpose of the campaigns is to interest the general public in astronomical phenomena. However, we have noticed that we might be able to extract some scientific results from these reports because of its huge numbers, for example, over 5,000. Therefore we tried to derive the hourly rate of meteor showers from accumulated data of some campaigns, which resulted in the success described in this paper.

2 Report Form

We recommend participants in the campaigns monitor the night sky more than 10 minutes when observing meteors by naked-eye. The participants are also recommended to report their results via the internet. We use a very simple form of questionnaire for this report because the main purpose is the outreach to the general public, including children. The participants in the campaigns answer questions about observation epoch, observation time duration, number of counted meteors, distinction of meteor shower from sporadic meteors, location, and so on.

M. Sato (✉) • J. Watanabe • NAOJ Campaign Team
National Astonomical Observatory of Japan; 2-21-1, Osawa, Mitaka, Tokyo, 181-8588, Japan. Phone: +81-422-343966; Fax: +81-422-343810; E-mail: Mikiya.Sato@nao.ac.jp

About observation epoch, participants are asked to choose the range of observed hour, for example, before 21h, 21-22h, 22-23h, ... 3-4h, after 4h on every day within the campaign period. About observation time duration, the choices are prepared as follows: less than 10 minutes, 11-20 minutes, 21-30 minutes, 31-40 minutes, 41-50 minutes, 51-60 minutes. The number of counted meteors is also divided into nine levels which are 0, 1, 2, 3-5, 6-10, 11-20, 21-30, 41-50 and more than 51. Although each report is not as precise as those coming from the skilled visual observers, the huge number of reports gives us good reason to look into the data in detail on the scientific aspect.

3 Method of Analysis

We try to analyze the collected data in order to derive activity profiles of each meteor shower. Because we set the discrete steps in our campaign, we have uncertainty in the actual observation time for each participant. We adopted the median value of each step when we analyzed data. For example, in case of the range of 11-20 minutes for the time of observation duration, we considered that it was 15.5 minutes in average. We applied the same way in the case of the meteor numbers; if the report of the number of observed meteors is the range of 6-10, we regarded this data as 8. The derived hourly rate (*HR*) is expressed as

$$HR = \Sigma\, (Nm * Nn) \,/\, \Sigma\, (Dm * Dn) * 60,$$

where *Dm* (minutes) is the median of observation duration time, *Dn* is the number of corresponding reports collected within the specified time epoch, *Nm* is the median of the number of counted meteors, and *Nn* is the number of the corresponding reports. We could remove contribution of the sporadic meteors on the basis of the judgment of each participant in the report.

4 Results

We analyzed data collected during four campaigns: Geminids in 2007 and 2009, Perseids in 2008 and Orionids in 2009. The following figures show the results plotted together with the data obtained by skilled Japanese observers (NMS; Nippon Meteor Society) for comparison. It is clear that the time profiles of the meteor showers in the campaigns are similar to those obtained by skilled observers. In order to show the similarity, the vertical axis of the NMS data in each figure is multiplied by one fourth or one fifth, of which the values are shown in the vertical axis in the right of the figure. This factor is thought to originate from the poor observational condition in the participants in the campaign. Most of the participants are in the large city or urban area where they have heavy light pollution in general.

In the case of the Geminids in 2007 (Figure 1), the derived *HR* of the campaign was about one-fifth of the data of NMS (Uchiyama 2007), while the time profile of the activity is similar to the NMS. On the other hand, the derived value of the *HR* in the campaign was one-fifth. This corresponds to the difference of a limiting magnitude of the observational condition between the campaign and NMS corresponds to 2.3 magnitude as population index (r) = 2.0 (IMO 2007).

In the case of the Perseids in 2008 (Figure 2), time variation of the hourly rate deduced from the campaign was also very similar to NMS (Uchiyama 2008), especially on August 12-13. The derived value of the *HR* was about one-fourth of the data of NMS. This corresponds to the difference of a limiting magnitude 1.9 magnitude as $r = 2.1$ (IMO 2008).

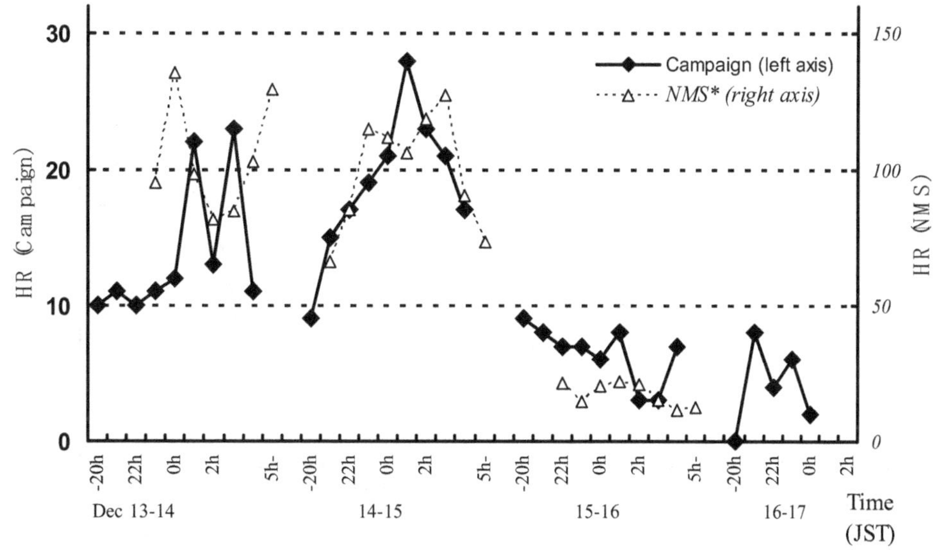

Figure 1. Hourly Rate of Geminids in 2007. The solid line with diamond marks is the results of our campaign and the dashed line with triangle marks is the results of the NMS (Nippon Meteor Society, Uchiyama 2007).

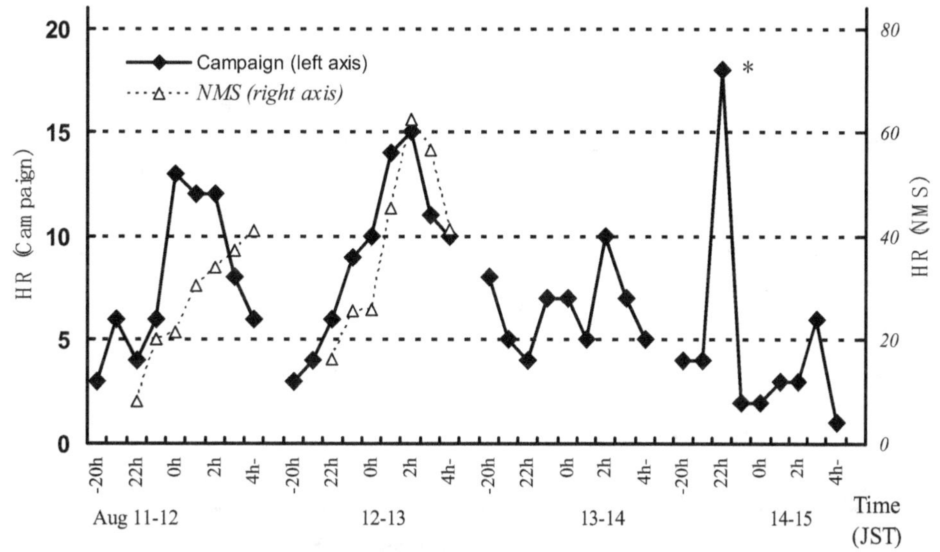

Figure 2. Hourly Rate of Perseids in 2008. The solid line with diamond marks is the results of our campaign and the dashed line with triangle marks is the results of the NMS (Uchiyama 2008). *The number of data was very few. (n = 2)

In the case of the Orionids in 2009 (Figure 3), the derived *HR* of campaign was also about one-fourth of the data of NMS (Iiyama 2009) like the case of the Perseids in 2008. The corresponding difference of a limiting magnitude is thought to be 2.0 magnitude when we apply the population index as $r = 2.0$ (IMO 2009). It should be noted that the time variation of the activity derived from our campaign seems to be smoother than the one by the NMS. Although this is mainly due to the huge number of reports, about 7,000, it may imply that the result by the huge number of observers may be better than that performed by a small number of skilled observers. We need further careful discussion on this point in the future.

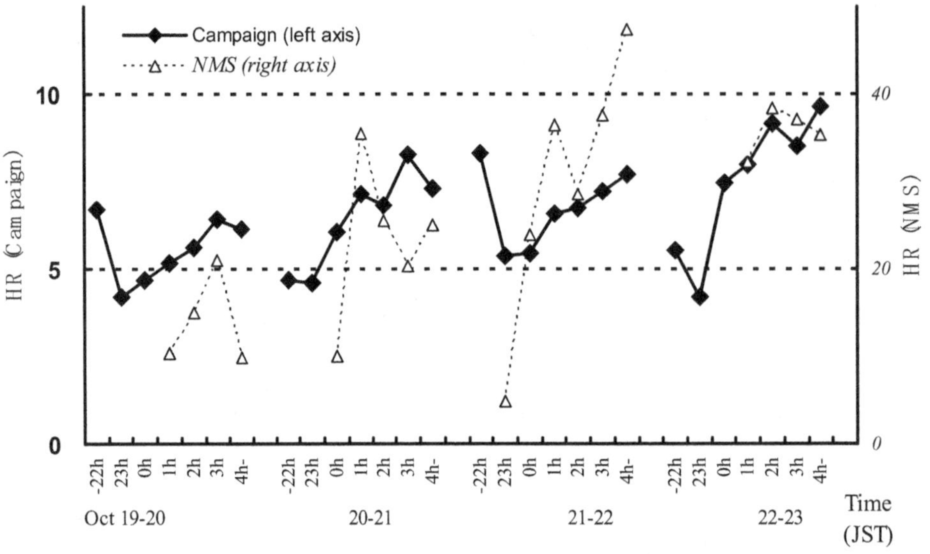

Figure 3. Hourly Rate of Orionids in 2009. The solid line with diamond marks is the results of our campaign and the dashed line with triangle marks is the results of the NMS (Iiyama 2009).

In case of the Geminids in 2009 (Figure 4), the derived *HR* of campaign was one-fifth of the data of NMS (Uchiyama 2009) from December 11 to 14. However, it changed to about one-seventh of the data of NMS from December 14 to 15. This corresponds to the variation of the limiting magnitude from 2.3 to 2.8, when we assume the population index is $r = 2.0$ (IMO 2009 No.2). The reason for this change may be due to the change of the sky condition of participating observers who reported to the campaign.

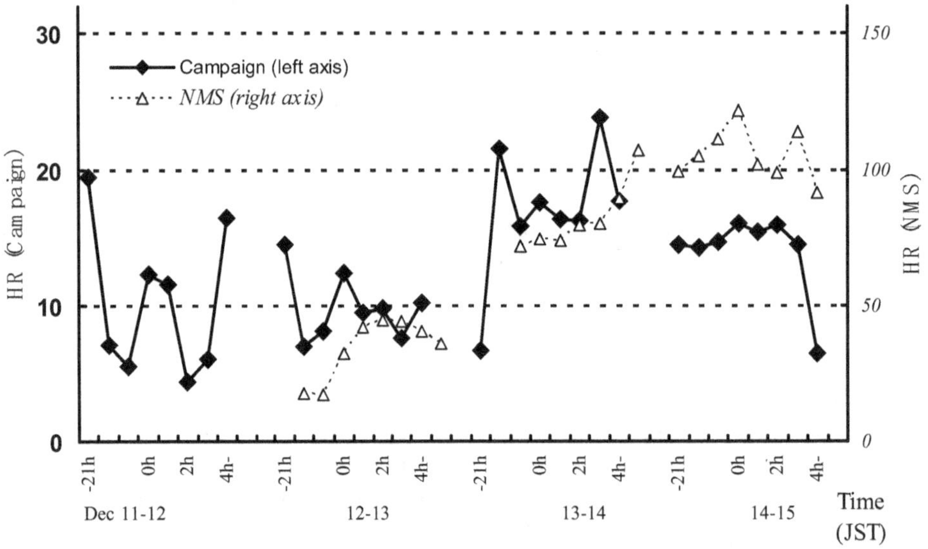

Figure 4. Hourly Rate of Geminids in 2009. The solid line with diamond marks is the results of our campaign and the dashed line with triangle marks is the results of the NMS (Uchiyama 2009).

5 Conclusion

We analyzed the data collected in public campaigns for four meteor showers, and confirmed that the derived time variation of the activities of meteor showers is very similar to those obtained by skilled visual observers.

On the other hand, the derived values of the *HR* in the campaigns are about one-fifth (Geminids in 2007 and 2009, except for from December 14 to 15) or about one-fourth (Perseids in 2008 and Orionids in 2009) compared to the data of the NMS. This is mainly due to poor observational sites for participants in the campaign, and probably partly due to immature skill of participants in the campaign. The difference of the limiting magnitude is estimated to be 1.9 ~ 2.3, as the average observational condition between the campaigns' participants and skilled observers. Even if we should have such difference, it is clear that we have a potential to extract scientific results from such outreach programs related to the meteor showers mainly due to the huge number of reports.

References

O. Iiyama, Report of visual observation on Octover in 2009, Astron. Circ., J. Nippon Meteor Soc., 810, 4 (2009)

IMO, Geminids 2007: Visual data quicklook, http://www.imo.net/live/geminids2007/ (2007)

IMO, Perseids 2008: Visual data quicklook, http://www.imo.net/live/perseids2008/ (2008)

IMO, Orionids 2009: Visual data quicklook, http://www.imo.net/live/orionids2009/ (2009)

IMO, Geminids 2009: Visual data quicklook, http://www.imo.net/live/geminids2009/ (2009 No.2)

S. Uchiyama, Quick results of Geminids in 2007, http://homepage2.nifty.com/s-uchiyama/meteor/shwr-act/12gemact/gem-act.html, 4 (2007)

S. Uchiyama, Quick results of Perseids in 2008, http://homepage2.nifty.com/s-uchiyama/meteor/shwr-act/08peract/per-act.html, 3 (2008)

S. Uchiyama, Quick results of Geminids in 2009, http://homepage2.nifty.com/s-uchiyama/meteor/shwr-act/12gemact/gem-act.html, 1 (2009)

Observations of Leonids 2009 by the Tajikistan Fireball Network

G. I. Kokhirova • J. Borovička

Abstract The fireball network in Tajikistan has operated since 2009. Five stations of the network covering the territory of near eleven thousands square kilometers are equipped with all-sky cameras with the Zeiss Distagon "fish-eye" objectives and by digital SLR cameras Nikon with the Nikkor "fish-eye" objectives. Observations of the Leonid activity in 2009 were carried out during November 13-21. In this period, 16 Leonid fireballs have been photographed. As a result of astrometric and photometric reductions, the precise data including atmospheric trajectories, velocities, orbits, light curves, photometric masses and densities were determined for 10 fireballs. The radiant positions during the maximum night suggest that the majority of the fireball activity was caused by the annual stream component with only minor contribution from the 1466 trail. According to the PE criterion, the majority of Leonid fireballs belonged to the most fragile and weak fireball group IIIB. However, one detected Leonid belonged to the fireball group I. This is the first detection of an anomalously strong Leonid individual.

Keywords observations · fireball · atmospheric trajectory · radiant · orbital elements · light curve · density · porosity

1 Introduction

Leonids are a well known meteor shower capable of producing meteor storms around November 17. The parent body is comet 55P/Tempel-Tuttle. Complex observations of Leonids were performed both by ground-based and aircraft facilities during 1998-2002 and in 2006 in connection with the high activity of the shower at this period. Owing to extensive observational data, very important results were obtained which significantly complemented meteor physics and dynamics and physical properties of cometary meteoroids. For the first time, extraordinary high beginning altitudes of the luminosity of the Leonid meteors were registered, among which some reaching the limit of almost 200 km, and are a result of both physical-chemical features of Leonid meteoroids and conditions of ablation at such altitudes (Spurny et al. 2000a, Spurny et al. 2000b, Koten et al. 2006).

According to several authors (Vaubaillon et al. 2005, Maslov 2007, Lyytinen and Nissinen 2009), high activity of the Leonids was predicted also in 2009.

In this work, the results of the photographic observations of the meteor shower Leonids in 2009 in Tajikistan are presented.

G. I. Kokhirova (✉)
Institute of Astrophysics of the Academy of Sciences of the Republic of Tajikistan. E-mail: kokhirova2004@mail.ru

J. Borovička
Astronomical Institute of the Academy of Sciences of the Czech Republic, Ondřejov Observatory

2 Observational Data

The photographic observations of the Leonids activity in 2009 were carried out during November 13-21, by the fireball network which consists of 5 stations situated in the south part of the Tajikistan territory and covering the area of near eleven thousands square kilometers (Babadzhanov and Kokhirova 2009b). The mutual distances between them range from 53 to 184 km. All stations of the network are equipped with all-sky cameras with the Zeiss Distagon "fish-eye" objectives (f = 30 mm, D/f = 1:3.5) using sheet films 9×12 cm and by digital SLR cameras "Nikon D2X" and "Nikon D300" with the Nikkor "fish-eye" objectives (f = 10.5 mm, D/f = 1:2.8).

As a result of observations, 16 Leonid fireballs have been photographed, from which 9 were registered on the night of maximum activity of November 17/18. Among all, 3 fireballs have been photographed from five stations, 1 – from four, 2 – from three, 7 – from two, and 3 – from one station.

The time of fireball appearance was determined by the method of combination of fireball images obtained by fixed and guided cameras, or by the digital fireball image. During the maximum night, double station video observations were performed simultaneously (Koten et al., in preparation). For six fireballs reported here, more precise times of appearance could be extracted from the video tapes. Here we present precise data of only 10 photographed fireballs for which the coordinates of radiants, heights, velocities, light curves, and orbital elements were determined. The geometrical conditions for the other three double-station fireballs were not good enough to compute reliable trajectories. Fireball photographs were measured using the Ascorecord device. Digital fireball images were measured using the Ascorecord measuring software "FISHSCAN" developed by J.Borovička for measurements of scanned photographs of fireballs registered by all-sky cameras.

Astrometric reduction procedures are the same as that used by the European Fireball Network, which allows determination of the position of an object at any point of photographic frame with the precision of one arc minute or better (Borovička et al. 1995, Babadzhanov et al. 2009).

3 Atmospheric Trajectories

The basic parameters of atmospheric trajectories of fireballs are given in Table 1, which contains the following data: the number of the fireball; the number of stations whose fireball photographs were involved in reduction; the type of camera which registered a fireball; date, the time of the fireball passage in UT;

- $L_{\odot}$ is the longitude of the Sun corresponding to the time of the fireball passage (J2000.0);
- v_B and v_E are the velocities at the beginning and at the end of the luminous trajectory;
- h_B and h_E are the beginning and the terminal heights of the luminous trajectory above the sea level;
- l is the total length of the luminous trajectory;
- M_P is the maximum absolute magnitude of the fireball;
- m_∞ is the initial mass of the meteoroid;
- m_E is the terminal mass of the meteoroid;
- PE is the empirical end height criterion for fireballs; the type of fireball according to Ceplecha and McCrosky (1976) classification. The standard deviations given for the beginning and the terminal points reflect the precision in computing the heights and positions of fireballs in the atmosphere. In Table 1, FC means fireball camera and DC – digital ones.

Table 1. Data of the atmospheric trajectories of the fireballs.

Fireball No.	TN171109A	TN171109B	TN171109C	TN171109D	TN171109E	TN171109F	TN171109G	TN131109	TN191109	TN211109
Number of stations, type of camera	5 stations FC 2 stations DC (5 total)	5 stations FC	3 stations FC 1 station DC (3 total)	2 stations DC	2 stations DC	4 stations FC	5 stations FC	2 stations FC 1 station DC (2 total)	2 stations FC 2 stations DC (3 total)	2 stations FC 1 stations DC (2 total)
Date, 2009	November 17	November 17	November 17	November 17	November 17	November 17	November 17	November 13	November 19	November 21
Time (UT)	20h39m09s ±10s	20h49m56s ±2s	21h10m25s ±2s	21h24m05s ±2s	22h17m14s ±2s	22h37m37s ±2s	23h35m27s ±2s	22h09m01s ±15s	22h14m41s ±15s	21h47m39s ±15s
$L^o_{\odot}$	235.504	235.511	235.526	235.535	235.572	235.590	235.627	231.535	237.589	239.590
v_∞ (km s^{-1})	71.84±0.05	72.07±0.17	71.45±0.38	71.77±0.18	71.71±0.60	71.59±0.53	70.57±0.49	71.80±0.16	70.38±0.36	71.78±0.03
h_B (km)	111.21±0.01	112.38±0.02	107.66±0.09	114.56±0.02	114.06±0.01	106.45±0.29	106.68±0.01	108.50±0.04	104.26±0.01	110.33±0.02
v_E (km s^{-1})	71.84±0.05	72.07±0.17	71.45±0.38	71.77±0.18	71.71±0.60	71.59±0.53	70.57±0.49	71.80±0.16	70.38±0.36	71.78±0.03
h_E (km)	91.03±0.00	91.51±0.04	91.05±0.07	98.91±0.02	77.84±0.01	89.04±0.28	87.01±0.01	91.14±0.03	91.05±0.02	98.20±0.01
l (km)	51.2	50.3	32.9	29.9	55.1	24.7	23.8	26.4	19.9	21.2
M_{max}	-7.2	-8.5	-8.3	-3.7	-3.4	-9.1	-8.3	-8.0	-6.3	-7.5
m_∞ (kg)	0.007	0.019	0.020	0.0002	0.00025	0.017	0.007	0.008	0.002	0.004
m_E (kg)	0	0	0	0	0	0	0	0	0	0
PE	-5.64	-5.90	-5.99	-5.76	-4.40	-5.98	-5.75	-5.82	-5.67	-6.34
Type	IIIA	IIIB	IIIB	IIIB	I	IIIB	IIIB	IIIB	IIIA	IIIB

Note that for all fireballs it was impossible to determine decelerations along the trajectories reliably. The cameras are not particularly suitable for studying velocities of very fast meteors like Leonids, since the shutter frequency is relatively low (12–15 breaks per second). In some cases we had to rely on 3 or 4 shutter breaks. Therefore, only average velocities were computed and were assumed to be equal to the initial velocities.

One digital camera was equipped with symmetrical two-blade shutters rotating with the frequency 370 rotations per minute in front of the objective. In the fireball cameras, the shutter is placed near the focal plane.

4 Photographic Beginning and End Heights of Visible Fireball Trajectories

It is undoubted now that the limit of beginning heights of photographic high-velocity meteors reaches 200 km. This fact was confirmed due to observations of the Leonids storm and outbursts during 1998-2002. Use of the more sensitive than photographic techniques provided a large number of meteors registered at the beginning heights between 130-200 km (see, e.g., Spurny et al. 2000a, Spurny et al. 2000b, Koten et al. 2006).

Our observational equipment does not allows us to record meteors at such heights because for the film's sensitivity I = 125 ISO units the limiting magnitudes of registration of meteors is equal to about -4 magnitudes. While, as was shown by Spurny et al. (2000a) and Koten et al. (2006), a brightness of meteors at heights above 130 km is more than 0 magnitude, as a rule.

The range of beginning heights of fireballs under investigation photographed by all-sky cameras is between 112-104 km. On observations from the same point it is revealed that the beginning height registered by the digital camera is equal to 128-114 km. This difference is caused by greater sensitivity of the digital camera. The standard range of terminal heights is 98-87 km for all-sky cameras and is practically the same for digital ones. One case of terminal height of 77.8 km was fixed only by digital camera.

From all-sky photographic records of Leonid fireballs Shrbeny and Spurny (2009) obtained the value 111 ± 5 km for beginning height for the range of maximum absolute magnitudes from -3 to -14, and concluded that this is the limiting altitude of all Leonids registration by the all-sky cameras.

Spurny et al. (2000b), investigating photographic and TV heights of high-altitude Leonid meteors (Hb > 116 km), found that photographic beginning height of a meteoroid weakly depends on its initial mass or maximum absolute magnitude. But they revealed relatively strong correlation on end heights, namely, very bright Leonid meteors, and consequently with greater mass, penetrate more than 20 km deeper than the faintest ones.

We plotted the same graphs using our data (Figures 1 and 2). The greatest magnitude and initial mass of described fireballs are $Mmax$ = -9.0 and m_∞ = 0.02 kg i.e. our data represents a half of the data range used by Spurny et al. (2000b).

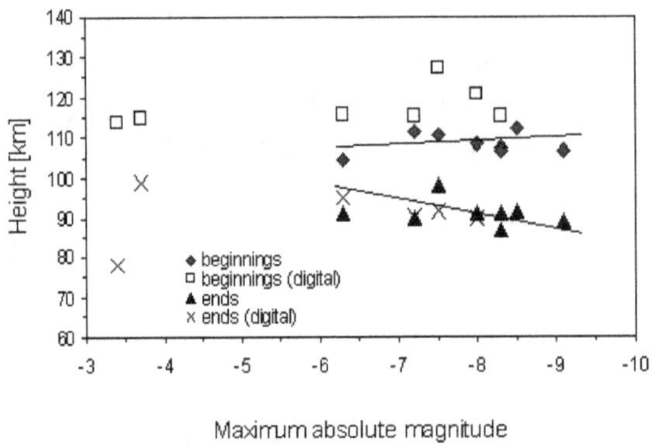

Figure 1. The Leonid beginning and terminal heights as a function of maximum absolute magnitude.

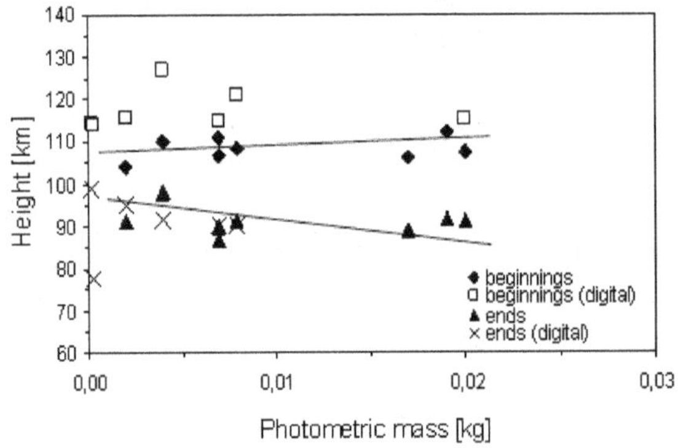

Figure 2. The Leonid beginning and terminal heights as a function of initial mass.

Nevertheless, gradual dependences of beginning and terminal heights on maximum absolute magnitude and initial mass can be seen clearly. However, the fireball TN171109E with maximal magnitude –3.4 and initial mass of only 2.5×10^{-4} kg was quite anomalous in this respect because it penetrated to the terminal height of 77.8 km, much deeper than more massive bodies.

5 Radiants and Heliocentric Orbits of Fireballs

Table 2 gives the results of determination of the coordinates of radiants and heliocentric orbits of the Leonid fireballs with their standard deviations. Here:

- α_R, δ_R are the right ascension and declination of the apparent radiant of fireball at the time of observation;
- z_R is the zenith distance of the apparent radiant;
- Q_p is the convergence angle between two planes (for multi-station fireballs the largest angle from all combinations of planes);
- v_∞ is the initial (preatmospheric) velocity;
- α_g, δ_g are the right ascension and declination of the geocentric radiant of fireball in J2000.0 equinox;

- v_g is the geocentric velocity;
- v_h is the heliocentric velocity;
- a, e, q, Q, ω, Ω, i are the orbital elements in J2000.0 equinox.

The results of determination of the coordinates of radiants of Leonid fireballs photographed during November 13-21, 2009, in dependency on longitude of the Sun, are illustrated in Figure 3 and compared with previously published radiant drifts. Using only our data, the daily radiant drift was found to be $\Delta\alpha = 0.78^{\circ}$ and $\Delta\delta = -0.53^{\circ}$. Maximum activity of Leonids occurred on the night of November 17/18 at the Solar longitude near 235.55°. The enhanced activity was predicted to be produced by two meteoroid trails ejected from the parent comet in 1466 and 1533, respectively. The annual Leonid shower was expected to peak approximately at the same time but with much lower activity.

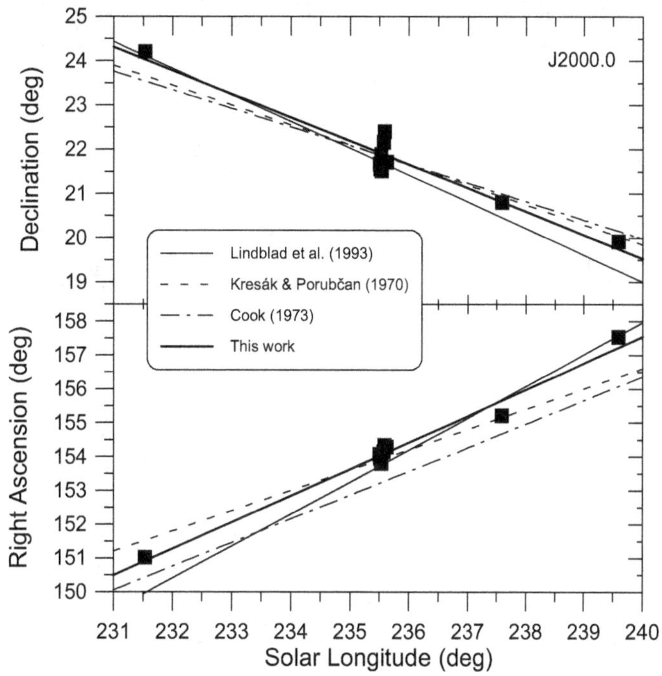

Figure 3. Drift of Leonid radiant as a function of Solar longitude. Our observations are compared with three published drifts as quoted in the book of Jenniskens (2006). Linear fit to our data is also shown. All coordinates are given in the equinox J2000.0.

Figure 4 shows the radiant positions of Leonids observed that night together with the predicted radiants for the 1466 and 1533 trails (Vaubaillon et al. 2009), the radiant of the annual shower according to various authors, and the so-called filament circle along which the radiants were spread during 2006 Leonids (Jenniskens et al. 2008). The radiants of two Leonids (D and F) have too large error to judge their origin. The radiant C, with moderate error, lies in between the annual radiant and the 1466 trail. Quite precise radiants A, B, and G lie closer to the annual shower or to the filament circle. Radiant E is the only one, which can be attributed with some confidence to the 1466 trail. None of the seven fireballs can be firmly attributed to the 1533 trail.

Table 2. Coordinates of radiants and heliocentric orbits of the fireballs.

Fireball No.	TN171109A	TN171109B	TN171109C	TN171109D	TN171109E	TN171109F	TN171109G	TN131109	TN191109	TN211109
α^o_R	153.87±0.04	153.77±0.02	153.70±0.14	153.73±1.11	154.17±0.01	154.42±0.58	154.51±0.03	151.04±0.09	155.26±0.07	157.51±0.07
δ^o_R	21.76±0.01	21.65±0.03	21.96±0.21	21.58±1.03	22.18±0.01	22.44±1.81	21.73±0.06	24.23±0.07	20.85±0.03	19.99±0.02
$cos\ z_R$	0.390	0.412	0.503	0.523	0.655	0.705	0.828	0.658	0.663	0.572
Q^o_p	74	55	25	5	33	17	74	51	62	58
α^o_g	154.07±0.04	153.95±0.02	153.81±0.15	153.81±1.12	154.14±0.02	154.33±0.59	154.29±0.03	151.02±0.09	155.21±0.07	157.54±0.07
δ^o_g	21.66±0.01	21.56±0.03	21.89±0.21	21.52±1.04	22.16±0.01	22.40±1.83	21.71±0.06	24.21±0.08	20.81±0.03	19.92±0.02
v_∞ (km s^{-1})	71.84±0.05	72.07±0.17	71.45±0.38	71.77±0.18	71.71±0.60	71.59±0.53	70.57±0.49	71.80±0.16	70.38±0.36	71.78±0.03
v_g (km s^{-1})	70.64±0.05	70.87±0.17	70.26±0.38	70.59±0.18	70.56±0.60	70.47±0.54	69.50±0.50	70.66±0.16	69.22±0.37	70.61±0.04
v_h (km s^{-1})	41.35±0.05	41.56±0.17	40.99±0.38	41.26±0.25	41.34±0.60	41.31±0.60	40.24±0.50	41.64±0.16	39.85±0.36	41.23±0.04
a (AU)	10.44±0.49	13.15±2.80	7.73±2.11	9.61±2.12	10.40±6.04	10.05±5.66	5.05±1.15	14.91±3.38	4.28±0.60	9.21±0.28
e	0.906±0.004	0.925±0.016	0.873±0.035	0.897±0.023	0.905±0.055	0.902±0.055	0.805±0.044	0.934±0.015	0.770±0.032	0.893±0.003
q (AU)	0.984±0.000	0.984±0.000	0.985±0.001	0.985±0.004	0.985±0.000	0.984±0.003	0.983±0.000	0.983±0.000	0.986±0.000	0.984±0.000
Q (AU)	19.90±0.98	25.32±5.59	14.48±4.23	18.23±4.24	19.81±12.08	19.11±11.32	9.12±2.30	28.85±6.76	7.58±1.20	17.44±0.56
ω^o	171.78±0.13	172.13±0.11	172.81±0.58	172.46±3.78	172.48±0.30	172.26±3.19	170.94±0.35	170.80±0.30	173.85±0.29	172.98±0.21
Ω^o	235.50±0.00	235.51±0.00	235.53±0.00	235.53±0.00	235.57±0.00	235.59±0.00	235.63±0.00	231.53±0.00	237.59±0.00	239.59±0.00
i^o	162.35±0.03	162.63±0.06	162.09±0.35	162.72±1.77	161.53±0.11	161.01±2.89	161.94±0.14	160.04±0.13	162.80±0.09	163.07±0.06

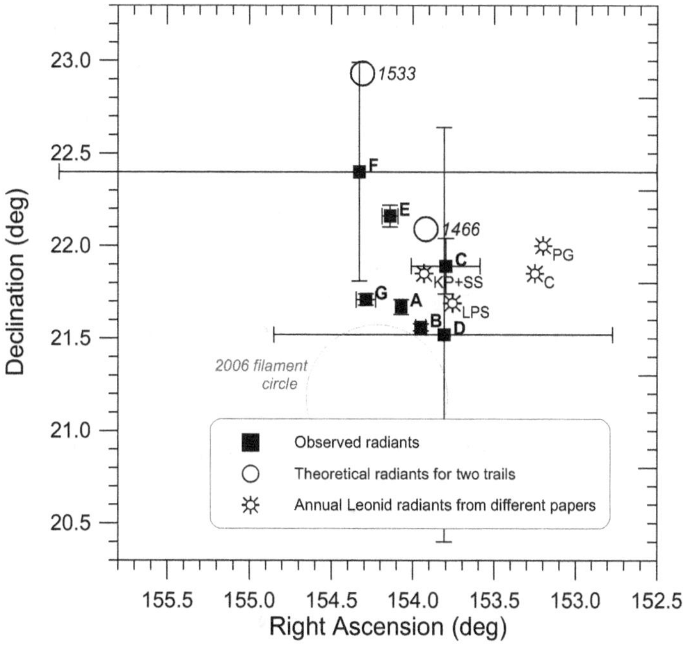

Figure 4. Leonid radiants during the maximum on November 17, 2009. The observed radiants are shown with their errors and compared with theoretical radiants for the 1466 and 1533 trails (Vauballion et al. 2009), with the annual Leonid radiant at Solar longitude 235.55° according to Cook (1973) (C), Kresák and Porubčan (1970) (KP), Lindblad et al. (1993) (LPS), Porubčan and Gavajdová (1994) (PG), and Shrbený and Spurný (2009) (SS), radiant almost identical to (KP). The filament circle as observed in 2006 (Jenniskens et al. 2008) is also shown. All coordinates are given in the equinox J2000.0.

The mean geocentric radiant of Leonid fireballs on November 17/18, 2009 is $\alpha = 153.66^{\circ} \pm 0.17^{\circ}$ and $\delta = 22.11^{\circ} \pm 0.31^{\circ}$, and is very close to the mean radiant values of Leonid fireballs in 1998 $\alpha = 153.63^{\circ}$, $\delta = 22.04^{\circ}$ for $L_{\odot} = 235.1^{\circ}$ (Betlem et al. 1999) and in 1999–2006 $\alpha = 153.6^{\circ} \pm 0.4^{\circ}$, $\delta = 22.0^{\circ} \pm 0.4^{\circ}$ for $L_{\odot} = 235.1^{\circ}$ (Shrbeny and Spurny 2009).

6 Light Curves of Fireballs

The photometry of Leonid fireballs was performed by the method developed for photographs taken by the Czech fish-eye camera (Ceplecha 1987). This method allows determine a brightness of fireball with the photometric precision of ±0.2 stellar magnitudes in the whole field to a zenith distance to 70°. Negatives, where fireball images have the best quality and the greater number of breaks, were used for photometry. The photometry of two fireballs observed only by the digital cameras was performed with the FISHSCAN program.

Maximum absolute magnitudes and initial photometric masses are given in Table 1. The maximum absolute magnitude ranges between –3.4 and -9.1, the masses are between 0.2 and 20 grams. The typical observed light curve of the fireball TN171109B is presented in Figure 5. We also present the light curve of deeply penetrating fireball TN171109E in Figure 6. All registered fireballs have smooth light curves with no significant flares. Almost all curves have asymmetric shape and the maximum points shifted towards to the end of luminosity.

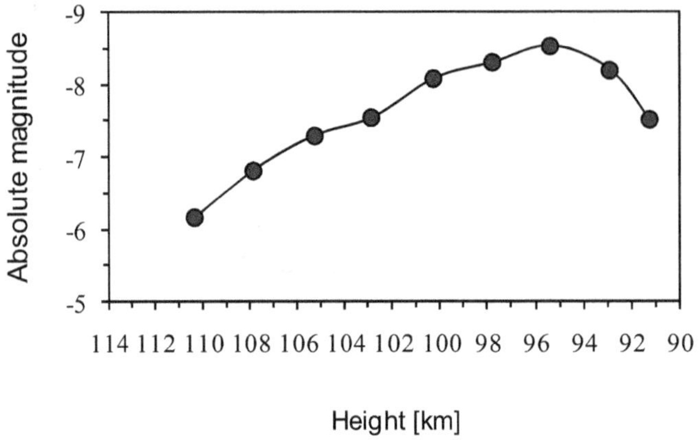

Figure 5. Observed light curve of Leonid fireball TN171109B.

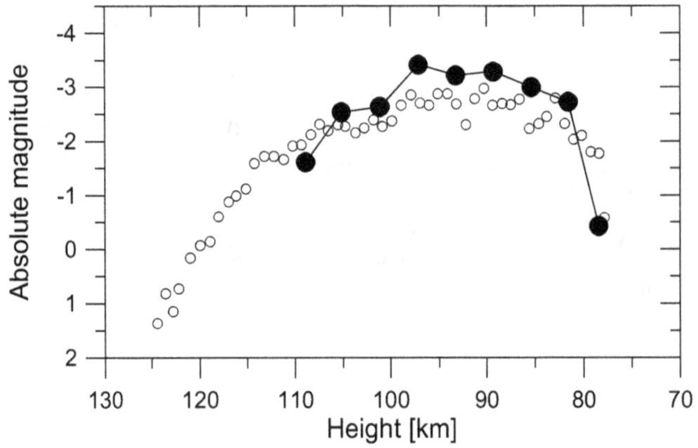

Figure 6. Observed light curve of Leonid fireball TN171109E. The empty circles are approximate magnitudes from the spectral video camera.

7 Physical Properties of Leonid Meteoroids

The values of the *PE* criterion given in Table 1 and calculated by the following expression:

$$PE = \lg \rho_E - 0.42 \lg m_\infty + 1.49 \lg v_\infty - 1.29 \lg \cos z_R,$$

where ρ_E – is the air density (g/cm^3) at the h_E – the terminal height of the fireball visible trajectory, indicate the penetration ability of a meteoroid; m_∞ is given in grams and v_∞ in km/s. For the majority of fireballs the *PE* values are typical for the fireballs of type IIIB according to Ceplecha and McCrosky (1976) classification or they lie close to the IIIA/IIIB boundary (*PE* = –5.70). The fireballs of group IIIB are produced by the meteoroids with the lowest bulk density equal to $\delta = 0.2$ g/cm^3, and represent the weakest cometary material. The fireball TN171109E was classified as type I, which is the absolute exception among Leonids and quite unusual for fireballs on cometary orbits. Type I fireballs are normally associated with stony meteoroids of density about 3.5 g/cm^3. The existence of different fireball types among the Leonid fireballs was also confirmed by Shrbeny and Spurny (2009). They recognized fireballs corresponding to types II, IIIA, and IIIB according to the *PE* criterion and made a conclusion on non-homogeneity of the parent comet.

Babadzhanov and Kokhirova (2009a) on the basis of photographic observations of Leonids determined mean bulk density equal to $\delta = 0.4 \pm 0.1$ g/cm^3, and mean mineralogical density of $\delta m = 2.3 \pm 0.2$ g/cm^3 of these meteoroids. Using the relation between these densities, the porosity of Leonid meteoroids was calculated to p = 83%. These confirm the very porous and fragile (weak) structure of the Leonid meteoroids. It turned out that density and porosity of Leonid meteoroids are very similar to those of Draconid meteoroids, which also were found to be porous aggregates of constituent grains with bulk density of $\delta = 0.3$ g/cm^3 and porosity of p = 90% (Borovicka et al. 2007).

The value of mean bulk density $\delta = 0.2$ g/cm^3 of Leonid meteoroids under investigation obtained according to the calculated values of *PE* criterion and fireball type, is in good agreement with mentioned results of investigation of density and porosity of cometary meteoroids. The nature of TN171109E with likely much larger bulk density is puzzling in this context. Nevertheless, small strong constituents penetrating much deeper than the majority of the meteoroid were observed in Leonids before (Spurný et al. 2000a, Borovička and Jenniskens 2000). TN171109E is the first case where a whole Leonid meteoroid was so strong that it was classified as type I meteoroid.

8 Conclusions

As a result of photographic observations by the Tajikistan fireball network during November 13-21, 2009, 16 Leonid fireballs were registered, from which 9 fireballs were captured at the night of maximum on November 17/18. This number confirms the forecasted enhanced activity of Leonids in 2009.

The results of determination of precise atmospheric trajectories, velocities, initial masses and orbits of 10 Leonid fireballs are presented in this study.

The daily radiant drift of Leonids was found to be $\Delta\alpha = 0.78^\circ$ and $\Delta\delta = -0.53^\circ$. The radiant positions during the maximum night suggest that the majority of the fireball activity (i.e. the majority of flux of Leonid meteoroids larger than 0.2 g) was caused by the annual stream component with only minor contribution of the 1466 trail. According to the *PE* criterion, the majority of Leonid fireballs belonged to the most fragile and weak fireball group IIIB, corresponding to the meteoroid mean bulk density of about 0.2 g/cm^3 and porosity of 80–90%. However, one detected Leonid of a size of about 5 mm belonged to the fireball group I and likely had a bulk density of few g/cm^3. This is the first detection of an anomalously strong Leonid individual.

Acknowledgments

The authors would like to express gratitude to specialists of the Institute of Astrophysics of Tajik Academy of Sciences M.I.Gulyamov, A.Sh. Mullo-Abdolov, A.O.Yulchiev, U.Kh.Khamroev, S.P.Litvinov, and to Dr. P. Koten (Astronomical Institute of the Academy of Sciences of the Czech Republic) who participated in the observations and data reduction. This work was supported by the International Science and Technology Centre Project T-1629.

References

P.B. Babadzhanov, G.I. Kokhirova, J. Borovička, P. Spurny, Photographic observations of fireballs in Tajikistan, Solar System Research **43** No. 4, 353-363 (2009)

P.B. Babadzhanov and G.I. Kokhirova, Densities and porosities of meteoroids, Astron. & Astrophys. **495**, Issue 1, 353-358 (2009a)

P.B. Babadzhanov and G.I. Kokhirova, Photographic fireball networks. Izvestiya Akad. Nauk Resp. Tajikistan **2(135)**, 46-55 (2009b)

H. Betlem, P. Jenniskens P., J. van't Leven, et al., Very precise orbits of 1998 Leonid meteors. Meteoritics & Planetary Sciences **34**, 979-986 (1999)

J. Borovička and P. Jenniskens, Time resolved spectroscopy of a Leonid fireball afterglow. Earth, Moon and Planets **82-83**, 399–428 (2000)

J. Borovička, P. Spurny, J. Keclikova, A new positional astrometric method for all-sky cameras. Astron. & Astrophys. Suppl. Ser. **112**, 173-178 (1995)

J. Borovička, P. Spurny, P. Koten, Atmospheric deceleration and light curves of Draconid meteors and implications for the structure of cometary dust. Astron. & Astrophys. **473**, 661-672 (2007)

Z. Ceplecha, Geometric, dynamic, orbital and photometric data on meteoroids from photographic fireball networks. Bull. Astron. Inst.Czechosl. **38** No. 4, 222-234 (1987)

Z. Ceplecha and R.E.J. McCrosky, Fireball end heights - A diagnostic for the structure of meteoric material. J. Geophys. Res. **81** No. 35, 6257-6275 (1976)

A.F. Cook, A Working List of Meteor Streams. In Evolutionary and Physical Properties of Meteoroids (eds. C.L. Hemenway, P.M. Millman, A.F. Cook), NASA SP-319, 183–191 (1973)

P. Jenniskens, Meteor Showers and their Parent Comets. Cambridge Univ. Press, New York, 790 p. (2006)

P. Jenniskens et al., Leonids 2006 observations of the tail of trails: Where is the comet fluff? Icarus **196**, 171-183 (2008)

P. Koten, P. Spurny, J. Borovička, S. Evans et al., The beginning heights and light curves of hight-altitude meteors. Meteoritics & Planetary Sciences **41** No. 9, 1305-1320 (2006)

L. Kresák and V. Porubčan, The dispersion of meteors in meteor streams. I. The size of the radiant areas. Bull. Astron. Inst.Czechosl. **21** No. 3, 153-170 (1970)

B.A. Lindblad, V. Porubčan, and J. Štohl, The orbit and mean radiant motion of the Leonid meteor stream. In Meteoroids and their Parent Bodies (eds. J. Štohl and I.P. Williams). Slovak Acad. Sci. Bratislava, 177–180 (1993)

E. Lyytinen and M. Nissinen, Predictions for the Leonid 2009 from a technically dense model. WGN **37:4**, 122-124 (2009)

M. Maslov, Leonid predictions for the period 2001-2100. WGN **35:1**, 5-12 (2007)

V. Porubčan and M. Gavajdová, A search for fireball streams among photographic meteors. Planetary and Space Science **42**, No. 2, 151-155 (1994)

L. Shrbeny and P. Spurny, Precise data on Leonid fireballs from all-sky photographic records. Astron. & Astrophys. **506**, 1445-1454 (2009)

P. Spurny, H. Betlem, K. Jobse, P. Koten, J. Van't Leven, New type of radiation of bright Leonid meteors above 130 km. Meteoritics & Planetary Sciences **35**, 1109-1115 (2000a)

P. Spurny, H. Betlem, J. Van't Leven, P. Jenniskens, Atmospheric behavior and extreme beginning heights of the thirteen brightest photographic Leonid meteors from the ground-based expedition to China. Meteoritics & Planetary Sciences **35**, 243-249 (2000b)

J. Vaubaillon, F. Colas, L. Jorda, A new method to predict meteor showers, II: Application to Leonids. Astron. & Astrophys. **439**, 761-770 (2005)

J. Vaubaillon, P. Atreya, P. Jenniskens, J. Watanabe, M. Sato, M. Maslov, D. Moser, B. Cooke, E. Lyytinen, M. Nissinen, and D. Asher, Leonid Meteors 2009. Central Bureau Electronic Telegram no. 2019 (2009 November 16)

CHAPTER 2:

ASTEROIDS AND METEOR SHOWERS: CASE OF THE GEMINIDS

Multi-Year CMOR Observations of the Geminid Meteor Shower

A.R. Webster • J. Jones

Abstract The three-station Canadian Meteor Orbit Radar (CMOR) is used here to examine the Geminid meteor shower with respect to variation in the stream properties including the flux and orbital elements over the period of activity in each of the consecutive years 2005 – 2008 and the variability from year to year. Attention is given to the appropriate choice and use of the *D*-criterion in the separating the shower meteors from the sporadic background.

Keywords meteor · orbital elements · radar · *D*-criterion

1 Introduction

Located near Tavistock, Ontario (43.26N, -80.77E) and operating at a frequency of 29.85 MHz, the three-station Canadian Meteor Orbit Radar (CMOR) has been in place for over a decade accumulating a considerable amount of data relating to meteor orbits, sporadic and shower (Jones et al, 2005). Here, observations of the Geminid meteor shower are used from an extended four year period (2005 – 2008) to cover the full range of solar longitude over which there is significant activity. The shower is known for its consistent return each year and the objective here is to look for variability in the waxing and waning stages in a given year and from year-to-year.

2 Observational Data

The radar is a back-scatter system and, aside from the occasional down-time for maintenance or weather events, operates continuously with a wide-angle all-round view of the sky. While sporadic meteors are widely spread in elevation and azimuth, the position of a detected shower meteor is governed by the shower radiant direction resulting in an effective "echo-line" on which the observed meteor lies (Kaiser, 1960). As the radiant rises, passes through transit and sets, the echo-line moves with it in a perpendicular fashion and with a minimum range which increases with the radiant elevation. As a result of this motion, the observed radar echoes move in range over the period when the radiant is above the horizon leaving a characteristic range-time "signature"; this is illustrated in Figure 1 for the Geminid shower. It will be noted that from the latitude of the radar site, this signature covers a total period of about 16 hours with a gap of about 3 hours centred on transit time.

In developing and applying the analysis routines, data from the year 2008 were first used over the anticipated period of significant activity, 251° to 267° in Solar Longitude (S.L.); the routines were then applied to the years 2005 – 2007 to complete the picture. The approach taken is illustrated in

A. R. Webster (✉) • J. Jones
Meteor Physics Group, The University of Western Ontario, London, ON. N6A5B9, CANADA. E-mail: awebster@eng.uwo.ca

Figure 2. The first filter employed (rather generous) restrictions on the values of Right Ascension (*RA*), Declination (*Dec*) eccentricity (*e*) and semi-major axis (*a*). The final selection of Geminid meteors made use of the *D*-criterion.

The application of the first filter to the 2008 data is shown in Figure 3, where the "range-time' signature of the Geminids is apparent, as is the peak in shower activity around 261° S.L.

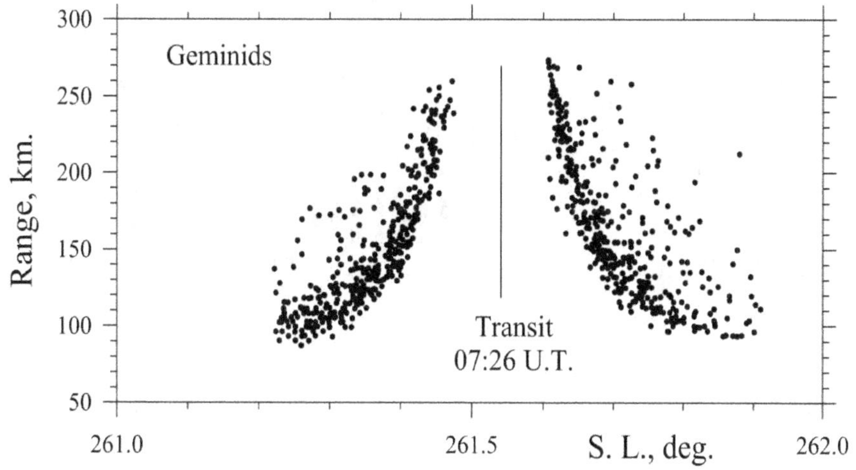

Figure 1. The "range-time" signature of the Geminid shower; the sharp minimum range will be noted.

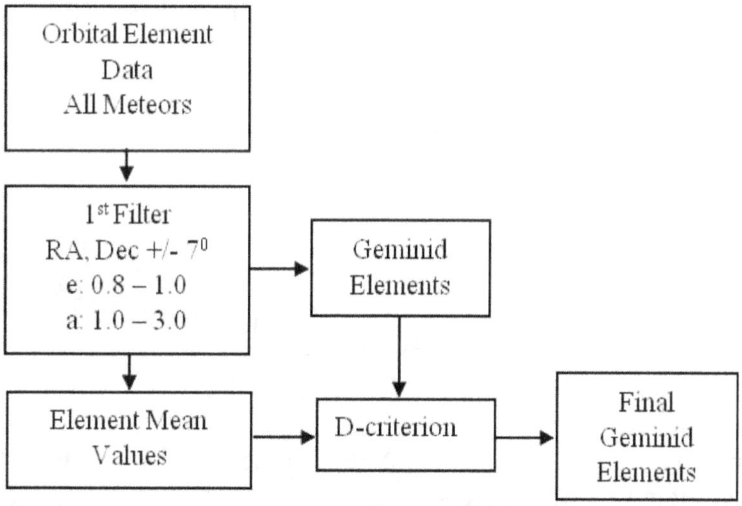

Figure 2. Extraction of Geminid meteors from the total observed.

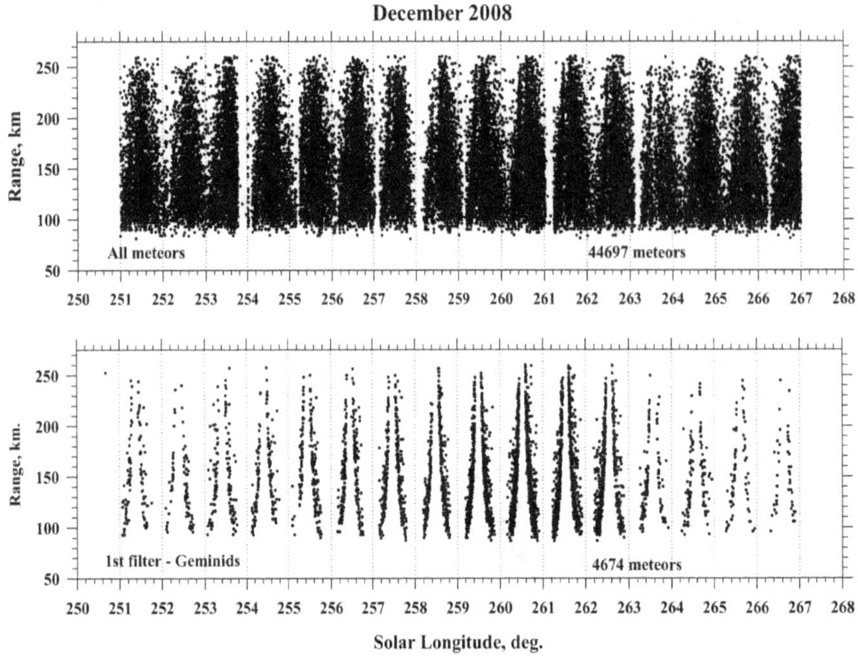

Figure 3. Total observed meteor echoes over the period 251° to 267° in S.L. in 2008 (top) and those extracted by the 1st filter (bottom).

The limits imposed in this first cut were deliberately made fairly wide to ensure that a high fraction of the Geminids present were selected, in the expectation that some sporadic meteors would be included. With this in mind, the application of the oft-used *D*-criterion was thought to be appropriate in reducing this contamination. The three versions based on the 5 orbital elements, q, e, i, ω, Ω commonly used were examined; Southworth and Hawkins (1963), Drummond (1981) and Jopek (1993) shown below (D_{SH}, D_D and D_J respectively), i.e.,

$$D_{SH}^2 = (e_1 - e_2)^2 + (q_1 - q_2)^2 + \left(2\sin\frac{I_{21}}{2}\right)^2 + \left(\frac{e_1 + e_2}{2}\right)^2 \left(2\sin\frac{\Pi_{21}}{2}\right)^2 \qquad (1a)$$

$$D_D^2 = \left(\frac{e_1 - e_2}{e_1 + e_2}\right)^2 + \left(\frac{q_1 - q_2}{q_1 + q_2}\right)^2 + \left(\frac{I_{21}}{180^0}\right)^2 + \left(\frac{e_1 + e_2}{2}\right)^2 \left(\frac{\Theta_{21}}{180^0}\right)^2 \qquad (1b)$$

$$D_J^2 = (e_1 - e_2)^2 + \left(\frac{q_1 - q_2}{q_1 + q_2}\right)^2 + \left(2\sin\frac{I_{21}}{2}\right)^2 + \left(\frac{e_1 + e_2}{2}\right)^2 \left(2\sin\frac{\Pi_{21}}{2}\right)^2 \qquad (1c)$$

where Π_{21} and Θ_{21} involve i, ω, and Ω. Application in turn of these to the data from the 1st filter results in the D values shown in Figure 4. The reference values used for the orbital elements were the mean values of the accepted meteors except for the longitude of the ascending node where the solar longitude at the time of occurrence is appropriate.

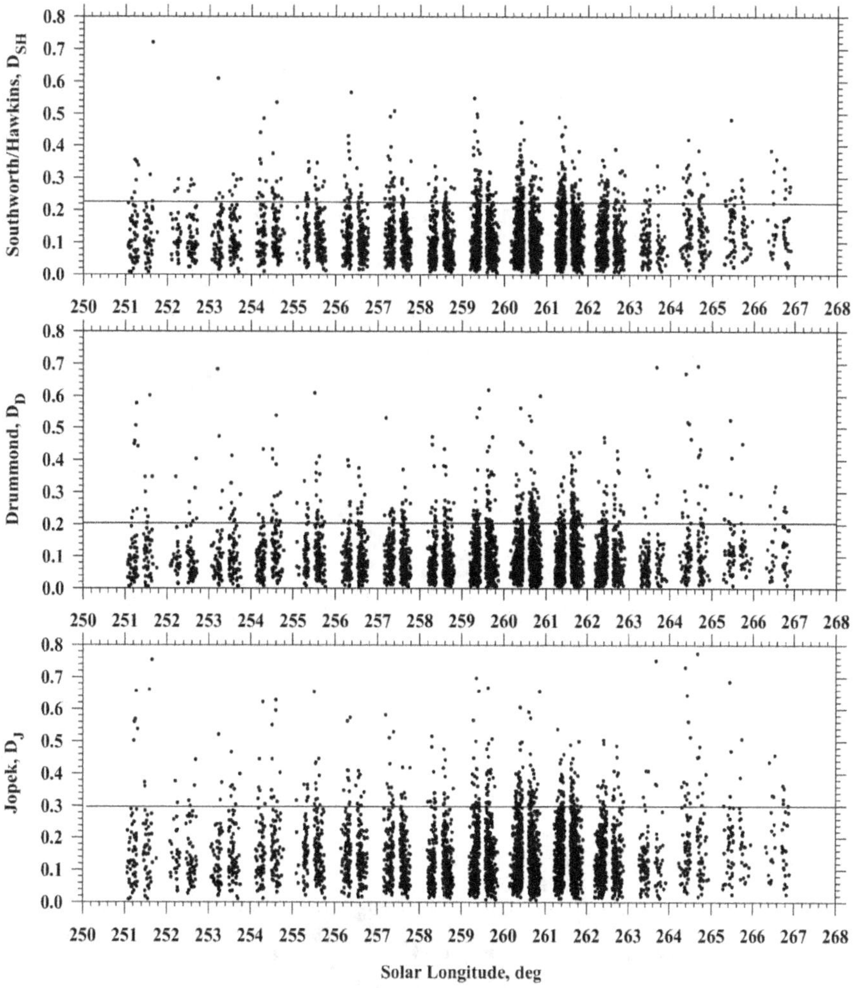

Figure 4. The *D*-criteria values as applied to the meteor orbital elements after the application of the 1st filter. The 90% cut-off values of $D_{SH} = 0.24$, $D_D = 0.21$ and $D_J = 0.30$ will be noted (see text and Figure 5.).

As can be seen in Figure 4, while the distributions are similar, the appropriate cutoff value to be used would be somewhat different. A better idea of this may be obtained from Figure 5 showing the differential and cumulative distributions for each of the criteria. In deciding what value of *D* to use for accepting the data, visual examination of Figure 4 suggests that at the time of the Geminid maximum, a significant number of shower meteors have *D* value higher than that normally used in this kind of application. Further, the waxing and waning of the activity in Figure 4 suggests that most of the meteors belong to the Geminid shower. Given the evidence in Figures 4 and 5, it was decided to apply a value of 0.21 to the Drummond data corresponding to the acceptance of ~90% of the meteors. This resulted in the reduction of presumed Geminids from 4674 to 4272 (Figure 6).

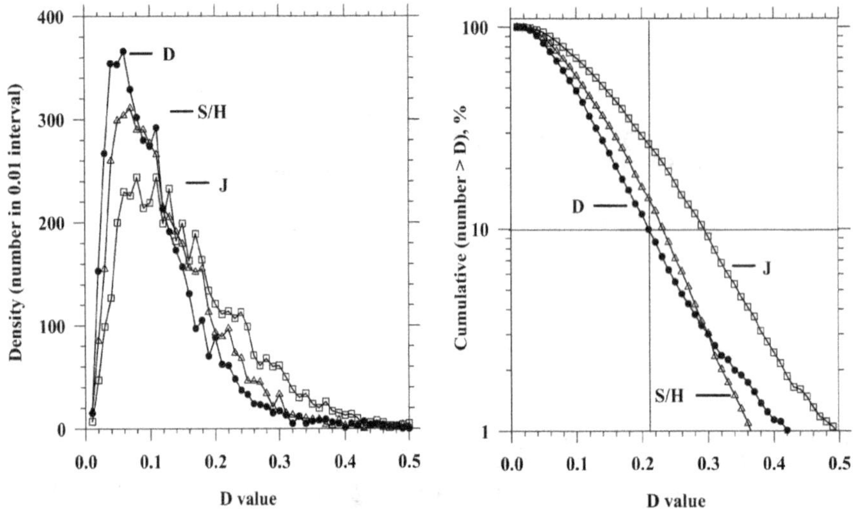

Figure 5. The differential (left) and cumulative (right) distributions of the three *D*-criteria D_{SH}, D_D and D_J.

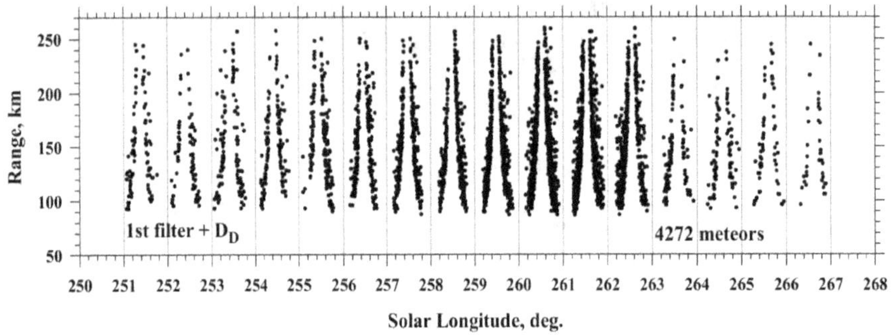

Figure 6. The range-time distribution of selected meteors after applying the 1st filter and the Drummond *D*-criterion with $D_D = 0.21$ cut-off (2008 data).

These remaining 4272 meteors in 2008 were assumed to represent a good estimate of the total observable Geminids with little contamination from other sources. The resulting echo rate, that is the total number of Geminid meteors seen by the radar over the period of significant activity, is presented in Figure 7, expressed in terms of the rate before and after transit and the total for a given night's observation. It will be remembered that the effective observing periods amounted to about 6.5 hours each before and after transit and the numbers presented represent a total for these periods.

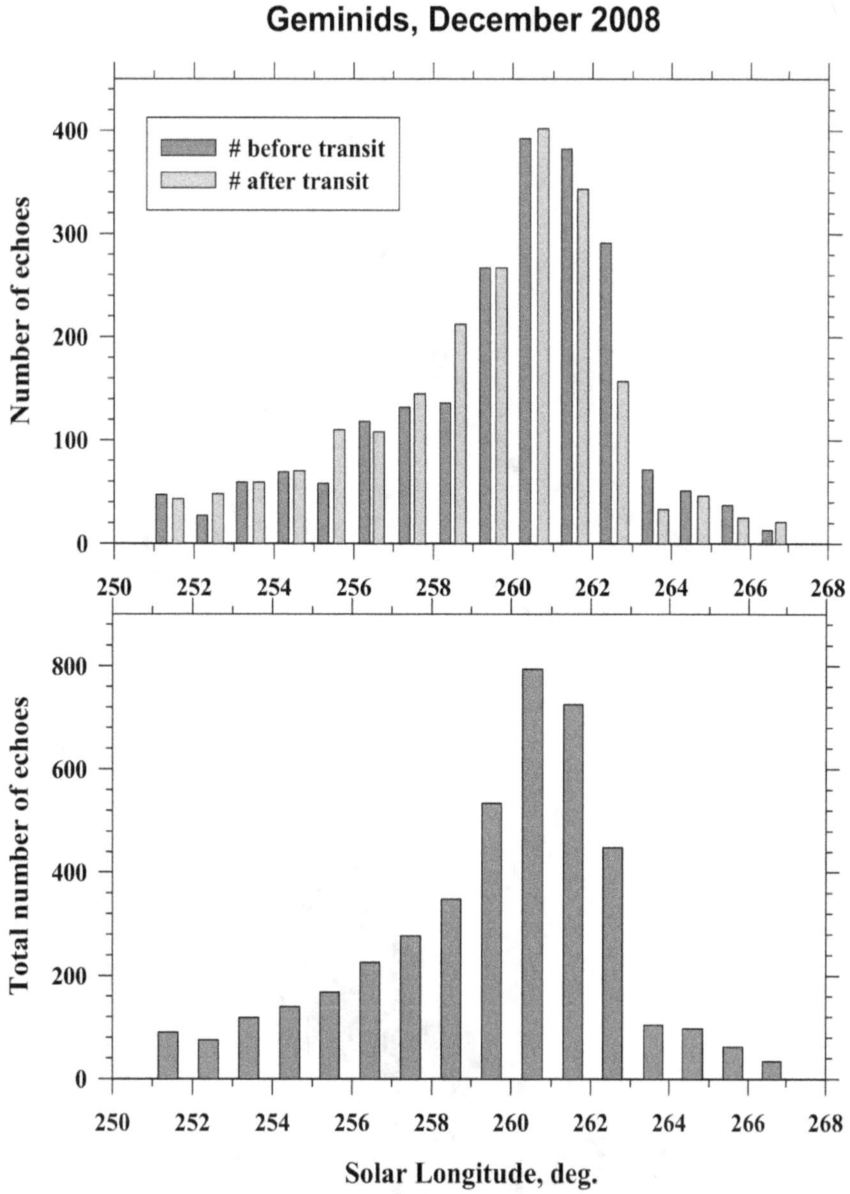

Figure 7. The echo rate seen by the radar over the period of significant activity; the rate before and after transit (top) and the total rate for each night (bottom).

The same routines were then applied to the data from the years 2005-2007 and the results consolidated into the activity shown in Figure 8. Again, for clarity, the total results for each night also are presented here. Since the transit time repeats every year, the fractional 0.25 day in the year causes a regression in the transit about 0.25° in Solar Longitude from year-to-year resulting in the "filling-in" seen in Figure 8. The classic rise to a maximum at about 261° in S.L. followed by a rapid fall in activity is apparent with little in the way of fluctuations. The residual activity at both ends of the observing period appears to be genuine.

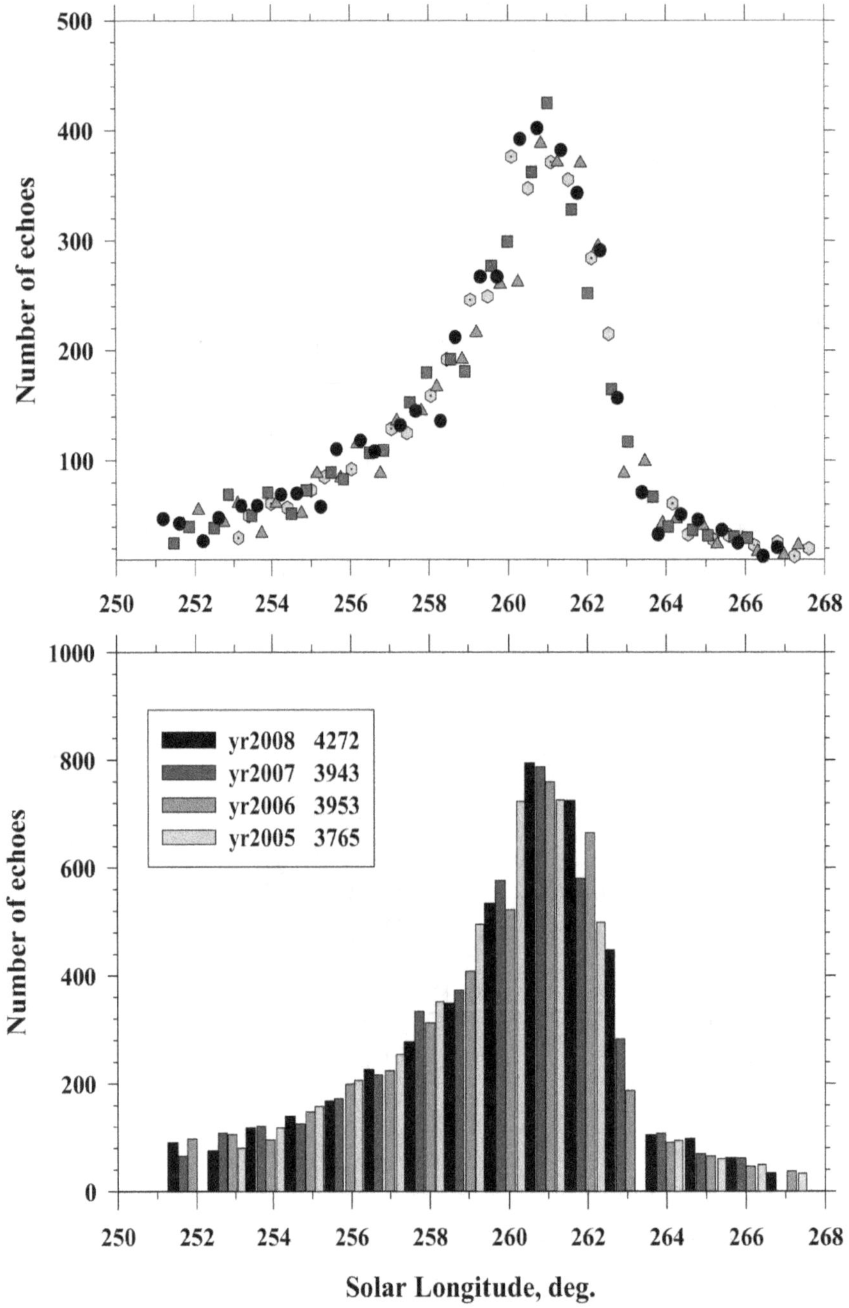

Figure 8. The activity of the Geminid shower over the four year period 2005 – 2008 showing: (top) the individual rates before and after transit; (bottom) the total number on a given night in each year for clarity.

The remarkable consistency from year-to-year is evident; it will be noted also that results are missing for 3 days in 2005, but were they available and in line with the trend, a further 200 or so would be added to the 2005 total.

The next step was to look at the variations in the various stream parameters including the orbital elements, velocities etc. All of these were available for each of the 15933 Geminid meteors selected, and linear regression was applied to plots of each parameter versus Solar Longitude. Examination of Figure 8 suggests that activity peaks at about $SL = 261°$ and this was used as the reference point. Figure 9 gives an example of this procedure showing the variation in orbital inclination. Similar results of this exercise for all the parameters are summarized in Table 1; the quoted uncertainties are standard errors.

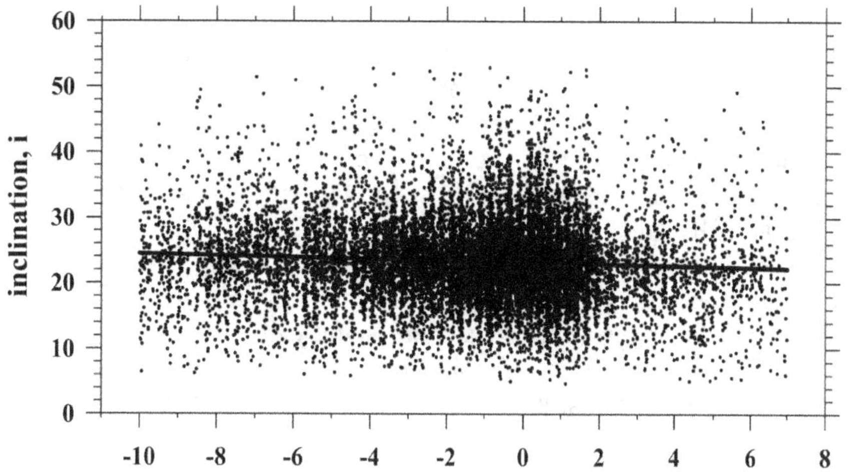

Figure 9. The variation in orbital inclination with Solar Longitude with $SL = 261°$ as the reference. The linear regression line is shown. All the 15933 selected Geminid meteors over the 4 year period are included.

Table 1. Mean Values and Variations with Solar Longitude
$y = b_0 + b_1*(SL\text{-}261)$

y	b_0	b_1
Semi-major axis, ***a*****,** AU.	**1.426** ± 0.003	+0.003 ± 0.001
Eccentricity, ***e***	**0.8964** ± 0.0003	+0.0007 ± 0.0003
Inclination, ***i*****,** deg.	**23.13** ± 0.05	-0.13 ± 0.02
Argument of perihelion, ***ω*****,** deg.	**324.9** ± 0.04	-0.06 ± 0.01
Right Ascension, deg.	**112.64** ± 0.02	+1.07 ± 0.01
Declination, deg.	**31.93** ± 0.02	-0.18 ± 0.01
Geocentric, v_g, km/s	**34.35** ± 0.05	-0.02 ± 0.02
Heliocentric, v_h, km/s	**33.79** ± 0.04	+0.01 ± 0.01

3 Discussion and Comments

The results presented here are part of the ongoing and continuous operation of CMOR over extended periods with stable properties. This allows confidence in comparative studies encompassing several years. As with any such system, there are uncertainties in the measured quantities, but the extensive numerical data gathering properties of CMOR allow meaningful answers to be drawn.

Examining the data from a 4 year period, with separated samples before and after transit, allows ~ 8 samples per degree in Solar Longitude. Although not entirely unexpected, the consistency of the flux of Geminid meteors from year-to-year is notable as are the relatively smooth variations from day-to-day. Given that, there is a suggestion of fluctuations in activity around the peak at *SL* ~261° which may be consistent with the more frequently sampled results presented by Rendtel (2005) using visual observations. The residual activity at each end of the period in this study is believed to represent Geminid meteors; a separate study using CMOR suggests that such activity may extend from late November to early January (Brown et al, 2010).

The changes in the orbital elements over the duration of the shower are notably small. For example, given the evidence in the literature for decreasing magnitude distribution exponent, generally associated with the Poynting-Robertson effect, a more significant increase in the semi-major axis, *a*, might be expected as the Earth moves from the inside to the outside of the stream.

The *D*-criterion has been, and is, used extensively in looking for connections between bodies orbiting the Sun and the three versions considered here have been use with differing cut-off values depending on the observing system used. In his paper, Drummond suggested values of $D_D = 0.105$ and $D_{SH} = 0.25$ in linking meteor streams and parent bodies based on the visual, photographic and radar data presented by Cook (1973) and Marsden (1979). Williams and Wu (1993) used the Drummond version with D_D again equal to 0.105. Galligan (2001) investigated the three criteria using the AMOR system in New Zealand and suggested a 90% recovery using $D_{SH} = 0.20$, $D_D = 0.18$ and $D_J = 0.23$. It might be remarked that different magnitude ranges can be involved in such studies which may influence the effectiveness; for example, AMOR has a limiting magnitude of around +13.5, CMOR of ~ +8.5 with visual and photographic usually brighter than ~ +6.0. We believe that the choice depends on the system, the interactions being studied and the quality of the data and that the use of $D_D = 0.21$ is appropriate here.

A further version of the *D*-criterion was introduced by Valsecchi et al (1999) which has found much favour in some applications. Instead of using the five orbital elements for comparison, the geocentric velocity (speed and direction in Earth oriented coordinates) is used. This is particularly useful when the data is available as direct, rather than derived, measurements. In the case of CMOR, all of the elements are derived from interferometric and time-delay measurements, though we are looking into this approach and developments. It is noted that Galligan (above) also considered this method and found it to be comparable and preferable in some circumstances.

Acknowledgements

The authors would like to acknowledge the many helpful discussions with those involved with the operation and data handling of CMOR, P. Brown, M. Campbell-Brown, Z. Krzemenski and R. Weryk, and the substantial support from the NASA Meteoroid Environment Office. Thoughtful insights from discussions with G. Valsecchi at Meteoroids 2010 are also acknowledged.

References

Brown, P., Wong, D., Weryk, R. and Wiegert. P., 2010, Icarus, **207(1)**, 66-81.

Cook, A.F., 1973, *Evolutionary and Physical Properties of Meteoroids*, U.S. Gov. Printing Office, NASA SP-319, 183-319.

Drummond, J.D., 1981, *Icarus*, **45**, 545-553.

Galligan, D.P., 2001, *Mon. Not. R. Astron. Soc.*, **327**, 623-628.

Jopek, T.J., 1993, *Icarus*, **106**, 603.

Jones, J., Brown, P., Ellis, K.J., Webster, A.R. Campbell-Brown, M.D., Krzemenski, Z. and Weryk, R.J. : 2005, *Planetary and Space Science*, **53**, 413 – 421.

Kaiser, T.R., 1960, *Mon. Not. R. Astron. Soc*, **121**, 3, 284–298.

Marsden, B.G., 1979, *Catalogue of Cometary Orbits*, 3rd ed., Cent. Bureau Astron. Telegrams, I.A.U., SAO, Cambridge, Mass.

Rendtel. J., 2004, *Earth, Moon and Planets*, **95 (1-4)**, 27-32, DOI10.1007/sl1038-004-6958-5.

Southworth, R.B. and Hawkins, G.S., 1963, *Smithson. Contrib. Astrophys.*, **7**, 262-285.

Valsecchi, G.B., Jopek, T.J. and Froeschle, C., 1999, *Mon. Not. R. Astron. Soc.*, **304**, 743-750.

Williams, I.P. and Wu, Z., 1993, *Mon. Not. R. Astron. Soc.*, **262**, 231-248.

The Distribution of the Orbits in the Geminid Meteoroid Stream Based on the Dispersion of their Periods

M. Hajduková Jr.

Abstract Geminid meteoroids, selected from a large set of precisely-reduced meteor orbits from the photographic and radar catalogues of the IAU Meteor Data Center (Lindblad et al. 2003), and from the Japanese TV meteor shower catalogue (SonotaCo 2010), have been analyzed with the aim of determining the orbits' distribution in the stream, based on the dispersion of their periods P . The values of the reciprocal semi-major axis $1/a$ in the stream showed small errors in the velocity measurements. Thus, it was statistically possible to also determine the relation between the observed and the real dispersion of the Geminids.

Keywords meteoroid · meteor showers · meteoroid streams

1 Introduction

One of the most intense annual meteor showers, Geminids are produced by a meteoroid stream unusual in having small orbits with aphelia well inside the orbit of Jupiter and perihelia close to the Sun. The Geminid's parent body, asteroid (3200) Phaethon, with a perihelion distance of only 0.14 AU and semi-major axis 1.27 AU, appears to be an inactive cometary nucleus (Gustafson 1989, Beech et al. 2003). The Phaethon's active period was determined by Gustafson (1989) as not more than 2000 years ago. This is in agreement with the age of the meteoroid stream, calculated dynamically, and which corresponds to a few thousand years (Ryabova 1999, Beech et al. 2002). The model for the formation of the Geminid meteor stream was developed by Fox and Williams (1982). Later, Williams and Wu (1993) produced a theoretical model showing that meteoroids ejected from Phaethon could have evolved, under the influence of planetary perturbations and radiation pressure, into Earth crossing orbits. The orbits of the Geminid meteoroids with aphelia far inside the orbit of Jupiter lead to the fact that the gravitational effects of the other outer planets are negligible. Furthermore, there have not been any close encounters significantly affecting their orbits during at least the last ten thousand years (Ryabova 2007). Thus, the orbital elements of most stream meteoroids vary little; furthermore, the spread in these elements is approximately invariant with the passage of time (Jones and Hawkes, 1986). Therefore, the structure of the Geminid meteoroid stream is dominated by the initial spread of meteoroid orbits. Ryabova (2001, 2007) developed a model explaining the two branches of the stream as being formed by the disintegration of the parent body, due to differences in orbital parameters of the individual particles ejected from the parent body before and after perihelion. The small perihelion distance may cause an intense thermal processing, which affects the physical properties of the meteoroids (Beech et al. 2003) and the higher density of Geminids, in comparison with other meteoroids (Babadzhanov and Konovalova, 2004).

M. Hajduková Jr. (✉)
Astronomical Institute of the Slovak Academy of Sciences, 84504 Bratislava, Slovakia. E-mail: astronomia@savba.sk

The present paper, based on a statistical analysis of a large set of precisely-reduced meteor orbits, shows the dispersion in the orbital elements of Geminid meteoroids for different mass ranges of the particles. For the analysis, data from the photographic and radar catalogues of the IAU Meteor Data Center (Lindblad et al. 2003) were used. Among the 4,581 photographic orbits, 385 meteoroids belonging to the Geminid meteor shower were identified using Southworth-Hawkins *D*-Criterion for orbital similarity (Southworth and Hawkins, 1963) and fulfilling the condition $D_{SH} \leq 0.20$. Similarly, we applied a limiting value of $D_{SH} = 0.25$ to 62,906 radar orbits and obtained 887 Geminids. The photographic data in the MDC catalogues are limited to the mass range of 10^{-4} kg (3^m) and radar data to 10^{-7} kg (5^m); for more powerful radars to 10^{-9} kg (15^m). To cover a broad mass range of the particles, quality orbits from the reduced database of 8,890 meteoroid orbits (Vereš and Tóth, 2010) of the Japanese TV meteor shower catalogue (SonotaCo 2009) were also used, giving 1,442 Geminids for the limiting value of $D_{SH} = 0.20$, detected mostly up to +2 magnitude.

2 Observed Dispersion of Orbital Elements

It is obvious that examination of the structure of meteoroid streams by means of the period of the individual particles is possible only for the short period meteoroid streams. The meteoroid streams with long periods of several decades to centuries, e.g. Lyrids, Perseids, Orionids, Leonids and Eta Aquarids, have heliocentric velocities close to the parabolic limit. The observational errors of those meteor streams greatly exceed the real deviations from the parent comet's orbit. Given that errors in the heliocentric velocity are a significant source of uncertainty in semi-major axes determination, it should be mentioned that errors in velocity determination in the IAU MDC can reach the value $v_H \sim 10$ km s^{-1}. The errors differ both for individual catalogues and for individual meteor showers. The largest spread was found for the Perseids from the catalogues with a lower precision, reaching values of 10 − 15 km s^{-1} (Hajduková 1993, 2007). But this is certainly not the case with the Geminids, the mean heliocentric velocity of which is only 36.6 km s^{-1}. The values of the reciprocal semi-major axis in this stream show small errors in the velocity measurements. The different precision of measurements, depending on the observation technique as well as on the quality of observations, causes a natural spread in the orbital elements. Figure 1 shows the dispersion in eccentricities, perihelion distances and semi-major axes. For the sake of comparison, we also plotted the orbital element of Geminid's parent body, which was obtained using the computer program Dosmeth (Neslušan et al. 1998).

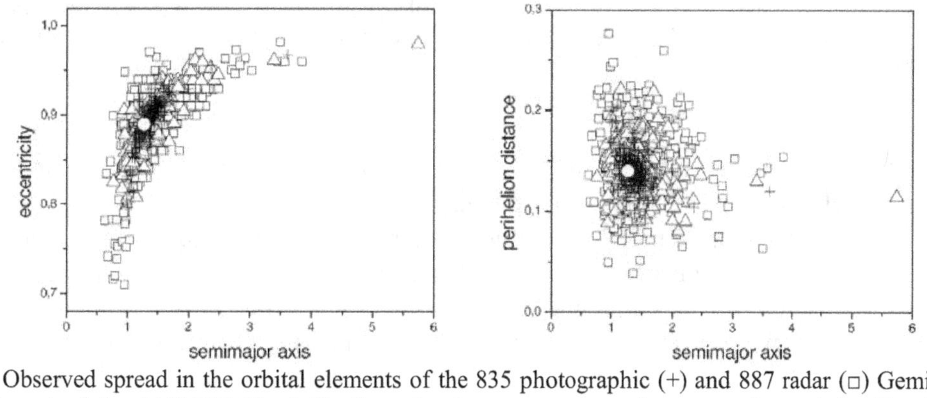

Figure 1. Observed spread in the orbital elements of the 835 photographic (+) and 887 radar (□) Geminids of the IAU MDC, and of the 1442 TV Geminids from the Japanese meteor shower catalogue (Δ). For the sake of comparison, we also plotted the orbital elements of the Geminid's parent body (○).

The observed dispersion of the orbit periods is shown in Figure 2 (left), separately for all three investigated data, obtained by different techniques. The mean period of the Geminids was found to be 1.59 and 1.48 years, derived from the photographic and video orbits, with a standard deviation of 0.37 and 0.24 respectively. The mean period of the fainter particles from the radar observation is 1.69 years, but the period determination from individual orbits varies from 0.53 to 7.54. It is clear that we are not dealing with a stream all of whose meteors have exactly the same period, but obviously the last observed spread in the values exceeds the real deviations.

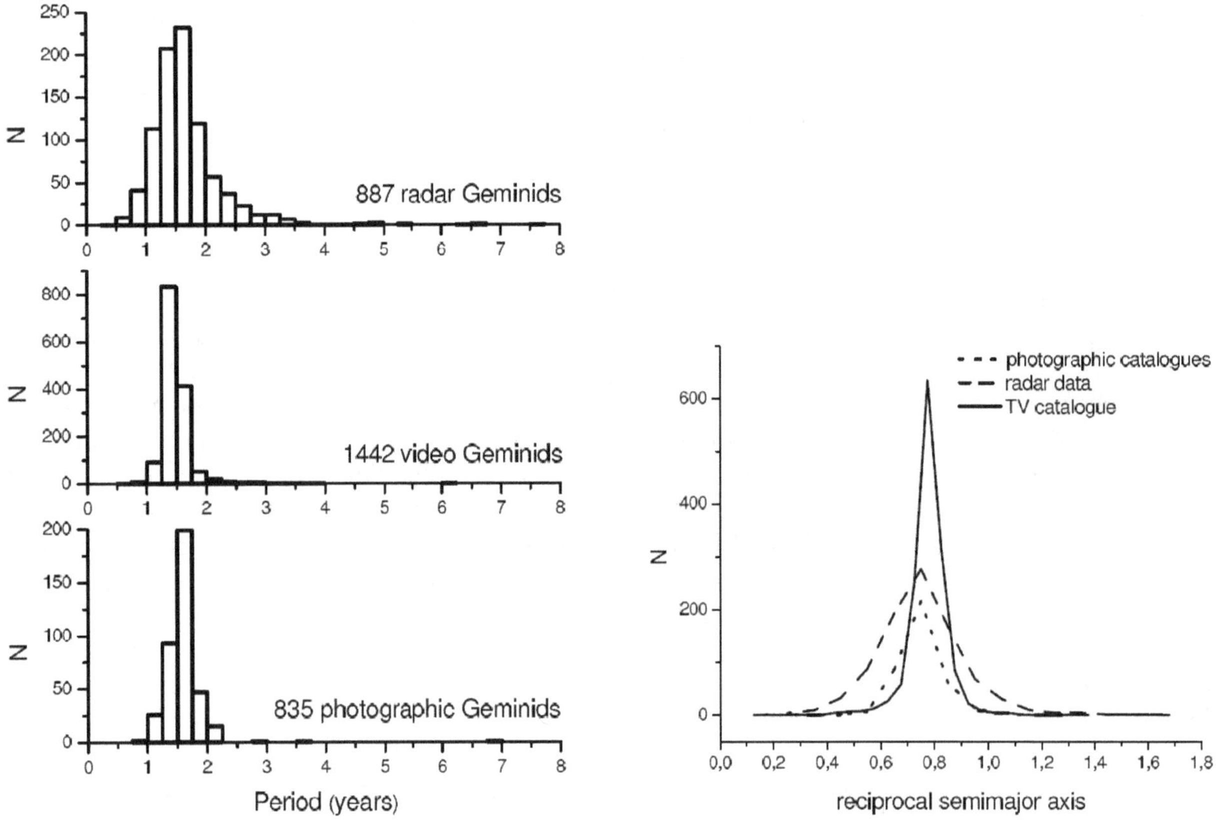

Figure 2. Comparison of observed dispersion of the period of revolution (left) and of the reciprocal semimajor axes (right) of Geminids from three different sets of data in term of mass particles, obtained using different techniques and different measurement methods. The observed dispersion is greater for radar Geminids in comparison with both other sets of data.

A complete study about the real dispersion of orbital periods in meteor streams was made by Kresák (1974), which showed that the observed dispersion of the semi-major axes involves the real orbital dispersion plus errors, which are greater by a factor of 10^4 for the orbits of the meteoroids than in the case of well-determined cometary orbits. Porubčan (1984), in his study of the dispersion of the orbital elements of meteor orbits, analyzing 153 photographic Geminids determined the mean orbital period at 1.66 years. The widely-observed dispersion is also seen in distributions of the reciprocal semi-major axes (Figure 2 right). The radar data in general are of a lower precision, which is obvious from the greater spread in the values of the semi-major axes in comparison with both other catalogues, in which the precision is comparable. The observed dispersion of the semi-major axis, defined by the standard deviation, is 0.079 for the photographic and 0.158 for the radar Geminids from the IAU MDC. The smallest standard deviation of 0.071 was derived from the Japanese video data, probably because we used a strict selection (Vereš and Tóth, 2010) of high quality video meteor orbits.

3 The Accuracy of the Semi-major Axes and their Dispersion

We tried to estimate the real dispersion of the semi-major axis within the meteor stream by comparing the observed dispersions in different catalogues of orbits, where the observational errors are different. However, for each observation technique, there are different sources of errors, which produce the observed dispersion in semi-major axis determination. On the basis of this fact, we chose in our analysis the median a_M as the most representative value of semi-major axis a, because the arithmetic mean value a is strongly affected by extreme deviations caused by gross errors. It was shown (Kresáková 1974) that the medians of $(1/a)_M$ in several major meteor showers do not differ from those of their parent comets beyond the limits of statistical uncertainty. The dispersion of the semi-major axis within the meteor stream is described by the median absolute deviation Δ_M in term of $1/a$: $\Delta_M\,(1/a) = |\,(1/a)_{1/2} - (1/a)_M\,|$, where $(1/a)_{1/2}$ are limiting values of the interval, which includes 50 percent of all orbits. The probable range of uncertainty is determined by $\pm\, n^{-1/2}\Delta_M\,(1/a)$, where n is the number of the meteor orbits used for the median determination $(1/a)_M$. For the sake of comparison, we also derived the deviation of the median $1/a$ from the parent body: $\Delta(1/a)_{Ph} = |\,(1/a)_M - (1/a)_{Ph}\,|$, where the $(1/a)_{Ph}$ is the reciprocal semi-major axis of Geminid's parent body Phaethon.

The results of our analysis are shown in Table 1 as in Figure 3. Table 1 summarizes the numerical results obtained separately for the three different sets of Geminids. The mean value, the standard deviations and the median semi-major axis a are listed in the first part of the Table. The second part contains the mean value, the standard deviations and the median reciprocal semi-major axis $1/a$. The median absolute deviation Δ_M in term of $1/a$, and the deviation of the median $1/a$ from the parent body, are listed in the last part of the Table. For comparison, we also list the chosen orbital elements from 3200 Phaethon.

Table 1. Numerical data obtained separately for the three different sets of Geminids observed by different techniques. n – number of meteors; $\bar{a}$, $\overline{1/a}$ – the mean values; σ_a, $\sigma_{1/a}$ – the standard deviations; a_M , $(1/a)_M$ – the median a, $1/a$ respectively; $\Delta_M\,(1/a)$ – the median absolute deviation; $\Delta(1/a)_{Ph}$ – deviation of the median $1/a$ from the parent body.

	$\bar{a}$	a_M	σ_a	$\overline{1/a}$	$(1/a)_M$	$\sigma_{1/a}$	$\Delta_M\,(1/a)$	$\Delta(1/a)_{Ph}$
$n_{phot} = 835$	1.361	1.356	0.180	0.744	0.737	0.079	0.040	0.049
$n_{tv} = 1442$	1.302	1.285	0.185	0.777	0.778	0.071	0.029	0.008
$n_{rad} = 887$	1.402	1.351	0.343	0.749	0.740	0.159	0.081	0.047
Phaethon	1.271			0.787				

The dispersion, described by the median absolute deviation Δ_M in terms of $1/a$ obtained from the photographic, video and radar catalogues, are 0.040, 0.029 and 0.081 AU^{-1} respectively. This corresponds to a deviation of ±0.01 years for the Geminid's period obtained from the precise photographic measurements. This is in agreement with a study by Kresáková (1974), which analyzed meteor orbits obtained from the most precise double-station photographic programs; it was shown that the dispersion of the 157 analyzed Geminids is moderate and the period can be put into narrow limits, between 1.62 and 1.64 years.

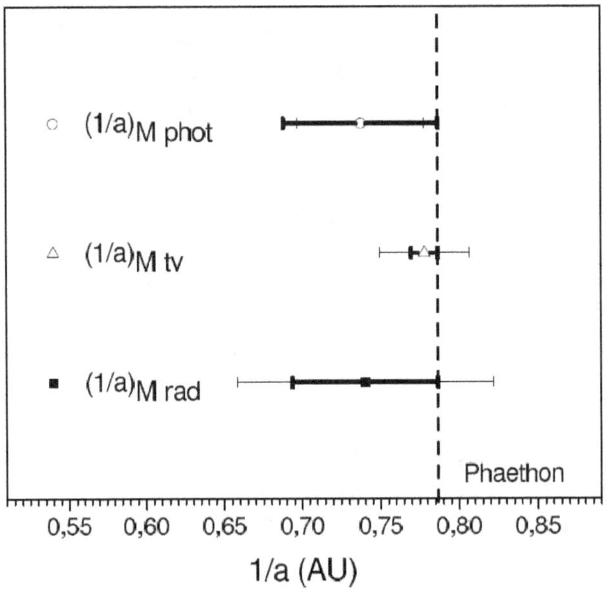

Figure 3. Dispersion in terms of $1/a$ for Geminids observed by three different techniques. Bold line – the deviation of the median $1/a$ from the parent body $\Delta(1/a)_{Ph}$; thin line – the absolute median deviation Δ_M in terms of $1/a$; vertical line – Phaethon.

The deviation of the median reciprocal semi-major axis from the parent body, obtained from Japanese video orbits, is only 0.008 AU^{-1}, whereas for the orbits from the IAU MDC catalogues, it is approximately five times greater. For the video and radar orbits, $\Delta(1/a)_{Ph}$ is considerably smaller than Δ_M $(1/a)$, but for photographic orbits, it is slightly bigger.

4 Conclusions

The analysis of a sufficient number of meteor orbits of chosen catalogues of meteors observed with different techniques allowed us to estimate the dispersion of semi-major axes within the Geminid meteor stream. It was shown that the dispersion differs considerably between the three different sets of data in terms of the different masses of the particles. This may be a consequence of different measurement errors for different observation techniques, as well as of different dispersions in the orbital elements for particles belonging to different mass ranges. The dispersion was found to be higher for small particles obtained by radars in comparison with the results of video and photographic observations of large meteoroid particles. It was found that the real dispersion of the Geminids is at least 2 times smaller than indicated by the observations, based on all three investigated catalogues. The deviations in terms of $1/a$ determined from the investigated catalogues range from ± 0.029 to ± 0.081 AU^{-1}. This corresponds to a deviation of ± 0.01 years for the Geminid's period obtained from the precise measurements and of ± 0.02 years using data of lower accuracy.

Acknowledgements

This work was supported by the Scientific Grant Agency VEGA, grant No 0636.

References

Babadzhanov, P. B. and Konovalova, N. A.: 2004, *Astron. Astrophys.* **428**, 241
Beach, M.: 2002, *Mon. Not. R. Astron. Soc.* **336**, 559
Beach, M.: 2003, *Meteoritics and Planetary Science* **38**, No.7, 1045
Fox, K. and Williams, I. P.: 1983, *Mon. Not. R. Astron. Soc.* **205**, 1155
Gustafson, B. A. S.: 1989, *Astronom. Astrophys.* **225**, 533
Hajduková, M. Jr.: 2007, *Earth, Moon and Planets* **102**, Issues 1-4, 67
Hajduková, M. Jr.: 1994, *Astronom. Astrophys.* **288**, 330
Jones, J. and Hawkes, R. L.: 1986, *Mon. Not. R. Astron. Soc.* **223**, 479
Kresák, L. and Kresáková, M.: 1974, *Bull. Astron. Inst. Czech.* **25**, No.6, 336
Kresáková, M.: 1974, *Bull. Astron. Inst. Czech.* **25**, No.4, 191
Lindblad, B., Neslušan, L., Porubčan, V. and Svoreˇn, J.: 2003, *Earth, Moon, Planets* **93**, 249
Neslušan,L., Svoreň, J., Porubčan, V.: 1998, *Astron. Astrophys.* **331**, 411
Porubčan, V.: 1978, *Bull. Astron. Inst. Czech.* **29**, No.4, 218
Ryabova, G. O.: 1999, *Solar System Research* **33**, 258
Ryabova, G. O.: 2001, *Proc. Meteoroids 2001 Conf. ESA Pub. Div., Noordwijk*, 77
Ryabova, G. O.: 2007, *Mon. Not. R. Astron. Soc.* **375**, 1371
Southworth, R. R. and Hawkins, G. S.: 1963, *Smithson. Contr. Astrophys.* **7**, 261
SonotaCo: 2009, *WGN, Journal of the IMO* **37**, 55
Vereš, P. and Tóth, J.: 2010, *WGN, Journal of the IMO* **38**, 1
Williams, I. P. and Wu, Z.: 1993, *Mon. Not. R. Astron. Soc.* **262**, 231

CHAPTER 3:

SPORADIC AND INTERSTELLAR METEOROIDS

Inferring Sources in the Interplanetary Dust Cloud, from Observations and Simulations of Zodiacal Light and Thermal Emission

A.C. Levasseur-Regourd • J. Lasue

Abstract Interplanetary dust particles physical properties may be approached through observations of the solar light they scatter, specially its polarization, and of their thermal emission. Results, at least near the ecliptic plane, on polarization phase curves and on the heliocentric dependence of the local spatial density, albedo, polarization and temperature are summarized. As far as interpretations through simulations are concerned, a very good fit of the polarization phase curve near 1.5 AU is obtained for a mixture of silicates and more absorbing organics material, with a significant amount of fluffy aggregates. In the 1.5-0.5 AU solar distance range, the temperature variation suggests the presence of a large amount of absorbing organic compounds, while the decrease of the polarization with decreasing solar distance is indeed compatible with a decrease of the organics towards the Sun. Such results are in favor of the predominance of dust of cometary origin in the interplanetary dust cloud, at least below 1.5 AU. The implication of these results on the delivery of complex organic molecules on Earth during the LHB epoch, when the spatial density of the interplanetary dust cloud was orders of magnitude greater than today, is discussed.

Keywords interplanetary dust · light scattering properties · thermal properties · atmospheric entry · comets · asteroids · meteoroids

1 Introduction

The question of the origin of the dust particles that are permanently replenishing the interplanetary dust cloud, thus allowing the appearance of the zodiacal light, has been extensively discussed all over the past years. Before the 1980s, the main source was assumed to be the dust released by active cometary nuclei in the interplanetary dust cloud (Whipple, 1955). In 1983, the detection of asteroidal bands and cometary trails by the Infrared Astronomical Satellite (IRAS) has allowed some authors to estimate that the main source was dust released by asteroidal collisions or disruptions (see e.g. Sykes and Greenberg, 1986). While minor sources of dust, such as dust from Jupiter and Saturn systems and dust of interstellar origin, have also been detected by Ulysses, Galileo and Cassini spacecraft (see e.g. Grün et al., 2001 and references therein; Taylor et al., 1996), the main source of interplanetary dust in the Earth environment has remained an open question.

It is most likely that the sources of most meteor streams are comet nuclei and that those of most meteorites are asteroidal fragments. Nevertheless, it is difficult to estimate whether comets or asteroids predominantly contribute to the zodiacal cloud, even in the vicinity of the Earth, and finally to know

A. C. Levasseur-Regourd (✉) • J. Lasue
UPMC (Univ. Paris 6), UMR 8190, BC 102, 46-45, 4ème, 4 place Jussieu, 75252 Paris Cedex 05, France. Phone: +33 1 4427 4875; Fax: +33 1 4427 3776; E-mail: aclr@aerov.jussieu.fr
Lunar and Planetary Institute, 3600 Bay Area Blvd., Houston, TX 77058, USA
LANL, Space Science and Applications, ISR-1, Mail Stop D-466, Los Alamos, NM 87545 USA

what are the sources of sporadic meteors and of micrometeorites. These questions are all the more important that the interplanetary dust cloud, even if assumed to be stationary, is likely to undergo numerous evolution processes, e.g. with fragmentation, weathering and partial sublimation of its dust particles. We will propose some answers through an approach that relies upon inversion of observations of the near-Earth zodiacal light and zodiacal thermal emission, and upon interpretations through numerical simulations. Finally, we will compare our results with those obtained for cometary dust and for the interplanetary dust through other approaches, and assess their implication for the delivery of carbonaceous compounds to the early Earth.

2 Results Derived From Observations

Observations from Earth's orbit in the visual and near infrared domains allow for the detection of the so-called zodiacal light and zodiacal thermal emission (see e.g. Levasseur-Regourd et al., 2001 and references therein). The zodiacal light is a faint veil of solar light, brighter towards the Sun and the near-ecliptic invariant plane of the solar system. The zodiacal thermal emission is the most prominent component of the light of the night sky in the 5 to 100 μm region, at least away from the galactic plane.

2.1. Near-Earth Zodiacal Light and Zodiacal Thermal Emission

The zodiacal light actually originates in the scattering of solar light by dust particles. The sharp increase of its brightness *Z*, towards the Sun and the invariant plane, indicates an increase in the space density of the interplanetary dust cloud, which forms a thick disk around the Sun. A slight enhancement in brightness, the gegenschein, also takes place in the anti-solar region; it corresponds to a backscattering effect. As expected from the scattering of randomly polarized solar light in an optically thin medium, the zodiacal light is partially linearly polarized. The polarization *P* is defined as the ratio of the difference to the sum of the brightness components respectively perpendicular and parallel to the scattering plane; it is slightly negative in the gegenschein region.

The brightness *Z* (in W m^{-2} sr^{-1} μm^{-1}) and the polarization *P* (in percent), as determined as functions of the helio-ecliptic latitude and ecliptic longitude, after correction for the invariant plane inclination (e.g. Leinert et al., 1998; Levasseur-Regourd et al., 2001), provide an estimation of the foreground noise induced by the zodiacal light, together with an optimization of the epochs of observations of faint extended astronomical objects. The zodiacal thermal emission, whose maximum is slightly above 10 μm, as observed from the Earth environment, corresponds to a temperature of about 250 K along the line-of-of sight. In the very near infrared domain, by 0.8 to 1.2 μm, the thermal emission is still negligible and the scattered light prevails. For larger wavelengths, observation of the thermal emission (which is isotropic) provides an easier detection of local heterogeneities than brightness emission, as recently illustrated by the detection from Spitzer spacecraft of the dust trail of comet 67P/Churyumov-Gerasimenko, the target of the Rosetta mission (Kelley et al., 2008).

2.2. Data Inversion and Local Results

Since the concentration and the temperature of the dust are changing significantly with the solar distance R, the local brightness and thermal emission are expected to vary along the line-of-sight for Earth or near-Earth based observations. Besides, it cannot be assumed that the interplanetary dust cloud is homogeneous and that the properties of the dust (e.g. albedo, size distribution) are the same everywhere in the cloud. The brightness, as well as its perpendicular and parallel components, and the thermal emission are thus integrals that need to be, at least partially, inverted. A rigorous inversion is feasible, for a line-of-sight tangent to the direction of motion of the observer and for the section of the line-of-sight where the observer is located. This approach has, up to now, provided bulk values of some local properties in the vicinity of the Earth (Table 1). To retrieve local information in regions that are not located on the orbit of the Earth, inversion mathematical methods, leading to comparable results, have been independently initiated by Dumont and Levasseur-Regourd (1988) and by Lumme (2000).

Table 1. Parameters relevant to the local properties of the interplanetary dust particles and their dependence with distance to the Sun R (0.3 to 1.5 AU range) in the near-ecliptic invariant plane (adapted from Levasseur-Regourd et al., 2001): Linear polarization P at 90° phase angle, temperature T, geometric albedo A and space density.

Parameter	Heliocentic gradient	Comment
$P_{90°}(R)$	$30\ R^{+0.5 \pm 0.1}$ (%)	Evolution of local polarization
$T(R)$	$250\ R^{-0.36 \pm 0.03}$ (K)	Not a perfect black-body
$A(R)$	$A_0\ R^{-0.34 \pm 0.05}$	Evolution of geometric albedo
Space density(R)	$10^{-17} R^{-0.93 \pm 0.07}$ (kg m^{-3})	Most likely $1/R$

One result is related to the shape of the local polarimetric phase curve (see Fig. 11 in Levasseur-Regourd et al., 2001). At 1.5 AU from the Sun in the invariant plane, it is smooth, with a slight negative branch, an inversion angle in the 15° to 20° range and a positive branch with a maximum of about 30 percent. This trend indicates that the scattering particles are irregular with a size greater than the wavelength of the observations, i.e. about 1 μm; it also suggests, assuming that the Umov empirical law is valid, that the particles have quite low an albedo. Another key result is related to the variations with the solar distance R (between 0.3 and 1.5 AU) of some local properties, which approximately follow power laws. The trend obtained for the local polarization at 90° phase angle, a ratio independent upon the concentration (see Fig. 5 in Levasseur-Regourd et al., 1991), establishes that the interplanetary dust cloud is heterogeneous, i.e. that the intrinsic properties of the dust vary with R. Since the dust particles spiral towards the Sun under Poynting-Robertson drag (or are blown away by solar radiation pressure), it can be assumed that the intrinsic properties vary with time and that the dust particles suffer a significant temporal evolution.

3 Interpretation Through Numerical Simulations

3.1 Zodiacal Light Results

Results need to be interpreted through appropriate simulations, with tentatively realistic assumptions about the size distribution, the composition and the structure of the particles (Levasseur-Regourd et al., 2007; Lasue et al., 2007). The size distribution may be assumed to be similar to that derived from in-situ measurements by Grün et al. (2001), showing a size distribution with a few branches following power-laws. We have approximated this size distribution with power-laws of index about -3 for sizes below 20 μm and about -4.4 for larger sizes. A predominance of silicates, with an average complex refractive index of about (1.62 + 0.03i) at 550 nm, and absorbing organic molecules or carbon, with an average complex refractive index of about (1.88 + 0.1i) at 550 nm, has been suggested from an analysis of previous studies of IDPs and micrometeorites by Lasue et al. (2007). The particles may either be compact, as expected for fragments resulting from asteroidal collisions and for some cometary dust, or constituted of aggregates, as expected for other cometary dust particles (as confirmed by Stardust mission, see also paragraph 4.1).

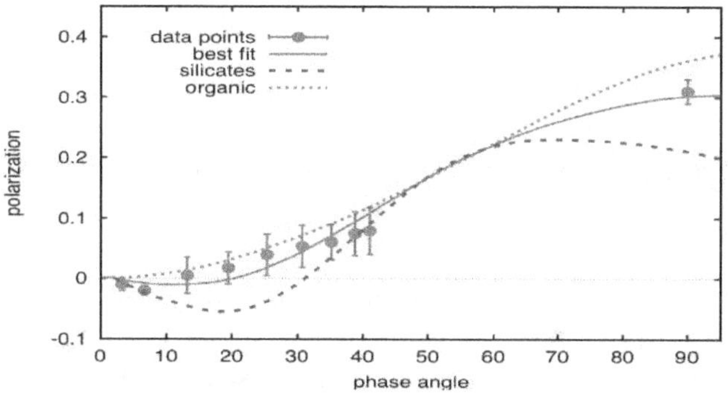

Figure 1. Best fit for the local polarimetric observations at 1.5 AU near the ecliptic. The dashed curve corresponds to non-absorbing silicates, the dotted curve to absorbing organic material. The solid curve is the best fit obtained by mixing 40% of organics and 60% of silicates in mass. (adapted from Lasue et al. 2007)

A combination of T-matrix calculations for small particles and ray-tracing simulations for larger particles is used to compute the light scattering from a cloud of dust particles built up of prolate spheroids and fractal aggregates of them. The best fit to the observational results constraints, at 1 AU in the invariant plane, the particles composition to 25-50% of organics in mass, and conversely to 75-50% of silicates in mass. The best estimate of the contribution of aggregated dust particles, simulated by irregular aggregates of spheroids randomly oriented, correspond to -at least- 20% of aggregates in mass (Lasue et al., 2007). This in turn, as extrapolated from the bulbous to single track ratio from the Stardust aerogel analyses (35% of bulbous tracks; Hörz et al., 2006; Burchell et al., 2008), would correspond to at least 50% in mass for the contribution of dust particles from comets.

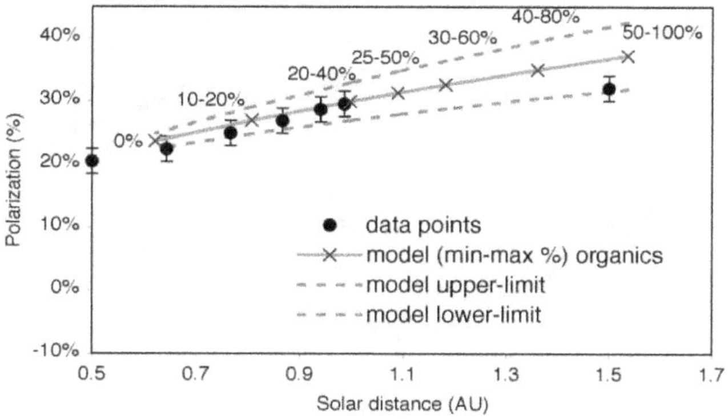

Figure 2. Interpretation of the decrease in polarization observed for the near-ecliptic zodiacal dust between 1.5 and 0.5 AU through an evolution of organics contribution. The results suggest the sublimation of the organics present in the particles.

3.2 Thermal Emission Results

The temperature variations with R, as deduced from the observations, do not follow a black-body relationship. This certainly indicates particular properties of the zodiacal dust cloud. The thermal equilibrium temperature of dust particles can be computed by equating the incident and emitted light integrated over a large range of wavelengths, λ, (typically from 0.1 to 1000 μm). At a distance R (in AU) from the Sun, this is obtained by solving the expression:

$$\left(\frac{r}{R}\right)^2 \int_0^\infty B(\lambda, T_S) Q_{abs}(a,\lambda) d\lambda = \varsigma \int_0^\infty B(\lambda, T) Q_{abs}(a,\lambda) d\lambda \quad (1)$$

where r is the radius of the Sun, $B(\lambda,T)$ the Planck function, T_S the solar surface temperature, ζ, the ratio of the emitting surface over $\pi a^2/4$, with a the diameter of the emitting particle and $Q_{abs}(a, \lambda)$ the absorption efficiency of a particle with a given optical index (see, e.g. Kolokolova et al., 2004).

The temperature variation with R (for R varying between 0.5 AU and 1.5 AU) of the dust particles is calculated by taking the absorption and emission properties of compact (spheroids) and irregular aggregates (aggregates of spheroids) dust particles with optical indices ranging from low absorbing silicates to highly absorbing carbonaceous compounds. The optical indices are taken to be those of astronomical silicates (Draine & Lee 1984) and refractive organic material (Li & Greenberg 1997). The behavior of the temperature for large particles (size > 100 μm) is always close to the black-body approximation. Only highly absorbing and small particles show a significantly different behavior. The variation with the solar distance is very dependent on the optical properties and size of the particles and less on the actual shape of the particles. The best estimate for the observed variation of temperature (Table 1) corresponds to small particles (effective radius < 2 μm) constituted of highly absorbing carbonaceous compounds such as organics or carbon as shown in Figure 3. Figure 3 also shows the thermal gradient with the solar distance for spheres and spheroids, indicating that the actual shape of the particle does not significantly modify the thermal behavior of the particles between non-absorbing silicates and absorbing organic compounds.

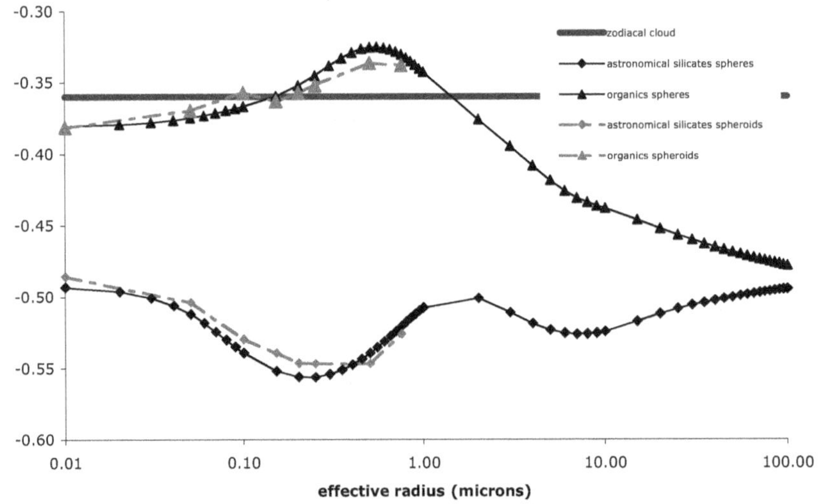

Figure 3. Calculations of the temperature gradient between 0.5 AU and 1.5 AU for two shapes of grains (spheres in black and spheroids in blue) as a function of the equivalent volume size of the grains and for the two different compositions relevant to the interplanetary dust cloud. (adapted from Lasue et al. 2007)

3.3 Significance of the Previous Results

To summarize, the local values derived from observational results, i.e. polarization, geometric albedo, temperature, indicate that, in the near-ecliptic invariant plane and in the 0.5-1.5 AU solar distance range, the dust cloud is heterogeneous and that the dust particles do not behave as black-bodies; they suggest that the dust properties change with time, as most of the particles spiral towards the Sun under Poynting-Robertson drag. Interpretation of the results obtained for the zodiacal light and the zodiacal thermal emission through robust numerical simulations favours the presence of both silicates and organics, with a steady decrease of the organics contribution. While the simulations require a significant amount of aggregates (most likely of cometary origin), it may be added that the *1/R* law derived for the increase of space density with decreasing solar distance is precisely what would be expected for dust particles under Poynting-Robertson drag in their formation region; in the above-mentioned region, significant amounts of cometary dust are actually ejected from active cometary nuclei, while it is unlikely that significant amounts of dust are released by asteroidal collisions.

4 Discussion and Conclusion

4.1 Comparison with Cometary Dust Properties

In-situ Vega and Giotto missions to comet Halley have revealed previously unsuspected properties of the dust ejected by the nucleus of this famous comet. From the dust mass spectrometer on-board Vega, the major constituents have been found to be silicate minerals and organic refractory materials (so-called CHON from their constitutive elements), both in comparable proportions (Kissel et al., 1986). From the optical probe and the dust impact detector on-board Giotto, the dust density has been estimated to be of about 100 kg m^{-3} (Levasseur-Regourd et al., 1999; Fulle et al., 2000). More recently, Stardust mission has provided some ground truth about the structure of the dust collected in comet Wild 2 coma, though the presence of both compact particles and fragile aggregates (Hörz et al., 2006).

As far as remote polarimetric observations are concerned, numerical simulations of the numerous observations of comets Halley and Hale-Bopp, through an approach similar to that described in 3.1, have allowed us to suggest that the dust particles present in the coma of these two comets consist of aggregates and some compact particles, with a percentage in mass of 40-65% of silicates and, conversely, of 60-35% of organics (Lasue et al., 2006; Lasue et al., 2009). In that work, the amount of aggregates present in the comae of comets Hale-Bopp and Halley was estimated to be at least respectively 18% and 10% in mass. We have mentioned in section 2.1 that 35% of the particles collected by Stardust were aggregates. Assuming that aggregate particles originate only from comets, such values would imply that from 50% up to 100% of the particles -both aggregates and compact- present in the zodiacal cloud would be of cometary origin. Experimental simulations have been also attempted to fit the polarimetric observations of comets. They also favour the presence, in addition to some compact silicates, of fluffy aggregates of silicates and carbonaceous compounds (Hadamcik et al., 2007). Finally, the presence of fragile low-density aggregates in the comae of various comets demonstrates that the aggregates noticed in the IDPs collected in the Earth stratosphere are of cometary origin.

4.2 Comparison with Recent Dynamical Studies

Nesvorny et al. (2010) have recently presented a new zodiacal cloud model based on the orbital properties and lifetimes of comets and asteroids, and on the dynamical evolution of dust after ejection, in order of determining the relative contributions of asteroidal and cometary material to the zodiacal cloud. The authors conclude that about 90% of the observed mid-infrared zodiacal thermal emission is produced by particles ejected from Jupiter family comets and that about 10% is produced by Oort cloud comets and/or asteroidal collisions.

While their approach is completely different from ours, and is only constrained by IRAS observations, it is certainly interesting to point out that both approaches establish that particles of asteroidal origin cannot be claimed to be the major source of interplanetary dust. Besides, it may be noticed that the value of about 50% in mass that we obtain for the contribution of dust particles from comets to the zodiacal cloud is likely to be underestimated. Dust particles of cometary origin are indeed, while their spiral towards the Sun under Poynting-Roberstson drag, most likely to suffer some evaporation of dark carbonaceous compounds, as well as some collisions, and thus to get more compact and comparable to particles of asteroidal origin. Finally, Nesvorny et al. (2010) estimate that the inner zodiacal cloud was at least 10^4 times brighter during the Late Heavy Bombardment epoch and derive the amount of primitive dark dust material that could have accreted on terrestrial planets. Taking into account the characteristic structure (with irregular grains and fluffy aggregates) of the particles of cometary origin, as already pointed out in Levasseur-Regourd et al. (2006), we will now carefully investigate this critical topic.

4.3 Implication for Earth Delivery of Carbonaceous Compounds

The theory of meteoritic ablation during atmospheric entry, including the effects of thermal radiation, heat capacity and deceleration for solid particles, has been described in a number of publications (e.g. Jones and Kaiser, 1966). In general, the thermal equilibrium of the particle is given by:

$$\frac{1}{2}\Lambda\rho_a v_\infty^{\ 3} A_{proj} = A_{tot}\varepsilon\sigma_S\left(T_s^4 - T_e^4\right) + \frac{4}{3}\pi r^3 \rho_m c_s \frac{dT_m}{dt} \tag{2}$$

where Λ is the heat transfer coefficient, ρ_a the density of the atmosphere, v_∞ the entry velocity of the particle, A_{proj} the projected surface of the particle, A_{tot} the total surface of the particle, ε the emissivity of the particle, σ_S the Stefan constant, T_s the surface temperature of the particle, T_e the environment temperature (atmosphere), r the equivalent radius of the particle (quantity for which $4\pi r^3/3$ equals the volume of the particle), ρ_m the density of the particle, c_s the specific heat of the meteoric substance, T_m the mean temperature of the particle, and t the time. This expression determines the relationship between the heat transfer from the atmospheric molecules to the particle and the light emission and heating of the particle.

As a first approximation, the transfer heat coefficient and the emissivity can be assumed to be equal to unity (Jones and Kaiser, 1966). Moreover, if the particle is small enough, typically with r less than tens of microns, then its temperature is always uniform (Murad, 2001) and the rightmost term of the equation (2) can be ignored. The equation (2) simplifies to:

$$\rho_{a1} \approx \xi \times \frac{\sigma_S T_b^4}{v_\infty^{\ 3}} \tag{3}$$

where $\xi = A_{tot} / A_{proj}$. In the case of spherical particles, $\xi = 4$, and assuming the evaporation temperature is about 2.1×10^3 K (Öpik, 1958), then evaporation of a particle that enters the atmosphere at 30 km s^{-1} starts at 101 km of altitude. Knowing that the ratio ξ can be 1.7 times higher for the case of typical spheroidal particles (oblate with a ratio of semi major axes of 2) and up to $\sim\pi$ for the case of aggregated fractal particles (Meakin and Donn 1988), this equation gives values for the altitude of evaporation of about 97 km and 93 km respectively for the same entry velocity.

However, the deceleration of the particle due to the collisions with the atmosphere molecules should also be taken into account. Assuming that the molecules stick to the particle and thus transmit all their momentum to the particle, the conservation of momentum implies:

$$v = v_\infty \exp\left[-\frac{3H\rho_a}{4R\rho_m \cos(\chi)}\right] \tag{4}$$

where H is the typical height of the atmosphere and χ the angle of the entry trajectory with respect to the zenith. Substituting this expression in equation (2) gives the expression for which the temperature obtained is maximal to be:

$$T_{\max}^4 = \frac{1}{\xi} \times \frac{4\Lambda R\rho_m v_\infty^3 \cos(\chi)}{18e\sigma_S \varepsilon H} \tag{5}$$

with e the natural base of logarithms. From this equation, the critical radius of the particles that can enter the atmosphere of Earth without being completely ablated can be determined. We have already seen that the shape parameter ξ can range from 4 for spherical particles to 4π for aggregated particles. The effect of the shape of the particles on the equilibrium temperature reached during atmospheric entry can be seen in Figure 4, assuming an entry velocity of 30 km s^{-1}. While the radius for which spherical particles reach the ground without being ablated is about 4.7 μm (Jones and Kaiser, 1966), the largest equivalent volume radius of irregularly shaped particles can reach up to 15 μm.

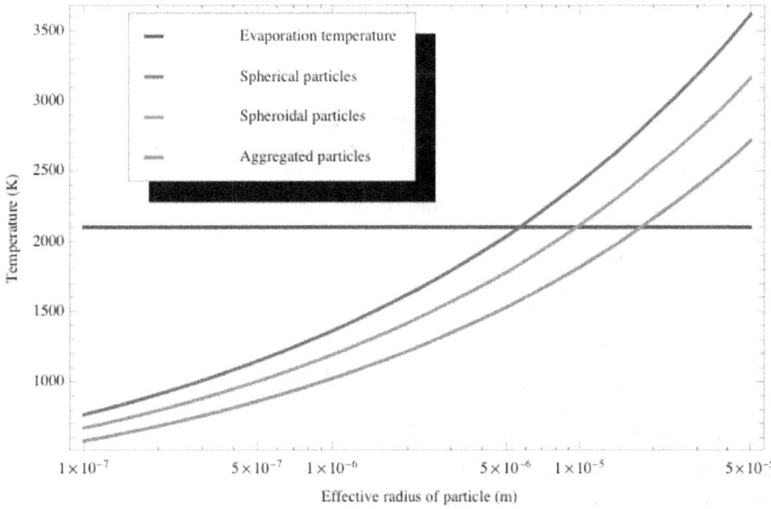

Figure 4. Maximum equilibrium temperatures for particles entering the Earth atmosphere at 30 km s^{-1}. The horizontal line corresponds to the temperature of sublimation of meteoritic materials suggested by Öpik (1958) of 2.1×10^3 K. The increase in size for the more efficiently decelerated particles (spheroids and aggregates) is obvious.

All parameters staying the same, irregularly shaped particles and fluffy aggregates can bring up to $\sim\pi^3$ more material in volume without being ablated to the Earth's surface than compact spherical particles. Cometary dust particles are therefore ideal candidates to bring carbonaceous compounds for seeding life on early Earth.

5 Conclusions

The long-standing controversy debated in the interplanetary dust community, around the relative contributions to the interplanetary dust cloud of dust resulting from asteroidal collisions and dust ejected by comet nuclei seems now about to be closed, with evidence for a major contribution of particles of cometary origin in the inner solar system and in the vicinity of the Earth, as established from their morphology (significant amount of aggregates), their composition (significant amount of organics) and their region of formation (inner solar system). It may thus be suggested that, not only meteor streams, but also sporadic meteors and micrometeorites, have mostly a cometary origin.

While more precise zodiacal observations are expected in a near future from Akatsuki spacecraft during its cruise between the Earth and Venus, a key implication of these conclusions is related to the early evolution of the solar system. During the LHB epoch, while the spatial density of dust in the interplanetary dust clouds was orders of magnitude greater than nowadays, the structure of dust particles originating from comets has quite likely favoured the survival of organics during their atmospheric entry.

Acknowledgments

We acknowledge partial funding from CNES. This is J. Lasue LPI contribution number ####.

References

M. J. Burchell and 11 co-authors MAPS **43**, 23 (2008)

B. T. Draine, H.M. Lee ApJ **285**, 89 (1984)

R. Dumont, A.C. Levasseur-Regourd A&A **191**, 154 (1988)

M. Fulle, A.C. Levasseur-Regourd, N. McBride, E. Hadamcik AJ **119**, 1968 (2000)

E. Grün, M. Baguhl, H. Svedhem, H.A. Zook In *Interplanetary dust*, Ed. by E. Grün, B. Gustafson, S. Dermott and H. Fechtig (Springer, 2001), p. 295

E. Hadamcik, J.B. Renard, F.J.M. Rietmeijer, A.C. Levasseur-Regourd, H.G.M. Hill, J.M. Karner, J.A. Nuth Icarus **190**, 660 (2007)

F. Hörz, and 43 colleagues Science **314**, 1716 (2006)

J. Jones, T.R. Kaiser MNRAS **133**, 411 (1966)

M.S. Kelley, W.T. Reach, D.J. Lien Icarus **193**, 572 (2008)

J. Kissel, and 18 colleagues Nature **321**, 280 (1986)

L. Kolokolova, M.S. Hanner, A.C. Levasseur-Regourd, B. Gustafson In *Comets II* Ed. by M. Festou, H.U. Keller, H.A. Weaver (Univ. Arizona Press, Tucson, 2004), p. 577

J. Lasue, A.C. Levasseur-Regourd JQSRT **100**, 220 (2006)

J. Lasue, A.C. Levasseur-Regourd, N. Fray, H. Cottin A&A **473**, 641 (2007)

J. Lasue, A.C. Levasseur-Regourd, E. Hadamcik, G. Alcouffe Icarus **199**, 129 (2009)

C. Leinert and 14 colleagues A&AS **127**, 1 (1998)

A.C. Levasseur-Regourd, J.B. Renard, R. Dumont In *Origin and evolution of interplanetary dust* Ed. by A.C. Levasseur-Regourd and H. Hasegawa (Kluwer, The Netherlands, 1991), p.131

A.C. Levasseur-Regourd, N. McBride, E. Hadamcik, M. Fulle A&A 348, 636 (1999)
A.C. Levasseur-Regourd, I. Mann, R. Dumont, M.S. Hanner In *Interplanetary dust* Ed. by E. Grün, B. Gustafson, S. Dermott and H. Fechtig (Springer, Berlin, 2001), p. 57
A.C. Levasseur-Regourd, J. Lasue, E. Desvoivres, Origin of Life and Evolution of Biosphere, **36**, 507 (2006)
A.C. Levasseur-Regourd, T. Mukai, J. Lasue, Y. Okada PSS **55**, 1010 (2007)
A. Li, J.M. Greenberg A&A, **323**, 566 (1997)
K. Lumme In *Light scattering by non spherical particles* Ed. by M.I. Mishchenko, J.W. Hovenier and L.D. Travis (Academic Press, San Diego, 2000), p. 555
P. Meakin, B. Donn ApJ. 329, L39 (1988)
E. Murad In *Meteoroids 2001 conference*, Ed. by B. Warmbein (ESA-SP-495, The Netherlands, 2001), p. 229
D. Nesvorny, P. Jenniskens, H.F. Levison, W. Bottke, D. Vokrouhlicky, M. Gounelle ApJ **713**, 816 (2010)
E.J. Öpik *Physics of meteor flight in the atmosphere* (Dover publication Inc., Mineola, New York, USA, 1958)
M.V. Sykes, R. Greenberg Icarus **65**, 51 (1986)
A.D. Taylor, W.J. Baggaley, D.I. Steel Nature **380**, 323 (1996)
F. Whipple ApJ **121**, 750 (1955)

Origin of Short-Perihelion Comets

A. S. Guliyev

Abstract New regularities for short-perihelion comets are found. Distant nodes of cometary orbits of Kreutz family are concentrated in a plane with ascending node 76° and inclination 267° at the distance from 2 up to 3 a.u. and in a very narrow interval of longitudes. There is a correlation dependence between q and cos I concerning the found plane (coefficient of correlation 0.41). Similar results are received regarding to cometary families of Meyer, Kracht and Marsden. Distant nodes of these comets are concentrated close three planes (their parameters are discussed in the article) and at distances 1.4; 0.5; 6 a.u. accordingly. It is concluded that these comet groups were formed as a result of collision of parent bodies with meteoric streams. One more group, consisting of 7 comets is identified. 5 comet pairs are selected among sungrazers.

Keywords short-perihelion comets · meteor streams · split comets

1 Kreutz Cometary Family

The Kreutz cometary family is quite a mysterious phenomenon in the solar system. The strength of this family, by rate of comets discovered during last years, might be estimated as tens of thousands. Hence, Kreutz comets form a singular belt around the Sun. Meanwhile, research on Kreutz comets, essentially, covers observation of individual objects of this class. This system is studied in insufficient detail. The reason for this is that the system is quite young and quickly replenishes.

There are some explanations concerning an origin of short-perihelion comets of the Kreutz family. However it is impossible to consider any of them as comprehensive one. It might be possible to consider conventionally that these comets are fragments one or several large proto-comet nucleus. The version about disintegration proves to be true even when some Kreutz comets sometimes break up to separate parts during astronomical observations.

We present and comment some new regularities of considered system in the present book. They were not known earlier. These regularities, in our opinion, might give a sufficient basis for revision of the discussed origin's mechanism concerning to Kreutz comets or bring essential updates in this mechanism, at least.

According of the catalogue by Marsden and Williams (2008) and Minor Planet Electronic Circulars for 2008-2009, the number of long-period comets with parameters close to values

$$q = 0.006a.e.;\ e = 1;\ \omega = 80°;\ \Omega = 0°;\ i = 144°$$

is equal to 1502 (as of early 2010).

A. S. Guliyev (✉)
Shamakhy Astrophysical Observatory, Academy of Sciences of Azerbaijan. Phone: +9940503325958; Fax. +99412 4975268; E-mail: ayyub54@yahoo.com

Primary viewing of Kreutz comets shows that their perihelions are not concentrated chaotically around a certain center. There is absolutely other way for the better description of perihelion distribution. Perihelion of comets are located along some arch of the celestial sphere. Before making comments on this feature of Kreutz comets, we have to make a substantiation of this assumption. If each point of perihelion with parameters (L_i, B_i) is present as a material point on a surface of a certain sphere, then coordinates (L, B) of the of inertia center of this sphere will be determined from expressions:

$$Nk \cos L \cos B = \Sigma\cos L_i \cos B_i$$
$$Nk \sin L \cos B = \Sigma\sin L_i \cos B_i$$
$$Nk \sin b = \Sigma\sin B_i ,$$

where N and k are number of perihelion and level of inferred concentration, accordingly. Calculations for 1502 points give following values:

$$L= 282°.82;\ B = 35°.06;\ R = 0.992$$

As a residual dispersion it is possible to consider value $\Sigma\sin^2\theta_i$, where θ_i are angular distances of perihelion from point (L, B).

$$S_p = \Sigma sin^2\theta_i = 12.33 \qquad (1)$$

Now let us consider a working hypothesis about perihelion location along the big circle of celestial sphere with parameters Ω' (ascending node) and I' (inclination). Calculations made by us give following values

$$I' = 37°.48 \qquad \Omega' = 171°.32 \qquad (2)$$

A residual dispersion in this case will be S_{res} = 5.24. This is almost twice less, than *(1)*.

It was found other plane with parameters

$$I'= 76°.34;\ \ \Omega'= 267°.15 \qquad (3)$$

concerning which distant nodes of Kreutz comets orbits have maximum in the interval 2 – 3 a.u. (Figure 1). It is close to the normal distribution with the maximum near 2.5 a.u in the interval of 0-5 a.u. (Hereinafter in the analysis are used overlapping on an axis abscissa each other intervals).

In addition angular sizes of distant nodes (DN) concerning a plane (3) have a sharp maximum in a narrow interval of longitude (Figure 2). These features of the distant nodes theoretically can be explained by two reasons: 1. Comets are generated by a planet body moving in the plane (3) and on distance nearby 2.5 a.u.; 2. There is an unknown meteoric stream in this plane and in the distance near 2.5 a.u., which is the reason of smashing Kreutz comets.

The first explanation seems to be extremely improbable as there is no similar body among known asteroids. If even it existed in the solar system, the mechanism of generation cometary nucleys by them would be not clear. Therefore it is evident to decide in favor of the second mechanism. It seems quite logical and explains almost all features of considered Kreutz comets.

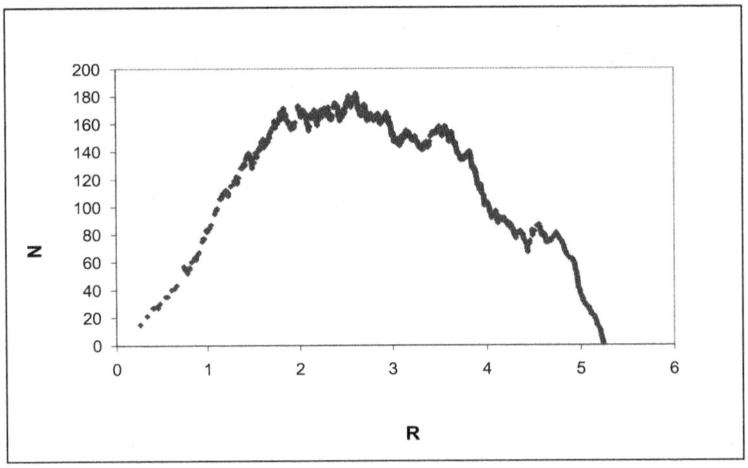

Figure 1. Distribution of distant nodes of Kreutz comets regarding to the plane (3) in the interval up to 5.3 a.u.

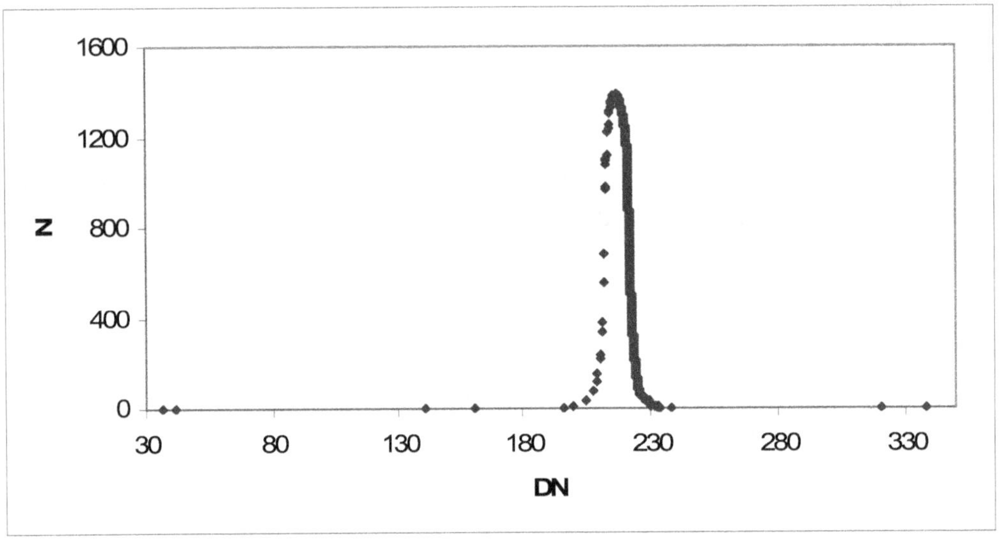

Figure 2. Distribution of distant nodes (DN) longitudes of Kreutz comets relative (3).

It is reasonable to make the following hypothesis on the origin of studied comets. Huge proto-comet nuclei, appearing in the inner part of the solar system at first, have fallen into unknown meteoric stream. It has got a lot of cracks. These cracks in a combination with tidal influence of the Sun have led to disintegration of proto-comet nuclei on to finer fragments. Fragments have fallen in the same meteoric stream at their next returning to perihelion and have got sets of impacts and cracks which lead to their secondary splitting, etc.

2 Meyer Group of Short – perihelion Comets

Under Meyer group of comets we will mean comets with parameters, varying around values:

$$q = 0.036a.e.;\ e = 1;\ \omega = 57°;\ \Omega = 73°;\ i = 73°$$

The number of such long-period comets, as of early 2010, was 100.

Results of our calculations and analyses show that the assumption of concentration along the plane

$$I' = 53°.69;\quad \Omega'_c = 11°.07 \qquad (4)$$

describes real distribution of perihelion better, than the similar assumption regarding to some point ($S_p = \Sigma \sin^2 B_i' = 0.265$). Ninety percent of points are concentrated in the field of ±4° regarding the plane (4)

Calculations show, that there is one more plane with parameters

$$I' = 84°.68;\quad \Omega' = 270°.87 \qquad (5)$$

near which distant nodes of cometary orbits have significant concentration in the interval 1.1 – 1.4 a.u. (Figure 3).

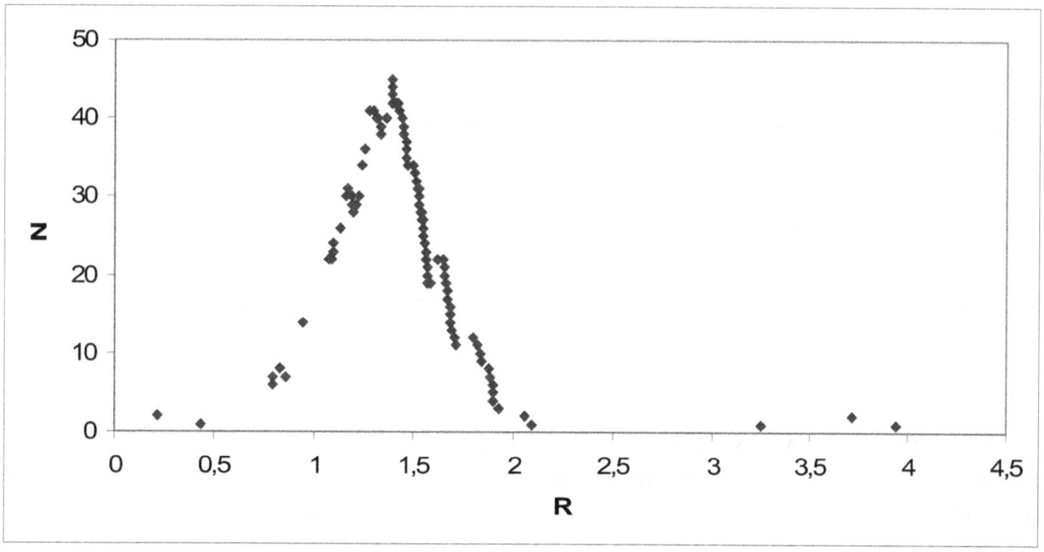

Figure 3. Distribution of distant nodes of Meyer comets regarding to the plane (5)

These features in combination with correlation between q and $\cos I$ (coefficient of correlation is equally to -0.3) give a basis to put forward the following hypothesis. One of the long-period comets having parameters

$$I = 72°.8;\ \Omega = 72°.6;\ q = 0.036$$

and appearing in the inner part of solar system for the first time at passage of the zone with parameters R ~1.4 a.e.; $I' = 84°.68$; $\Omega' = 270°.87$ has got powerful jets of a meteoric stream. The orbit of comet had an inclination to this plane about 150°. Therefore a head-on collision occurred, i.e. impacts of meteoric particles on comet nuclei were powerful. As a result, comet nucleus has collapsed on to many fragments.

3 Kracht and Marsden Cometary Groups

Analogical results have been obtained concerning the cometary groups of Kracht and Marsden. First of them has following characteristics

$$q = 0.045 a.e.;\ e = 0.98;\ \omega = 59°;\ \Omega = 44°;\ i = 13°$$

and contains 35 comets (2010). It is established at first that distant nodes of these comets are concentrated near the plane

$$I' = 24°.08;\quad \Omega' = 104°.51$$

and in the interval of the distance 0.4 – 0.6 a.u. There is a sharp concentration of distant nodes on longitude in this case too.

Group of Marsden has following characteristics

$$q = 0.050 a.e.;\ e = 0.98;\ \omega = 24°;\ \Omega = 79°;\ i = 27°,$$

and contains 32 comets (2010). Calculations show that perihelion of these comets are concentrated near the plane

$$I' = 10°.21;\quad \Omega' = 359°.60$$

At the same time we have found that distant nodes of these comets are concentrated near the plane

$$I' = 89°.50;\quad \Omega' = 101°.22$$

and in the distances from 3 up to 8.7 a.u.

In the opinion of the author, these two groups have been formed as a result of comet-meteor stream collisions, too.

4 New Group of Sungrazers and Other Splitted Comets

The author has analyzed features of 63 sporadic short-perihelion comets by own methods described in the book. A new group was identified among them. It contains 7 comets (C/2007 K19, C/2006 L7, C/2007 L12, C/2005 L10, C/2006 M6, C/2007 M6, C/1997 M5). Perihelion of these comets are concentrated near the plane with parameters:

$$I' = 53°.9;\quad \Omega' = 222°.1.$$

Five pairs among short-perihelion comets are selected except this group: C/2002 V5 and C/1996 V2; C/2004 U2 and C/2005 M3; C/2005 D1 and C/2007 C12; C/2000 V4 and C/2001 T5; C/2008 S2 and C/2004 X7. Probably they are fragments of splitted comets.

References

B.G.Marsden and G.V.Williams, 2008. Catalogue of Cometary Orbits, 17th ed. 195p.

Identification of Optical Component of North Toroidal Source of Sporadic Meteors and its Origin

T. Hashimoto • J. Watanabe • M. Sato • M. Ishiguro

Abstract We succeeded to identify the North Toroidal source by optical observations performed by the SonotaCo Network, which is a TV observation network coordinated by Japanese amateurs. This source has been known only for radar observations until now. The orbits of the optical meteors in the North Toroidal source are relatively large eccentricity and semi-major axis, compared with those of the radar meteors. In this paper, we report the characteristics of this North Toroidal source detected by optical observations, and discuss the possible origin and evolution of this source.

Keywords sporadic source · North Toroidal · optical method

1 Introduction

The major six sources of sporadic meteors were discovered mainly by radar observations: Helio (H) and Antihelio (HA), South and North Apex components (SA/NA), and South and North Toroidal (ST/NT). Due to the high efficiencies realized in modern radar technologies, high resolution and sensitive observations have been carried out on these sporadic sources (Campbell-Brown 2008). On the other hand, optical data has not been enough to study these sources until now. Especially, Toroidal sources have never been identified by optical method. In this paper, we report the first identification of the North Toroidal sources among the data obtained by the TV observation network coordinated by Japanese amateurs. We also report the characteristics of the orbits of meteors belonging to the NT source, and discuss the possible origin and evolution of this source.

2 Observational Material

We analyzed data collected by SonotaCo network, which is the coordinated monitoring observation network of automated detection for bright meteors or fireballs among amateur astronomers (SonotaCo 2009). We selected the meteors by using analysis software, UFOOrbit ver. 2.11 for securing well-determined orbits with the following conditions: length of the trail $\geq$ 1.5 degrees, the angle of the intersection of two apparent passes of trails' extension $\geq$ 10 degrees. The total number of the selected samples is 13,275. Among them, 5,341 meteors are judged to belong to 20 major meteor showers using

T. Hashimoto (✉)
The Nippon Meteor Society 1-28-1, Kinugaoka, Hachioji, Tokyo, 192-0912, Japan , E-mail: thashi@din.or.jp

J. Watanabe • M. Sato
National Astonomical Observatory of Japan 2-21-1, Osawa, Mitaka, Tokyo, 181-8588, Japan

M. Ishiguro
Department of Physics and Astronomy, Seoul National University, 599 Gwanak-ro, Gwanak-gu, Seoul 151-742, Republic of Korea

the analysis software UFOAnalyzer Ver. 2. The rest of 7,934 meteors are thought to be sporadic meteors. The radiants of these 7,934 meteors are plotted in Figure 1.

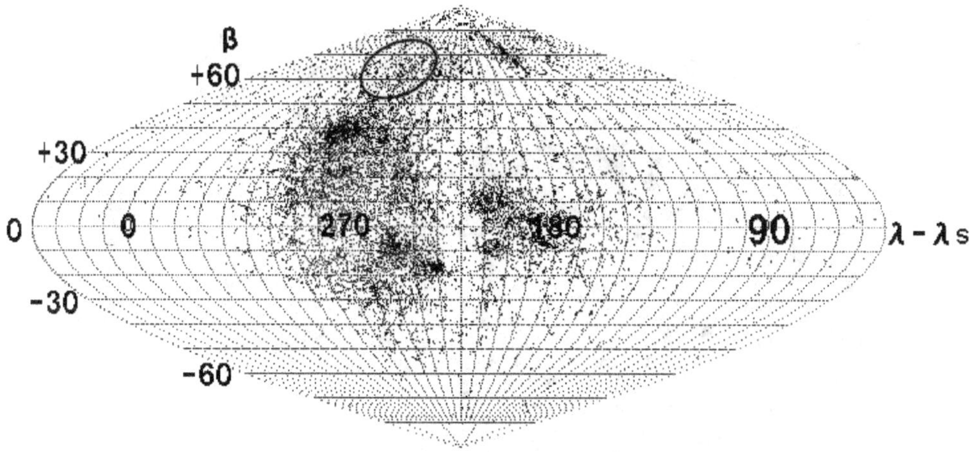

Figure 1. Distribution of radiant points of 7,934 sporadic meteors.

While it is clear that there are concentrations corresponding to HA, and SA/NA, there is also a weak concentration at around $\lambda - \lambda_{sun} = 230 \sim 290$ degrees, and $\beta = +50 \sim +80$ degrees. This area corresponds to the NT source determined by radar observations. There are 410 meteors with radiants are located in this area.

3 Characteristics of Optical NT Meteors

Assuming these meteors belong to the NT source, we analyzed the characteristics of these meteors in order to compare to radar NT meteors. Due to the optical monitor, these meteors are relatively bright, including the fireball-class. Figure 2 shows the absolute magnitudes of detected optical NT meteors. This means that the original size of the optical NT meteoroids is larger than that of radar NT meteoroids.

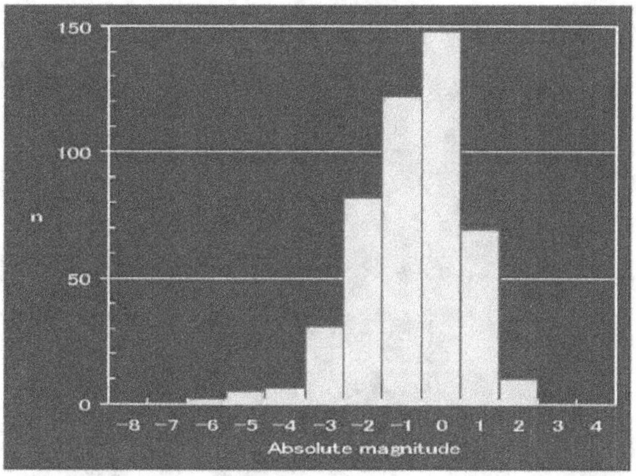

Figure 2. Absolute magnitude of optical NT meteors.

The orbital elements of the optical NT meteors are also different from radar NT meteors. Figures 3 and 4 indicate the distribution of their eccentricities and semi-major axes, respectively. Each figure contains the value of the radar NT meteors studied by Jones and Brown (1993) for comparison.

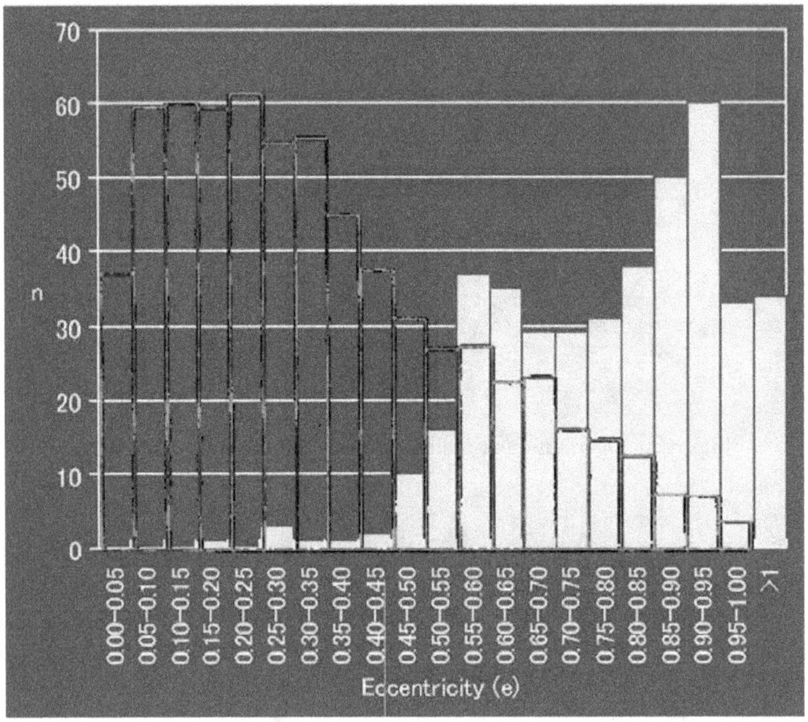

Figure 3. Distribution of eccentricity of optical NT meteors.

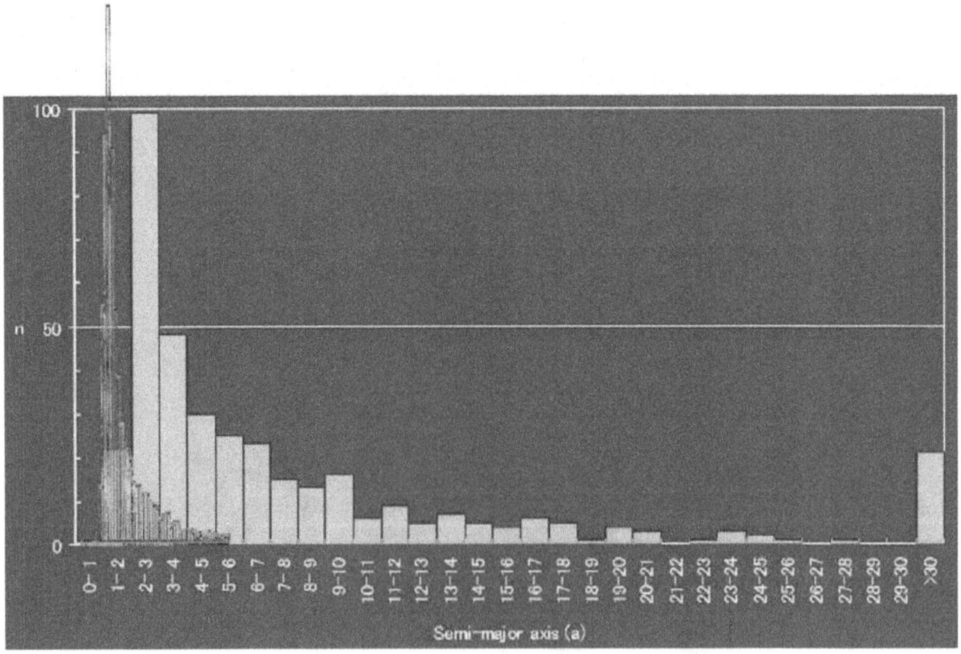

Figure 4. Distribution of semi-major axis of optical NT meteors.

The optical NT meteors have a more eccentric orbit with larger semi-major axis than the radar NT meteors. On the other hand, the inclination is not so different from radar NT meteors, as shown in Figure 5.

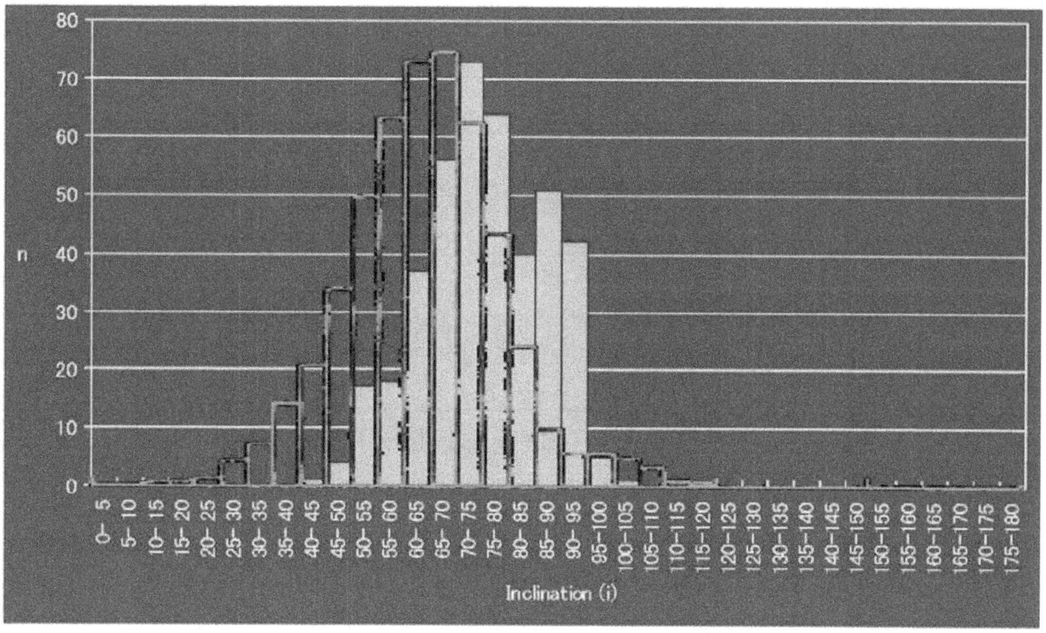

Figure 5. Distribution of inclination of optical NT meteors.

4 Origin of the NT

It is clear that the orbits of the NT meteors depend on their size such that larger eccentricity and semi-major axis with larger meteoroids, while the inclination is similar. This situation gives us a strong implication regarding the orbital evolution of the NT meteoroids. Because of the P.R. effect, the smaller meteoroids change their orbits faster than the larger ones. Even if the orbits of all the meteoroids of different size are the same initially as large eccentricity and semi-major axis, the orbits of smaller meteoroids shrink into smaller and circular orbits more rapidly than larger meteoroids. The NT meteoroids are thought to be a stage on the way of such orbital evolution. If so, we speculate that the parent object or objects should have been close to the orbit of larger-size meteoroids, namely large eccentricity and relatively large semi-major axis of more than a few A.U.

The theoretical evolutional tracks of the orbits of the NT meteoroids can be plotted in the *a-e* diagram. Within this diagram, the evolutional track depends strongly on the initial orbit, and not on the size of meteoroids. Smaller meteoroids evolve along the track into the smaller and circular orbits faster than larger meteoroids. Therefore, the observed distribution of the orbits of optical and radar NT meteoroids should be located in the one evolutional track if the origin is the same. It is important to find out any appropriate evolutional track which passes through the both observed components of the NT. Our preliminary trials show that two possible groups of evolutional tracks are plausible. One is a group of large-*e* & small-*a* orbits, and the other is that of large-*e* & large-*a* orbits. Figures 6 and 7 show the *a-e* diagrams with evolutional tracks of the two groups, respectively.

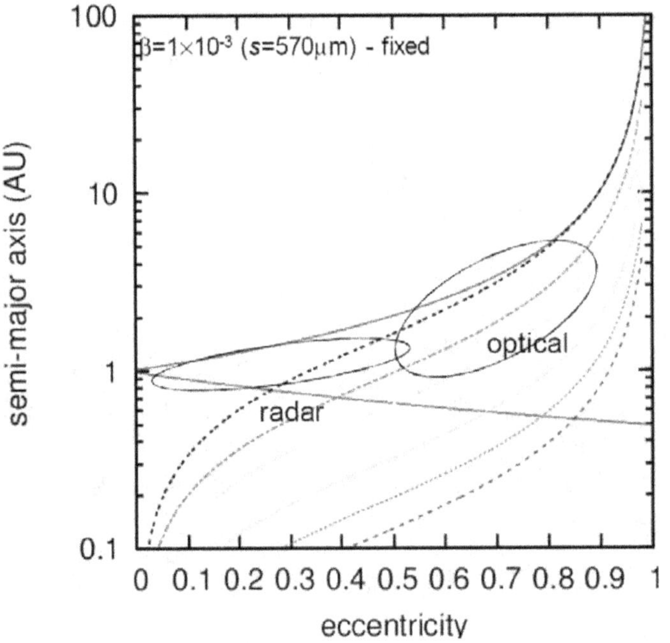

Figure 6. *a-e* diagram of the possible evolutional track of the meteoroids from large-*e* and large-*a* orbits.

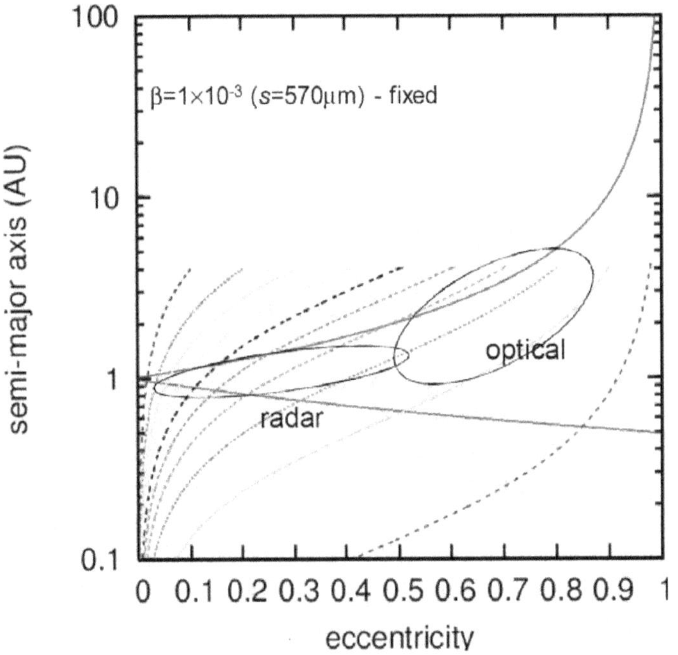

Figure 7. Same as Figure 6, but from large-*e* and small-*a* orbits.

The former group is suggesting high-inclined short-period comets, which was already suggested by Wiegert (2008). Although Wiegert et al. (2009) also tried to simulate the NT from the near-Earth asteroids, it seems to be impossible to explain the origin of the large-*a* meteoroids detected in optical NT, if we assume the origin of the NT is only one object.

However, it should be noted that there is an annual variation of the NT source. In our sample, the number of the optical NT increased in autumn and winter. Recent detailed study of the radar NT meteors by Campbell-Brown & Wiegert (2009) clearly shows that the NT has several components of different orbital characteristics. This suggests that the NT source has been originated from several different parent objects. Anyway, no definite candidate has been identified yet. Further studies should be needed to clarify the origin of the NT source.

5 Conclusion

We identified the optical component of the North Toroidal source that the size of the meteoroids is larger than that detected by the radar method. These larger NT meteoroids have different orbital characteristics; larger eccentricity and semi-major axis than those of the radar NT meteoroids. This strongly suggests the orbital evolution of the meteoroids in the NT source by the P.R. effect. One of the possible parent(s) of the NT source should have larger eccentricity and semi-major axis of a few or much larger values.

References

Campbell-Brown, M.D., Wiegert, P., Seasonal Variations in the north toroidal sporadic meteor source, Meteoritics & Planet. Sci., 44, 1837-1848 (2009)

Campbell-Brown, M.D., High resolution radiant distribution and orbits of sporadic radar meteoroids, Icarus, 196, 144-163 (2008)

Jones, J., and Brown, P., Sporadic meteor radiant distributions: orbital survey results, Mon. Not. Roy. Astron. Soc., 265, 524-532 (1993)

SonotaCo, A meteor shower catalog based on video observations in 2007-2008, WGN, Journal International Meteor Organization, 37, 55-62 (2009)

Wiegert, P., The dynamics of low-perihelion meteoroid streams, Earth, Moon and Planets, 102, 15-26(2008)

Wiegert, P., Vaubaillon, J., and Campbell-Brown, M., A dynamical model of the sporadic meteoroid complex, Icarus, 201, 295-310 (2009)

Distributions of Orbital Elements for Meteoroids on Near-Parabolic Orbits According to Radar Observational Data

S. V. Kolomiyets

Abstract Some results of the International Heliophysical Year (IHY) Coordinated Investigation Program (CIP) number 65 "Meteors in the Earth Atmosphere and Meteoroids in the Solar System" are presented. The problem of hyperbolic and near-parabolic orbits is discussed. Some possibilities for the solution of this problem can be obtained from the radar observation of faint meteors. The limiting magnitude of the Kharkov, Ukraine, radar observation program in the 1970's was +12, resulting in a very large number of meteors being detected. 250,000 orbits down to even fainter limiting magnitude were determined in the 1972-78 period in Kharkov (out of them 7,000 are hyperbolic). The hypothesis of hyperbolic meteors was confirmed. In some radar meteor observations 1 – 10% of meteors are hyperbolic meteors. Though the Advanced Meteor Orbit Radar (AMOR, New Zealand) and Canadian Meteor Orbit Radar (CMOR, Canada) have accumulated millions of meteor orbits, there are difficulties in comparing the radar observational data obtained from these three sites (New Zealand, Canada, Kharkov). A new global program International Space Weather Initiative (ISWI) has begun in 2010 (http://www.iswi-secretariat.org). Today it is necessary to create the unified radar catalogue of near-parabolic and hyperbolic meteor orbits in the framework of the ISWI, or any other different way, in collaboration of Ukraine, Canada, New Zealand, the USA and, possibly, Japan. Involvement of the Virtual Meteor Observatory (Netherlands) and Meteor Data Centre (Slovakia) is desirable too. International unified radar catalogue of near-parabolic and hyperbolic meteor orbits will aid to a major advance in our understanding of the ecology of meteoroids within the Solar System and beyond.

Keywords meteors · meteoroids · meteor orbits · meteor radar · hyperbolic meteors

1 Introduction

In a series of publications (Kolomiyets and Kashcheyev 2005, Kolomiyets 2002, Andreyev et al. 1993) the authors have identified a set of meteor orbits, with $e \geq 1$, of meteor sporadic background based on the Kharkov radar observations, which they named "hyperbolic meteors" similar to previous publications (Vsekhsvyatskiy 1978; Shtol 1970) based on analogous data. The Kharkov radar orbital data from the 1970s has proven to be extremely promising for finding the real hyperbolic orbits, as they were statistically many in terms of volume and uniformity, there have been twenty-four-hour and round off the annual cycles of observations were weaker meteors between masses 10^6 – 10^9 kg, which are important for the building of the Meteor engineering distribution models (Dikarev et al. 2001). In addition, these data were obtained as a result of carefully designed and carefully executed multi-year monitoring experiment (Kashcheyev and Tkachuk 1980), using the Meteor automated radar system (MARS) of the Kharkov National University of radio electronics (KhNURE), which was recognized at that time to be the best in the world (Fedynskiy et. al. 1976, Kashcheyev 1977, Kashcheyev et al. 1977,

S. V. Kolomiyets (✉)
Kharkiv National University of Radioelectronics, Lenin ave., 14, Kharkiv, 61166, Ukraine. E-mail: s.kolomiyets@gmail.com

Voloshchuk et. al. 1984). Hyperbolic meteors were recorded and continue to be recorded by other meteor radar and optical observations (Kramer et. al. 1986), and in "in situ" experiments (Weidensehilling 1978, Grün et al. 2001). The information on hyperbolic orbits is currently available as the new 2003 version at the International Astronomical Union Meteor Data Center IAU (MDC), provided by scientists from Slovakia (Hajdukova 2008; Hajdukova and Paulech 2006). Nevertheless, data on hyperbolic orbits that are available to scientists in print is very heterogeneous and not always meaningful for the categorical conclusions. Part of it are the consequence of errors (Hajduk 2001). In addition to that the real hyperbolic meteor complex has a naturally compound structure. The theories of the origin of hyperbolic meteor orbits near the Earth orbit and in the Solar System are still ambiguous and contradictory (Meisel et al. 2002a,b; Janches et al. 2001; Grun and Landgraf 2000; Kramer et al. 1998; Belkovich and Potapov 1985; Kazantsev 1998; Vsekhsvyatskiy 1978). The majority of scientists do not contradict the reality of hyperbolic meteor orbits altogether, but at the same time it is becoming increasingly attractive to research the emergence of new information and new submissions on this issue. As a rule the number of meteor orbits with the eccentricities much greater than 1 is very small, both theoretically and experimentally (~ 1%). Thanks to scanty statistics the problem of hyperbolic identities meteors ($e \geq 1$)is actually a problem near-parabolic orbits meteoroids ($e \sim 1$). The set of near-parabolic orbits of meteoroids is the most dynamic part of meteor substance of the Solar System. This orbital series is statistically far richer than the set of hyperbolic meteor orbits only and its properties and characteristics are the keys to solving both problems of hyperbolic meteor orbits, and other problems of cosmology and cosmogony of the Solar System. (Lebedinets 1980, 1990; Rietmeijer 2008, Drolshagen et. al. 2008, Suggs et. al. 2008, Chapman 2008).

2 The Kharkov (Ukraine) Meteor Radar Data

The final test of the validity of a theory has always been an experiment. The 1972-1978 Kharkov meteor radar data mentioned above was the result of a carefully designed and performed at the highest level experiment. During the radar observations of faint meteors in Kharkov, special attention was paid to the regularity, continuity and stability of the sensitivity of the surveillance equipment. The scheduling of observations was designed such that the observing cycles were distributed more or less evenly throughout the year. For example, during 1975, 29 observing cycles, ranging five to eight days, took place and, as a result, over 54,000 orbits of meteoroids were determined. The monitoring, carried out in times when main meteor showers were absent, with few exceptions (for ex., Geminids and Quadrantids), allows observation of prevalently the sporadic meteor background. Therefore the derived distribution of meteors was hardly influenced by meteoroids of main showers and characterized mainly sporadic meteor complex. In the 1972-1978 MARS of the KHNURE (Kharkov) registered about 250 thousand radiants, velocities and orbits of small meteoroids. The limiting magnitude of the Kharkov radar observation program in the 1970s was +12m (faint meteors). Parameter distributions of small meteoroid orbits registered in Kharkov were constructed. Variations of those distributions with time, seasons, and factors of selectivity were taking into account. Thus, the empirical model of the meteor substances from radar data in Kharkov between masses $10^6 - 10^9$ kg with mass parameter $s = 2$ was formed. Some of the properties and characteristics of this model were published (Kashcheyev and Tkachuk 1979, Tkachuk 1979). As a guide to the Kharkov meteor orbital empirical model, based on monitoring data of the 1972-1978, the selective catalogue of 5,317 meteors of up to +12 magnitude (Kashcheyev and Tkachuk, 1980) can be used. It demonstrates in brief all the characteristics of the model, the parameters, the methodology and peculiarities of radar observations (Kashcheyev et al. 1967, Tkachuk 1974). It contains

5,317 orbits, registered in Kharkov during the 1975, out of total record of 54,000 orbits.

Some characteristics of the Kharkov empirical model of orbital distributions of meteoroids using radar observations from 1975 in Kharkov are shown in Figure 1, where the dashed lines represent the set of elliptical orbits, available in the catalogue of Kashcheev and Tkachuk (1980), and the solid lines represent the set of hyperbolic orbits, selected by Kolomiyets (Kashcheyev et al. 1982). Meteoroid number distributions are plotted versus three orbit elements: perihelion distance, inclination and perihelion argument. The author listed nearly 1,000 meteor hyperbolic orbits with eccentricities close to 1, based on the 1975 data obtained in Kharkov. Their orbital distributions and some other facts support the existence of "hyperbolic meteors" (Kolomiyets 2001).

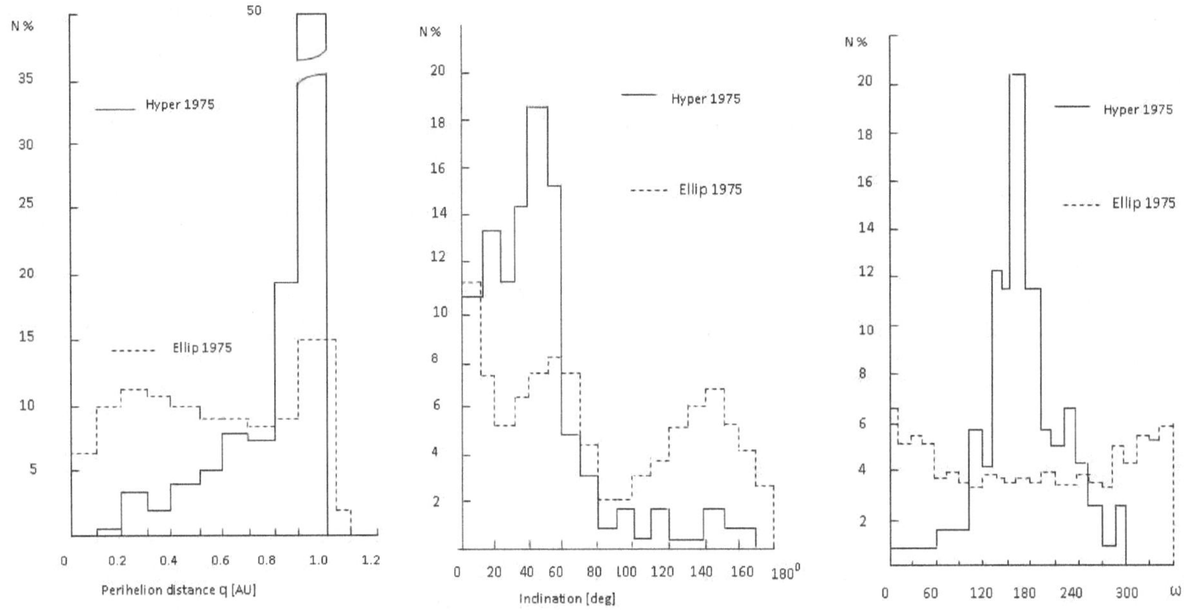

Figure 1. Left: Histograms of the number of orbits N (in %) depending on the perihelion distance q (in AU). Middle: inclination (in degrees) and right: argument of perihelion ω (in degrees) for two types of orbits with different values of the eccentricity: elliptical (dashed lines) and hyperbolic(solid lines).

In Table 1 we show an example of the data on hyperbolic and near-parabolic orbits of meteoroids registered on July 12-13, 1975 by radar method in Kharkov. During the 1990s, registered meteor data from 1972-1978, including the velocities, radiant coordinates and orbits, have been recalculated and put into electronic format. On the basis of this electronic database, the more sophisticated model of the meteor complex near the Earth's orbit for elliptical orbits of meteoroids (for stream and sporadic components) of faint meteors was constructed. A detailed description of the specified database and its thorough analysis for elliptic orbits is presented by Voloshchuk et al. (1995, 1996, 1997). In this analysis we did not include the hyperbolic orbits of meteoroids. Now the KhNURE scientists have the possibility to use the re-calculated meteor orbit database of the 1972-1978 dataset when they perform meteor research in the KhNURE. For the analysis of distributions of hyperbolic and near-parabolic orbits of meteoroids according to radar observations during the period 1972-1978, the author also used the recalculated KHNURE electronic database.

Table 1. Example of data on the hyperbolic and near-parabolic orbits of meteoroids from Kharkov (36.90E, 49.40 N) radar observations program 1975 (July 12-13). The columns are: (*H*:*M*) – hour and minute; V_g – geocentric velocity; V_h –heliocentric velocity; (β', λ') – radiant heliocentric coordinates; E_s' – radiant elongation from the Sun; *e* – eccentricity and σ_e – the standard deviation of eccentricity; *q* – perihelion distance; *p* – orbit parameter; *i* – inclination; ω – perihelion argument; Ω – longitude of ascending node; $\pi = \omega + \Omega$ – longitude of perihelion; ($R_{\Omega 1}$, $R_{\Omega 2}$) – nodes radius vectors.

H:*M*	V_g	V_h	β'	λ'	E_s'	$e \pm \sigma_e$	*q*	*p*	*i*	ω	Ω	π	$R_{\Omega 1}$	$R_{\Omega 2}$
July, 12														
02:09	41±2.2	50±1.9	40	240	120	1.78±0.19	0.81	2.3	48	226	109	336	–	1.02
04:45	54±2.8	56±2.6	70	215	95	2.63±0.34	1.01	3.7	71	188	109	297	–	1.02
04:53	40±2.1	44±1.9	52	239	112	1.25±0.18	0.88	1.9	59	220	109	330	–	1.02
05:03	46±2.4	59±2.6	46	174	73	2.91±0.34	0.96	3.7	49	157	109	267	–	1.02
05:06	67±3.4	43±3.4	28	342	121	1.15±0.26	0.76	1.6	145	238	109	348	4.07	1.02
05:08	65±3.3	49±3.2	36	313	136	1.45±0.23	0.55	1.4	118	257	109	6	1.99	1.02
05:40	44±2.3	42±2.0	68	232	101	1.12±0.19	0.98	2.1	71	201	109	311	–	1.02
06:10	59±3.0	41±3.0	21	313	148	0.99±0.08	0.28	0.5	136	296	109	46	0.38	1.02
06:46	59±3.0	56±3.2	55	272	122	2.35±0.37	0.80	2.7	78	226	109	336	–	1.02
07:43	55±2.8	51±3.1	60	272	118	1.83±0.32	0.84	2.4	80	223	109	333	–	1.02
07:49	67±3.4	43±3.3	43	38	76	1.20±0.32	0.97	2.1	134	155	109	265	–	1.02
08:02	39±2.1	48±3.0	46	232	111	1.62±0.30	0.90	2.4	51	215	109	325	–	1.02
09:10	58±3.0	52±4.3	37	281	122	1.94±0.46	0.80	2.3	84	227	109	337	–	1.02
11:54	39±2.1	44±1.6	-0	138	28	1.06±0.05	0.26	0.5	0	243	289	173	1.02	0.35
July, 13														
02:05	42±2.1	44±1.0	21	264	113	1.09±0.07	0.33	0.7	42	287	110	37	0.53	1.02
03:40	41±2.2	41±1.7	63	236	111	0.97±0.15	0.95	1.9	68	210	110	321	11.93	1.02
05:26	59±3.1	67±3.2	57	229	105	4.15±0.49	0.97	5.0	60	199	110	310	–	1.02
05:31	62±3.2	43±3.1	16	318	147	1.04±0.10	0.31	0.6	147	290	110	41	0.46	1.02
08:37	67±3.4	49±3.6	43	328	125	1.57±0.32	0.74	1.9	123	236	110	347	14.32	1.02
15:29	35±1.9	46±1.6	9	152	42	1.25±0.10	0.52	1.2	14	96	110	207	1.33	1.02
15:57	31±1.7	41±1.4	28	158	53	0.98±0.09	0.65	1.3	36	106	110	217	1.78	1.02
16:08	50±2.6	71±2.5	15	168	59	4.22±0.35	0.8	4.4	17	142	110	253	–	1.02
16:23	34±1.9	54±1.7	18	170	61	2.14±0.18	0.84	2.6	21	138	110	249	–	1.02

2.1 Empirical Model of Orbital Distributions of Meteoroids with Near-parabolic Orbits According to the Kharkov Radar Data

Celestial bodies are moving around the Sun in curves of the second order, which are the conic sections with the Sun in one of the foci. The orbital elements are p, e, ω, Ω, i, τ, where p is the orbital parameter, e is the eccentricity, ω is the argument of perihelion, Ω is the longitude of ascending node, i is the inclination and τ is the time registration. These elements are called Kepler's elements and they determine the orbit of any type, elliptical $e < 1$, parabolic $e = 1$ or hyperbolic $e > 1$.

The author presents here the empirical model of orbital meteoroids complex for near parabolic orbits of faint meteors. This model is based on the observational data obtained by the MARS radar system in 1972-1978 in Kharkov. The model is presented in the form of distributions of numbers of orbits versus the orbital elements, perihelion distance q, inclination i and argument of the perihelion ω, for different types of orbits and different eccentricity values. As an important informative source, the distributions of the number of orbits versus geocentric and heliocentric velocities were also constructed. The model is constructed in such a way that one can compare a specific orbit-registered-size meteoroid samples that represent sets of orbits, which are close to the exact parabolic orbit, for both elliptical and hyperbolic orbits. That is, the selection of orbit was based on the approximation to the exact parabola in varying degrees. Depending on the degree of approximation the selections were called classic, close or average. These approximations had the following criteria. Classic selection for elliptical site of orbits (approaching the parabola from one side) was performed according to $0.9 < e < 1.0$, and hyperbolic test for site of orbits (approaching the parabola from the other side) by criterion $1.0 < e < 1.1$. Close approximation had $0.99 < e < 1.0$ for elliptical orbits, and $1.0 < e < 1.01$ for hyperbolic orbits. Average approximation criterion was $0.95 < e < 0.98$ for elliptical orbits, and $1.1 < e < 2.35$ for hyperbolic orbits. The set of the distributions (the empirical model) gives a clear representation of behavior of a meteoric orbital complex near a parabolic limit $e = 1$ (Figures 2-6).

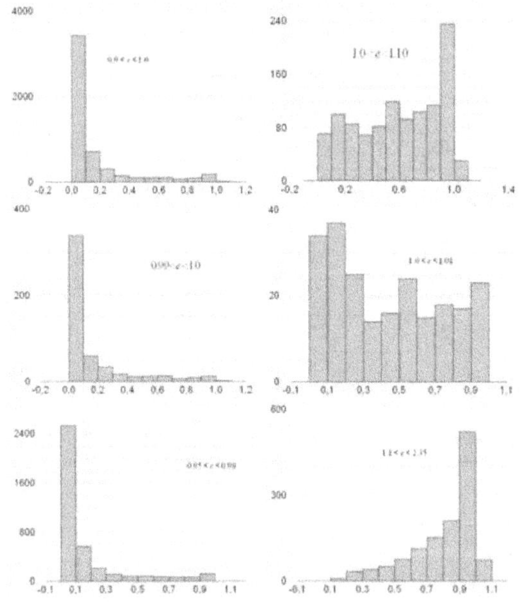

Figure 2. Histograms of the number of orbits N with different values of eccentricity e vs. perihelion distance q (in AU) for two types of near-parabolic orbits, elliptical (left column) and hyperbolic (right column).

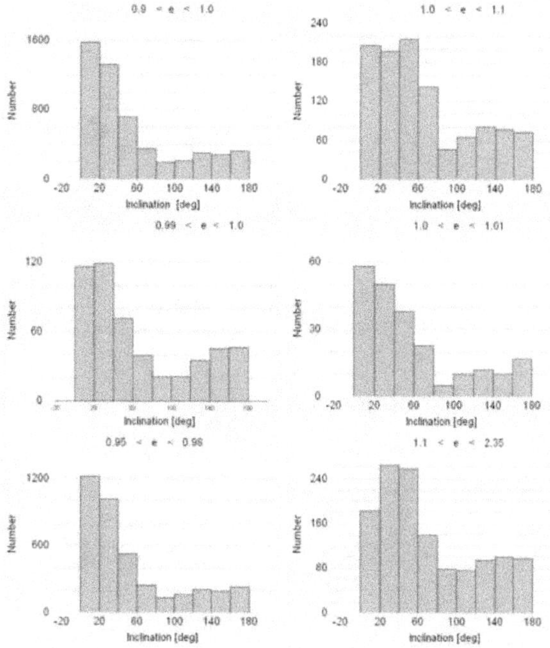

Figure 3. Number of orbits N with different values of eccentricity e vs. inclination i (in degrees) for elliptical (left column) and hyperbolic (right column) orbits.

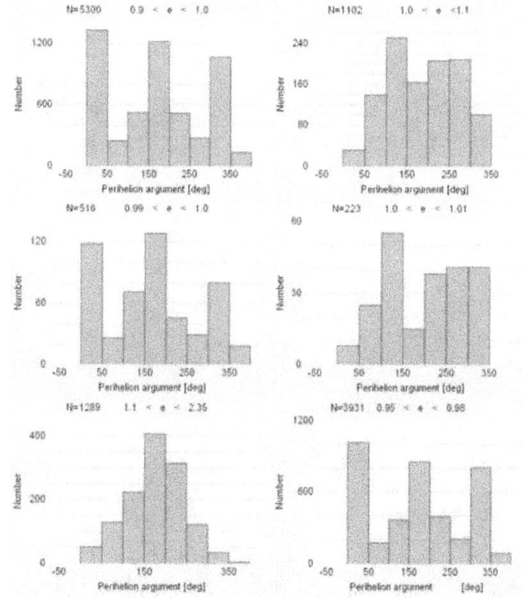

Figure 4. Number of orbits N with different values of eccentricity e vs. perihelion argument ω (in degrees) for elliptical (left column) and hyperbolic (right column) orbits.

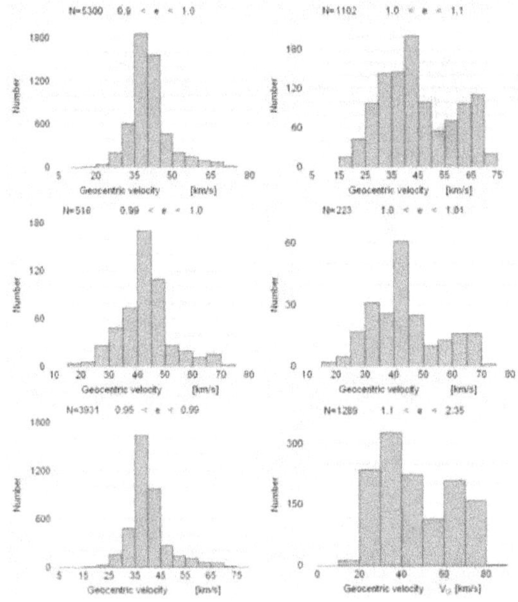

Figure 5. Histograms of the number of orbits N with different values of eccentricity e vs. geocentric velocity V_g for elliptical (left column) and hyperbolic (right column) orbits.

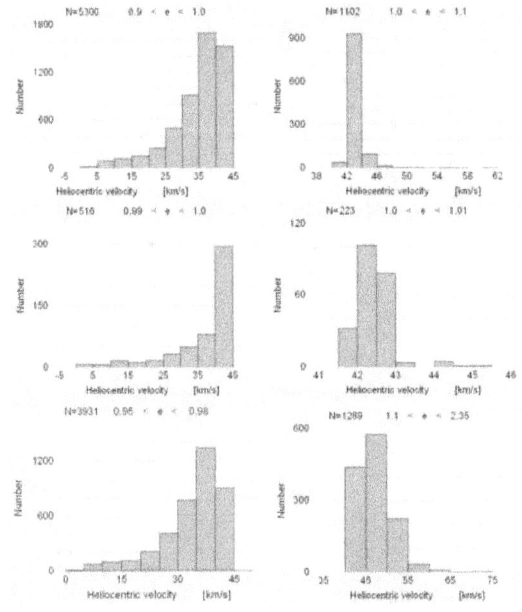

Figure 6. Histograms of the number of orbits N with different values of eccentricity e vs. heliocentric velocity V_h for elliptical (left column) and hyperbolic (right column) orbits.

3 Small-size Orbits of Meteoroids Near the Earth's Orbit

In the studies of hyperbolic meteors, the meteoroids on hyperbolic and near-parabolic orbits are mostly regarded as newcomers from distant regions of the Solar System and even from interstellar space (Baggaley 2005, Weryk and Brown 2005, Meisel et al. 2002a, b; Hawkes et. al. 1998). The fact that part of the hyperbolic and parabolic orbits complex can be formed and replenished by the component with small-size orbits in the nearby space between the Sun and the Earth's orbit is largely ignored. The recent sharp increase in interest in small bodies in the Solar System is undoubtedly due to the immediate opportunity to observe the Sun-grazing comets thanks to SOHO/LASCO and STEREO/SECCHI programs carried out over the past thirteen years. Spectacular images of comets, recorded on the disk of the Sun special satellites are available online (http://sungrazer.nrl.navy.mil/index.php) and are exciting to everyone. Comets grazing the Sun have been known for a very long time as the Kreutz comets. The working hypothesis of the origin of the Kreutz comets is the ongoing disintegration of one giant comet (Marsden 1967), and today there is some additional data to it (Guliyev 2010). These sungrazing comets are one of the specific parent sources of meteoroids with small-size orbits. The second specific parent source of meteoroids with small-size orbits is the Aten, Apollos and Amor streams that cross the Earth's orbit (AAA-asteroids).

An asteroid is considered a Near Earth Asteroid (NEA) when it comes to within 1.3 AU of Earth. A NEA is called a Potentially Hazardous Asteroid (PHA) when its orbit comes within 0.05 AU of the Earth's orbit and its absolute magnitude becomes $H < 22$ mag (i.e., its diameter is $D > 140$ m). The estimated total population of PHAs is ~ 25, 000 (http://neo.jpl.nasa.gov/ca). At the same time it is estimated that 32% of the total number of NEAs are Amors, 62% are Apollos and 0.6% are Atens.

The meteoroids-asteroids population discovered by A.K. Terent'yeva (Galibina and Terent'yeva 1981) is known as the Eccentrides. A table presenting the sample of orbital elements of some of the Eccentrides (Simonenko et al.1986) is shown in Table 2.

Table 2. Orbital elements of some of the Eccentrides. Columns N_2, N_3 are as in Simonenko et al. (1986). Other column names are as in Table 1.

N	N_2	N_3/name	e	a	q	Q	Ω	ω	i
1	4	6096	0.62	0.61	0.23	1.0	113	176	139
2	15	10573	0.87	0.54	0.07	1.0	191	177	135
3	19	11855	0.77	0.61	0.14	1.1	42	349	10
4	20	11941	0.79	0.62	0.13	1.1	44	13	47
5	38	231	0.75	0.57	0.14	1.0	260	354	34
6	39	11041	0.85	0.56	0.09	1.0	210	353	9
7	43	4473	0.94	0.53	0.03	1.0	177	3	17
8	51	1954XA	0.35	0.78	0.51	1.1	190	57	4
9	52	Hathor	0.45	0.84	0.46	1.2	211	40	6
10	53	Ra-Shalom	0.44	0.83	0.47	1.2	170	356	16

Eccentrides were defined as groups of small bodies in the Solar System with the smallest orbits ($a < 1$ AU) of medium or large eccentricity whose aphelion is near the Earth's orbit ($Q < 1.15$ AU). From existing meteors' and bolides' photographic data, Simonenko et al. (1986) has selected fifty Eccentrides. Three asteroids of the Atens team were also selected as Eccentrides (2340 Hathor, 2100 RA-Shalom and 1954 HA), although Hathor and RA-Shalom have an aphelion distance of $Q \sim 1.2$ AU.

Out of these objects, seven deserve special attention as a specific group having the most eccentric orbits (Simonenko et al. 1986). These orbits, projected on the ecliptic plane, are shown in Figure 7 (the orbital elements are presented in Table 2).

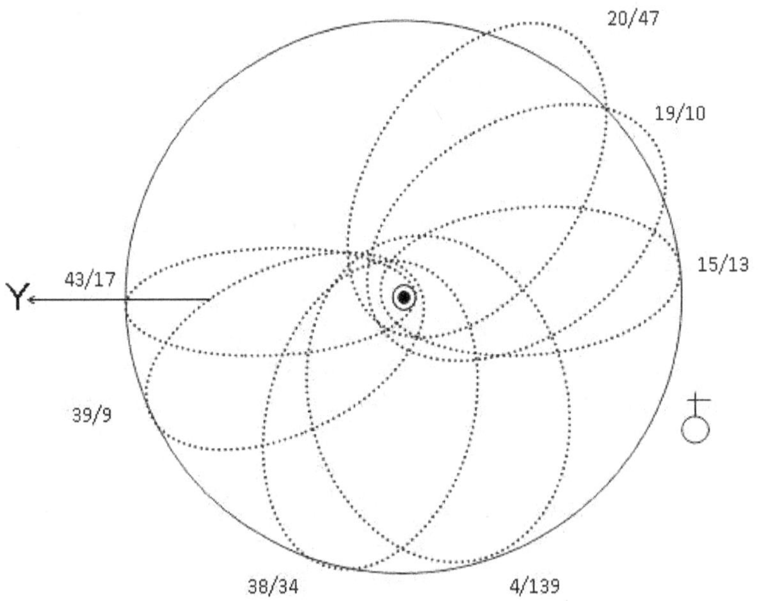

Figure 7. Seven Eccentrides with most eccentric orbits, projected on the ecliptic plane (orbital elements presented in Table 2). The numbers next to the aphelions are object numbers and their orbital inclinations, respectively (Simonenko et al. 1986).

According to Levin et al. (1981), at least 10% of the meteorites on Earth come from the population that has very small-size orbits, located entirely within the orbit of the Earth (such as, for example, Mauch, Murray, Old Peschanoe, Gorlovka and Vashugal). This class of meteorites has attracted the special attention of researchers, since they belong to the source of potentially dangerous objects for the Earth.

As the most dynamic component of the Solar System, meteoroids on near-parabolic orbits and orbits with very high eccentricities are a valuable source of information either about their progenitors, or about the place and mechanism of their formation. For example, from the Kharkov database of near-parabolic orbits it is possible to select a set of orbits with the aphelions that are characteristic for the Eccentrides. Figure 8 shows the distribution of near-parabolic orbits of sporadic meteoroids with the same aphelion distances Q as for Eccentrides. Using the streaming component (5160 orbits) of the Kharkov meteor electronic database (Voloshchuk et al. 1996, 1997, 1998), Voloshchuk et al. (2002), while calculating the probability of collision between the Earth and the parent bodies of meteor streams, has found that the most dangerous are the parent bodies whose corresponding meteor orbits have an aphelion distance of 1 AU. The authors selected 100 of the most potentially dangerous meteor streams, whose parent bodies may fall on Earth. Almost all of their orbits are the Eccentridestype. A table with examples from this list of the Eccentrides with $0.9 < e < 1$ (i.e. near-parabolic) is given in Table 3, where N_2 is a number in the list of meteoroids of the Kharkov Meteor database (ordered according to the likelihood of the stream falling on the Earth). This factor, identified above for the sporadic meteors of the Eccentrides-type, of the very low values of the perihelion distance q ("the Sungrazing orbits"), has also been identified in 12 meteor streams selected as the Eccentrides.

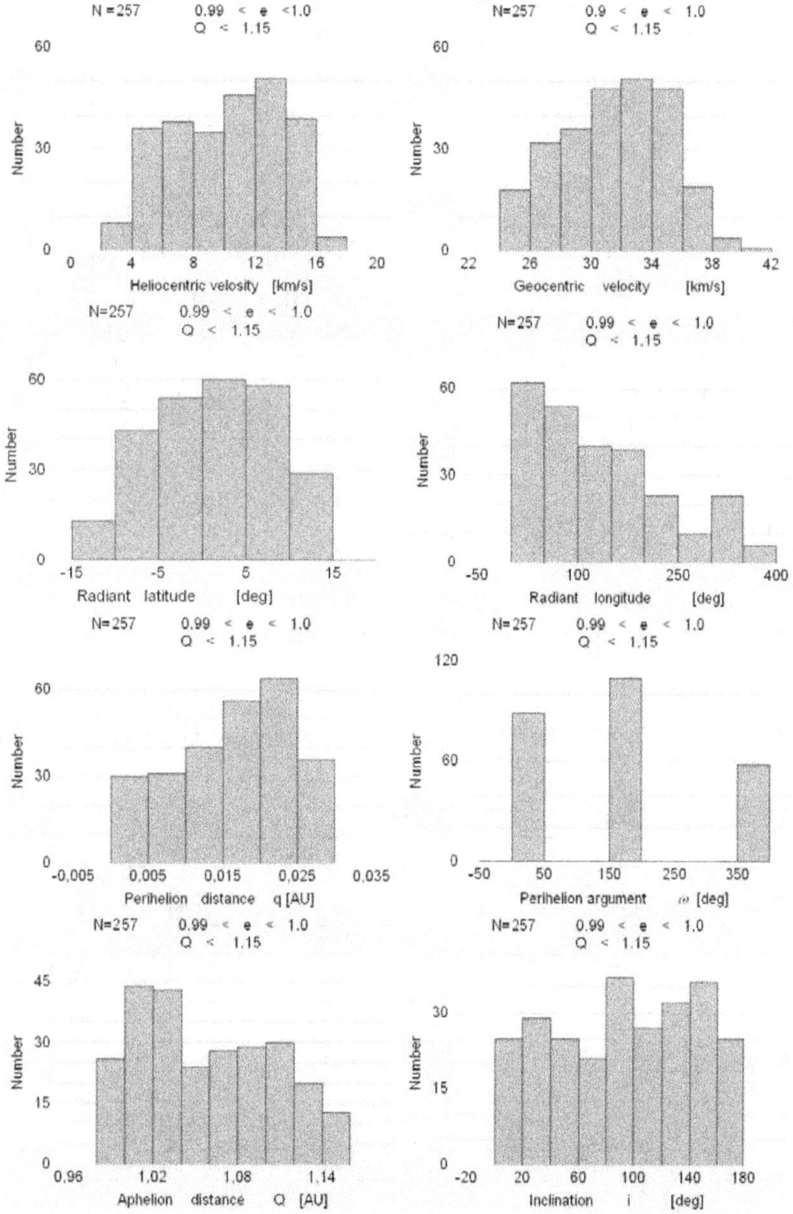

Figure 8. Distribution of the number of near-parabolic orbits of sporadic meteoroids with aphelion distance Q. The labeling of x and y-axes is the same as in Figs. 2-6 for Eccentrides. V_h is heliocentric velocity, V_g is geocentric velocity, (β, λ) are radiant latitude and longitude in ecliptic system, q and Q are perihelion and aphelion distance, ω is the argument of the perihelion, and i is the orbital inclination.

Table 3. Parameters of some streams according the KhNURE data (Eccentrides-type with $e > 0.9$) that have the highest probability of colliding with the Earth. N_2 is the number from the list of 100 dangerous streams, N_3 is the number in the KhNURE catalogue, Members - quantity, e is the eccentricity, i is the inclination, q is the perihelion distance and Q is the aphelion distance (Voloshchuk et al. 2002).

N	N_2	N_3	*Members*	e	i	q	Q
1	11	855	7	0.983	32.5	0.008	0.99
2	15	1167	10	0.98	172.9	0.001	1.02
3	16	3621	10	0.914	148.2	0.045	1.00
4	24	2807	13	0.958	163.3	0.022	1.02
5	31	313	13	0.943	41.4	0.029	1.00
6	36	4333	21	0.906	137.7	0.049	0.99
7	42	3175	18	0.923	160.2	0.041	1.03
8	45	3155	13	0.966	110.2	0.018	1.01
9	51	4123	8	0.943	159.1	0.030	1.03
10	62	3981	13	0.935	77.1	0.034	1.00
11	65	2596	7	0.996	155.2	0.002	1.04
12	98	3530	9	0.910	27.4	0.049	1.04

4 World Radar Data Resources of Hyperbolic Orbits

Main modern holders of world radar data resources of orbits of meteoroids are specified in Table 4. From Table 4 it can be seen that the r`esource-monitoring data on near-parabolic and hyperbolic orbits of meteoroids is quite impressive.

Table 4. World data resources of hyperbolic orbits: data, the methodology and the nominal parameters of meteoric automatic radar systems MARS, CMOR, and AMOR.

Country	Ukraine	Ukraine	Canada	New Zealand	Puerto Rico
Radar name	MARS	MARS	CMOR	AMOR	Arecibo meteor radar
Radar type	VHF	VHF	HF/VHF SKiYMET	HF/VHF SKiYMET	UHF, HPLA
Method	Impulse-diffraction, mirror reflect	Impulse-diffraction, mirror reflect	Impulse-diffraction, mirror reflect	Impulse-diffraction, mirror reflect	Not mirror reflection
Frequency	22.38 MHz	31.1 MHz	29.85 MHz	26.2 MHz	430 MHz
City	Kharkov	Kharkov	Tavistock, ON	Banks Peninsula	Arecibo
LAT	49.4 N	49.4 N	43.3 N	43.2 S	18.3 N
LON	36.9 E	36.9 E	80.8 W	172.5 E	66.8 W
Period	1967-1971	1972-1978	2002-2004	1995-1999	1997-1999, 2002
Enter data	ATC	ATC	ATC	ATC	Head echo
Record / Holding	Oscillograph / photofilm	Computer/paper tape/ Electronic (with 1996)	Computer / Electronic	Computer / Electronic	Computer / Electronic
Orbits	~90,000	~250,000	>1,000,000	~500,000	~50,000
Magnitude or size	+8m / +12m	+12m	+8m	+8m / +13m	< 20 – 100 μm
Hyperbola content	Didn't search	1-3%	1-10%	1-3%	~2%

There are radars in New Zealand and Canada providing extensive observation results (reported by Baggaley et al. 2001, Weryk and Brown 2005). The Advanced Meteor Orbit Radar (AMOR) is located near Banks Peninsula on the South Island in New Zealand (172.6E, 43.6S). The Canadian Meteor Orbit Radar (CMOR) is located near Tavistock, Canada (80.8W, 43.3N). The CMOR has accumulated over one million meteor orbits. These meteor radars (AMOR and CMOR) are based on the commercially available SKiYMET system. The Kharkov meteor radar of 1970s (MARS) had some distinctions.

Difficulties exist in comparing the radar observation data obtained from these three sites (Banks Peninsula, New Zealand; Tavistock, Canada; Kharkov, Ukraine). Moreover, comparison of data collected by the above mentioned three stations with the classical meteor radar and the Arecibo radar data requires an even more complex approach (Pellinen-Wannberg 2001). This data is not published in full and is not accessible for the general use, neither it is transferred to the IAU MCD.

A new global program "International Space Weather Initiative" (ISWI) started in 2010 (http://www.iswisecretariat.org). Today it is necessary to create the general unified meteor radar orbit catalogue (with hyperbolic and near-parabolic orbits) in the framework of this new international program ISWI (or in any other way) with the collaboration of Ukraine, the USA, Canada, New Zealand, possibly Japan, and other countries. Both the IAU MDC (Slovakia) and the Virtual Meteor Observatory (the Netherlands) shall be used for creating this International Radar Catalogue.

5 Links to International Projects

5.1 International Heliophysical Year

This work was undertaken in the framework of the international project 2007-2009 International Heliophysical Year (Harrison et. al.2007, Davila et. al.2004). Meteor research was officially included as an IHY program under the title "Meteors, Meteoroids and Interplanetary Dust" only in 2007 (Kolomiyets and Slipchenko 2008). The principal mechanism for coordinating scientific activities for the IHY was the Coordinated Investigation Programs (CIPs). Information on research works in the scientific discipline "Meteors, Meteoroids, Dust" (Coordinator Svitlana Kolomiyets, Ukraine) of the IHY project is shown in Table 5.

Table 5. The meteor IHY 2007/9 Activities of the NIS (the Discipline: Meteor/Meteoroids/Dust). It has 7 Coordinated Investigation Programs: CIP 60, CIP 65, CIPs 72-76.

CIP	Program Title	Lead Proposer	Affiliation, *city*, country
CIP 60	Influence of Space Weather on Micrometeoroid Flux	Dr. Thomas Djamaluddin, Senior Researcher, Head of Center for Application of Atmospheric Science and Climate	National Institute of Aeronautics and Space (LAPAN), *Bandung*, Indonesia
CIP 65	Meteors in the Earth Atmosphere and Meteoroids in the Solar System	Dr. Svitlana Kolomiyets, Researcher, Meteor Radar Centre	Kharkov National University of Radioelectronics (KhNURE), *Kharkov*, Ukraine
CIP 72/65	Meteors in the Earth Atmosphere and Meteoroids in the Solar System	Prof. Oleg Belkovich	Kazan State University, Zelenodolsk branch, *Kazan*, Tatarstan, Russia
CIP 73/65	Meteors in the Earth Atmosphere	Prof. Nelly Kulikova	Obninsk State Technical University, *Obninsk*, Russia
CIP 74	Meteoroid-Atmosphere Interactions	Dr. Olga Popova, Senior Researcher (SR)	Institute for Dynamics of Geospheres of the Russian Academy of Sciences, *Moscow*, Russia
CIP 75	Meteoroid Streams: Origin, Formation, Observations	Prof. Galina Ryabova	Tomsk State University, *Tomsk*, Russia
CIP 76	Physical Properties of Meteoroids and Bolide-Meteorite-Asteroid Associations	Dr. Natalia Konovalova, Senior Researcher	Institute of Astrophysics, Tajik Academy of Sciences, *Dushanbe*, Tajikistan

The IHY, an international program of scientific collaboration in order to understand the external drivers of planetary environments, has come to the end. Many aspects of the IHY are continuing through the program International Cosmic Weather Initiative. As it was presented and discussed on February 18, 2009 at the meeting of the UN's COPUOS (United Nation Committee on the Peaceful Uses of Outer Space) Science and Technical Subcommittee (STSC), the ISWI is a 3-year plan (2010-2013). The study of the energetic events in the Solar System will pave the way for safe human space travel to the Moon and planets in the future, and may serve as an inspiration for the next generation of space physicists. To complement the ground-based data, a huge amount of data from space-based missions on the Earth and heliospheric phenomena is available. Support of local governments and institutions is needed for local scientists to participate in the analysis and interpretation of this data.

5.2 The Meteor Heritage of the Twentieth Century

One of the objectives of the IHY project and the coordinated research IHY CIP65 Meteors in Earth's atmosphere and meteoroids in the Solar System is to reflect the important role in the development of meteor studies during the previous similar worldwide program The International Geophysical Year 1957 (IGY). At the same time the CIP 65 draws the attention of the scientific community in a large reserve not only unpublished observation data and knowledge gained during the Soviet period in the meteor centers of the USSR, but also to the significant scientific publications of the meteor heritage of the former USSR, which continue to be available only in Russian. The huge amount of data and knowledge about meteors of scientific value was accumulated in the former USSR thanks to the rapid development of meteor science during the second half of the twentieth century, from realization of the IGY project in 1957-1959 (Lebedev and Sologub 1960). The linguistic barrier, along with other reasons, limits access of world meteor science to the sources of meteor information of the former Soviet Union. The meteor heritage of the NIS is also not available to every modern researcher of meteors. Without the knowledge and the experience of meteor centers of the former USSR, the modern re-searchers of meteors sometimes have to 'invent a bicycle all over again'. This, of course, impoverishes modern meteor science and, perhaps, slows the pace of its development. In Fig. 12 the table displays the main supervision centers of meteor studies in the former USSR that participated in the international IGY program, and where the powerful meteor scientific schools were subsequently developed. These centers keep the meteor heritage of the twentieth century of the former Soviet Union.

5.2.1 Historical Note

The IGY program played an important role in the development of science, and the meteor science, inter alia. The IGY was the largest and most extensive international scientific program of the 20th century on the world-scale with 69 countries participanting, whose most significant result was the launch of the first artificial satellite of the Earth (Sputnik). The IGY has established the institutions for international scientific collaboration, which continues to play an important role in modern scientific cooperation. One such structure is the International Data Centers (IDC) that were created to store the obtained information. The first data centers were established in the USA (Boulder, IDC A), the Soviet Union (Moscow, IDCB), the UK (Slough) and Japan. The IDCs collected the observational reports from participants in all sections of geophysics, including meteor data (activity numbers, etc.). The preparations for the IGY started in 1950, but the meteor program was introduced only after 1954. The founders of the IGY Meteor Program were Prof. D. Link, Prof. V. Guth and Prof. B. Lovell. The IGY meteor studies were supervised by the 22 Commission of thr International Astronomical Union (IAU) with Prof. Guth in charge. At the same time, the Special Committee for the IGY was established in the USSR and Prof. V. V. Fedynskiy was appointed as the head of the Soviet meteor program adapted to local conditions.

The main objective of the IGY was the research of solar-terrestrial connections, with the emphasis on understanding the ionosphere and near-Earth space. Rocket technology and radar techniques were the cornerstones of the IGY. These areas are directly connected to the studies of meteors in the Earth's atmosphere and of meteoroids in the Solar System. Meteors as a research area were included in section V "Ionosphere" of the IGY program under the title "Ionosphere. Meteors". The main reason for the progress in IGY meteor studies was the implementation of the radar method. This is reflected in the table in Table 6.

Table 6. Participants of the IGY-1957 meteor program (section V Ionosphere. Meteors) in the USSR. Meteor observations: *R* radar, *Ph* photographic, *V* visual (Fedynskiy 1962).

№	City, number	φ	λ	*H* *m*	Scientific institute/Republic of the USSR/Head	Program IGY number
1	Ashkhabad (C126)	37° 56'	58° 24'	200	Astrophysical Laboratory of the Institute of Physics and Geophysics AS / Turkmen SSR / Sadykov, Ya.F., Astapovich, S.I.	R, Ph, V N696
2	Kazan	55° 47'	49° 07'	80	Astronomical observatory named Engelgart of the Kazan University / Tatarstan / Russian SFSR / Kostylyov, K.V.	R N233
3	*Kiev*	50° 27'	30° 30'	185	Astronomical observatory of the Kiev University named Shevchenko / Ukrain. SSR / Bogorodskikh, A.F.	R, Ph N320
4	*Odessa*	46° 29'	30° 46'	50	Astronomical observatory of the Odessa University / Ukrain. SSR / Tsesevich, V.P.	R, Ph, V N680
5	Stalinabad (Dushanbe) (C115)	38° 34'	68° 46'	820	Institute of Astrophysics AS Tajik SSR / Tajik SSR / Babadzhanov, P.B.	R, Ph, V N680
6	Tomsk	56° 29'	84° 59'	120	Tomsk Polytechnical Institute / Russian SFSR / Fialko, Ye.F.	R N224
7	*Kharkov* (B141)	50° 90'	36° 14'	140	Kharkov Polytechnical Institute / Faculty of Radioengineering / Ukrain. SSR / Kashcheyev, B.L.	R N358

All meteor centers of the Soviet Union that performed the IGY observation program had to carry out radar observations. In the former USSR a great importance has been given to the fulfillment of the IGY meteor program with allocation of public funds (the main initiative and the general management was performed by Prof. V.V. Fedynskiy). During the existence of the USSR, the research on meteors, both in specified centers (see Table 6) and some other establishments, has been actively sponsored at the highest level (as is a rule for large international projects). In the second half of the twentieth century, the experimental meteor radar-tracking supervisions, lead by Kharkov, were considered as one of the best in the world.

With the purpose of preservation and the development of meteoric knowledge in view of a meteor heritage of the former Soviet Union, it is necessary to establish a sponsored program for the accumulation of Soviet meteor study results of the NIS. The first implementation of such a program can be the establishment in Kharkov, Ukraine, of the first piloted center of preservation and development of meteor knowledge of the former Soviet Union on the basis of the KhNURE. KhNURE possesses access to the basic part of the meteor scientific heritage of the former USSR due to the fact that it is one of the oldest meteor radar centers of the former USSR.

Other countries also face problems in the preservation of the meteor scientific potential of the 20th century, especially for NIS. In the 20th century, the amount of data was so great that the researchers were unable to cope with its handling, especially since the existing computer facilities were inadequate. In the 21st century, new levels of information processing may allow processing of data from previous years with modern methods. This also applies to the meteor data that were preserved in WDCs (Boulder, USA; Moscow, Russia; Slow, UK, and Japan). Finding, extracting and translating meteor observation data of the past to modern media could fill up the Slovakia international meteor data centre. This also applies to the meteor data recorded in the sixties on 35-mm film everywhere in the world.

6 Conclusions

- This work was undertaken in the framework of international projects 2007-2009 International Heliophysical Year.
- Received in the KhNURE, distributions of parameters of a class of near-parabolic and hyperbolic meteoric orbits on the Kharkov data of radar-tracking supervision of 1972-1978 represent an empirical model of an observable sporadic complex of meteor orbits of this class.
- Separate attention is deserved with an observable complex of meteoric orbits of the small sizes (e.g. the Eccentrides, the Sungrazing group).
- The problem of near parabolic/hyperbolic orbits is not solved yet.
- There are facts supporting the reality of "hyperbolic meteors". Scientists haven't enough published uniform hyperbolic orbital data.
- There are difficulties in comparing the radar observation data obtained from 4 sites (Banks Peninsula, New Zealand; Tavistock, Canada; Kharkov, Ukraine; Arecibo, Puerto Rico).
- Today it is necessary to create the common unified radar catalogue, maybe, in the frame of the international program ISWI, maybe other ways, with collaboration of the Ukraine, the USA, New Zealand, Canada, Slovakia (IAU MDC), the Netherlands (Virtual meteor radar observatory), Japan, etc. in addition to the major advances in our understanding of the ecology of meteoroids within the Solar System and beyond it.
- There is dormant meteor data in the Meteor Centers of the IGY and WDCs.
- It is necessary to create international meteor centers of the NIS for preserving meteor heritage, outreach and to promote meteor research, for example, with a pilot center located in Kharkov.

Acknowledgements

This work was undertaken in the framework of international projects of the 2007-2009 International Heliophysical Year, discipline "Meteors meteoroids, dust", CIP 65. Svitlana V. Kolomiyets is grateful for the support from the Meteoroids 2010 LOC, the Kharkov National University of Radioelectronics, the Ukrainian Astronomical Association, the International Charitable Fund of Olekcandr Feldman, The Fund of Oleksandr Feldman for her participation in the Meteoroids 2010 Conference. The author would like to express her gratitude to Dr. M. Safonova for considerable help with English editing.

References

1. G.V. Andreyev, B.L. Kashcheyev and S.V. Kolomiyets, Hyperbolic meteoroid flax near Earth orbit, in Abstracts of IAU Symposium 160: Asteroids, Comets, Meteors (1993, Belgirate, Italy) (1993), p.12.
2. W.J. Baggaley, S.H. Marsh and R.G.Bennett, Feature of the enhanced AMOR facility: the advanced meteor orbit radar, in Proceedings of the Meteoroids 2001 Conference (2001, Kiruna, Sweden), ed. by B.Warmbein, (ESA Publication Division, ESTEC, Noordwijk, the Netherlands), (2001), pp.387-392.
3. W.J. Baggaley, Interstellar dust in the Solar System, in Modern Meteor Science An Interdisciplinary View, ed. By R. Hawkes, I. Mann and P. Brown (Springer, Dordrecht, 2005), pp. 197-209.
4. O.I.Belkovich and I.N. Potapov, Expected distribution of some of the orbits elements of interstellar particles in the solar system. Astronomicheskiy Vestnik 19, N3, 206-210 (1985), in Russia.
5. C.R. Chapman, Meteoroids, meteors, and the Near-Earth Object impact hazard, in Advances in Meteoroid and Meteor Science, ed. by J.M. Trigo-Rodriguez et. al. (Springer Science, 2008), pp.417-424 (2008). doi:10.1007/978 – 0 – 387 – 78419 – 958.
6. J.M. Davila, A.I. Poland and R.A. Harrison, IHY: A Pro- gram of Global Research Continuing the Tradition of Previous International Years. Adv. Space Res. 34, 2453-2458 (2004).
7. V. Dikarev, E. Grun, M. Landgraf, W. J. Baggaley and D. Galligan, Interplanetary dust model: from micron- sized dust to meteors, in Proceedings of the METEOROIDS 2001 Conference (2001, Kiruna, Sweden), ed. by B.Warmbein (ESA Publication Division, ESTEC, the Netherlands) (2002), pp.237-239.
8. G. Drolshagen, V. Dikarev, M. Landgraf, H. Krag and W. Kuiper, Comparison of meteoroids flux models for Near Earth Space, in Advances in Meteoroid and Meteor Science, ed. by J.M. Trigo-Rodriguez et. al. (Springer Science, 2008), pp. 191-197. doi:10.1007/978 – 0 – 387 – 78419 – 927
9. V.V. Fedynskiy, in Ionospherical Researches (Meteors), N8, Results of researches of the program of the IGY1957-1958-1959, ed. by V.V. Fedynskiy (Publishing house of the Academy of Science of the USSR, Moscow, 1962), pp. 5-6, in Russian.
10. V.V. Fedynskiy, B.L. Kashcheyev, Yu.I. Voloshchuk and A.A. Diykov, Radar observations of meteors with application of the automated systems. Vestnik Akademii Nauk SSSR, N10, Moscow (1980), p.89-94, in Russian.
11. I.V. Galibina and A.K Terentiyeva, Evolution of meteoroid orbits by age-old disturbances. Astronomicheskiy Vestnik, XV, N3, 180-186 (1981), in Russian.
12. E. Grun and M. Landgraf, Collisional consequences of big interstellar grains. JGR 105, 10, 291-297 (2000).
13. E. Grun, S. Kempf, H. Kruger, M. Landgraf and R. Srama, Dust astronomy: a new approach to the study of interstellar dust, in Proceedings of the Meteoroids 2001 conference (2001, Kiruna, Sweden), ed. by B.Warm- bein (ESA Publication Division, ESTEC, Noordwijk, the Netherlands, 2001), pp.651-662.
14. A.S. Guliyev, Origin of Short-perihelion Comets (Elm-poazerbaydzganski nauka, Baku, Azerbaijan, 2010).
15. M. Hajdukova, On the very high velocity meteors, in Proceedings of the Meteoroids 2001 conference, (2001, Kiruna, Sweden), ed. by B. Warmbein (ESA Publication Division, ESTEC, the Netherlands) (2001), pp. 557-559.
16. M. Hajdukova and T.Paulech, Hyperbolic and interstellar meteors in the IAU MDC radar data. Contrib. Astron. Obs. Skalnate Pleso, 37, 1830 (2007).
17. M. Hajdukova, Meteors, in the IAU Meteor Data Center on hyperbolicc orbits, in Advances in Meteoroid and Meteor Science, ed. by M. Trigo-Rodriguez et. al. (Springer Science, 2008), pp. 67-71 (2008). doi: 10.1007/978 – 0 – 78419 – 910.
18. R.A. Harrison, A. Breen, B. Bromage and J. Davila, 2007: International Heliophysical Year. Astron&Geophys. 46, 3.27-3.30 (2007).
19. R. Hawkes, T. Close and S.Woodworth, Meteoroids from outside the Solar System, in Proceeding of the METEOROIDS 1998 Conference (1998, Tatranska Lomnica), pp.257-263.
20. D. Janches, D.D. Meisel and J.D. Mathews, Orbital properties of the Arecibo micrometeoroids at earth interception. Icarus 150, 206-218 (2001).
21. B.L. Kashcheyev, V.N.Lebedinets and M.F. Lagutin, Meteoric phenomena in the earth's atmosphere, in Soviet Geophysical Committee Results of researches on international geophysical projects of Meteor Investigations. Interdepartmental geophysical committee of the Science Academy Presidium Academy of Science of the USSR, N2 (1967, Sovietskoye Radio, Moscow, USSR), in Russian.
22. B.L. Kashcheyev, State and prospects of development of radiometeor research, in Soviet Geophysical Committee Results of researches on international geophysical projects of Meteor Investigations. Interdepartmental geophysical committee of the Science Academy Presidium Academy of Science of the USSR (Soviet Geophysical Committee), N4 (1977) (1977, Sovietskoye Radio, Moscow, USSR), pp.5-10, in Russian.

23. B.L. Kashcheyev, Yu.I. Voloshchuk, A.A. Tkachuk, B.S. Dudnik, A.A. Diakov, V.V. Zhukov and V.A. Nechitailenko, Meteor automated radar system, in Soviet Geophysical Committee Results of researches on international geophysical projects of Meteor Investigations. Interdepartmental geophysical committee of the Science Academy Presidium Academy of Science of the USSR, N4 (1977, Sovietskoye Radio, Moscow, USSR), pp.11-61, in Russian.
24. B.L. Kashcheyev and A.A. Tkachuk, Distribution of orbital elements of minor meteoric bodies, Space Physics Problems, the Republican interdepartmental scientific collection (1979, Vyshchya shkola, Kiev) issue 14, (1979), pp.44-51, in Russian.
25. B.L. Kashcheyev and A.A. Tkachuk, in Results of Radar Observations of Faint Meteors: Catalogue of Meteor Orbits to +12m, (Soviet Geophysical Committee of the Academy of Sciences of the USSR, Moscow, 1980), p. 232, in Russian.
26. B.L. Kashcheyev, A.A. Tkachuk and S.V. Kolomiyets, On the problem of hyperbolic meteors, in Space Physics Problems, the Republican interdepartmental scientific collection (1982, Vyshchya shkola, Kiev), issue 17, pp.3- 15 (1982), in Russian.
27. A.M. Kazantsev, The possibility to detect interstellar meteoroids, Kinematics and Physics of Celestial Bodies. 82-88 (1998), in Russian.
28. S.V. Kolomiyets, Interstellar particle detection and selection criteria of meteor streams, in Proceedings of the Meteoroids 2001 conference (2001, Kiruna, Sweden), ed. by B.Warmbein (ESA Publication Division, ESTEC, Noordwijk, the Netherlands), pp.643-650 (2001).
29. S.V. Kolomiyets, Structure of the meteoroids complex with about parabolic and Hyperbolic orbits near the Earth, according to data of the KHNURE catalogue, in Proceedings of the Conference on Asteroids, Comets, Meteors (2002, Berlin, Germany), ed. B. Wambein (ESA Publication Division, ESTEC, the Netherlands 2002), pp. 237-239.
30. S.V. Kolomiyets and B.L.Kashcheyev, Complex of meteoroid orbits with eccentricities near 1 and higher, in Modern Meteor Science An Interdisciplinary View, ed. By R. Hawkes, I. Mann and P. Brown (Springer, Dordrecht, 2005), pp.229-235.
31. S.V. Kolomiyets and M.I. Slipchenko, The meteors, meteoroids and interplanetary dust program of the International Heliophysical Year 2007/9, in Advances in Meteoroid and Meteor Science, ed. by J.M. Trigo- Rodriguez et. al. (Springer Science, 2008), pp. 305 307. doi:10.1007/978 – 0 – 387 – 78419 – 943
32. E.N. Kramer, I.S. Shestaka and A.K. Markina, in Catalogue of Meteor orbits from photographic observations1957-1983, ed. by B.L. Kashcheyev (Soviet Geophysical Committee of the USSR Academy of Sciences, Moscow, 1986).
33. E.N. Kramer, A.K. Markina and L.Ya. Skoblikova, Statistical Distribution of Interstellar Meteoric Matter and the Probability of Its Collision with the Earth. Astronomicheskiy vestnik 32, N3, 277-283 (1998), in Russian.
34. E.N. Kramer, Dust particles, dust aggregates and other interstellar meteoroids, in Abstracts of International Conference on METEOROIDS (1998, Tatranska Lomnica), pp. 440-441.
35. T.S. Lebedev and V.B. Sologub, Contribution of scientists of the Ukraine to investigation in accordance with the international geophysical year program, in Information bulletin of the Presidium of the Academy of Sciences of the Ukrainian SSR, N2, 3-31 (1960), in Russian.
36. V.N. Lebedinets, Dust in the upper atmosphere and space, in The meteors, (IEM, Leningrad, 1980), in Russian.
37. V.N. Lebedinets, Interplanetary organic matter the key to unsolved problems of interaction between cosmic dust and atmosphere. Interdepartmental geophysical committee of the Science Academy Presidium Academy of Science of the USSR (Soviet Geophysical Committee, Moscow, 1990), in Russian.
38. B.Yu. Levin and A.N. Simonenko, On the Implausibility of a Cometary Origin for Most ApolloAmor Asteroids. Icarus 47, 487491 (1981).
39. B.G. Marsden, The sungrazing comet group. A.J. 72 (9), 1170-1183 (1967). doi:10.1086/11396.
40. D.D. Meisel, D. Janches and J.D. Mathews, Extrasolar micrometeors radiating from the vicinity of the local interstellar bubble. Ap.J. 567, 323-341 (2002).
41. D.D. Meisel, D. Janches and J.D. Mathews, The size distribution of Arecibo interstellar particles and its implication. Ap.J. 579, 895-904 (2002).
42. A. Pellinen-Wannberg, The high power large aperture radar method for meteor observations, in Proceedings of the Meteoroids 2001 Conference (2001, Kiruna, Sweden), ed. by B.Warmbein, (ESA Publication Division, ESTEC, Noordwijk, the Netherlands), (2001), pp.443-450
43. F.J.M. Rietmeijer, Natural variation in comet-aggregate meteoroid compositions, in Advances in Meteoroid and Meteor Science, ed. by J.M. Trigo-Rodriguez et. al. (Springer Science, 2008), pp. 461-471 (2008). doi:10.1007/978 – 0 – 387 – 78419 – 962.
44. J. Shtol, On the problem of hyperbolic meteors. Bull. Astron. Inst. Czechosl. 21, 10-17, (1970).
45. R.M. Suggs, W.J. Cooke, R.J. Suggs, W.R. Swift and N. Hollon, The NASA Lunar impact monitoring program, in Advances in Meteoroid and Meteor Science, ed. by J.M. Trigo-Rodriguez et. al. (Springer Science, 2008), pp.293- 298 (2008). doi:10.1007/978 – 0 – 387 – 78419 – 941.

46. A.N. Symonenko, A.K.Terenteva and I.V. Galibina, Meteoroid within the Earth's orbit: kscentrid. Astronomicheskiy Vestnik XX, N1, 61-75 (1986), in Russian.
47. A.A. Tkachuk, Selectivity of faint meteors, Space Physic Problem, the Republican interdepartmental scientific collection, the Republican interdepartmental scientific collection (1974, Vyshchya shkola, Kiev), issue 9 (1974), pp.92-98, in Russian.
48. A.A. Tkachuk, Daily variation of the orbital parameters of observered radiometeos. Space Physics Problems, the Republican interdepartmental scientific collection (1979, Vyshchya shkola, Kiev) issue 14, 52-61 (1979), in Rus- sian.
49. Yu.I. Voloshchuk, B.L. Kashcheyev and V.G. Kruchi- nenko, Meteors and a Meteoric Substance, (Naukova Dumka, Kiev, 1989), in Russian.
50. Yu.I. Voloshchuk, B.L. Kashcheyev and V.A. Podolyaka, A meteor complex near the Earths orbit: sporadic background, stream, and association: I. The technique for selection of streams and associations from a large sample of individual meteor orbits. Solar System Research 29, N5, 382-391 (1995).
51. Yu.I. Voloshchuk and B.L. Kashcheyev, The meteor complex near the Earths orbit: sporadic background, stream, and association. Solar System Research 30, N6, 480-498 (1996).
52. Yu.I. Voloshchuk, A.V.Vorgul, and B.L. Kashcheyev, The meteor complex near the Earths orbit: sporadic back- ground, stream, and accotiation: III. Sources of Stream and sporadic meteoric bodies. Solar System Research 31, N4, 306-329 (1997).
53. Yu.I. Voloshchuk, B.L. Kashcheyev, S.V. Kolomiyets and N.I. Slipchenko, The information about possible orbits of NEOS from the results of long-term radar observations of meteors in the Laboratory of Radioengineering of KHNURE, in Proceedings of the Conference on Asteroids, Comets, Meteors (2002, Berlin, Germany), ed. By B. Wambein, (ESA Publication Division, ESTEC, Noordwijk, the Netherlands), (2002), pp.825-828.
54. S.K. Vsekhsvyatskiy, Eruptive processes in Galaxy and hyperbolic meteors Space Physics Problems, the Republican interdepartmental scientific collection, issue 13, 141, (1978, Vyshchya shkola, Kiev), in Russian.
55. S.J. Weidensehilling, The distribution of orbits of cosmic particles detected by Pioneers 8 and 9. Geophys. Res. Lett. 5, N7, 606-608, (1978).
56. R.J. Weryk and P. Brown, A search for Interstellar meteoroids using the Canadian Meteor Orbit Radar (CMOR), in Modern Meteor Science An Interdisciplinary View, ed. by R. Hawkes, I. Mann and P. Brown (Springer, Dordrecht, 2005), pp. 221-227

Preliminary Results on the Gravitational Slingshot Effect and the Population of Hyperbolic Meteoroids at Earth

P. A. Wiegert

Abstract Interstellar meteoroids, solid particles arriving from outside our Solar System, are not easily distinguished from local meteoroids. A velocity above the escape velocity of the Sun is often used as an indicator of a possible interstellar origin. We demonstrate that the gravitational slingshot effect, resulting from the passage of local meteoroid near a planet, can produce hyperbolic meteoroids at the Earth's orbit with excess velocities comparable to those expected of interstellar meteoroids.

Keywords meteors · meteoroids · interstellar material · orbital dynamics

1 Introduction

The search for interstellar meteoroids is complicated by contamination of the sample by the abundant meteoroids originating within our own Solar System. Meteoroid velocity is frequently used as a filter to distinguish between these two samples, with velocities above the hyperbolic limit at the Earth's orbit taken as being interstellar in origin. This criterion is based on the assumption that meteoroids on hyperbolic orbits do not originate within our Solar System.

However, there are processes at work in our Solar System that certainly produce unbound meteoroids. One of these is the so-called gravitational slingshot, whereby a meteoroid or other particle passing near a planet can exchange energy and momentum with it. Such interactions should produce hyperbolic meteoroids at the Earth's orbit that are of a purely local origin. In order to distinguish these from true interstellar meteoroids, an understanding of the properties and fluxes of such meteoroids is needed.

Meteoroids ejected from other solar systems are expected to enter the Solar System with excess velocities typical of the velocity dispersion of stars in the solar neighborhood, about 20 km/s. Since energy and not velocity is conserved, they would arrive at Earth with a velocity near $(20^2 + 42^2)^{1/2} \approx 46.5$ km/s where 42 km/s is the escape velocity from the Sun calculated at the Earth's orbit. The presence of this excess velocity has been the traditional hallmark searched for when one looks for extra-solar meteors.

2 Review

Whether of an interstellar nature or not, hyperbolic meteors have been reported in the past, having been observed both in space and at the Earth. Spacecraft dust detectors aboard the Ulysses, Galileo and Helios

P. A. Wiegert (✉)
Dept. of Physics and Astronomy, The University of Western Ontario, London Ontario CANADA. Phone.: +1-519-661-2111 ext. 81327; Fax: +1-519-661-3283; E-mail: pwiegert@uwo.ca

spacecraft (Grün et al 1993; Krüger et al 2007) have detected very small (10^{-18} - 10^{-13} kg) grains moving at speed above the local solar system escape velocity and parallel to the local flow of interstellar gas. These particles are too small to be detected as meteors at the Earth, sizes > 10^{-10} kg may be required for this. These larger particles have also been reported to have a significant hyperbolic component. Between 0.2% and 22% of meteors observed at the Earth by various surveys, optical and radar-based, have shown a hyperbolic component according to a recent review by Baggaley et al (2007). Conversely, other work (Hajduková and Paulech 2007; Hajduková 2008) has shown that many hyperbolic meteors may only appear so as the result of measurement errors. For example, many of the hyperbolic meteors are associated with shower radiants or the ecliptic plane, unlikely associations for interstellar meteors. As a result observations of hyperbolic meteors in the Earth's atmosphere remain somewhat controversial. The problem rests on the velocity, the key signature of an interstellar origin, but which often has an uncertainty (~10%) which is of the same order as the effect one is trying to detect.

The question of the nature of hyperbolic meteors and the possible presence of interstellar meteoroids within our Solar System is an interesting one, but here we address the question of whether or not hyperbolic meteors could be produced within our own Solar System, in particular by the gravitational slingshot effect.

3 Model

In this preliminary work, we simply consider the well-known problem of two-dimensional gravitational scattering of meteoroids off a moving planet. The planets are all considered to be on circular coplanar orbits, with the meteoroids moving within this same plane. A proper treatment relevant to our Solar System will require considering the full three-dimensional scattering problem, but the simple two dimensional problem provides us with initial insight into the broad strokes of the result.

We consider the scenario depicted in Figure 1 below. The planet is moving to the left with a velocity V. The meteoroid arrives with speed v, direction ϕ and impact parameter y, all measured in the heliocentric frame. The arrival velocity is assumed to be less than the local solar escape velocity at the scattering planet. After scattering off the planet, the meteoroid departs with a new velocity v_f and direction ϕ_f. If this final velocity places the meteoroid on an unbound orbit but one which will cross that of the Earth before leaving the Solar System, we conclude that it constitute an observable hyperbolic meteoroid of local origin.

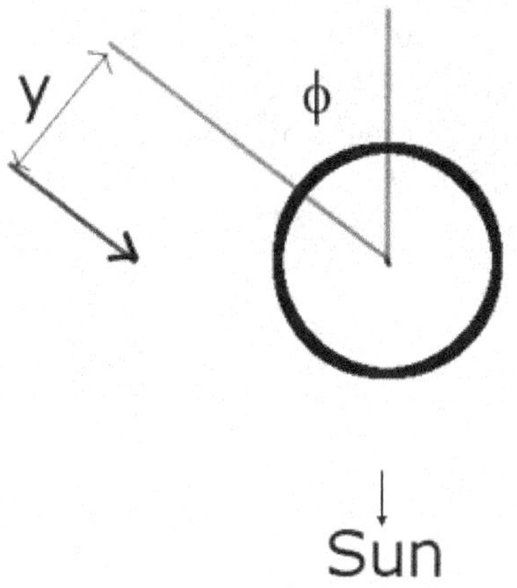

Figure 1. The angle ϕ and the impact parameter y are defined as shown, in the heliocentric frame.

For the purposes of this study, we assume that all the planet are bombarded by meteoroids arriving from all directions, with all possible impact parameters and (bound) velocities, and ask what fraction of these would become observable hyperbolic meteors at the Earth.

The results for meteoroids arriving at a particular planet with a particular speed can be summarized in a single figure displaying the scattering results for a range of arrival direction and impact parameter, here taken on a 100x100 grid. Figure 2, for example, shows the result of meteoroids arriving at Jupiter with a heliocentric velocity of 1.4 times the local circular velocity. A substantial fraction of these objects, indicated by the black area in the figure, leave Jupiter on hyperbolic Earth-crossing orbits. Of course, having arrived at Jupiter on nearly-unbound orbits (the local escape speed is $2^{1/2} \approx 1.414$ times the circular velocity), many of these meteoroids are close to the parabolic limit and thus are relatively easy to scatter onto hyperbolic orbits. Lower arrival velocities (Figures 3 to 5) produce fewer hyperbolic meteoroids, as would be expected.

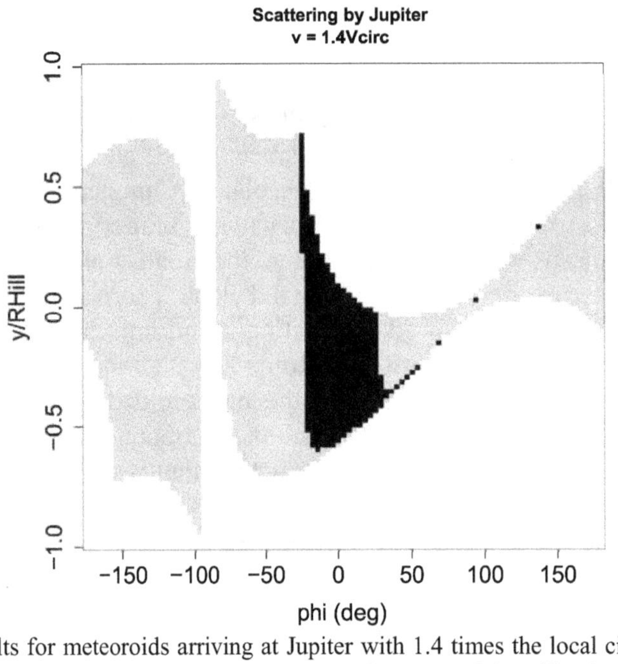

Figure 2. Scattering results for meteoroids arriving at Jupiter with 1.4 times the local circular velocity. Phi is the arrival direction ϕ and y is the impact parameter, as a fraction of the size of the Hill sphere. Grey indicates particles which leave on hyperbolic heliocentric orbits but do not cross the Earth's orbit, black indicates particles scattered onto hyperbolic Earth-crossing orbits.

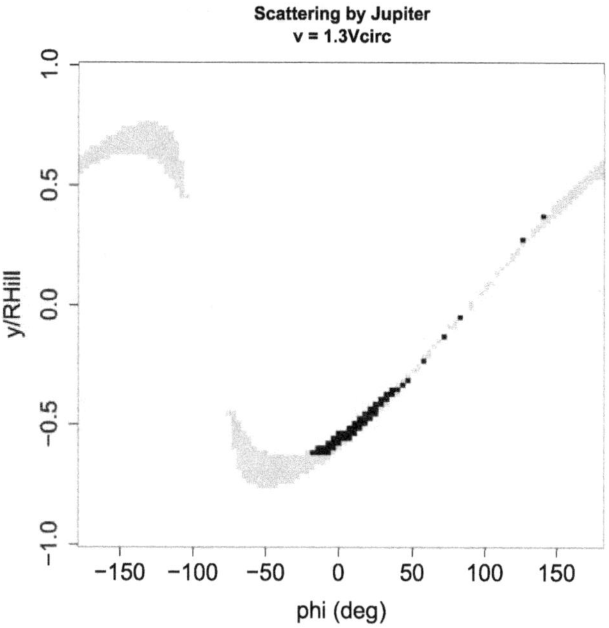

Figure 3. Scattering results for meteoroids arriving at Jupiter with 1.3 times the local circular velocity.

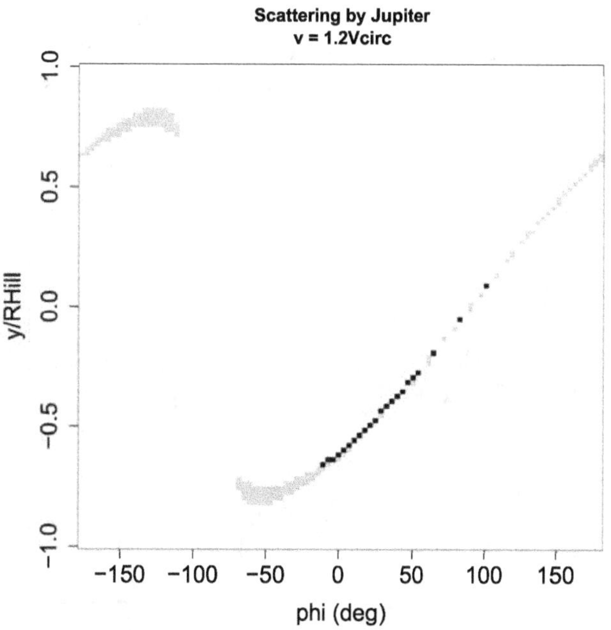

Figure 4. Scattering results for meteoroids arriving at Jupiter with 1.2 times the local circular velocity.

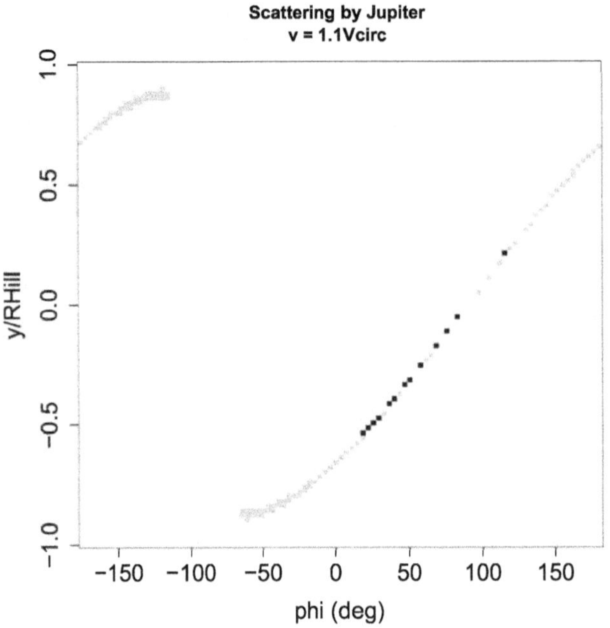

Figure 5. Scattering results for meteoroids arriving at Jupiter with 1.1 times the local circular velocity.

The distribution of velocities that these meteoroids would have measured should they happen to impact the Earth is displayed in Figure 6. This figure collects all the hyperbolic meteoroids produced during the simulations used in the production of Figures 2 to 5, and displays the excess velocity that would be observed at Earth. Most of the hyperbolic meteoroids are just above the hyperbolic limit, but there are some which can reach excess velocities of a few km/s, just what is expected of interstellar meteoroids. Thus we cannot conclude that hyperbolic meteoroids are necessarily of interstellar origin.

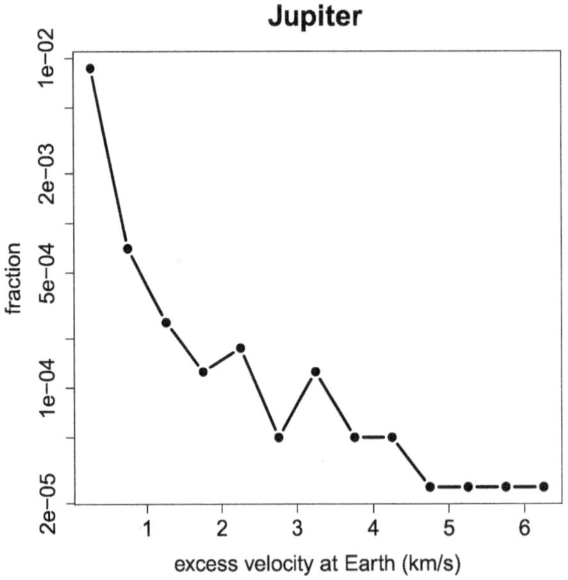

Figure 6. Distribution of excess velocities measured at the Earth for hyperbolic meteoroids of Figures 2 to 5. Fraction is relative to the total number of meteoroids simulated.

The other planets are also capable of producing hyperbolic meteoroids. Mercury and Mars are the least efficient due to their low masses, and are not plotted amongst the following figures, which illustrate the velocity distribution produced from a similar consideration of Saturn (Figure 7), Uranus (Figure 8), Neptune (Figure 9) and Venus (Figure 10). These planets are all much less efficient than Jupiter and produce hyperbolic meteoroids that almost exclusively arrive at Earth with excess velocities below 1 km/s.

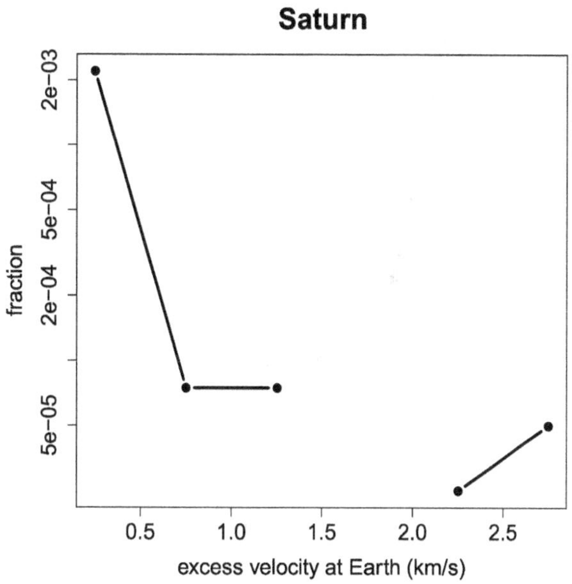

Figure 7. Distribution of excess velocities measured at the Earth for the hyperbolic meteoroids scattered by Saturn. Missing points indicate those arrival velocities which are not produced by any of the initial conditions considered.

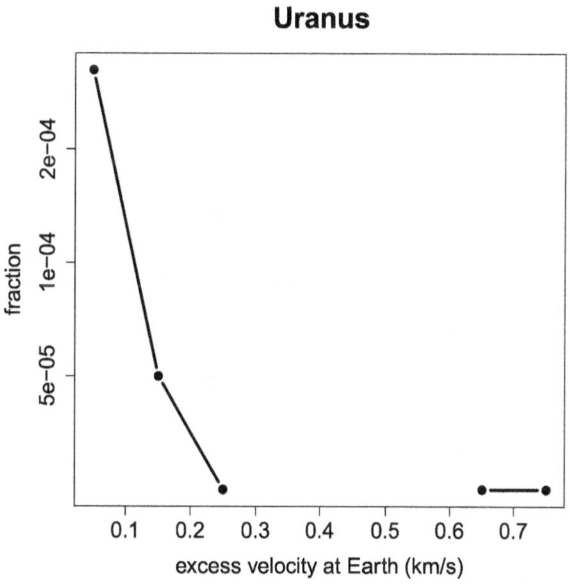

Figure 8. Distribution of excess velocities measured at the Earth for the hyperbolic meteoroids scattered by Uranus.

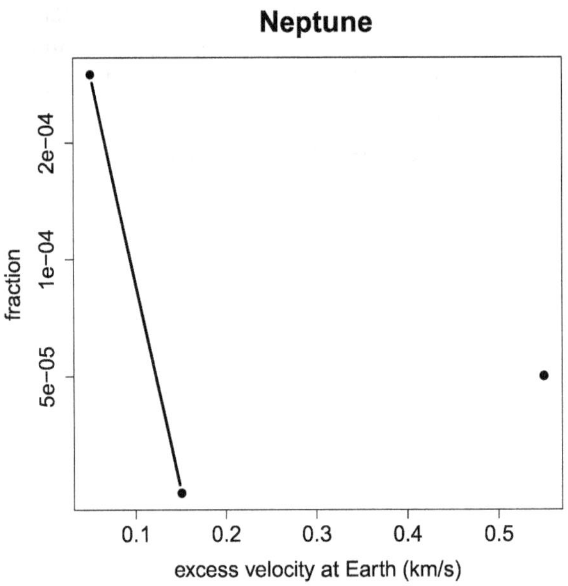

Figure 9. Distribution of excess velocities measured at the Earth for the hyperbolic meteoroids scattered by Neptune.

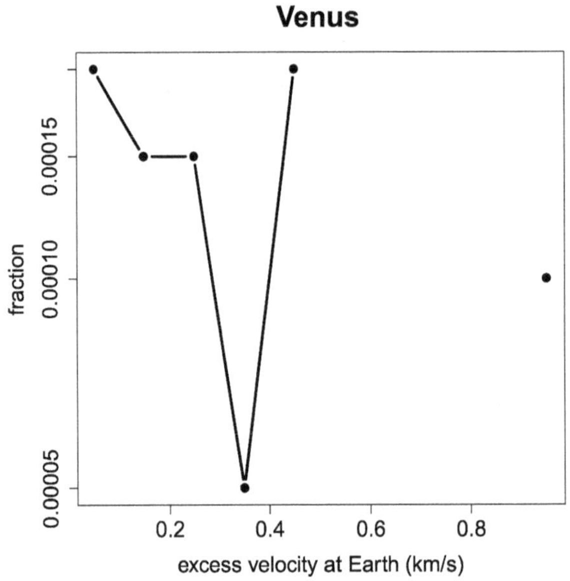

Figure 10. Distribution of excess velocities measured at the Earth for the hyperbolic meteoroids scattered by Venus.

4 Conclusions

The gravitational slingshot effect can produce meteors with hyperbolic heliocentric velocities measured at Earth that originate wholly within our Solar System. Though our study here is far from exhaustive, we have found that hyperbolic are most easily produced by Jupiter from meteoroids with near-parabolic orbits. The majority have small (< 1 km/s) excess velocities but some can exceed 5 km/s. Thus we conclude that hyperbolic excess velocities, even of a few km/s, are not unequivocal signatures of an interstellar nature.

Future work would involve extending these results to full three-dimensional scattering, which we are currently undertaking. In addition, estimates of the flux of gravitationally scattered meteoroids at the Earth would be of great value. However, this calculation will require the determination of the meteoroid environments of the planets first, as the production of hyperbolic meteoroids depends sensitively on both the speed and direction with which the meteoroids approach the scattering planet, and the relative populations of such meteoroids is not yet known.

Acknowledgements

This work was performed in part with support from the Natural Sciences and Engineering Research Council of Canada.

References

Baggaley WJ, Marsh SH, Close S (2007) Interstellar Meteors. Dust in Planetary Systems 643:27–32

Grün E, Zook HA, Baguhl M, Balogh A, Bame SJ, Fechtig H, Forsyth R, Hanner MS, Ho-ranyi M, Kissel J, Lindblad B, Linkert D, Linkert G, Mann I, McDonnell JAM, Morfill GE, Phillips JL, Polanskey C, Schwehm G, Siddique N, Staubach P, Svestka J, Taylor A (1993) Discovery of Jovian dust streams and interstellar grains by the ULYSSES spacecraft. Nature 362:428–430, DOI 10.1038/362428a0

Hajduková M (2008) Meteors in the IAU Meteor Data Center on Hyperbolic Orbits. Earth Moon and Planets 102:67–71, DOI 10.1007/s11038-007-9171-5

Hajduková M Jr, Paulech T(2007) Hyperbolic and interstellar meteors in the IAU MDC radar data. Contributions of the Astronomical Observatory Skalnate Pleso 37:18–30

Krüger H, Landgraf M, Altobelli N, Grün E (2007) Interstellar Dust in the Solar System. Space Science Reviews 130:401–408, DOI 10.1007/s11214-007-9181-7, 0706.3110

CHAPTER 4:

METEOROID IMPACTS ON THE MOON

Lunar Meteoroid Impact Observations and the Flux of Kilogram-sized Meteoroids

R. M. Suggs • W. J. Cooke • H. M. Koehler • R. J. Suggs • D. E. Moser • W. R. Swift

Abstract Lunar impact monitoring provides useful information about the flux of meteoroids in the hundreds of grams to kilograms size range. The large collecting area of the night side of the lunar disk, approximately 3.8×10^6 km^2 in our camera field-of-view, provides statistically significant counts of the meteoroids striking the lunar surface. Over 200 lunar impacts have been observed by our program in roughly 4 years. Photometric calibration of the flashes observed in the first 3 years along with the luminous efficiency determined using meteor showers and hypervelocity impact tests (Bellot Rubio et al. 2000; Ortiz et al. 2006; Moser et al. 2010; Swift et al. 2010) provide their impact kinetic energies. The asymmetry in the flux on the evening and morning hemispheres of the Moon is compared with sporadic and shower sources to determine their most likely origin. These measurements are consistent with other observations of large meteoroid fluxes.

Keywords impact flash · lunar impact · meteoroid flux

1 Introduction

Video observations of the Moon during the Leonid storms in 1999 and 2001 (Dunham et al. 2000; Ortiz et al. 2000, 2002) confirmed that lunar meteoroid impacts are observable from the Earth. One probable Geminid impact was observed from lunar orbit by Apollo 17 astronaut Dr. Harrison Schmitt (NASA 1972). NASA's Marshall Space Flight Center (MSFC) began routine monitoring of the Moon in June 2006 with multiple telescopes following our first detection in November 2005 (Cooke et al. 2006 and 2007). Of the more than 175 impacts observed in the first 3 years, 115 of them have been used to determine the flux of impactors in the 0.1 to 10s of kilogram size range. This flux is compared with other measurements in section 5 and the correlation of the observations with meteor showers and sporadic is examined in section 4.

2 Observation and Analysis Process

The observations are carried out at the Automated Lunar and Meteor Observatory located on-site at the

R. M. Suggs (✉)
NASA, Space Environments Team, EV44, Marshall Space Flight Center, Huntsville, AL 35812, USA. E-mail: rob.suggs@nasa.gov

W. J. Cooke • H. M. Koehler • R. J. Suggs
NASA, Space Environments Team and Meteoroid Environment Office, EV44, Marshall Space Flight Center, Huntsville, AL 35812, USA

D. E. Moser
MITS Dynetics, Space Environments Team, EV44, Marshall Space Flight Center, Huntsville, AL 35812, USA

W. R. Swift
Raytheon/MSFC Group, Space Environments Team, EV44, Marshall Space Flight Center, Huntsville, AL 35812, USA

MSFC near Huntsville, Alabama (latitude 34.66 north, longitude 86.66 west) and at a remotely controlled observatory near Chickamauga, Georgia (34.85 north, 85.31 west). The instrument complement has changed somewhat over time beginning with a 10 inch (254 mm) diameter Newtonian reflector for the initial observations then two Meade RCX400 14 inch (355mm) diameter telescopes with Optec 0.33x focal reducers and StellaCam EX or Watec 902H2 Ultimate monochrome video cameras. Both cameras use the same Sony HAD EX ½ inch format CCD. The effective focal length is approximately 923mm giving a horizontal field of view of 20 arc minutes covering approximately 4×10^6 square km or 12% of the lunar surface (see Figure 1). In 2008, one of the 14 inch telescopes was replaced with a Ritchey Chretien Optical Systems 20 inch (0.5 m) telescope with the focal reducer adjusted to give approximately the same field of view as the 14 inch instruments. The limiting stellar magnitude at the 1/30 second frame rate is approximately 12. The video from the cameras is digitized using a Sony GV-D800 digital tape deck and sent by Firewire to a personal computer where it is recorded on the hard drive for subsequent analysis.

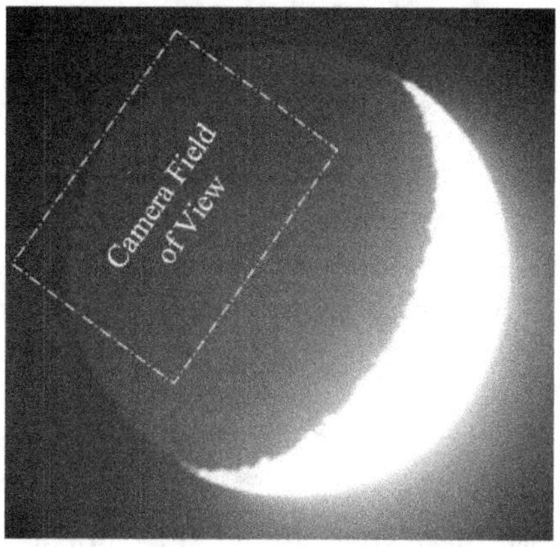

Figure 1. Camera field of view and orientation.

The observations of the night portion of the Moon are made when the sunlit portion is between 10% and 50% illuminated. This occurs on about five nights and five mornings per month. No observations are attempted during phases less than 10% since the time between twilight and moon rise or set is too short. Observations are not made during phases greater than 45 - 50% because the scattered light from the sunlit portion of the Moon is too great and masks the fainter flashes. Large lunar albedo features are easily visible in the earthshine and are used to determine the approximate location of the impacts on the lunar surface.

The recorded video is analyzed using two custom programs. LunarScan (available at http://www.gvarros.com) was developed by Peter Gural (Gural 2007). The software finds flashes in the video which are statistically significant (as described in Suggs et al. 2008) and presents them to a user who determines if they are cosmic ray impacts in the detector, sun glints from satellites between the Earth and the Moon, or actual meteoroid impacts. By requiring that a flash be simultaneously detected in two telescopes, cosmic rays and electronic noise can be ruled out. Five of the detected impacts were observed with only one telescope early in the program but only flashes which spanned more than two video frames and showed a proper light curve (abrupt brightness increase followed by gradual decay) were counted. There have also been a few impacts independently observed by amateur astronomers using 8 inch (200 mm) telescopes (Varros 2007; Clark 2007). For short flashes where satellite motion might not have been detectable, custom software was used to check for conjunctions with Earth orbiting satellites whose orbital elements are available in the unclassified satellite catalog (www.space-track.org). Since there is some probability that orbital debris or a classified satellite not listed in this catalog could cause such a short flash, a remotely controlled observing station was constructed in northern Georgia about 125 km from MSFC. This allows parallax discrimination between impact flashes and sun glints from manmade objects, even at geosynchronous altitude. After 3 years of operation of the remote observatory only one candidate flash due to orbital debris has been seen that could have been mistaken for an impact and that one showed orbital motion upon closer inspection. Whenever the weather doesn't allow operation of the remote observatory, temporally short flash images are enhanced and closely examined for any sign of motion with respect to the lunar surface.

After detection and confirmation, another computer program, LunaCon, is used to perform photometric analysis (Swift et al. 2007). Background stars are used as photometric references to determine the observed luminous energy of the flashes. Since a reference star is unlikely to be in the frame during a flash, the earthshine on the Moon is used as a transfer standard thereby correcting for first order extinction. LunaCon also displays graphics showing the lunar surface brightness, contrast between the lunar surface and space next to the limb, lunar elevation angle, lunar surface area in the field of view, and other data quality diagnostics as a function of time during the night. These displays make it obvious when clouds pass, twilight is contaminating the observations, the Moon drifts in the field of view, and atmospheric extinction is extreme. Using this information, time spans of clear weather and good data quality were determined for use in the calculations of observation time necessary for flux calculations. Flashes outside of these time spans were not used in the analysis reported here. Photometric accuracy is estimated to be approximately ± 0.5 magnitudes.

3 Observational Results

Using the photometric quality criteria described above, 115 impacts were observed during periods of consistent photometric quality. By plotting the histogram of number of flashes per magnitude bin (Figure 2), we determined that our completeness limit was approximately 10th magnitude (Johnson-Cousins R band) and there were 108 flashes brighter than that. These were included in the dataset for further analysis.

Magnitude Distribution

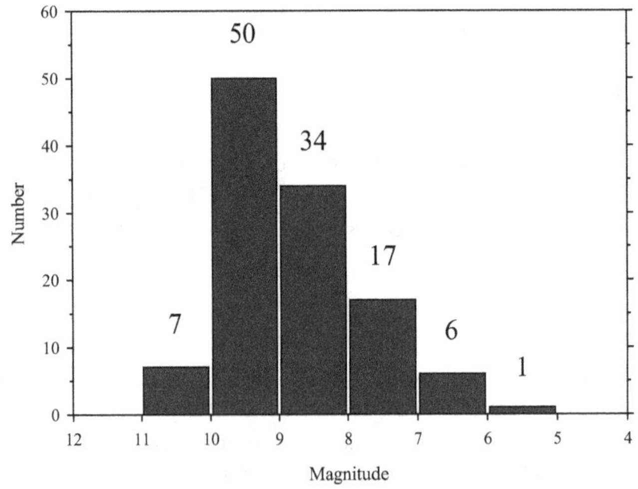

Figure 2. Histogram of flash magnitudes showing completeness to approximately magnitude 10

Calculating the flux to this completeness limit:

$$\text{Flux} = 108 \text{ impacts} / (212.4 \text{ hours} * 3.8\times10^{6} \text{ km}^{2}) = 1.34\times10^{-7} \text{ km}^{-2} \text{ hr}^{-1}$$

To compare with other estimates of meteoroid fluxes, the limiting kinetic energy corresponding to the limiting magnitude of our observations must be determined. We observe the intensity of the impact flash in our camera passband. The ratio of the optical energy and the impact kinetic energy is the luminous efficiency η.

$$\eta = \frac{optical\ energy\ in\ passband}{\frac{1}{2}\, m\, v^2}$$

where m is the mass of the impactor and v is its velocity. The luminous efficiency is a function of velocity and has been determined using laboratory measurements at low velocities (Swift et al. 2010) and using several meteor showers (Bellot Rubio et al. 2000 for Leonids and Moser et al. 2010 for Geminids, Lyrids and Taurids). The luminous efficiencies determined from laboratory and shower observations have been assimilated into a single expression by Swift et al. (2010) for the passband of the cameras used in our observations

$$\eta_{cam} = 1.5 \times 10^{-3}\, e^{-\left(\frac{9.3}{v}\right)^2}$$

Using this expression and the velocities of the various showers associated with the observations we estimated the mass at our completeness limit to be approximately 100 grams.

The impact asymmetry between the western (left, leading) and eastern (right, trailing) hemispheres evident in Figure 3 is real and when corrected for hours of observation amounts to a ratio of 1.45:1. The explanation for this asymmetry is addressed in the next section.

Figure 3. Impact flashes observed between June 2006 and June 2009 and culled for use in this analysis. Continuous monitoring was from April 2006 to the present.

4 Modeling

Our initial explanation for the asymmetry was this: observations of the western hemisphere occur leading up to first quarter phase when the observed portion of the Moon is exposed to both the Apex and Anthelion sporadic meteoroid sources. The eastern hemisphere is observed following last quarter phase when the Apex source is only visible from the farside of the Moon thus no Apex meteoroids can impact the portion of the Moon we are observing. The Apex source's flux is lower than the Antihelion's but the velocities are higher so the impact kinetic energy at a given mass would be higher. Thus the limiting mass would be lower and more meteoroid impacts would be visible. This seemed like a reasonable explanation but modeling of the asymmetry using the Meteoroid Engineering Model (McNamara et al. 2004) showed that the ratio would be 1.02:1 rather than the observed 1.45:1 so sporadics could not be the dominant source of the impacts. This result was confirmed by similar calculations by Wiegert (private communication).

Shower meteoroids then were a more likely explanation for the observed impacts and the expected rates and hemispheric asymmetry were calculated to test this hypothesis. Figure 4 shows the temporal variation of impacts compared with shower peaks.

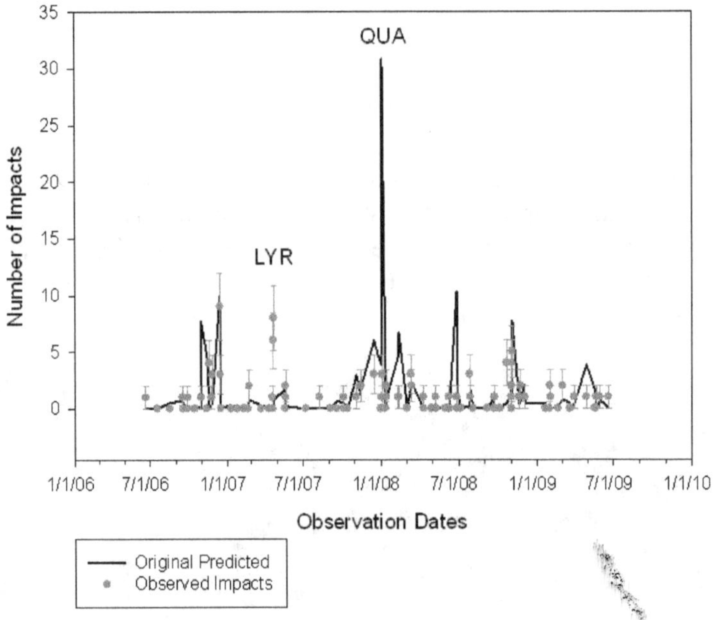

Figure 4. Impact flash distribution versus time compared with meteor showers. The red points are the observed rates with error bars representing the square root of the number of impacts per bin. The black curve is the impact rate calculated from observed values of zenithal hourly rate at the Earth. See text for discussion of this calculation.

The predicted flash rate was calculated using the reported shower zenithal hourly rates (ZHR), speed, and population/mass index. Knowledge of the camera energy threshold, combined with the shower speed and the luminous efficiency (Swift et al. 2010), enables the computation of the limiting mass for each shower. This may then be used with the ZHR (corrected for the lunar location) and the population and mass indices to obtain a flux. The predicted rate is obtained by multiplying this flux by the fraction of the observed lunar surface visible from the shower radiant. There are obviously uncertainties in the photometry and other quantities, so these were used to constrain the adjustment of the energy threshold, which was varied until a best fit with the observed Geminid rate was achieved. The Geminids were chosen because 1) they are the strongest annual shower in terms of rates, and thus 2) they have the best determined mass and population indices. The resulting impact rate was plotted for comparison with the observed rates (Figure 4).

There is a clear correlation between the observed and predicted rates. Some of the weaker showers, such as the June Bootids, JBO, do not correlate as well due to their small zenithal hourly rate and poorly determined mass index. The shallow mass indices for showers relative to the steeper one for sporadics means that there are relatively more large particles in the showers. This fact alone argues that observed lunar impacts are dominated by shower meteoroids. Sporadic source populations are less likely to contain larger particles but they do contribute to the overall observed rate. Since we are observing impacts from meteoroids larger than 10^{-1} kg and visual and video observers (from which the population indices and ZHRs are derived) have limiting masses around 10^{-7} to 10^{-5} kg, we are extrapolating over several decades in mass to estimate the impact rate we observe. It is remarkable that the rates match as closely as seen in Figure 4. The mass indices for two showers had to be adjusted to get a better match. Figure 5 shows that the calculated impact rate for the Quandrantids (QUA) was too high and for the Lyrids (LYR) was too low. A better fit was obtained for the 2007 Lyrids when its population index was changed from 2.9 to values of 2.5, 2.3, and 2.6 for the dates of April 21, 22, and 23, respectively. This shallower distribution increased the number of larger meteoroids to better match those impacts we observed. The 2008 Quadrantids had a reported population index of 2.1 which overestimated the number of large meteoroids by a factor of 10. When the population index was adjusted to 2.6, a better match with our observations was obtained. Figure 4 has these adjustments included while Figure 5 does not.

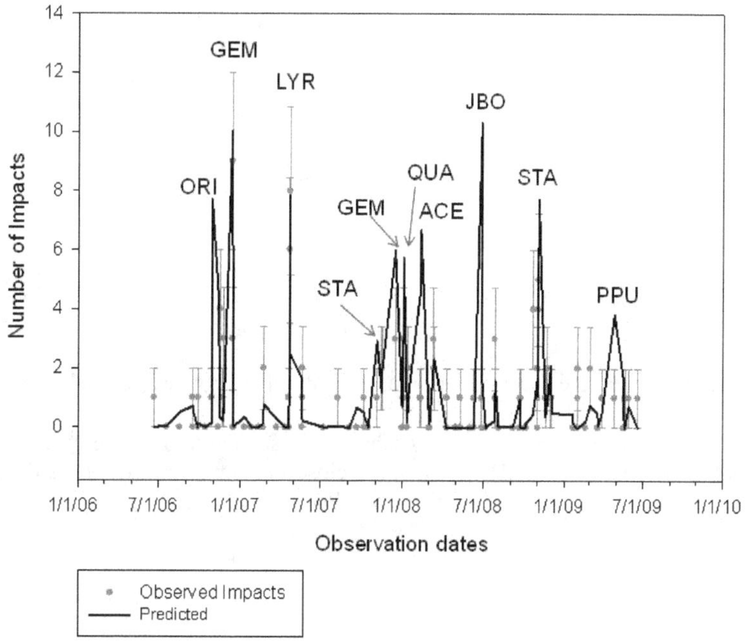

Figure 5. Impact flash distribution versus time compared with meteor showers using observed ZHRs and population indices from the International Meteor Organization (http://www.imo.net/data/visual). The symbols are similar to those in figure 4. Adjustment of the population indices for the Quadrantids (2.1 to 2.6) and Lyrids (2.9 to ~2.5) yielded the better fit seen in Figure 4.

Using these adjusted rates for the meteor showers gives a predicted hemispheric asymmetry during our observation periods of 1.57 compared to the observed ratio of 1.45:1. This is compelling evidence that shower meteoroids, including those from minor showers, dominate the observed impacts.

5 Flux Comparison

The observed flux of meteoroids with impact energies greater than our completeness limit was compared with fluxes determined by other techniques for larger objects. Figure 6 plots the flux determined using impact observations with those determined by all-sky fireball cameras, infrasound of meteor entries, lunar craters, satellite observations of fireballs, and telescopic observations of near-earth asteroids (Silber et al. 2009). The comparison is very favorable including the slight downturn from the power-law fit observed by the fireball network.

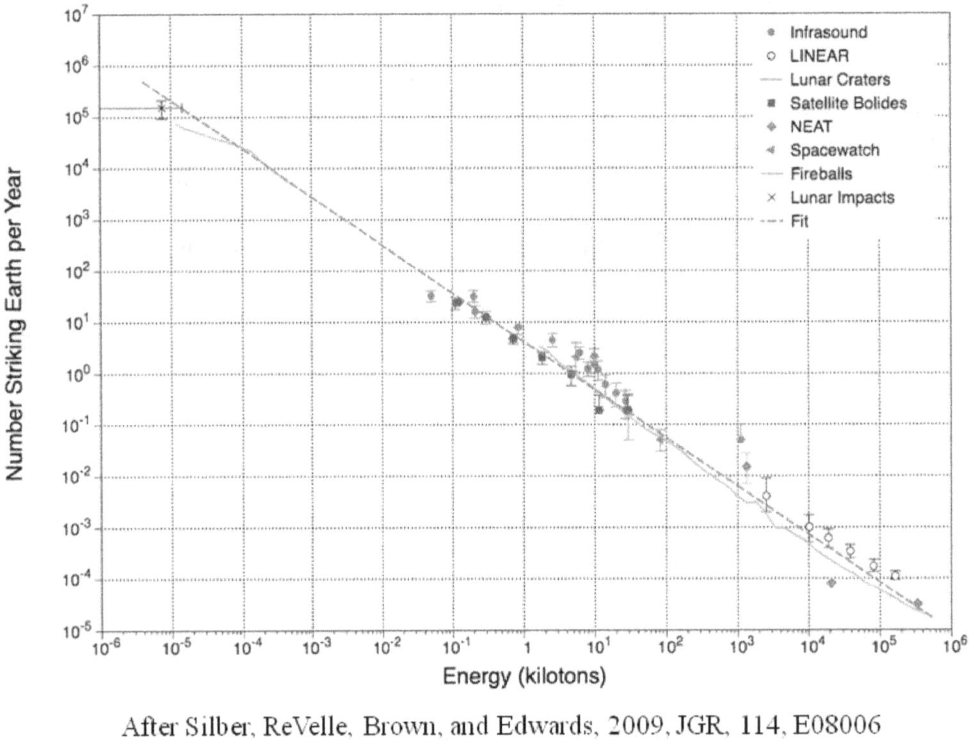

After Silber, ReVelle, Brown, and Edwards, 2009, JGR, 114, E08006

Figure 6. Number of meteoroids striking the Earth each year versus the impact energy in kilotons of TNT. Our measurement is to the extreme upper left. The cyan curve closest to our measurement is determined from all-sky camera observations of fireball meteors in the Earth's atmosphere.

6 Conclusions

MSFC's 4 years of routine lunar impact monitoring has captured over 200 impacts. Data from the first 3 years of operation were analyzed to investigate the source of the meteoroids, their flux, and the observed hemispheric asymmetry. It was found that shower meteoroids dominate the environment in this size range and explain the evening/morning flux asymmetry of 1.45:1. The observed flux of meteoroids larger than 100 g impacting the Moon is consistent with fluxes determined by all-sky fireball meteor cameras. With sufficient numbers of impacts, this technique can potentially help determine the population index for some showers in a size range not normally measured.

Future plans include performing detailed calculations to investigate the observed concentration of impacts on the trailing hemisphere limb. Observations will be continued to build up number statistics to improve our understanding of meteoroids in this size range. A dichroic beamsplitter system is under construction to allow simultaneous observations with visible and near-infrared cameras with our 20 inch (0.5 m) telescope now located in southern New Mexico. This arrangement allows 1 telescope to be used to detect and confirm impacts and allows temperature measurements of the impact flash. Observations supporting robotic lunar seismic and dust investigation missions are also planned.

Acknowledgements

The authors wish to acknowledge the meticulous and dedicated support of the following observers who helped record the video: Leigh Smith, Victoria Coffey, and Richard Altstatt. Thanks also to Anne Diekmann who performed some observations and a portion of the analysis. This work was partially supported by the NASA Meteoroid Environment Office, the Constellation Program Office, and the MSFC Engineering Directorate.

References

Bellot Rubio, L.R., Ortiz, J.L, Sada, P.V,: 2000, "Luminous efficiency in hypervelocity impacts from the 1999 lunar Leonids", *Astrophys. J.* **542**, L65-L68.

Clark, D.: 2007, private communication.

Cooke, W.J., Suggs, R.M., Suggs, R.J., Swift, W.R., and Hollon, N.P.: 2007, "Rate And Distribution Of Kilogram Lunar Impactors", Lunar and Planetary Science XXXVIII, Houston, Texas, LPI, Paper 1986.

Cooke, W.J., Suggs, R.M., and Swift, W.R.: 2006, "A Probable Taurid Impact On The Moon", Lunar and Planetary Science XXXVII, Houston, Texas, LPI, paper 1731.

Dunham, D. W., Cudnik, B., Palmer, D.M., Sada, P.V., Melosh, J., Frankenberger, M., Beech, R., Pelerin, L., Venable, R., Asher, D., Sterner, R., Gotwols, B., Wun, B., Stockbauer, D.: 2000, "The First Confirmed Videorecordings of Lunar Meteor Impacts.", Lunar and Planetary Science Conference XXXI, Houston, Texas, LPI, Paper 1547.

Gural, P.: 2007, "Automated Detection of Lunar Impact Flashes", 2007 Meteoroid Environments Workshop, NASA MSFC, Huntsville, Alabama.

McNamara, H., Suggs, R., Kauffman, B., Jones, J., Cooke, W., and Smith, S.: 2004, "Meteoroid Engineering Model (MEM): A Meteoroid Model for the Inner Solar System", *Earth, Moon, and Planets* **95**, 123-139.

Moser, D.E., Suggs, R.M., Swift, W.R., Suggs, R.M., Cooke, W.J.: 2010, "Luminous Efficiency of Hypervelocity Meteoroid Impacts on the Moon Derived from the 2006 Geminids, 2007 Lyrids, and 2008 Taurids", this issue.

NASA, December 1972. "Apollo 17 air-to-ground communications transcript", http://www.jsc.nasa.gov/history/mission_trans/AS17_TEC.PDF , page 455.

Ortiz, J.L., Aceituno, F.J., Quesada, J.A., Aceituno, J., Fernandez, M., Santos-Sanz, P., Trigo-Rodriguez, J.M., Llorca, J., Martin-Torres, F.J., Montanes-Rodriquez, P., Palle, E.: 2006, "Detection of Sporadic Impact Flashes on the Moon: Implications for the Luminous Efficiency of Hypervelocity Impacts and Derived Terrestrial Impact Rates", Icarus, **184**, 319-326.

Ortiz, J.L., Quesada, J.A., Aceituno, J., Aceituno, F.J., and Bellot Rubio, L.R.: 2002, Astrophysical Journal, **576**, 567-573.

Ortiz, J.L., Sada, P.V., Bellot Rubio, L.R., Aceituno, F.J., Aceituno, J., Gutierrez, P.J., Thiele, U.: 2000, "Optical detection of meteoroidal impacts on the Moon", *Nature* **405**, 921-923.

Silber, E.A, ReVelle, D.O., Brown, P.G., and Edwards, W.N.: 2009, Journal of Geophysical Research, **114**, E08006.

Suggs, R.M., Cooke, W.J., Suggs, R.J., Swift, W.R., and Hollon, N.:2008, "The NASA Lunar Impact Monitoring Program", Earth Moon Planets, **102**, 293.

Swift, W.R., Suggs, R.M., Cooke, W.J.: 2007, "Algorithms for Lunar Flash Video Search, Measurement, and Archiving", this issue.

Swift, W.R., Moser, D.E., Suggs, R.M., and Cooke, W.J.: 2010, "An Exponential Luminous Efficiency Mode for Hypervelocity Impact into Lunar Regolith", this issue.

Varros, G.: 2007, private communication.

Wiegert, P.: 2008, private communication.

An Exponential Luminous Efficiency Model for Hypervelocity Impact into Regolith

W. R. Swift • D. E. Moser • R. M. Suggs • W. J. Cooke

Abstract The flash of thermal radiation produced as part of the impact-crater forming process can be used to determine the energy of the impact if the luminous efficiency is known. From this energy the mass and, ultimately, the mass flux of similar impactors can be deduced. The luminous efficiency, η, is a unique function of velocity with an extremely large variation in the laboratory range of under 6 km/s but a necessarily small variation with velocity in the meteoric range of 20 to 70 km/s. Impacts into granular or powdery regolith, such as that on the moon, differ from impacts into solid materials in that the energy is deposited via a serial impact process which affects the rate of deposition of internal (thermal) energy. An exponential model of the process is developed which differs from the usual polynomial models of crater formation. The model is valid for the early time portion of the process and focuses on the deposition of internal energy into the regolith. The model is successfully compared with experimental luminous efficiency data from both laboratory impacts and from lunar impact observations. Further work is proposed to clarify the effects of mass and density upon the luminous efficiency scaling factors.

Keywords hypervelocity impact · impact flash · luminous efficiency · lunar impact · meteoroid

1 Introduction

The impact of meteoroids on the lunar surface is accompanied by a brief flash of light, detectable with small telescopes from the ground, Figure 1. These impact flashes have been successfully observed on the Moon by Earth-based telescopes during several showers (e.g. Dunham et al., 2000; Ortiz et al., 2000; Cudnick et al., 2002; Ortiz et al., 2002; Yanagisawa & Kisaichi, 2002; Cooke et al., 2006; Yanagisawa et al., 2006, Cooke et al., 2007; Suggs et al., 2008a,b; Yanagisawa et al., 2008) and for sporadic meteoroids by a campaign conducted by the NASA Marshall Space Flight Center (MSFC) since early 2006. Although the initial shock wave from a hypervelocity impact produces a significant high temperature plasma and blackbody flash lasting on the order of microseconds as the shock wave passes through the material this is generally buried below the regolith surface and not readily observable, Figure 2 lower (Ernst and Schultz, 2007). Also obscured and/or quenched by the regolith is the plasma and vapor plume observed from impacts into solid surfaces, Figure 2 upper, as modeled in early lunar impact models (Melosh et al., 1993; Nemtchinov et al., 1998). What is observed at video rates by terrestrial telescopes is the secondary blackbody radiation from the cooling hot debris thrown upwards in

W. R. Swift (✉)
Jacobs ESTS Group/Raytheon, NASA/Marshall Space Flight Center, Huntsville, AL, 35812 USA. E-mail: wesley.r.swift@nasa.gov

D. E. Moser
MITS/Dynetics, NASA/Marshall Space Flight Center, Huntsville, AL, 35812 USA

R. M. Suggs • W. J. Cooke
Meteoroid Environment Office, NASA/Marshall Space Flight Center, Huntsville, AL, 35812 USA

the initial moments of crater formation. Since the optical energy of such flashes can be readily measured telescopically, it is highly desirable to be able to estimate the energy of the meteoroid impact given the luminous efficiency η of the event. The concern then is how the luminous efficiency scales with the velocity, mass, and density of the impactor.

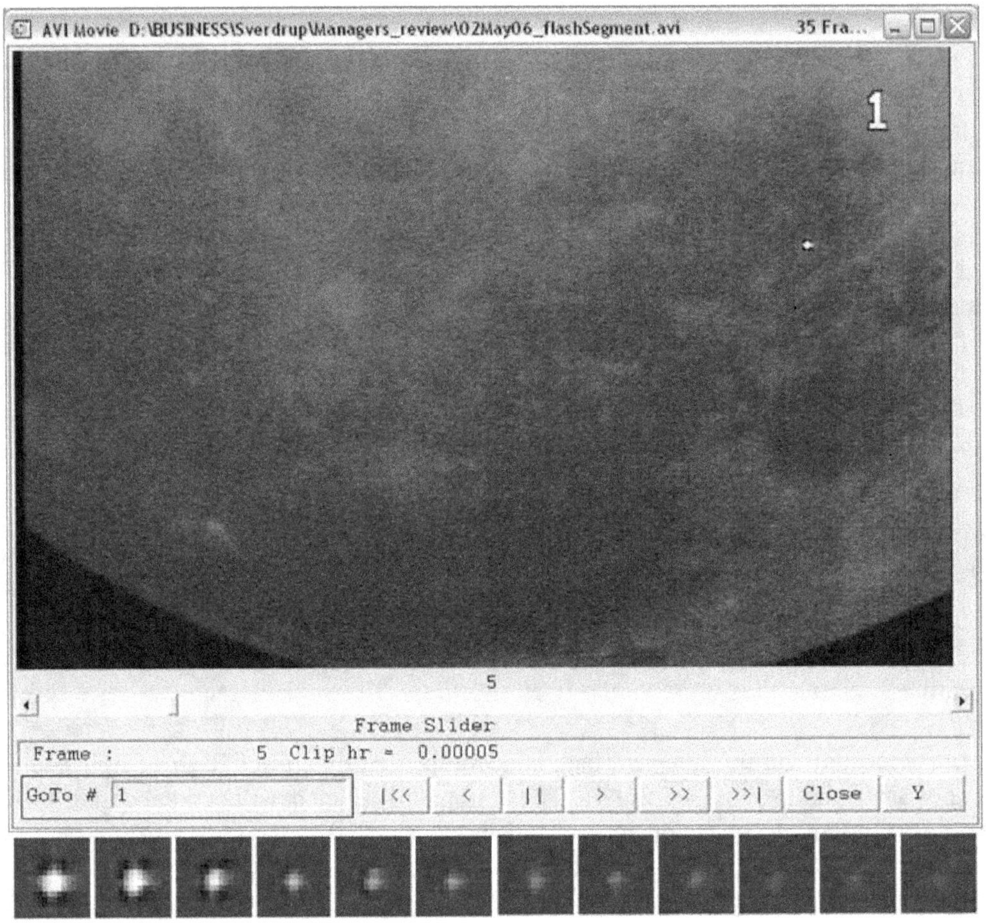

Figure 1. Lunar impact as seen on May 2, 2006 with a 254mm aperture telescope at 30 frames/second. The lower sequence shows a magnified view of the flash decay versus frame. This impact is one of the brighter impacts observed to date.

Similarly, in light gas gun experiments into pumice and lunar simulant, Figure 2, there is often a very brief (microsecond) high temperature spike recordable by high speed photodiodes (Ernst and Schultz, 2004, 2007). This early-time spike is followed over the next tenth(s) of a second by a slowly decaying secondary production of light from the hot ejecta. Moderately fast ejecta particle trails are quite evident in video rate (1/30 second) images of gas gun tests as is the cooling of the ejecta from frame to frame. Although the first video field after impact is usually the brightest, localized initial shock heating is not readily apparent in the hot ejecta dominated image. High speed camera images of lab tests (not shown) also show the primary source of illumination to be hot ejecta moving up, away from the impact rather than primary emissions from the shock wave propagating down into the target. Due to the much longer time period of these secondary emissions, their total output is significantly larger than the brief but intense shock and plasma emissions. This is especially true since most of the prompt emissions are hidden beneath the impactor and the particulate target surface.

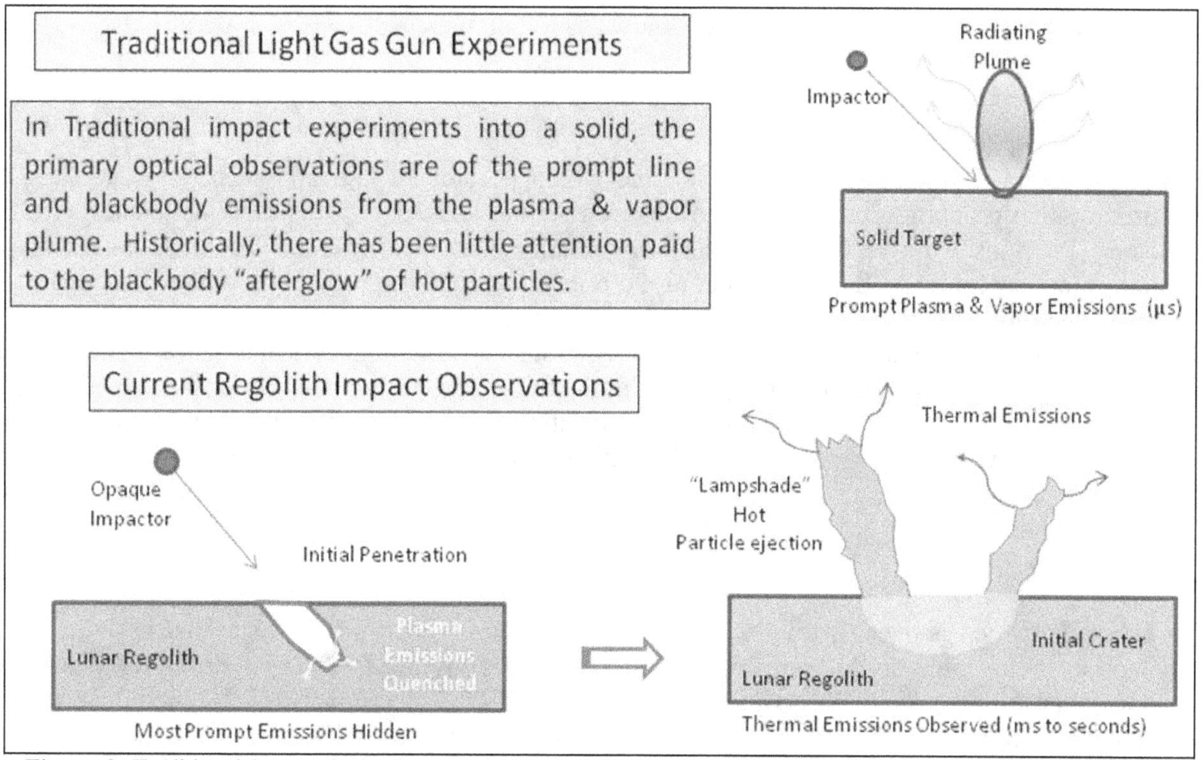

Figure 2. Traditional hypervelocity impact observations compared with impact into regolith. The emissions are thermal in nature and much longer lasting.

A series of light gas gun experiments were conducted at the Ames Vertical Gun Range (AVGR) in which a Pyrex® glass bead was shot into JSC-1a lunar regolith simulant (McKay et al., 1997; Zeng et al., 2010) at various angles and velocities. It was a relatively simple matter to calculate the luminous efficiency of light gas gun experiments since the mass, material properties, and velocity of the impactor were precisely known and the flash intensity readily measured. A problem arose when one attempted to correlate this luminous efficiency with velocity over the small range of velocities (< 7 km/s) available to the technique. The increase of luminous efficiency with velocity between 2 km/s and 6 km/s was so steep that polynomial fits extrapolate to unrealistic ($\eta > 1$) values well before the usual meteoroid velocities, V_m, of some tens of km/s. Furthermore, if curves analogous to conventional impact crater dimension scaling with exponents of V^1 to V^2 (Holsapple, 1993) are plotted through the luminous efficiency versus velocity data (almost vertical) they appear orthogonal (almost horizontal) to the data from these experiments. This implies the existence of additional phenomena that scales quite differently from conventional impact crater dimension scaling.

In order to determine an appropriate model of impact luminous efficiency versus impact velocity, it is useful to briefly examine the internal energy produced by the initial impact shock wave itself and early post shock conditions. One can then relate these conditions to the special case of the luminous efficiency of an impact into lunar regolith to obtain evidence leading to an appropriate model. Finally, this model will be compared to knowledge of the luminous efficiency from both light gas gun experiments and the growing database of lunar impact measurements.

2 Lunar Impact Luminous Efficiency

It is useful to estimate the kinetic energy of an impactor on the moon's surface from the total optical energy detected by a camera, E_λ, using a ratio known as the luminous efficiency, η_λ defined as:

$$\eta_\lambda \equiv E_\lambda / KE_{impactor} \tag{1}$$

where E_λ is defined as that energy at the source which is radiated into all space (4π steradians) as measured by that proportion received in the camera aperture and $KE_{impactor}$ is the kinetic energy of the impactor. Previous work has assumed surface radiation into 2π steradians (Swift et al. 2008) or radiation into 3π steradians (Belio Rubio et al. 2000). The geometric projection removes the effect of telescope aperture from the measurements leaving bandpass considerations unresolved. Initial assumptions that the radiation was from the early crater surface and thus into 2π steradians were abandoned when it was realized that the primary radiation was from free particles above the surface. E_λ is instrument specific, leading to the camera optical ratio, $O_c \equiv E_\lambda / E_t$, with an alternate definition of luminous efficiency, η_t or total luminous efficiency, based on total radiant energy, E_t

$$\eta_t \equiv E_t / KE_{impactor} = \sum_i m_i E_i' / KE_{impactor} \tag{2}$$

where the summation is over i particles of mass m_i and specific energy E'_i. Note that O_c is less than unity and is a function of the camera spectral response convolved with the declining blackbody emissions over the time of the observation. Improvements in the determination of O_c and the variation from camera to camera are underway but the distinctions between E_λ and E_t, are poorly defined. Note that, unlike the rate of thermal emissions, which is fourth power in temperature, E_t, is the integral over time and is almost linear in temperature since the thermal specific energy for each particle is the specific heat capacity, Cp, times the temperature change, ΔT, during emission, $E'_i = {}_{,}Cp\Delta T_i$. Unless otherwise defined, whenever η is mentioned it is usually safe to assume that η_λ is implied for the purpose of this paper.

NASA's Marshall Space Flight Center has been consistently monitoring the Moon for impact flashes produced by meteoroids striking the lunar surface since early 2006 (Cooke et al 2006). The 2006 Geminids, 2007 Lyrids, and 2008 Taurids, Table 1 below, produced a small but sufficient, sample of lunar impact flashes with which to perform a luminous efficiency analysis like that outlined in Bellot Rubio et al. (2000b). The analysis technique, discussed in detail by Moser et al. (2010), involves 'backing out' the luminous efficiency by relating the number of impacts expected on the Moon as a function of energy to the time integral of the flux of meteors of known size and the lunar area perpendicular to the shower radiant of known mass index, S. The resulting luminous efficiencies for the cameras used for the observations are shown in Table 1 with the published results of Bellot Rubio et al. (2000b) for the 1999 Leonids. Although their results are for a less sensitive camera and are based on the assumption of radiation into 3π steradians rather than 4π as assumed here, the results are consistent with the current determinations. Also shown are the results of hydrocode modeling of the 1999 Leonids by Artemieva et al. (2000, 2001). Although the agreement of this hydrocode model to the other results is entirely fortuitous, it is shown here for reference purposes. Expected errors are less than ± 20% for the camera dependant luminous efficiency. Note the almost constant luminous efficiency, η_λ, over these velocities.

Table 1. Luminous Efficiency from Lunar Impact Observations (Moser et al., 2010).

Shower	# Flashes	Obs. Time (hr)	*V* (km/s)	*S* (mass index)	η_λ
2008 Taurids	12	7.93	27	1.8	1.6×10^{-3}
2006 Geminids	12	2.18	35	1.9	1.2×10^{-3}
2007 Lyrids	12	10.22	49	1.7	1.4×10^{-3}
1999 Leonids*	5	1.5	71	2	2×10^{-3}*
1999 Leonids**	N/A	(model)	71	N/A	$1\times10^{-3}/2\times10^{-3}$

* Bellot Rubio et al. (2000) results for a different camera and slightly different geometry.
** Artemieva et al. (2000, 2001) hydrocode model results for densities 0.1 / 1.0 g/cm^3.

3 Light Gas Gun Camera Angle, Impact Angle and Velocity Experiments

A series of hypervelocity impacts into JSC-1a lunar regolith simulant at various angles and velocities were observed with the same video cameras used for lunar impact monitoring (Suggs et al. 2008b). Multiple cameras at three view angles were used in staring mode at the video rate of 29.97 frames per second. Their field of view, Figure 3 left, comprised the complete impact zone and the lenses were fitted with calibrated neutral density filters to obtain correct exposures. This contrasts with traditional light gas gun observations as illustrated in Figure 2, particularly in the time scale here of hundreds of milliseconds as opposed to hundreds of microseconds or less. Due to the long exposure sequence and good near IR sensitivity of the cameras, the hot ejecta from these impacts forms a cooling curve lasting multiple frames very similar to the bulk of the signals observed in lunar meteoroid impacts.

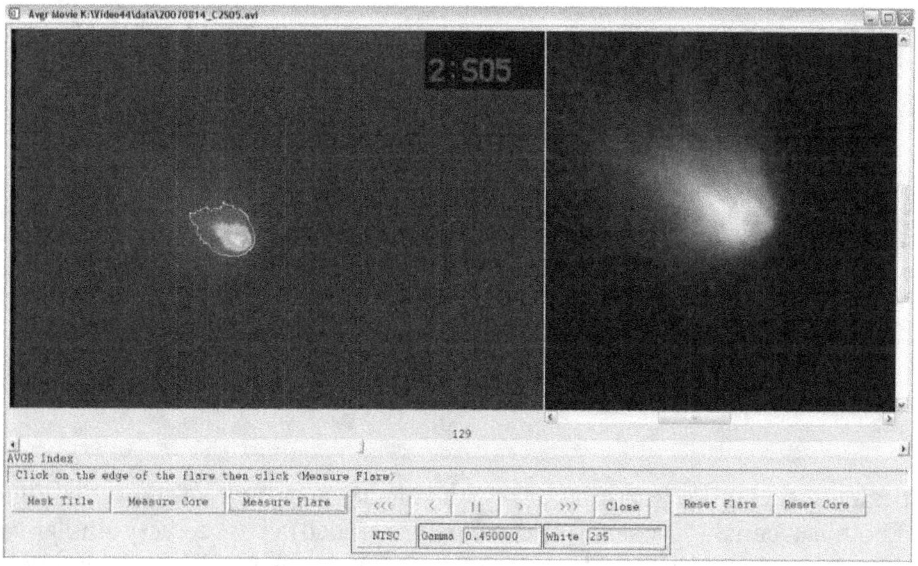

Figure 3. Software was written to semi-automatically determine the illuminated area and to compensate for background and video intensity scaling. The complete "encircled" image is in the false color image on the left while an enlarged view centered on the impact is to the right.

For these experiments, Pyrex® spheres 6.35mm in diameter and of mass 0.29 g were fired in vacuum at velocities from 2.4 km/s to 5.75 km/s at elevations of 15 to 90 degrees into a deep horizontal pan of JSC-1a lunar simulant. The cameras were mounted to observe at three angles: A) camera 2 with a 25mm lens used at f/10.84 was aimed near normal at 65 degrees elevation, 2.13m from impact, B) camera 3 with a 25mm lens used at f/12.04 was aimed at 33 degrees elevation, 1.75m from the impact and C) camera 5 with a 17mm lens used at f/4.0 viewed horizontally 1.3m from the impact. Cameras 2 and 3 were StellacamEX video cameras set at the gain used for lunar meteor impact observations. For these observations the cameras were fitted with Andover precision neutral density filters from optical density (OD) from OD 1.02 to OD 3.77. These dark filters were chosen to keep the extremely bright signals from saturating the images. Camera 5 was a Watec model 902-H2 Ultimate with the same charge coupled device (CCD), gain, and filters as the others. A parallel set of cameras fitted with photographic grade neutral density filters had radiation leaks in the IR so the data was discarded. Laboratory and stellar calibrations were used to determine the electron gain of these cameras and the published quantum efficiency curve, *QE(λ)*, for the Sony ICX248AL CCD was used to evaluate spectral response. The *QE* was used to convert from photon counts, which these cameras measure, to detected energy in order to determine η. Software was written in the Interactive Data Language (IDL) computer language, Figure 3, to isolate the flash area in each image, compensate for NTSC-J video scaling, measure the intensity, subtract backgrounds, and calibrate the results. The total emission meaning that from all illuminated pixels for all illuminated frames is used to calculate η as shown in Figure 4.

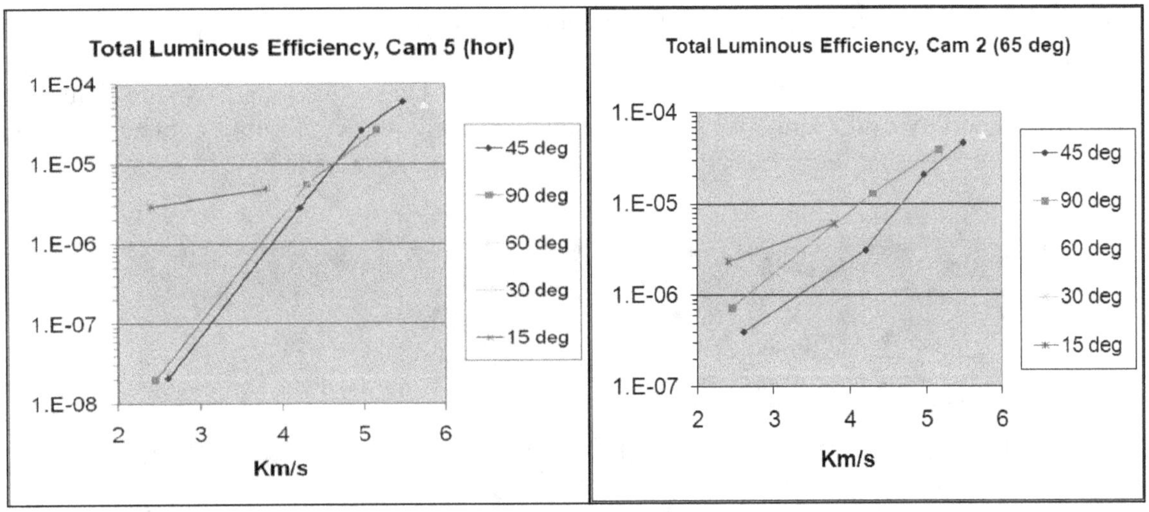

Figure 4. Total luminous efficiency of impacts of Pyrex into JAS-1 versus velocity and impact elevation. On the left is the horizontal view and on the right is the view from above. Note the convergence in both elevation and view angle near 5.5 km/s.

A brief examination of the variation of η with velocity and angle of impact in Figure 4, shows a convergence in both tangential (horizontal) and normal (overhead) views to very similar values at higher velocities for all angles of incidence. The low velocity enhancement of low angle impacts due to the "plowing up" of particles is evident as well as the negation of the effect at higher velocities. The low velocity, low angle of incidence η can be "compensated" to an equivalent η at normal incidence with a simple sine function of the impact angle that disappears above 4.4 km/s: , $\eta_c = \eta * Sin(i)\hat{}(4.4-MIN(4.4,v))$. One can see the effect of incidence compensation in Figure 5 where the normal data is shown as blue diamonds and the compensated normal data with yellow triangles. This compensation makes comparison with meteoroid impacts more realistic. The independence of luminous efficiency with angle of incidence at high velocities was also noted by Artemieva et al. (2000) and Nemtchinov et al. (1998). It is also a very convenient result for lunar impact observations since the impact angle is often unknown.

It is also desirable to correct for view angle, particularly since, due to gun emplacement, the normal view is not available. A useful viewing geometry, although inexact, is that of an oblate spheroid having a unit circle projection from above (normal) and an elliptical projection seen from any other angle. Development of this spheroid cross section model is straight forward. One lets the tangential view be approximated by a standard ellipse with unity half width a and half height b with area πab. The normal view is a circle with unit radius a and area πa^2 so that the tangential cross section ratio is b/a or just b. The height of the cross section of the spheroid viewed from angle θ is given by the radius in polar form of the ellipse where r, is given by $r^2 = a^2b^2 / (a^2 \sin^2\theta + b^2 \cos^2\theta)$. The area at view angle θ is πar so that the cross section ratio is simply r. Given experimental normal and tangential emission components at various velocities, their ratio can be used to determine the parameter, $b = 0.8V - 0.13$, a function of velocity which becomes unity (spherical) above 10.9 km/s. This has been used to correct the camera 2 data to the normal in Figure 5 prior to impact angle compensation. The primary lesson learned from this is that the surface intensity ellipse converges to a sphere and view angle effects are minimal for the higher velocities found in lunar meteoroid impacts: a very convenient result. Furthermore, it is the normal result from impact experiments that is to be compared with meteoroid impacts. A likely explanation is that at high impact velocities, most of each particle's emission is into free space significantly above the surface. This implies radiation into 4π steradians rather than 2π surface radiation or a compromise of 3π steradians (Bellot Rubio et al. 2000b).

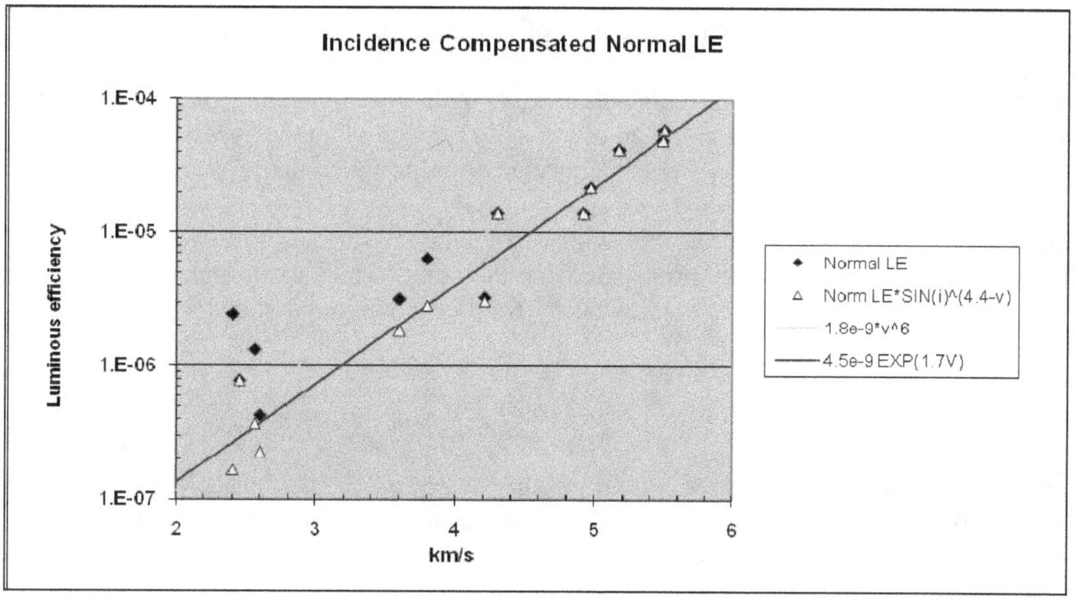

Figure 5. A trial fit compensating the luminous efficiency data for impact elevation was made for the vertical (normal) view. The normal, incidence compensated view is the one to use when comparing to meteoroid velocity lunar impacts.. Also shown is a power law velocity fits to V^6, light blue, and an exponential fit, dark line. The power law fit becomes absurd at meteoroid velocities giving $\eta > 1$ above 28.7 km/s.

Also shown in Figure 5 are trial fits to the incidence compensated η versus impact velocity data. As can be expected with a log-linear plot, a traditional power law fit appears curved while an exponential is a straight line fit to the data. The normal incidence data is approximated by a power law fit of V^6 which, unfortunately, becomes improbable at meteoroid velocities giving $\eta > 1$ above 28.7 km/s. It is also difficult to imagine a physical model with such an exponent of velocity covering three orders of magnitude change for a less than 3x change in velocity. Simple exponential functions, although a better fit over the range of the data, also become unlikely at meteoroid velocities implying an exponential form that is not simply direct with velocity as is the one shown here. These questions drive much of the discussions to follow.

A luminous efficiency error analysis was performed for the η determinations yielding an estimated one sigma precision of 21% in η. The largest contributors to the error are the camera distance, the electron gain, the effective *QE* and the average energy per photon. The distance is problematic since the emission plume is a dynamic, three-dimensional object and each pixel views a part of the image at a different distance. Note that if one doubles this error the final uncertainty will increase by about 27% to 30%. The electron gain uncertainty, e-/IU, is relatively small but can be reduced further with careful spectral calibration. The effective *QE* and energy per photon uncertainties are both due to incomplete understanding of how the CCD reacts to the color changes in images of rapidly cooling particles. Refinements for future experiments are possible which would significantly reduce the uncertainty although, due to the extremely large dynamic range of the η data (up to five orders of magnitude), the estimated precision is deemed sufficient for current purposes.

4 Impact of Shock Waves in Materials

A logical first step to determine the correct scaling of impact luminous efficiency versus impact velocity is to briefly examine the internal energy produced by the initial impact shock wave itself and early post shock conditions. Indeed, this is the approach used in hydrocode modeling of impacts (Nemtchinov, 1998; Artemieva, 2000, 2001). One can then relate these conditions to the special case of the luminous efficiency of an impact into lunar regolith to deduce an appropriate model. One starts with a review of the basics (Melosh 1989; Lyzenga 1980).

Impact of a hypervelocity projectile with a solid target surface, such as that of a particle of regolith, produces shock waves which propagate from the point of impact through the target. The shock wave speed in the target, U_s can be represented by the linear Hugoniot shock velocity relation in the notation of Melosh (1989):

$$U_s = C_b + S\,u_p. \tag{4}$$

Here C_b is the bulk speed of sound in the target, u_p is the particle speed and S is an experimentally determined material property. Coupling at impact is determined by comparing the shock impedance Z_s of the target and the impactor:

$$Z_s \equiv \partial pressure / \partial velocity = \rho_0 U_s \tag{5}$$

Then $$P_s = Z_s u_p = \rho_0 U_s u_p \tag{6}$$

Here ρ_0 is the initial target density and P_s is the pressure behind the shock wave. Note that, from Equation 6 above, the shock pressure is second order in u_p, which in direct impact experiments is the impact velocity. A few idealized special cases serve to introduce the role of shock impedance. Assume the target and impactor are the same size and $Z_{target} < Z_{impactor}$ then the impactor and target move together after impact at a reduced velocity. Similarly, if $Z_{\text{target}} > Z_{impactor}$ then the impactor bounces back from the target and target and impactor move in opposite directions. If both materials have the same shock impedance then the impactor will stop and the target will move away at the contact speed u_p. The extreme pressures P_s of the shock wave which give rise to acceleration of the target to u_p also give rise to irreversible effects which can include heating, thermal radiation, phase change, and decomposition. Due to the energy lost from the shock wave, U_s and thus u_p decline along the direction of propagation. This implies that, in a series of impacts, the energy transferred in each impact is some fraction of that of the preceding impact.

Early high pressure research (Walsh and Christian, 1955; McQueen et al., 1967) showed that solid materials under extreme pressure followed a pressure-volume curve characteristic of the material called the Hugoniot, Figure 6 (Lyzenga, 1980). Indeed, the determination of the Hugoniot for geophysical materials, (McQueen et al., 1967; Ahrens et al.,1969) is of central importance in planetary mantle investigations and drives much of the impact work to date. In a material which is transparent in the un-shocked state, shock temperature and shock velocity, V_s, can be measured by optical pyrometry. The work by Lyzenga (1980) and Lyzenga and Ahrens (1982) in which the primary thermal emissions from shocked transparent minerals are examined provides a useful introduction to the techniques involved. Shock emission techniques are further developed theoretically and experimentally by Svendsen et al. (1987) with attention paid to emissions from the shock interface. Of particular interest is the sensible (thermal) internal specific energy of the shocked state, which can be determined from the product of the change in volume times the change in pressure, $E' = \frac{1}{2}(V_0 - V_1)\Delta P$, as in Figure 6, since this energy gives rise to the observed primary and secondary thermal emissions. Although similar determinations for opaque materials such as lunar regolith are not as easily performed the same principles apply. Also note that the physical properties of the material, including shock impedance, melting point, heat of fusion, emissivity, etc. all tend to vary along the Hugoniot adding an interesting complexity to the problem.

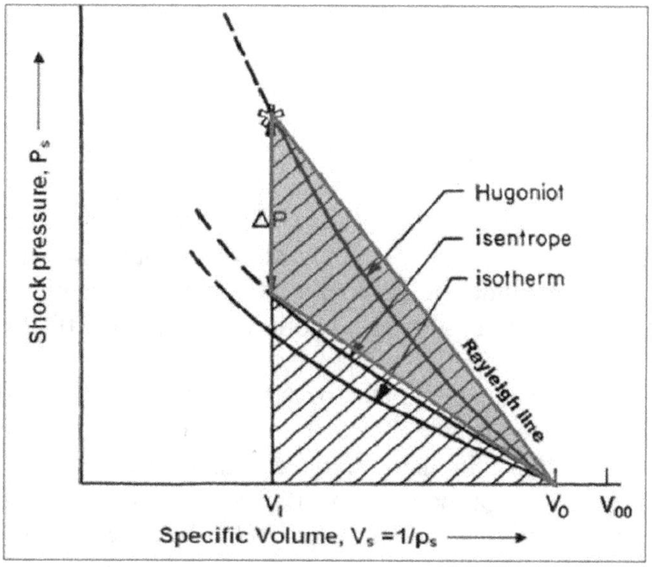

Figure 6. Simple Hugoniot compared with isotherm and isentrope of compression by Lyzenga (1980). Upon impact, a solid target is compressed along the Rayleigh line from V_o to V_1. Decompression after shock wave passage is at V_1 along ΔP followed by isentropic relaxation The total energy is given by the shaded area while the irreversible internal specific energy, the red portion, is $E' = \frac{1}{2}(V_0 - V_1)\Delta P$.

The sensible portion of this internal energy is expressed immediately as a temperature change giving rise to the primary thermal radiation observed in transparent shocked materials. Although the shock temperature with phase change is less than it would be without phase change, observed shock temperature ranges from 4000 K to 8000 K as measured by multi channel optical pyrometry. A fast response (5 ns) is required since sample thicknesses of approximately 3 mm result in emissions lasting about a third of a microsecond while the shock wave traverses the material. Such direct emissions are consistent with the brief initial spike observed in impacts into pumice (Ernst and Schultz, 2007) and lunar simulant by a transparent projectile but not an opaque one. Investigations have been performed by Ahrens et al. (1973) and Ahrens and Cole (1974) using lunar regolith returned by the Apollo missions to determine their shock properties. Similar work (Anderson and Ahrens, 1998, Schmidt et al., 1994) has also been done for chondritic meteorites where the porosity was found to be of particular importance. After relaxation, the remaining sensible energy and much of the phase change internal energy will be found in thermal form providing the cooler but still hot particles observed in a laboratory or lunar impact into granular materials.

It is desirable to compare these investigations to the observations of higher velocity meteoroid impacts on the moon (Ahrens and O'Keef, 1972) and indeed the material properties determined in the laboratory are used in hydrocode simulations which attempt to answer similar questions. For current purposes, it is sufficient to note the following:

- Passage of shock wave leaves energy in the target
- This residual shock energy is expressed as heat in the target
- Residual specific energy (heat) is traditionally expressed as V^2
- Remainder of shock wave energy is passed on as kinetic energy
- Target material becomes an impactor with reduced kinetic energy
- Powder targets imply multiple serial impacts within the target

5 Shock Waves in Porous Materials and Powders

The moon is covered with a thick layer of porous lunar regolith so lunar impact emissions are governed in a large part by the porosity of the target. In the usual model, porous materials are first compacted to a dense state prior to the initiation of the shock wave into the body of the material. Although this compaction occurs at pressures well below that of the shock wave, volume changes and $\Delta P \Delta V$ work can be a significant contributor to the post shock temperature of the bulk material (Dijken and DeHosson, 1994a). For experiments to determine the Hugoniot of some material this "interface" heating is an annoying artifact but for impact sintering to form exotic materials the effect does useful work (Dijken and DeHosson, 1994b).

The approach taken by Dijken and De Hosson (1994a, 1994b) for powder sintering by impact is particularly instructive in that they couch the effects in term of impactor velocity u_p and the ratio of solid to powder specific volume V_0/V_{00}. In their approach, they follow a path in the P-V plane that compresses at zero pressure from initial powder specific volume V_{00} to solid density V_0 then compress with V_0 constant to the constant internal specific energy $(E'-E'_0)$ curve giving the shock pressure P_s as the starting point for determining u_s. This implies an additional internal energy component of $(V_{00}-V_0)P_S$. In their development, the powder is viewed as initially separated planes of identical solid material which, by symmetry, leads to the equipartition of internal and kinetic energy. One can define a partition function B of energy in the target mass m_t between internal (thermal) and kinetic energy as follows:

$$KE_{impactor} = \frac{1}{2} m_i u_i^2 \geq (1-B) m_t (E' - E'_0) + B\ \mathrm{m_t u_t^2}/2 \qquad (7)$$

$$\text{Equipartition} \Rightarrow B = 1/2 \Rightarrow (E' - E'_0) = \frac{1}{2} \mathrm{u_t^2}$$

The simple equipartition approximation is shown to be particularly accurate (better than 5%) for loose powders with impactor velocities below 5 km/s when compared with data and more precise models (Dijken and DeHosson, 1994c). Lunar regolith (Ahrens and Cole 1974) with a bulk density of 1500 to 1800 kg/m^3 and a solid density averaging 3100 kg/m^3, has a relative powder density of 0.48 to 0.58, for which the above approximations are reasonable. The JSC-1a lunar regolith simulant (McKay et al. 1997) used in the above luminous efficiency determinations is by design very similar to the Apollo samples in these respects.

When one examines the internal energy effects of a sequence of impacts, Figure 7, each target particle becomes the impactor for the subsequent impact. From the equipartition assumption, $B = 1/2$ and the energy is quickly expended in the powder as internal (thermal) energy within a short distance from the initial penetration track. One can imagine a similar result when the effect is generalized to a branched chain series of impacts. Radiation, conduction, and plasma quenching, all lead to a rapid statistical distribution of this energy within the initial zone. Although the primary impactor can have impedance significantly different from the solid particles of the powder giving an initial ratio, B_0, different from the equipartition assumption, the serial impacts between like particles in the regolith predominate. In any case, it is clear that the impactor energy is thermalized very rapidly in the penetration phase of the impact into regolith. This view is confirmed by recent high speed camera results by Ernst et al. (2010) which show that in the first 50 μs the energy of the impactor is primarily confined to several impactor radii of the impact. This compact thermal reservoir leads to a useful macroscopic thermal approach to the problem of energy partitioning in the impact zone.

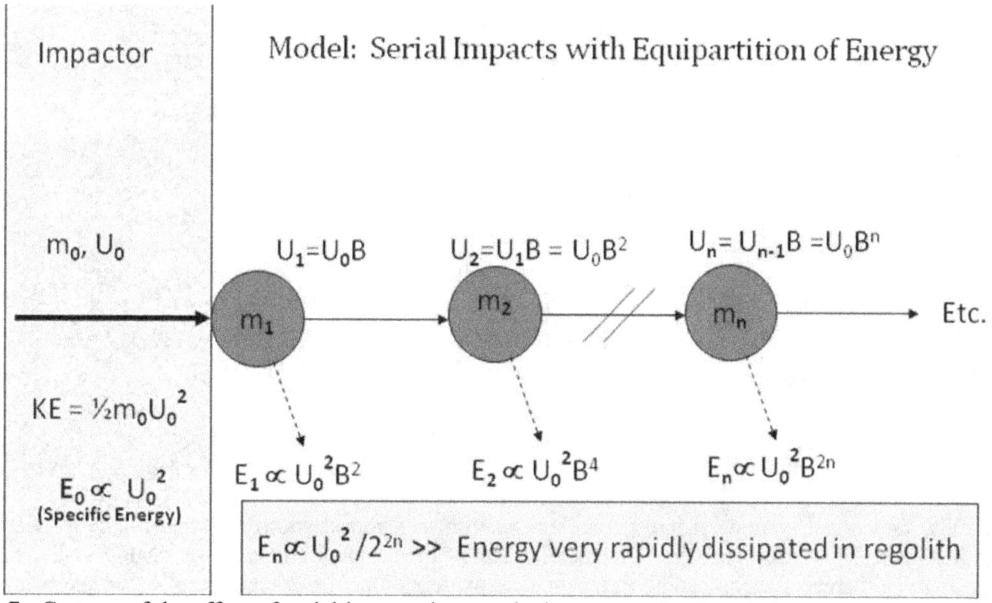

Figure 7. Cartoon of the effect of serial impacts in a particulate target. In the usual case, B = ½ corresponding to equipartition of energy. Note that the specific kenetic energy expressed by velocity U_n declines extremely rapidly.

6 A Statistical Physics Approach

The impact zone defines a thermal reservoir of many small but macroscopic particles thermally linked with one another. These are precisely the assumptions used in the development of the canonical probability distribution of the particle energy states, Figure 8. It is a small extension of the canonical representation of the energy of particle r, E_r, in Joules to the representation of that energy as an energy density, E'_r in J/Mol. Similarly, the temperature parameter, $\beta = 1/kT$, becomes $1/RT$ when expressed as an energy density. The ratio remains unchanged. Similarly, the specific energy of particle r can be expressed as $E'_r = V^2_r$ in J/kg and the specific energy of the impact zone thermal system can be expressed as $E'_r = V^2_m$ in J/kg where V_m is the impactor velocity and V_r is the specific energy equivalent velocity of state r. The resulting probability of a particle being in state r becomes

$$P_r = Ce^{-V_r^2 / V_m^2} \tag{8}$$

where C is the normalization constant. The energy density E'_T of any particular set of states, those states emitting visible radiation in this case, then becomes

$$E'_T = \sum_{r_{visible}} Ce^{-V_r^2 / V_m^2} \Delta V_r^2 \tag{9}$$

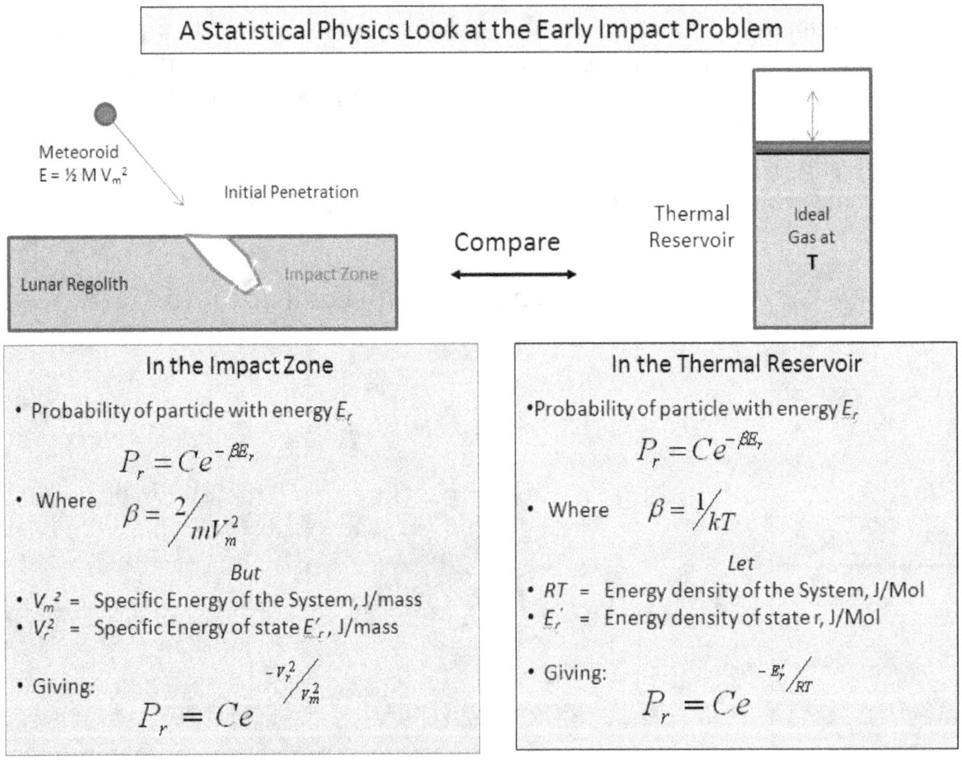

Figure 8. With hypervelocity impacts into particulate regolith, the impact specific energy is rapidly thermalised leading to a statistical physics approach. The specific energy of the impact is an exact analog of the canonical energy density of a thermal system leading to a canonical expression of the probability of a particle being in any particular energy state.

For the macro case of blackbody radiation the possible states, r, are numerous making V_r is essentially continuous allowing the summation in Equation 9 to be converted to an integral:

$$E'_T = C \int_{V_T^2}^{V_m^2} e^{-V_r^2 / V_m^2} d(V_r^2) \qquad (10)$$

Where the energy densities are left as velocities squared for clarity. In Equation 10 the velocity of the lower limit, V_T is that of the lowest detectable energy. If the problem were to determine the portion of the energy expended to melt the regolith, then this would be just the square root of the minimum energy density of the molten material. For the cameras it would be the velocity equivalent of the coolest visible blackbody radiator. In Figure 9 the fraction of photons collected from a blackbody emitter are plotted versus temperature for a typical camera used for lunar impact studies. From this it becomes evident that there is no defined threshold, V_T ,for the lower limit which would enable the integral in Equation 10 to be evaluated directly. One can, however, somewhat arbitrarily put a lower bound on the visible blackbody temperature of 1000K for a ΔT of about 900K for these silicon Vis/NIR cameras. From this one can set a lower bound on V_T of about 1.2 km/s.

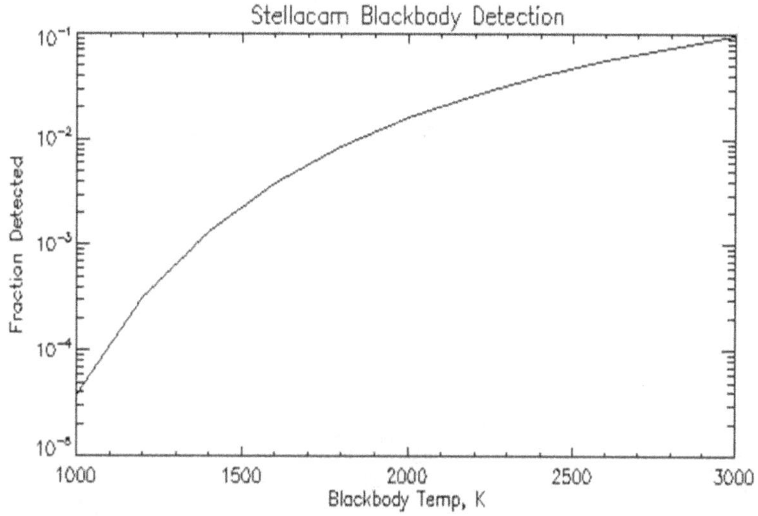

Figure 9. Fraction of blackbody emissions detected by the typical camera used for lunar impact flash detection. From this it is evident that there is no particular minimum detectable blackbody temperature. If 1000K is taken as a lower bound then the equivalent specific energy velocity, V_T would be about 1.2 km/s.

At this point we apply the Mean Value Theorem. When applied to Equation 10 the mean value theorem implies that:

there exists $V_C \in [V_T, V_m]$ such that $$E'_T = C(V_m^2 - V_T^2)e^{-V_C^2 / V_m^2} \qquad (11)$$

IF $V_m^2 << V_T^2$ then $$E'_T \cong CV_m^2 e^{-V_C^2 / V_m^2} \qquad (12)$$

Where an integration with a difficult limit, V_T, has been replaced with a characteristic velocity, V_c, and a simpler form in Equation 12. Note that for the usual case with an unresolved camera constant, O_c is lumped with the normalization constant, C. With E_T normalized as energy density E'_T and recognizing that $E'_m = V^2_m$ the luminous efficiency of the impact assumes a particularly simple form:

$$\text{since} \quad \eta = E_T/E_m = E'_T/V_m^2 \qquad \text{then} \quad \eta(V_m) = Ce^{-V_C^2/V_m^2} \qquad (13)$$

where one has two undetermined constants: a characteristic velocity V_C and a scaling factor C.

One can now use the luminous efficiencies determined from lunar impact observations in Table 1 with the light gas gun luminous efficiencies using the same cameras in Figure 5 to estimate the characteristic velocity V_C and scaling factor C. These results are shown in Figure 10, below. Also plotted for comparison are the historical luminous efficiency determinations of Bellot Rubio (2000) and Ernst and Schultz (2005). The data spans almost six orders of magnitude in η and ranges from just over 2 km/s to 71 km/s in velocity. Due to the form of Equation 13, it is immediately evident that the scaling factor is almost completely determined by the lunar impact data while the light gas gun data affects the critical velocity to a great extent. The lunar impact data yields a scaling factor estimate of $C = 1.5\times10^{-3} \pm 10\%$. Due to the wide range and natural variability of the light gas gun data various fitting techniques gave slightly different results with characteristic velocity fit ranging from 9 km/s to almost 11 km/s. From this it is estimated that the critical velocity, $V_c = 9.3$ km/s ± 10%.

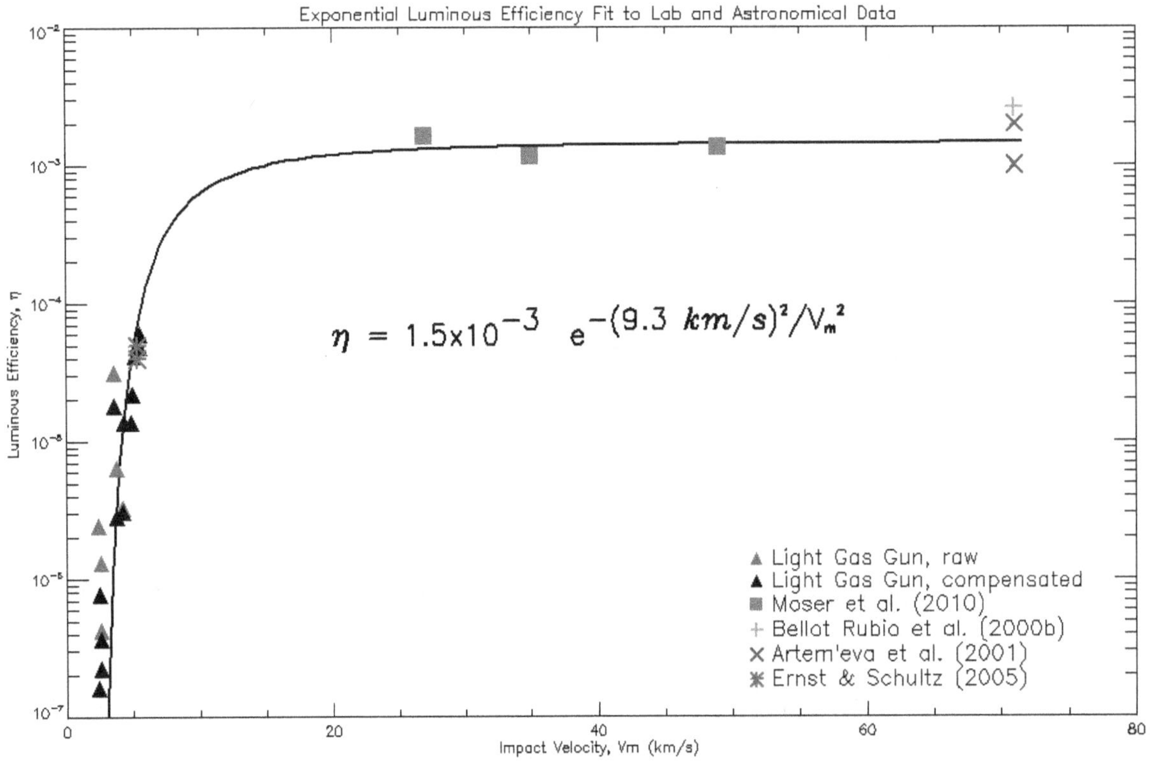

Figure 10. Lunar impact data from Table 1 is shown with light gas gun data from Figure 5 and historical data. The constants in Equation 13 are fit to the combined Table 1 and Figure 5 data. Characteristic velocity V_c is estimated to be 9.3 km/s and the scaling factor C is estimated to be 1.5×10^{-3}.

7 Conclusions

The luminous efficiency of hypervelocity impacts has been examined both in the laboratory and from observations of lunar meteoroid impacts. The luminous efficiency is a unique function of velocity with an extremely large variation with velocity in the laboratory range of 2 to 6 km/s, but a necessarily small variation with velocity in the meteoric range of 15 to 71 km/s. An exponential model of impact thermal emission efficiency is developed using fundamental principles of statistical physics which fits the combined laboratory and astronomical luminous efficiency data. This exponential model differs significantly from the polynomial models used to describe crater formation and dynamics. The model is valid for the early time portion of the process and focuses on the deposition of internal energy into the regolith which is subsequently observed as a bright blackbody flash. The model is compared with luminous efficiency data from laboratory impacts and from lunar impact observations. From these comparisons a critical velocity of 9.3 km/s and scaling factor of 1.5×10^{-3} are estimated. Further work to clarify the effects of mass and density of both the impactor and target upon the model is required. This model improves confidence in meteoroid mass estimates for lunar impacts and thus knowledge of the local space environment.

The unique energy partitioning approach embodied by luminous efficiency and this model can perhaps be extended to impact melting, another early time energy concern. Note that, since the melting point can be precisely known, Equation 10 can be evaluated directly. Although some melting is evident in light gas gun impacts into regolith it is not measurable while the light flash is. This is a possible starting point for future investigations.

Acknowledgements

Special thanks are due to the NASA Ames Vertical Gun Range crew for their efforts and to Dr. Peter Schultz of Brown University for help in coordinating those efforts. Work for this paper was supported in part by NASA contracts NNM05AB50C (Swift), NNM10AA03C (Moser) and the NASA Meteoroid Environments Office.

References

T. J. Ahrens, C. F. Petersen, and J. T. Rosenberg, Shock compression of feldspars, J. *Geophys. Res., 74*, 2727-2746 (1969)

T. J. Ahrens, and J. D. O'Keefe, Shock melting and vaporization of lunar rocks and minerals, *The Moon, 4*, 214-249 (1972)

T. J. Ahrens, J. D. O'Keefe, and R. V. Gibbons, Shock compression of a recrystallized anorthositic rock from Apollo 15, 4th Lunar Sci. Conf, Suppl. 4, *Geochim. Cosmochim. Acta, 3*, 2575-2590 (1973)

T. J. Ahrens, and D. M. Cole, Shock compression and adiabatic release of lunar fines from Apollo 17, Proc. 5th Lunar Science Conf., Suppl. 5, *Geochim. et Cosmochim. Acta, 3*, 2333-2345 (1974).

T. J. Ahrens, G. A. Lyzenga, and A. C. Mitchell, Temperatures induced by shock waves in minerals, In *High pressure research in geophysics,* edited by S. Akimoto and M.H. Manghnani, pp 579–594, Center for Academic Publications, Japan (1982).

W. W. Anderson, and T. J. Ahrens, Shock wave equations of state of chondritic meteorites, in *Shock Compression of Condensed Matter - 1997*, edited by S. C. Schmidt, et al., pp. 115-118, AIP Press, Woodbury, NY, 1998 (1997).

O. L. Anderson, and R. C. Liebermann, Sound velocities in rocks and Minerals, VESIAC State-of-the-Art Report 7885-4-X, *Inst. Sci. Techn.,* Univ. Michigan, Ann Arbor(1966).

N.A. Artemieva, V.V. Shuvalov, I.A. Trubetskaya, Lunar Leonid Meteors- Numerical Simulations, *Lunar and Planetary Science Conference XXXI*, paper 1402 (2000).

N.A. Artemieva, I.B. Kosarev, I.V. Nemtchinov, I.A. Trubetskaya, and V.V. Shuvalov, *SoSyR* **35**(3), 177-180 (2001a).

L. R. Bellot Rubio, J. L. Ortiz, and P. V. Sada, Observation And Interpretation Of Meteoroid Impact Flashes On The Moon, *Earth, Moon, and Planets, v. 82/83*, p. 575-598 (2000a).

L. R. Bellot Rubio, J. L. Ortiz, and P. V. Sada, Luminous Efficiency in Hypervelocity Impacts from the 1999 Lunar Leonids, *The Astrophysical Journal, 542*:L65-L68 (2000b).

M. B. Boslough, and T. J. Ahrens, Particle Velocity Experiments in Anorthosite and Gabbro, in *Shock Waves in Condensed Matter-1983*, edited by J. R. Asay *et.al.,* Elsevier Science Publishers (1984).

W. J. Cooke, R.M. Suggs, and W.R. Swift, A Probable Taurid Impact on the Moon, *Lunar and Planetary Science Conference XXXVII*, paper 1731 (2006).

W. J. Cooke, R.M. Suggs, W.R. Swift, and N.P. Hollon, Rate And Distribution Of Kilogram Lunar Impactors, *Lunar and Planetary Science Conference XXXVIII*, Paper 1986 (2007).

B. M. Cudnik, D. W. Dunham, D. M. Palmer, A. C. Cook, R. J. Venable, and P. S. Gural Ground-Based Observations of High Velocity Impacts on the Moon's Surface – The Lunar Leonid Phenomena of 1999 and 2001, *Lunar and Planetary Science Conference XXXIII*, Paper 1329 (2002).

B. M. Cudnick, D. W. Dunham, D. M. Palmer, A. C. Cook, R. J. Venable, and P. S. Gural The Observation and Characterization of Lunar Meteroid Impact Phenomena, *Earth, Moon and Planets 93*: 97–106 (2003).

D. K. Dijken, and J. Th. M. De Hosson Thermodynamic Model of the Compaction of Powder Materials by Shock Waves, *J. Appl. Phys. 75 (1)*, p.203 (1994a).

D. K. Dijken, and J. Th. M. De Hosson , Shock Wave Equation of State of Powder Material, *J. Appl. Phys. 75 (2)*, p 809 (1994b).

D. K. Dijken, and J. Th. M. De Hosson Shock Wave Velocity and Shock Pressure for low density powders: A Novel Approach, *Appl. Phys. Lett. 64 (7)*, p. 933 (1994c).

D.W. Dunham, B. Cudnick, D.M. Palmer, P.V. Sada, H.J. Melosh, M. Beech, R. Frankenberger, L. Pellerin, R. Venable, D. Asher, R. Sterner, B. Gotwols, B. Wun, and D. Stockbauer, *LPSC XXXI*, Paper 1547 (2000).

C. M. Ernst, and P. H. Schultz, Early Time Temperature Evolution of the Impact Flash and Beyond, *Lunar and Planetary Science Conference XXXV*, Paper 1986 (2004).

C. M. Ernst, and P. H. Schultz, Investigations of the Luminous energy and Luminous Efficiency of Experimental Impacts into Particulate Targets, *Lunar and Planetary Science Conference XXXVI*, paper 1475 (2005).

C. M. Ernst, and P. H. Schultz, Temporal and Spatial Resolution of the Early-Time Impact Flash: Implications for Light Source Distribution, *Lunar and Planetary Science Conference XXXVIII*, paper 2353 (2007).

C. M. Ernst, O. S. Barnouin, and P. H. Schultz, *Lunar and Planetary Science Conference XLI*, paper 1381 (2010).

D. E. Gault, E. D. Heitowit, and H. J. Moore Some Observations of Hypervelocity Impacts with Porous Media, *Proceedings of the Lunar Surface Materials Conference*, Academic Press, Boston, Massachusetts, 1964 (1963).

G. Heiken, D. Vaniman and B. French, *Lunar Sourcebook - A Users Guide to the Moon.* Cambridge University Press, Cambridge, UK (1991).

K. A. Holsapple and R. M. Schmidt, Point Source Solutions and Coupling Parameters in Cratering Mechanics, *J. Geophys. Res., 92, No. B7*, p. 6350 (1987).

K. A. Holsapple, The Scaling of Impact Processes in Planetary Sciences, *Ann Rev. Earth Planet. Sci, 21,* p. 333-373 (1993).

R. C. Liebermann and A. E. Ringwood, Elastic properties of anorthite and the nature of the lunar crust, *Earth Planet. Sci. Letters, 31*, 69-74 (1976).

G. A. Lyzenga, Shock Temperatures of Materials: Experiments and Applications to the High-Pressure Equation of State, Ph.D. dissertation, California Institute of Technology, Pasadena, California (1980).

D. S. McKay, J.L. Carter, W.W. Boles, C.C. Allen and J.H. Allton, JSC-1: A new Lunar Regolith Simu1lant, *Lunar and Planetary Science Conference XXIV*, 963 (1997).

R. C. McQueen, S. P. Marsh and J. N. Fritz, Hugoinoit equations of state of twelve rocks, *J. Geophys. Res., 72*, 4999-5036 (1967).

H. J. Melosh, *Impact Cratering, a Geological Process,* Oxford University Press, New York (1989)

H. J. Melosh, N. A. Artemieva, A. P. Golub, I. V. Nemchinov, V.V. Shuvalov, and I. A. Trubetskaya, Remote Visual Detection of Impacts on the Lunar Surface, *Lunar and Planetary Science Conference XXIV*, 975 (1993).

D. E. Moser, R.M. Suggs, W. R. Swift, R. J. Suggs, W. J. Cooke, A. M. Diekmann, H. M. Koehler, *Meteoroids 2010*, Breckenridge, CO (2010).

I.V. Nemtchinov, V. V. Shuvalov, N. A. Artemieva, B. A. Ivanov, I. B. Kosarev, and I. A. Trubetskaya, Light Impulse Created by Meteoroids Impacting the Moon, *Lunar and Planetary Science Conference XXIX*, 1032 (1998).

J.L. Ortiz, P.V. Sada, L.R. Bellot Rubio, F.J. Aceituno, J. Aceituno, P.J. Gutiérrez, and U. Thiele, *Nature* **405**, 921-923 (2000).

J.L Ortiz, J.A. Quesada, J. Aceituno, F.J. Aceituno, and L.R. Bellot Rubio, *ApJ* **576**, 567-573 (2002).

J. L. Ortiz, F. J. Aceituno, J. A. Quesada, J. Aceituno, M. Fernández, P. Santos-Sanz, J. M. Trigo-Rodríguez, J. Llorca, F. J. Martín-Torres, P. Montañés-Rodríguez, and E. Pallé, Detection of Sporadic Impact Flashes on the Moon: Implications for the Luminous Efficiency of Hypervelocity Impacts and Derived Terrestrial Impact Rates. *Icarus 184*, 319-326. (2006).

R. T. Schmitt, A. Deutsch, and D. Stöffler, Calculation of Hugoniot Curves and Post-Shock Temperatures for H- and L-Chondrites, *Lunar and Planetary Science Conference XXV*, p.1209 (1994).

R. M. Suggs, W. Cooke, R. J. Suggs, H. McNamara, W. Swift, D. Moser, and A. Diekmann, Flux of Kilogram-sized Meteoroids from Lunar Impact Monitoring., *Bulletin of the American Astronomical Society 40*, 455 (2008a).

R. M. Suggs, W. J. Cooke, R. J. Suggs, W. R. Swift, and N. Hollon, The NASA Lunar Impact Monitoring Program, *Earth, Moon, and Planets 102*, 293 (2008b).

R.M. Suggs, W.J. Cooke, H.M. Koehler, D.E. Moser, R.J. Suggs, and W.R. Swift, *Meteoroids 2010*, Breckenridge, CO (2010).

B. Svendsen, T. J. Ahrens, and J. D. Bass, Optical radiation from shock-compressed materials and interfaces, *High Pressure Research in Mineral Physics*, edited by M. Manghnani and Y. Syono, pp. 403-426, Terra Scientific Publishing Co., Tokyo (1987).

D. C. Swift, A. Seifter, D. B. Holtkamp, V. W. Yuan, D. Bowman, and D. A. Clark, Explanation of anomalous shock temperatures in shock-loaded Mo samples measured using neutron resonance spectroscopy,*Phys. Rev. B, vol. 77,* 092102 (2008).

W.R. Swift, R.M. Suggs, and W.J. Cooke, *Earth Moon Planets* **102**, 299-303 (2008).

Swift, W., Suggs, R., and Cooke, B. 2008. Algorithms for Lunar Flash Video Search, Measurement, and Archiving. *Earth, Moon, and Planets* **102**, 299.

J. M. Walsh and R. H. Christian, Equation of state of metals from shock wave measurements, *Phys. Rev., 97*, 1544-1556 (1955).

M. Yanagisawa and N. Kisaichi, Lightcurves of 1999 Leonid Impact Flashes on the Moon, *Icarus* **159**, 31–38 (2002).

M. Yanagisawa, K. Ohnishi, Y. Takamura, H. Masuda, Y. Sakai, M. Ida, M. Adachi, and M. Ishida, *Icarus* **182**(2), 489-495 (2006).

M. Yanagisawa, H. Ikegami, M. Ishida, H. Karasaki, J. Takahashi, K. Kinoshita, and K. Ohnishi, *M&PSA* **43**, Paper 5169 (2008).

X. Zeng,, C. He, H. A. Oravec, A. Wilkinson, J. H. Agui, and V. M. Asnani,. Geotechnical Properties of JSC-1A Lunar Soil Simulant. Journal of Aerospace Engineering 23, no. 2: 111 (2010).

Luminous Efficiency of Hypervelocity Meteoroid Impacts on the Moon Derived from the 2006 Geminids, 2007 Lyrids, and 2008 Taurids

D. E. Moser • R. M. Suggs • W. R. Swift • R. J. Suggs • W. J. Cooke • A. M. Diekmann • H. M. Koehler

Abstract Since early 2006, NASA's Marshall Space Flight Center has been routinely monitoring the Moon for impact flashes produced by meteoroids striking the lunar surface. During this time, several meteor showers have produced multiple impact flashes on the Moon. The 2006 Geminids, 2007 Lyrids, and 2008 Taurids were observed with average rates of 5.5, 1.2, and 1.5 meteors/hr, respectively, for a total of 12 Geminid, 12 Lyrid, and 12 Taurid lunar impacts. These showers produced a sufficient, albeit small sample of impact flashes with which to perform a luminous efficiency analysis similar to that outlined in Bellot Rubio et al. (2000a, b) for the 1999 Leonids. An analysis of the Geminid, Lyrid, and Taurid lunar impacts is carried out herein in order to determine the luminous efficiency in the 400-800 nm wavelength range for each shower. Using the luminous efficiency, the kinetic energies and masses of these lunar impactors can be calculated from the observed flash intensity.

Keywords hypervelocity impact · impact flash · luminous efficiency · lunar impact · meteoroid

1 Introduction

When a meteoroid strikes the Moon, a large portion of the impact energy goes into heat and crater production. A small fraction goes into generating visible light, which results in a brilliant flash at the point of impact that can be seen from Earth. The luminous efficiency, η, relates how much of the meteoroid's kinetic energy, *KE*, is converted into luminous energy, *LE*, in wavelength range, λ.

$$LE_\lambda = \eta_\lambda KE \quad (1)$$

The luminous efficiency plays a vital role in understanding observations and constraining models of the near-Earth meteoroid environment. Experiments into lunar regolith simulant at *low velocities* (2 to 6 km/s) have been performed at hypervelocity gun test ranges in order to determine η (Swift et al., 2010), but *high velocities* – meteoroid speeds, 18 to 71 km/s – are impossible to replicate in the laboratory using particle sizes typical of meteoroids. Scaling these low velocity luminous efficiency results to the

D. E. Moser (✉)
MITS/Dynetics, NASA/Marshall Space Flight Center, Huntsville, AL, 35812 USA. E-mail: danielle.e.moser@nasa.gov

R. M. Suggs • R. J. Suggs • W. J. Cooke • H. M. Koehler
Meteoroid Environment Office & Space Environments Team, NASA/Marshall Space Flight Center, Huntsville, AL, 35812 USA

W. R. Swift
ESTS/Raytheon, NASA/Marshall Space Flight Center, Huntsville, AL, 35812 USA

A. M. Diekmann
ESTS/Jacobs Technology, NASA/Marshall Space Flight Center, Huntsville, AL, 35812 USA

high velocity regime results in luminous efficiencies greater than 1 – a result that is completely unphysical. Numerical hydrocode simulations, like that of Nemtchinov et al. (1999), have mainly focused on particles of asteroidal composition moving at low speeds. There are limited simulations of high speed cometary particles impacting the Moon (e.g. Artem'eva et al., 2001).

Impact flashes have been successfully observed on the Moon by Earth-based telescopes during several showers (e.g. Dunham et al., 2000; Ortiz et al., 2000; Cudnick et al., 2002; Ortiz et al., 2002; Yanagisawa & Kisaichi, 2002; Cooke et al., 2006; Yanagisawa et al., 2006, Cooke et al., 2007; Suggs et al., 2008a,b; Yanagisawa et al., 2008). Observations of lunar impact flashes associated with meteor showers offer an opportunity to measure η at high velocities, since some properties of the impactors, like direction and speed, are known. This was first accomplished by Ortiz et al. (2000) and later detailed in Bellot Rubio et al. (2000a, b) for the 1999 Leonid lunar impact flashes.

The NASA Marshall Space Flight Center (MSFC) has routinely monitored the un-illuminated portion of the Moon for lunar impact flashes in the 400-800 nm range. As the Earth has witnessed several meteor showers in the past few years, so has the Moon. Since the beginning of our monitoring program in 2006, we have captured video of probable Leonid, Geminid, Lyrid, Quadrantid, Orionid, Bootid, Southern Delta-Aquariid, and Taurid meteoroid impacts on the Moon. Multiple lunar impact flashes were detected during the 2006 Geminids, 2007 Lyrids, and 2008 Taurids, allowing for a luminous efficiency analysis like that performed by Bellot Rubio et al. (2000b) for the 1999 Leonids.

This paper is organized as follows: in Section 2, an overview of the lunar impact monitoring program is given, with specifics regarding the data collected during the showers of interest. In Section 3, the luminous efficiency analysis is described, with the results for each shower presented and discussed in Section 4.

2 Observations

2.1 Lunar Impact Monitoring Program Overview

MSFC conducts observations from the Automated Lunar and Meteor Observatory (ALaMO) located in Huntsville, Alabama, USA (34.°66 N, 86.°66 W) and the Walker County Observatory (WCO) near Chickamauga, Georgia, USA (34.°85 N, 85.°31 W). The un-illuminated (earthshine) portion of the Moon is simultaneously observed with two identical Meade RCX-400 0.35 m diameter Cassegrain telescopes, online in June 2006 and September 2007, and one RCOS 0.5 m diameter Ritchey-Chrétien telescope, online in January 2008; two telescopes reside at the ALaMO with the remainder at WCO. The ALaMO telescopes are outfitted with focal reducers resulting in nearly identical 20 arcmin fields of view covering approximately 4×10^6 km^2 or about 10% of the lunar surface. ASTROVID StellaCamEX and Watec 902-H2 Ultimate monochrome CCD cameras (400-800 nm bandwidth) are employed to monitor the Moon. The interleaved, 30 fps video is digitized and recorded straight to hard-drive.

Impact flash detection and analysis is performed by two custom programs: LunarScan (Gural, 2007) and LunaCon (Swift et al., 2008). LunarScan software is used to detect impact flashes in the video. LunaCon determines flash magnitudes, time on target, photometric quality (including sky condition), and lunar area within the field of view. Candidate flash detections are those multi-pixel flashes simultaneously detected in two or more telescopes at the same selenographic location or those that are more than 1 frame (1/30 s), or two video fields (1/60 s each) in duration. Candidate flashes do not exhibit any motion from video field to field but do demonstrate a suitable light curve: a sudden brightness increase followed by a gradual decrease. These criteria rule out cosmic rays, electronic noise, and most sun glints from orbiting satellites. The WCO telescope, located about 125 km from the ALaMO, functions only to eliminate any additional satellite sun glints via parallax. Short flashes observed before the second observatory came online, or flashes not detected by this third telescope due to weather, viewing geometry, or equipment problems, are checked against the unclassified satellite catalog.

Observations of the un-illuminated portion of the Moon are typically conducted when sunlight illuminates between 10 and 50% of the Earth-facing surface. This yields a maximum of 10 observing nights per month. At illuminations greater than 50%, the scattered light overwhelms the video and faint flashes go undetected. Observing at illuminations less than 10% is considered an inefficient use of time and resources since the time between twilight and moon set or moon rise is very limited at these phases. Additional descriptions of the lunar impact monitoring program and analysis techniques are given in Suggs et al. (2008a,b) and Suggs et al. (2010).

2.2 Shower Data

The illumination criterion and weather conditions resulted in several nights of observations at/near the peak of the 2006 Geminids, 2007 Lyrids, and 2008 Taurids. Table 1 lists the observation dates coinciding with the showers, the telescopes employed, and the number of hours of data recorded that were of a consistent photometric quality. The 2006 Geminids, 2007 Lyrids, and 2008 Taurids were observed a total of 2.18 hrs, 10.22 hrs, and 7.93 hrs, respectively. Candidate flashes are associated with a shower if they occur within days of the shower peak and are located in an area on the Moon that is visible to the radiant. Visibility plots for each shower are shown in Figures 1.

Table 1. List of observing times during the 2006 Geminids, 2007 Lyrids, and 2008 Taurids. All times UT.

Date	Shower	Telescopes	Obs Timespan	Obs Time (hr)
14 Dec 2006	Geminids	two 0.35 m	08:30 – 09:29	0.98
15 Dec 2006	Geminids	two 0.35 m	09:12 – 10:24	1.20
20 Apr 2007	Lyrids	two 0.35 m	01:18 – 02:24	1.10
21 Apr 2007	Lyrids	two 0.35 m	01:16 – 03:18	2.03
22 Apr 2007	Lyrids	two 0.35 m	01:12 – 04:29	3.28
23 Apr 2007	Lyrids	two 0.35 m	01:11 – 05:00	3.81
02 Nov 2008	Taurids	0.5 m, two 0.35 m	00:04 – 00:47, 23:46 – 24:00	0.95
03 Nov 2008	Taurids	0.5 m, two 0.35 m	00:00 – 00:13, 00:30 – 01:33, 23:42 – 24:00	1.57
04 Nov 2008	Taurids	0.5 m, two 0.35 m	00:00 – 02:09, 23:42 – 24:00	2.45
05 Nov 2008	Taurids	0.5 m, two 0.35 m	00:00 – 02:58	2.96

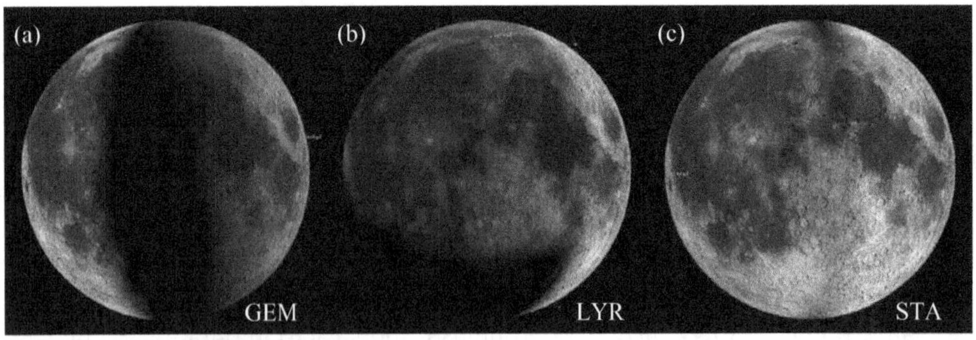

Figure 1. Shower visibility for the (a) 2006 Geminids, (b) 2007 Lyrids, and (c) 2008 Taurids. The colored portion indicates the area of the un-illuminated Moon visible to the radiant.

In all, 12 Geminid, 12 Lyrid, and 12 Taurid impacts were detected during periods of consistent photometric quality. (The data for an additional 8 Geminids, 3 Lyrids, and 2 Taurids detected during the monitoring period was of poor quality and is not considered here.) The details for each flash are given in Table 2.

Table 2. Details of the lunar impact flashes detected during the 2006 Geminds, 2007 Lyrids, and 2008 Taurids.

Shower [*obs time*]	ID	Date	Time (UT) ± 0.02s	Duration (ms)	R Mag	Lum. Energy, LE_{cam} (J)
Geminids	G01	14 Dec 2006	08:32:06.647	33	+9.2	5.6×10^4
[*2.18 hrs*]	G02	14 Dec 2006	08:32:51.993	50	+8.9	7.1×10^4
	G03	14 Dec 2006	08:39:57.155	17	+9.8	3.1×10^4
	G04	14 Dec 2006	08:46:01.957	17	+9.6	3.7×10^4
	G05	14 Dec 2006	08:50:36.200	33	+8.4	1.2×10^5
	G06	14 Dec 2006	08:51:20.562	17	+9.1	6.2×10^4
	G07	14 Dec 2006	08:56:42.837	17	+8.7	8.5×10^4
	G08	14 Dec 2006	09:00:22.142	33	+8.4	1.2×10^5
	G09	14 Dec 2006	09:03:32.851	33	+9.8	3.1×10^4
	G10	15 Dec 2006	09:15:14.040	33	+8.4	1.1×10^5
	G11	15 Dec 2006	09:17:39.336	17	+7.6	2.3×10^5
	G12	15 Dec 2006	09:53:28.464	83	+6.4	7.0×10^5
Lyrids	L01	20 Apr 2007	01:40:04.044	50	+7.8	2.1×10^5
[*10.22 hrs*]	L02	22 Apr 2007	01:15:05.616	67	+8.8	7.9×10^4
	L03	22 Apr 2007	01:15:43.956	33	+10.0	2.6×10^4
	L04[a]	22 Apr 2007	01:38:33.864	33	+8.0	1.6×10^5
	L05[b]	22 Apr 2007	03:12:24.372	67	+6.8	4.9×10^5
	L06	22 Apr 2007	03:52:37.182	17	+9.1	6.0×10^4
	L07	23 Apr 2007	01:15:54.547	17	+8.7	8.5×10^4
	L08	23 Apr 2007	02:23:21.361	50	+8.8	7.7×10^4
	L09	23 Apr 2007	04:08:48.755	50	+8.0	1.7×10^5
	L10	23 Apr 2007	04:40:45.912	33	+9.2	5.6×10^4
	L11	23 Apr 2007	04:42:34.781	83	+6.4	7.1×10^5
	L12	23 Apr 2007	04:59:57.557	50	+7.3	3.3×10^5
Taurids	T01	02 Nov 2008	23:48:39.996	50	+9.4	4.5×10^4
[*7.93 hrs*]	T02	03 Nov 2008	00:11:06.144	50	+7.9	1.9×10^5
	T03	03 Nov 2008	00:33:37.620	50	+9.1	6.0×10^4
	T04	03 Nov 2008	23:59:24.504	50	+8.7	9.0×10^4
	T05	04 Nov 2008	00:04:06.060	50	+8.9	7.2×10^4
	T06	04 Nov 2008	01:10:01.272	67	+8.1	1.5×10^5
	T07	04 Nov 2008	01:39:03.744	67	+6.3	7.8×10^5
	T08	05 Nov 2008	00:38:37.860	117	+7.4	2.9×10^5
	T09	05 Nov 2008	00:53:58.308	67	+8.5	1.1×10^5
	T10	05 Nov 2008	02:05:07.908	100	+7.3	3.0×10^5
	T11	05 Nov 2008	02:09:44.748	50	+9.3	4.9×10^4
	T12	05 Nov 2008	02:32:47.184	67	+8.1	1.5×10^5

[a] Also detected by independent observer Dave Clark in Houston, Texas, USA using a 0.2 m Schmidt Cassegrain telescope.
[b] Also detected by independent observer George Varros in Mt Air, Maryland, USA using a 0.2 m Newtonian telescope.

All of the events had durations between 17 and 117 ms and magnitudes between +10.0 and +6.3. Impact flash locations are shown in Figure 2. Figures 3 and 4 give a sample of impact flashes detected during each shower, shown as video stills of the impact flash on the Moon, and as a sequence of 1/30 s image squares, respectively.

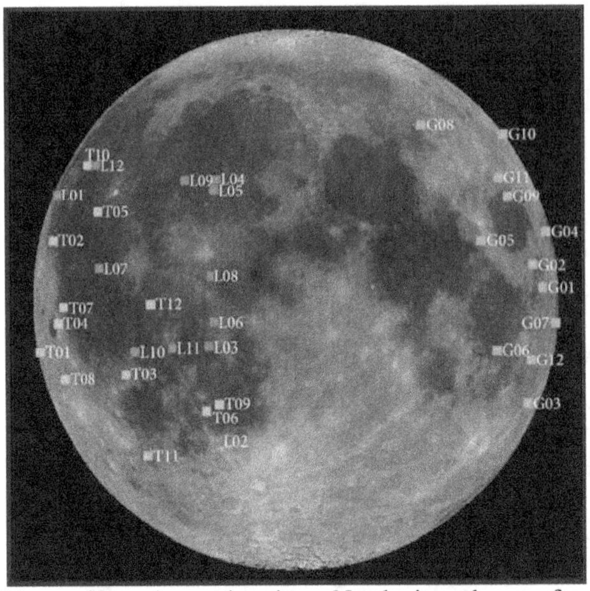

Figure 2. Observed lunar impact locations. Numbering scheme refers to Table 2.

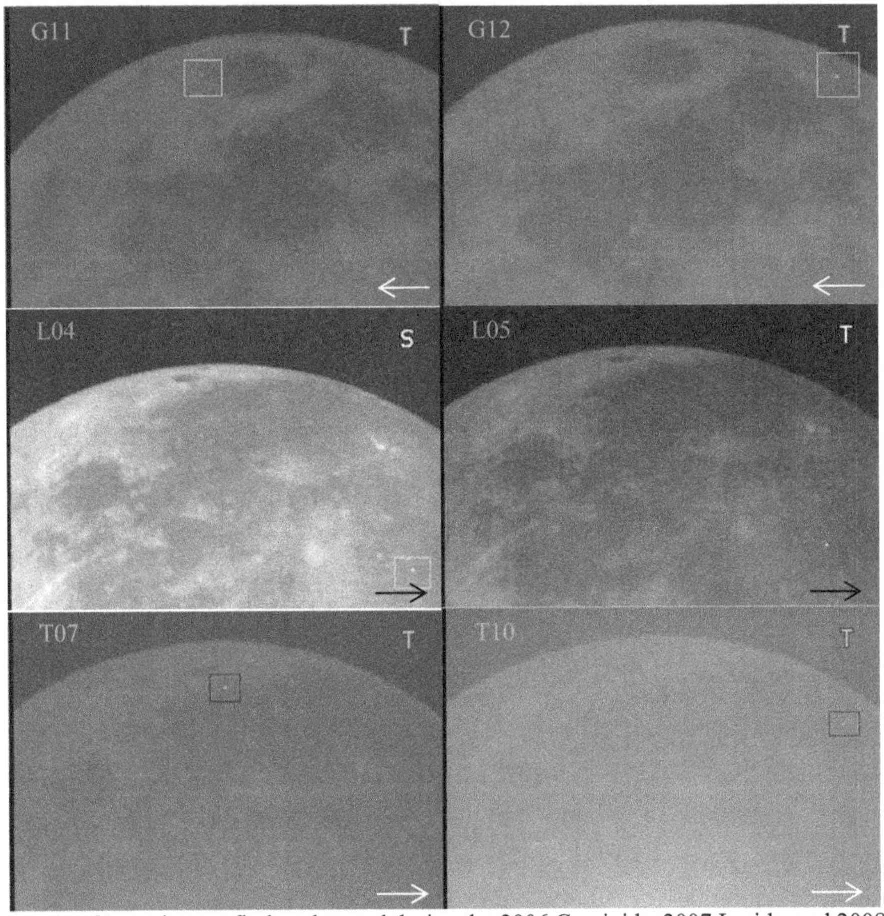

Figure 3. A sample of lunar impact flashes detected during the 2006 Geminids, 2007 Lyrids, and 2008 Taurids. Arrows indicate the direction of selenographic north. The numbering scheme refers to Table 2.

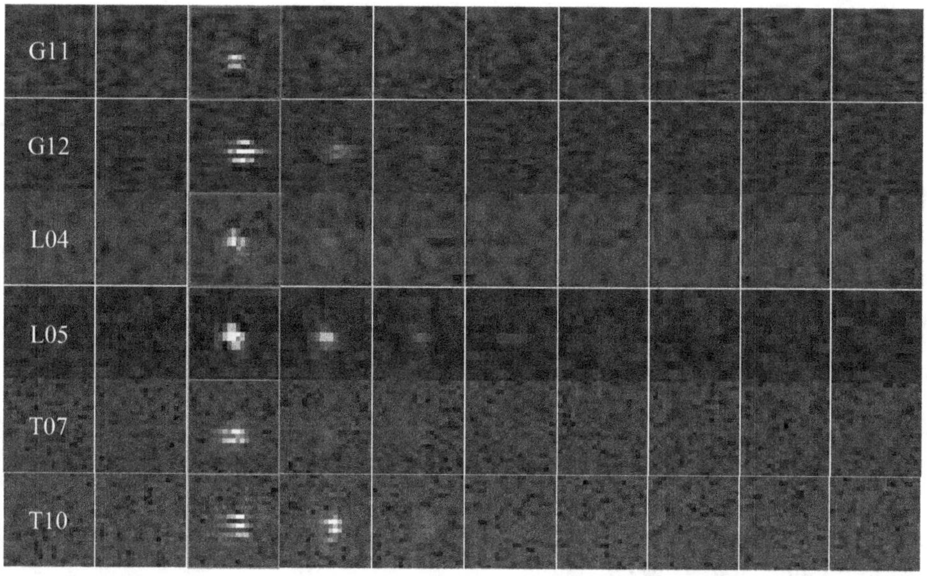

Figure 4. A sample of lunar impact flashes detected during the 2006 Geminids, 2007 Lyrids, and 2008 Taurids. The dimensions of each square in the series are about 35 x 35 arcseconds and each covers 1/30 s. The numbering scheme refers to Table 2.

The amount of sporadic contamination in this sample of meteoroids can be crudely calculated. Using the Grün sporadic flux model (Grün, 1985), and taking lunar shielding into account, it is estimated that roughly 3 of the 36 impact flashes may be caused by sporadic meteoroids as opposed to shower meteoroids. But there is no way to remove this contamination.

3 Luminous Efficiency Analysis

3.1 Theory

The technique for determining luminous efficiency incorporates the method first referenced by Ortiz et al. (2000) and then detailed by Bellot Rubio et al. (2000a, b). Their method is restated in this section and referenced in the text hereafter as BR2000. In addition to this method, an iterative process is used to determine the final luminous efficiency η, and is better suited to discussion alongside a description of the flux parameter inputs in Section 3.2.5.

The number of meteoroids that impact the Moon in time span t_1 to t_2 is

$$N = \int_{t_1}^{t_2} F(t) A_\perp(t) dt \qquad (2)$$

where $F(t)$ is the flux as a function of time, t, and $A_\perp(t)$ is the observed lunar area that is perpendicular to the meteor shower radiant also as a function of time.

The cumulative flux distribution of meteoroids of mass m is given by

$$F(m) = F(m_0)\left(\frac{m}{m_0}\right)^{1-s} \qquad (3)$$

where $F(m)$ is the flux of particles having mass greater than m, $F(m_0)$ is the flux of particles of known mass greater than mass m_0, and s is the mass index.

The masses of the meteoroids impacting the Moon are unknown. For an impactor of mass m and velocity V, the kinetic energy is $KE = ½\ m\ V^2$. Substituting this into Eq (3) gives a cumulative flux distribution as a function of kinetic energy.

$$F(KE) = F(m_0)\left(\frac{2KE}{V^2 m_0}\right)^{1-s} \qquad (4)$$

Solving Eq (1) for KE and substituting this into Eq (4) gives a cumulative flux distribution as a function of luminous energy, depending on the luminous efficiency η_λ in a particular wavelength range.

$$F(LE_\lambda) = F(m_0)\left(\frac{2LE_\lambda}{\eta_\lambda V^2 m_0}\right)^{1-s} \qquad (5)$$

Using Eq (5), Eq (2) becomes the number of lunar meteoroid impacts producing luminous energies greater than LE_λ in the time span t_1 to t_2.

$$N(LE_\lambda) = \left(\frac{2LE_\lambda}{\eta_\lambda V^2 m_0}\right)^{1-s} \int_{t_1}^{t_2} F(m_0,t) A_\perp(t)dt \qquad (6)$$

This result is comparable to Eq (4) of BR2000.

In short, the analysis technique involves 'backing out' the luminous efficiency by matching the number of impacts expected on the Moon to that actually observed. One of the difficult problems in using this technique alone derives from uncertainties in the various inputs, namely the flux and mass index. This is discussed in the next section.

3.2 Inputs

The inputs for Eq (6) in the 400-800 nm range are summarized in Table 3 and outlined in the following sub-sections.

Table 3. Input parameters for Eq (6) for the 2006 Geminids, 2007 Lyrids, and 2008 Taurids. The *average* area perpendicular to the radiant in the field of view is given, for illustration purposes.

Shower	V (m/s)	s	$F(m_0,t)$ (#/m²/hr)	m_0 (kg)	t_1, t_2 (hr)	$A_{\perp ave}$ (km²)	LE_{cam} (J)
Geminids	35000	1.9	*Suggs*	4.7×10^{-2}	*from*	3.2×10^6	*from*
Lyrids	49000	1.7	*et al.*	8.4×10^{-2}	*Table*	1.1×10^6	*Table*
Taurids	27000	1.8	*(2010)*	2.4×10^{-2}	*1*	3.6×10^6	*2*

3.2.1 Luminous Energy, LE_λ

The energy received at Earth [J/m²] is calculated using

$$\varepsilon_\oplus = \tau \cdot Flux_{0\lambda} 10^{-0.4 m_\lambda} \qquad (7)$$

where τ is the camera exposure time [s], $Flux_{0\lambda}$ is the flux [J/m²/s] from a zero magnitude star in the camera's wavelength range λ, and m_λ is the measured magnitude of the impact flash. Stellacam and

Watec cameras operate in the 400-800 nm range with a peak response approximated by the R passband. Flash photometry is performed utilizing local background stars in the video as reference and Vega is used as the calibration star with $Flux_{0R} = 3.39 \times 10^{-9}$ J/m^2/s. The exposure time of the camera is 0.0167 s.

The luminous energy at the Moon [J] is related to the energy received at Earth by

$$LE_{\lambda} = f\ \pi\ d^2 \varepsilon_{\oplus} \tag{8}$$

where f is a factor describing the distribution of the light ($f = 4$ for spherical emission into 4π steradians, $f = 2$ for hemispherical emission into 2π steradians, etc.) and d is the distance in meters between the impact flash on the Moon and the telescope on Earth. It is chosen that $f = 4$, since the radiating plume is most likely above the surface, created from hot meteoroid and regolith materials, and d is assumed a constant 3.84×10^8 m. The resulting luminous energies for each flash, including a correction factor to produce energies in the camera's passband, are seen in Table 2 as LE_{cam}. For more photometry details, see Swift et al. (2008).

This differs from the inputs in the BR2000 method in the choice of f (the compromise $f = 3$) and wavelength range (400-900 nm). In addition, the cameras used in their study peak in the visual range, whereas the cameras we use peak in the red-NIR.

3.2.2 Time Span, t_1 to t_2

Observing sessions typically run from moonrise to twilight (waning phases) or twilight to moonset (waxing phases). Only those times that are of a consistent photometric quality are used in the analysis. For each video, plots of lunar disk brightness and contrast versus time are examined. Any video segments that exhibit obvious cloud attenuation, a loss of contrast due to cirrus haze or fogged optics, a rapid change in extinction during moonrise or moonset, twilight, or obvious obstructions from the observatory dome or trees, are excluded. The time spans $t_1 - t_2$ used in this analysis are listed in Table 1.

3.2.3 Perpendicular Lunar Area, $A_{\perp}(t)$

During each observing session the Moon drifts slightly within the telescope's field of view, thereby changing the amount of lunar surface area detected. The LunaCon analysis software identifies and calculates the lunar area visible in the video. This is accomplished by first detecting the location of the limb within a video frame and solving for the center and radius of the lunar disc in image pixels. From the radius, the lunar area of the center pixel is calculated in square kilometers, and, knowing the radial distance of each pixel in the lunar image, a weight is applied for each pixel to compensate for spherical Moon effects (an image pixel near the limb contains more area than one near the center of the disc); pixels at the lunar limb with *extreme* weights are discarded. Summing over all the lunar pixels in the image with their appropriate weights yields the total lunar area. In this way, the lunar area within the field of view as a function of time is determined (Swift et al., 2008).

To determine the lunar area perpendicular to the shower direction within the field of view, $A_{\perp}(t)$, the area as a function of time determined by LunaCon is modeled as 1 million equal area cells. The area in each cell is multiplied by the cosine of the zenith angle of the radiant. Summing yields the total perpendicular lunar area within the field of view as a function of time. For illustration purposes, the average perpendicular area, $A_{\perp ave}$, for each shower is given in Table 3.

In comparison, the BR2000 method calculates $A_{\perp}$ using Monte Carlo simulations and it is considered a constant during the 90 min of Leonid observations they performed in 1999.

3.2.4 *Shower Parameters, V and s*

The speeds, V, and mass indices, s, for each shower are taken from the annual meteor shower tables compiled by the International Meteor Organization (IMO, 2006, 2007, 2008). Gravitational effects from the Earth and Moon are not considered in the velocity parameter as they are too small to be considered significant. The mass index characterizing the mass distribution of (small) shower meteors in the visual range may not be applicable to large particles. As there are no measurements of the mass index for these shower meteoroids in the lunar impactor size range, it is only possible to estimate s from shower observations; this makes s a rather uncertain parameter input. The speeds and mass indices used are listed in Table 3.

In comparison, the BR2000 method explored the luminous efficiency results from two different mass indices. The first was an extrapolated mass index from the 1999 terrestrial Leonid fireballs of the IMO Visual Meteor Database and the second was a constant $s = 2.0$.

Looking at the effects of varying s has not yet been done for the showers discussed here and is classified as future work.

3.2.5 *Flux Parameters, $F(m_0,t)$ and m_0*

To determine the flux parameters, the lunar impacts were first considered as an ensemble. The MSFC detected 115 lunar impact flashes in 212 hours of observing between 2006 and 2009, the majority of which are most probably produced by shower meteoroids (Suggs et al., 2010). We calculate an initial limiting magnitude and subsequently an initial limiting kinetic energy based on the ensemble lunar impact data, incorporating previously determined luminous efficiency values based on gun test work (Suggs et al., 2008b) and the 1999 Leonid work by Bellot Rubio et al. (2000). This, in turn, is used to calculate the number of impacts we should have detected, based on observed and historical IMO ZHR data, and given the lunar collecting area in the field of view, observing time, and the shower geometry. Matching the observed and expected number values requires adjustment of the luminous efficiency or limiting magnitude. As there is more uncertainty in the limiting magnitude, this value was adjusted to best fit IMO observations, resulting in a final limiting kinetic energy corresponding to a mass of 100 g moving at a speed of 25 km/s. The final limiting mass, m_0, for each shower yielding the equivalent final limiting kinetic energy is given in Table 3. The flux corresponding to this limiting mass is $F(m_0)$ and the data and time dependence is taken from the observed lunar impact flux, removing any impact flashes that have a magnitude fainter than the corresponding final limiting magnitude . For a more in depth discussion on the flux determination, see Suggs et al. (2010).

The procedure described above is just the first step in an iterative process. Using the 'final' limiting mass determined from the *ensemble* of lunar impacts, which incorporates an initial luminous efficiency estimate, a new luminous efficiency is calculated based on the energies of the *individual* lunar impact flashes using the technique outlined in Section 3.1. The new luminous efficiency is then used to compute a more accurate limiting energy, as in the above paragraph, and the process repeats until convergence.

The determination of the flux at the Moon in the original BR2000 method is quite different. The method scales the terrestrial flux for the 1966 Leonids by a factor of 4 and adopts the timing of the terrestrial 1999 Leonids, shifted to the Moon. Their fluxes are tied to the mass of a Leonid meteoroid producing a meteor of magnitude +6.5 on Earth.

Looking at the ensemble of lunar impacts and comparing it to observations on Earth, we have instead determined a lunar flux for each shower, $F(m_0, t)$, of particles with mass greater than the limiting mass, m_0. Fluxes are discussed in Suggs et al. (2010) and the limiting mass for each shower is listed in Table 3.

4 Results and Discussion

As stated previously, the analysis technique involves 'backing out' the luminous efficiency by matching the number of impacts expected on the Moon to that actually observed. The expected cumulative number of lunar meteoroid impacts, $N(LE_\lambda)$, producing luminous energies greater than LE_λ (as discussed in Section 3) for the 2006 Geminids, 2007 Lyrids, and 2008 Taurids is plotted in Figures 5 alongside the observed cumulative lunar impacts using two different energy binning schemes. Fig 5(a) shows

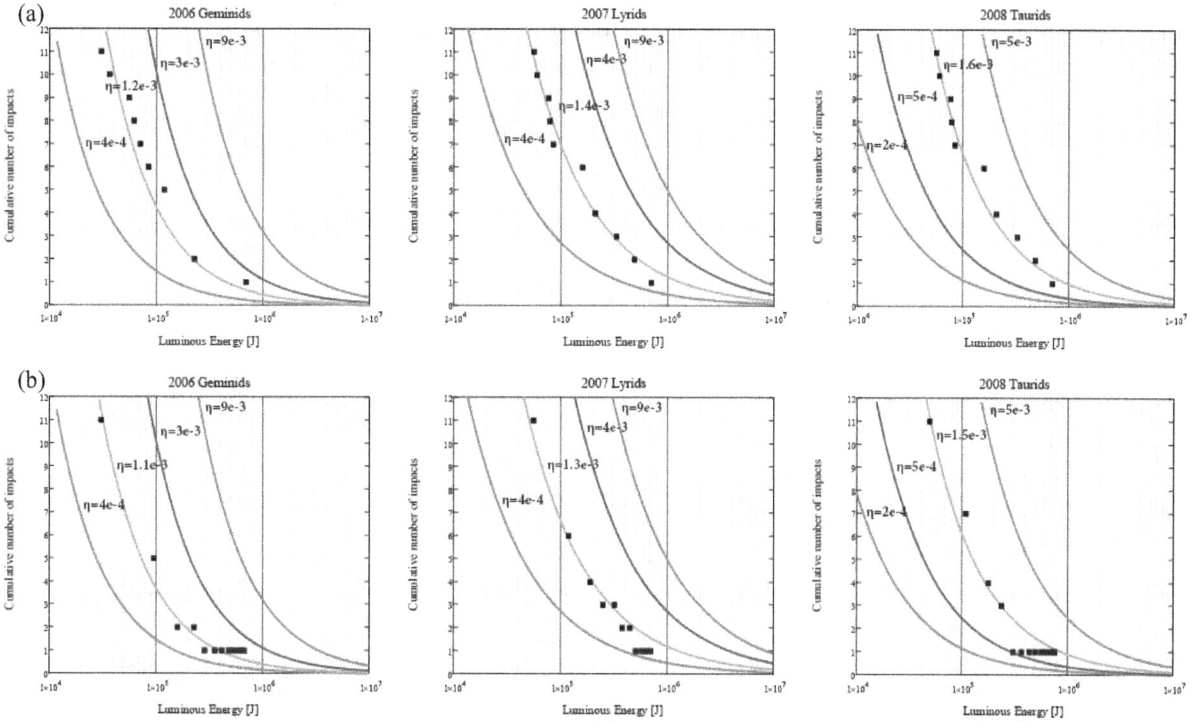

Figure 5. $N(LE_\lambda)$ vs LE_λ, the comparison between expected number of cumulative impacts (colored solid lines) and observed data (black squares) for two different energy binning schemes for the 2006 Geminids, and 2007 Lyrids, and 2008 Taurids. (a) No binning of luminous energy observed during the impact flash, (b) observed luminous energies are binned with bin size = 65,000 J. Wavelength λ is 400-800 nm.

the comparison between expected cumulative number of impacts at various values of luminous efficiency and observed number using almost no binning, since number statistics are poor. Fig 5(b) shows this same comparison with luminous energy bins set at 65,000 J. Binning using the two different schemes yields similar results for luminous efficiency in the 400-800 nm range, η_{cam}, as listed in Table 4 and illustrated in Figures 5.

Table 4 Calculated luminous efficiencies η_{cam} for the 2006 Geminids, 2007 Lyrids, and 2008 Taurids using two different binning schemes (a) and (b); listed in order of increasing velocity. An estimated impactor mass range corresponding to the flashes we detected for each shower is also calculated.

Shower	# Flashes	Obs. Time (hrs)	V (km/s)	s	η_{cam} (a)	η_{cam} (b)	Mass Range (kg)
2008 Taurids	12	7.93	27	1.8	1.6×10^{-3}	1.5×10^{-3}	0.09-1.4
2006 Geminids	12	2.18	35	1.9	1.2×10^{-3}	1.1×10^{-3}	0.04-0.99
2007 Lyrids	12	10.22	49	1.7	1.4×10^{-3}	1.3×10^{-3}	0.03-0.44
*1999 Leonids**	*5*	*1.5*	*72*	*1.83*	2×10^{-3}	*n/a*	*0.12-4.9*

* Bellot Rubio et al. (2000a, b), shown for reference. Results are from a different camera with a different λ range.

Errors in η_{cam} may be on the order of a few percent. The Bellot Rubio et al. (2000a, b) result for the 1999 Leonids is reproduced in Figure 6. A better agreement between the observed number of impacts and the expected number of impacts was found in this work than in Bellot Rubio et al. (2000a, b), as seen by a comparison of Figures 5 and 6, indicating a perhaps more reliable value of luminous efficiency.

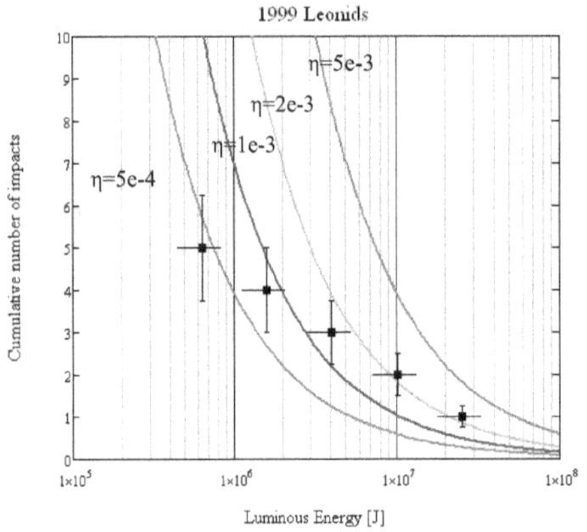

Figure 6. Results of Bellot Rubio et al. (2000b). $N(LE_\lambda)$ versus LE_λ adapted from Figure 2 of the same reference. Wavelength λ in this case is 400-900 nm. Compared to Figure 5, the observed data points do not fit the curves as well.

The luminous efficiency derived by Bellot Rubio et al. (2000a, b) for the 1999 Leonids is also listed in Table 4 for comparison purposes. It should be noted that this data was observed with cameras having a slightly different spectral response and sensitivity than the cameras in this study and a light distribution coefficient of $f = 3$ instead of 4; other differences in technique are outlined in Section 3.2. Despite these differences, the 1999 Leonid luminous efficiency is consistent with those of the 2006 Geminids, 2007 Lyrids, and 2008.

Luminous efficiency determinations at low speeds into lunar simulant JSC-1a have been made at the NASA Ames Vertical Gun Range employing the same cameras used to monitor the Moon (Swift et al., 2010). These values appear in Figure 7, along with the luminous efficiencies calculated in this paper. Also plotted for reference are previous results found in the literature. A fit to the lunar impact derived data from this paper and the hypervelocity gun test data from Swift et al. (2010) yields the following equation for luminous efficiency in the 400-800 nm wavelength range of our cameras

$$\eta_{cam} = 1.5 \times 10^{-3} e^{-(9.3)^2 / V^2} \tag{9}$$

where V is the speed of the impactor in km/s. The lunar impact data mainly controls the constant scaling factor in Eq (9) while the hypervelocity gun test data largely controls the number in the exponential. Findings from other data sources shown in Figure 7 – Bellot Rubio et al. (2000b) for the 1999 Leonids at 400-900 nm, Ernst & Schultz (2005) considering gun tests into powdered pumice at 340-1000 nm, and numerical hydrocode simulations by Artem'eva et al. (2001) for two different densities, 0.1 g/cm^3 and 1 g/cm^3 – are not considered in the fit, but the results seem to be quite consistent.

The range of estimated impactor masses is computed and given in Table 4, using the luminous efficiencies in binning scheme (a). Simulations by Artem'eva et al. (2001) indicate that luminous efficiency weakly depends (10-20%) on size of the impactor, while luminous efficiencies are twice as high as for low-density impactors. The dependence of luminous efficiency on impactor mass and/or density is left for future work.

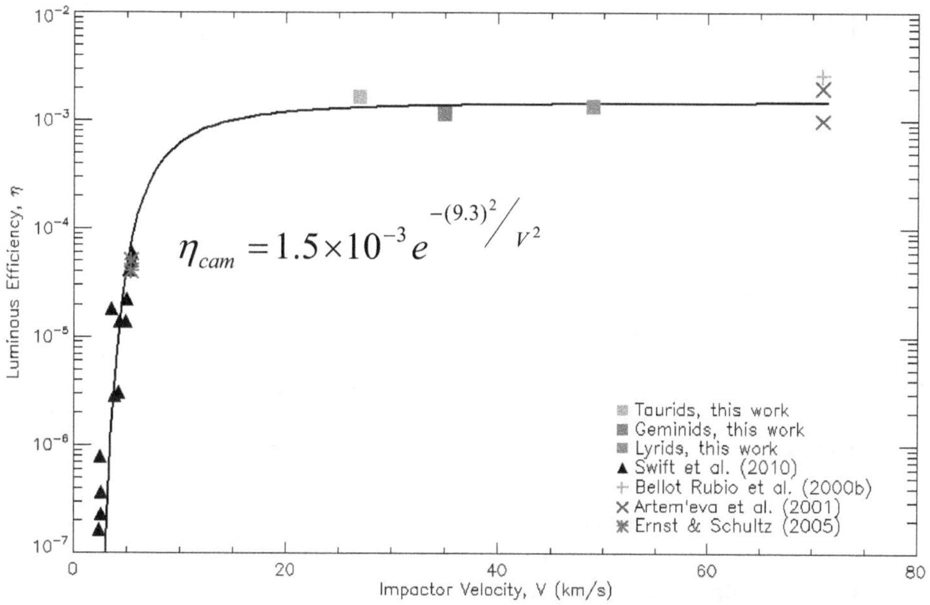

Figure 7. Plot of luminous efficiency versus velocity using several different methods. The data from gun tests into lunar regolith simulant from Swift et al. (2010) populates the low velocity end of the graph. At high meteoroid velocities (this work), the curve is relatively constant. The Bellot Rubio et al. (2000b) point from the 1999 Leonids and Ernst & Schultz (2005) point from gun tests into powdered pumice do not represent the spectral response of the cameras used in this study and were not used in the fit (solid black line); they are shown for comparison purposes only. The results of hydrocode simulations by Artem'eva et al. (2001) are also shown only for comparison purposes. The Bellot Rubio number has a correction applied to convert from the originally assumed $f = 3$ to $f = 4$.

5 Summary

Utilizing the technique of Bellot Rubio et al. (2000a,b), the best estimate for the luminous efficiency of lunar impacts involving the 2006 Geminid, 2007 Lyrid, and 2008 Taurid meteoroids is $\eta_{cam} = 1.2 \times 10^{-3}$, 1.4×10^{-3}, and 1.6×10^{-3}, respectively, in the 400-800 nm wavelength range of our cameras. These values are consistent with that found by Bellot Rubio et al. (2000) for the Leonid lunar impacts of 1999 and numerical simulations performed by Artem'eva et al. (2001). Number statistics are poor in all cases, however, and more observations are needed. It must be noted that η is highly dependent on the mass index though how much the determination of η varies with s is left to future work. Mass indices found in the literature and used in this analysis may not apply to the size range considered for lunar impacts. More work to determine mass indices for meteoroids larger than 100 g is needed.

Luminous efficiencies determined from lunar impact flash analyses are fairly constant at meteoroid speeds. Luminous efficiencies calculated as the result of hypervelocity gun tests into lunar simulant has revealed a large variation in η at low velocities. Luminous efficiency values imply impactor masses of roughly 30 to 1400 g. The dependence of luminous efficiency on impactor mass/density is also a topic of future work.

Acknowledgements

This work was partially supported by the NASA Meteoroid Environment Office, the Constellation Program Office, the MSFC Engineering Directorate, and NASA contracts NNM10AA03C and NNM05AB50C.

References

N.A. Artem'eva, I.B. Kosarev, I.V. Nemtchinov, I.A. Trubetskaya, and V.V. Shuvalov, *SoSyR* **35**(3), 177-180 (2001).
L.R. Bellot Rubio, J.L Ortiz, and P.V. Sada, *ApJ* **542**, L65-68 (2000a).
L.R. Bellot Rubio, J.L Ortiz, and P.V. Sada, *Earth Moon Planets* **82-83**, 575-598 (2000b).
W.J. Cooke, R.M. Suggs, W.R. Swift, *LPSC XXXVII*, Paper 1731 (2006).
W.J. Cooke, R.M. Suggs, R.J. Suggs, W.R. Swift, and N.P. Hollon, *LPSC XXXVIII*, Paper 1986 (2007).
B.M. Cudnick, D.W. Dunham, D.M. Palmer, A.C. Cook, R.J. Venable, and P.S. Gural, *LPSC XXXIII*, Paper 1329 (2002).
D.W. Dunham, B. Cudnick, D.M. Palmer, P.V. Sada, H.J. Melosh, M. Beech, R. Frankenberger, L. Pellerin, R. Venable, D. Asher, R. Sterner, B. Gotwols, B. Wun, and D. Stockbauer, *LPSC XXXI*, Paper 1547 (2000).
C.M. Ernst and P.H. Schultz, *LPSC XXXVI*, Paper 1475 (2005).
E. Grün, H.A. Zook, H. Fechtig, and R.H. Giese, *Icarus* **62**, 244-272 (1985).
P.S. Gural, *Meteoroid Environment Workshop*, NASA MSFC, Huntsville, AL (2007).
IMO, Meteor Shower Calendars, http://www.imo.net/calendar/2006 (2006).
IMO, Meteor Shower Calendars, http://www.imo.net/calendar/2007 (2007).
IMO, Meteor Shower Calendars, http://www.imo.net/calendar/2008 (2008).
I.V. Nemtchinov, V.V. Shuvalov, N.A. Artemieva, I.B. Kosarev, and I.A. Trubetskaya, *Int J Impact Eng* **23**, 651-662 (1999).
J.L. Ortiz, P.V. Sada, L.R. Bellot Rubio, F.J. Aceituno, J. Aceituno, P.J. Gutiérrez, and U. Thiele, *Nature* **405**, 921-923 (2000).
J.L Ortiz, J.A. Quesada, J. Aceituno, F.J. Aceituno, and L.R. Bellot Rubio, *ApJ* **576**, 567-573 (2002).
R.M. Suggs, W.J. Cooke, R.J. Suggs, W.R. Swift, and N. Hollon, *Earth Moon Planets* **102**, 293-298 (2008a).
R.M. Suggs, W.J. Cooke, R.J. Suggs, W.R. Swift, D.E. Moser, A.M. Diekmann, and H.A. McNamara, *BAAS* **40**, 455 (2008b).
R.M. Suggs, W.J. Cooke, H.M. Koehler, D.E. Moser, R.J. Suggs, and W.R. Swift, *Meteoroids 2010*, Breckenridge, CO (2010).
W.R. Swift, D.E. Moser, R.M. Suggs, and W.J. Cooke, *Meteoroids 2010*, Breckenridge, CO (2010).
W.R. Swift, R.M. Suggs, and W.J. Cooke, *Earth Moon Planets* **102**, 299-303 (2008).
M. Yanagisawa and N. Kisaichi, *Icarus* **159**(1), 31-38 (2002).
M. Yanagisawa, K. Ohnishi, Y. Takamura, H. Masuda, Y. Sakai, M. Ida, M. Adachi, and M. Ishida, *Icarus* **182**(2), 489-495 (2006).
M. Yanagisawa, H. Ikegami, M. Ishida, H. Karasaki, J. Takahashi, K. Kinoshita, and K. Ohnishi, *M&PSA* **43**, Paper 5169 (2008).

CHAPTER 5:

METEOR LIGHT CURVES AND LUMINOSITY RELATIONS

Constraining the Physical Properties of Meteor Stream Particles by Light Curve Shapes Using the Virtual Meteor Observatory

D. Koschny • M. Gritsevich • G. Barentsen

Abstract Different authors have produced models for the physical properties of meteoroids based on the shape of a meteor's light curve, typically from short observing campaigns. We here analyze the height profiles and light curves of ~200 double-station meteors from the Leonids and Perseids using data from the Virtual Meteor Observatory, to demonstrate that with this web-based meteor database it is possible to analyze very large datasets from different authors in a consistent way. We compute the average heights for begin point, maximum luminosity, and end heights for Perseids and Leonids. We also compute the skew of the light curve, usually called the *F*-parameter. The results compare well with other author's data. We display the average light curve in a novel way to assess the light curve shape in addition to using the *F*-parameter. While the Perseids show a peaked light curve, the average Leonid light curve has a more flat peak. This indicates that the particle distribution of Leonid meteors can be described by a Gaussian distribution; the Perseids can be described with a power law. The skew for Leonids is smaller than for Perseids, indicating that the Leonids are more fragile than the Perseids.

Keywords meteor light curves · physical properties · meteoroids · double-station observations

1 Introduction

The shape of the light curve of meteors can be used as an indicator for the physical properties of the underlying meteoroid particle. In general, a single solid grain would be expected to increase in brightness and stop emitting light at the end of its flight path at the point of maximum brightness. Fragile particles will start to disintegrate high up in the atmosphere, and the luminosity will be the sum of the light emitted around the individual particles. In the extreme case, a meteoroid will fragment very quickly and reach its highest magnitude early on in its light curve. As the individual particles ablate and slow down, the magnitude of the complete meteor will decrease slowly over its path.

Several authors have analyzed larger numbers of observational data, typically from observing campaigns of meteor streams (*e.g.* Fleming *et al.* 1993, Murray *et al.* 2000, Koten *et al.* 2004). Data from very few meteors was analyzed in very high detail *e.g.* by Jiang and Hu (2001) or Campbell-Brown and Koschny (2004). All of these analyses derive meteoroid physical properties from the shape of the

D. Koschny (✉)
European Space Agency, ESA/ESTEC, SRE-SM, Keplerlaan 1, Postbus 299, 2200 AG Noordwijk, The Netherlands. E-mail: Detlef.Koschny@esa.int

M. Gritsevich
Institute of Mechanics, Lomonosov Moscow State University, Michurinskii Ave. 1, 119192 Moscow, Russia; Faculty of Mechanics and Mathematics of the Lomonosov Moscow State University, Leniskie Gory, 119899 Moscow, Russia

G. Barentsen
Armagh Observatory, College Hill, Armagh BT61 9DG,United Kingdom

light curves. The underlying model is based on an idea by Öpik (1958) and was worked out in detail by Hawkes and Jones (1975) into the so-called dustball model. It assumes that meteoroids are composed of small grains held together by a low boiling point 'glue'. By heating up the meteoroid this glue is evaporated and the particles disintegrate. This model was detailed *e.g.* by Beech and Hargrove (2004). Koschny *et al.* (2002) have modeled the light curve of meteors based on assuming mechanical fragmentation; Campbell-Brown and Koschny (2004) use a detailed aerodynamical model adding a thermal fragmentation mechanism. In this paper we analyze data obtained in several meteor campaigns, but also one data set (from the Perseids 2009) found only by data mining within the Virtual Meteor Observatory (VMO). In addition to the light curve evaluation, one goal of this work was to assess the useability of the VMO for this task.

2 Input Data and Observational Setup

We have been using data stored in the openly available Virtual Meteor Observatory. The Virtual Meteor Observatory (VMO) is a data storage facility for a wide range of meteor data; see Koschny *et al.* (2008) and Barentsen *et al.* (2008a, 2008b). In the currently available beta version single station video meteor data of the International Meteor Organization until ~2007 has been ingested. In addition, double-station data of selected campaigns in the time span from 1997 to 2009 is available. The database can be queried remotely using SQL syntax (SQL = Structured Query Language). All queries used for the paper here are available and can be reused in exactly the same way once more data is available.

In this work we use so-called orbit data sets of the VMO. Most of them are derived from dedicated double-station observing campaigns, using image-intensified camera systems. One of the datasets was extracted by using the VMO functionality of finding potential double-station meteors. This query will go through the existing single-station data for a given time range, and read out the location and pointing direction of all cameras in the database. From the derived geometry, it will identify camera systems which look into the same volume in the atmosphere. Given a maximum delta time, the VMO will identify all observations which could possibly be the same meteor. A list of potential double-station meteors is then presented to the user. The user can select which one (or all) of the meteors should be used for computing orbits. The orbit computation is also done within the VMO using the software MOTS (Meteor Orbit and Trajectory Software, Koschny and Diaz del Rio, 2002).

The contents of such a data set is a list of orbits computed from the observations from two different stations, giving time and initial shower association of a meteor together with all the orbital elements and their associated error bars. Additionally computed information is the peak magnitude of the meteor, a derived photometric mass, velocities, height for the begin, peak brightness and end points, the apparent and geocentric radiant of the meteor, the zenith angle and convergence angle, and a flag whether the meteor started and ended in- our outside the field of view.

The cameras used to produce the data sets were either image-intensified cameras as described in Koschny *et al.* (2002) with field of views between 20° and 60° or, in the case of the Perseids 2009, a non-intensified Mintron camera with a 6 mm f/0.8 wide-angle lens yielding a field of view of 60°. The typical stellar limiting magnitude of the intensified cameras was between 5 and 6 mag; for the non-intensified camera of the Perseids 2009 data the limiting stellar magnitude was around 3 mag.

The following datasets were used:

(a) Perseids 1997 (data set name ORB-KOSDE-PER1997), using two intensified video cameras with 30 deg circular field of view, a faintest star of 6.5 mag, and using a total of 74 meteors;

(b) Perseids 2007 (data set name ORB-KOSDE-PER2007), again using intensified video cameras with 30 deg circular field of view, a faintest stellar magnitude of 6.5 mag, using a total 28 meteors;
(c) Perseids 2009 (ORB-KOSDE-PER2009), using one intensified video camera, one un-intensified camera, with only 13 orbits available.
(d) Leonids 2001 (ORB-BARGE-LEO2001), using two intensified video cameras with 15 x 12 and 33 deg field of view, with 64 available orbits.

From these datasets, we selected only meteors containing 8 or more magnitude datapoints.

3 Results

3.1 Height Distribution

For each dataset, we determined the average beginning height, height of peak light intensity, and average end height. The errors were computed by simply taking the standard deviation of the individual data points. To interpret these errors, one should keep in mind that the input data is quantized. Assume one meteor has precisely 8 data points, and it starts at 110 km and ends at 95 km (for a typical Perseid, see Figure 1). Then the 'quantization noise' already is about 15/8 km ~ 2 km. So each individual meteor height can not be determined more accurate than this simply due to the fact that we look at discrete video images. Table 1 shows the results.

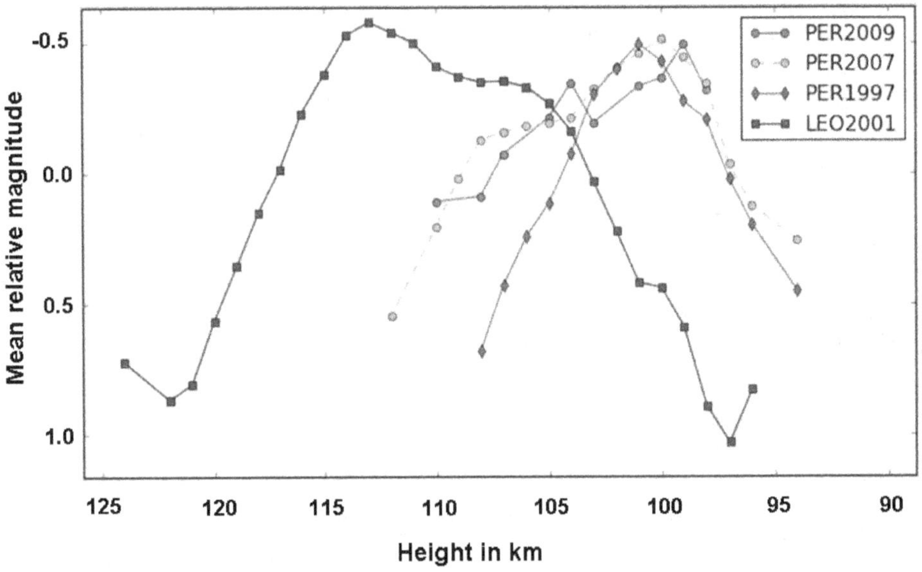

Figure 1. Average light curves from two different meteor showers. For a detailed explanation, see the text. No error bars are shown at the data points; the typical deviation between the mean value and the actual magnitude of a meteor in the given height bin is about 0.3 mag (indicated in the upper left area).

Table 1. Average heights for the beginning point, the point of peak brightness, and the end point for the different datasets. For comparison, the values from a very similar paper as this one (Koten 2004) are given in the same table. The last column lists the *F*-parameter which indicates the skew of the light curve.

Dataset	h_{begin} in km	h_{peak} in km	h_{end} in km	F
This work:				
PER1997	107.7 +/- 3.1	100.6 +/- 3.2	96.0 +/- 3.5	0.61 +/- 0.19
PER2007	113.8 +/- 3.7	102.4 +/- 3.7	97.5 +/- 2.9	0.68 +/- 0.21
PER2009	115.1 +/- 12.1	97.8 +/- 7.2	93.1 +/- 8.1	0.75 +/- 0.20
LEO2001	117.0 +/- 10.9	110.1 +/- 10.3	103.0 +/- 10.7	0.47 +/- 0.26
Koten 2004:				
PER1998-2001	113.9 +/- 2.4	104.4 +/- 2.9	96.0 +/- 4.1	0.535 +/- 0.010
LEO2000	120.0 +/- 3.5	106.9 +/- 3.8	96.5 +/- 3.7	0.498 +/- 0.014

3.2 Light Curves and *F*-parameter

We present a novel way to show the typical light curves of a meteor stream. For each individual meteor, we compute the average brightness in magnitudes. Each individual brightness measurement is converted to a relative brightness by subtracting the average value. We then take height bins of one kilometer and average all values in this bin. This results in a smooth curve which is an indication for the typical light curve behavior of a meteor stream, independent of the magnitude.

All Perseid years showed meteors starting between 108 and 113 km, ending between 94 and 97 km. The peak seems to shift slightly from ~102 km in 1997 to 97-99 km in 2007/2009. However, due to the small number of meteors analyzed one should be careful in giving this result too much significance. The important result is that all three years the light curve is clearly peaked, with the peak slightly behind the half length of the profile.

The Leonids begin much higher, the end height is close to the Perseid end height. The main difference to the Perseids is that the curve shows a flat-topped shape, i.e. between 115 km and 105 km the brightness as about constant. This can mean that either all meteors are really flat-topped, or that the peak height of the Leonids varies in this range in such a way that the average curve looks flat. Looking at several individual light curves the former seems to be the case. This is also consistent with Murray (2000) for their Leonids 1998 data but not for their 1999 data.

Note that by displaying the curve like this, any relation between e.g. end height and brightness will be hidden. The apparent increase in magnitude at the begin and end point are assumed to be artifacts, possibly by contamination due to sporadic meteors.

The *F*-parameter (Hawkes and Jones, 1975, Fleming 1993) is defined as

$$F = \frac{H_{begin} - H_{\max}}{H_{begin} - H_{end}}$$

where H_{begin} is the beginning height, H_{max} the height of peak brightness, and H_{end} the end height. The *F*-parameter was computed for each meteor individually. The result is shown in Figure 2 as a function of absolute magnitude (the peak magnitude normalized to a distance of 100 km) and shower. Obviously, the *F*-parameter is ill-defined for a flat-topped meteor light curve. Thus, interpreting of the Leonid results has to be done with care. Still, it can be seen that the values for the Leonids are in a different regime than those for the Perseids. For a given absolute magnitude, the Leonids show lower *F*-

parameters, i.e. the peak occurs earlier. In the average light curves shown in Figure 1 this is evident by the bump in the light curve for high altitudes; Table 1 shows the average F-parameter which is 0.47 for the Leonids but between 0.6 and 0.75 for the Perseids.

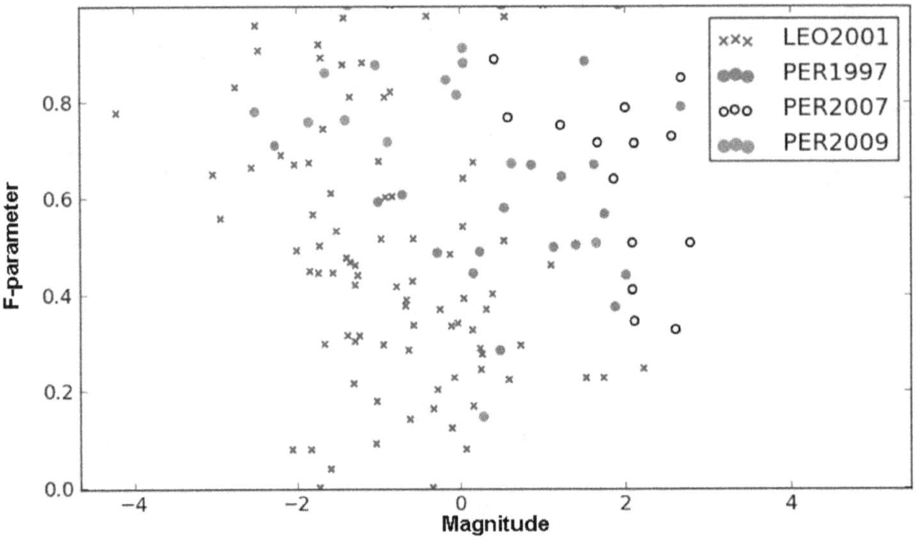

Figure 2. The F-parameter as a function of absolute magnitude for the different data sets. A value of 1 means that the meteor has its peak brightness at the end; 0.5 indicates that the light curve is symmetric. Average values are shown in Table 1, this Figure shows the trend between different meteor streams and the relation to the magnitude.

Both Leonids and Perseids show an 'empty region' in the lower left part of the diagram with the same tendency – there are no low F-values for brighter meteor, i.e. the brighter the meteor, the later it reaches its maximum.

3.3 Discussion

Beginning, peak, and end heights as reported here are in good agreement with other papers; for comparison we give the height data and the F-parameter as determined by Koten *et al.* (2004) in Table 1. The beginning height is an indication for the fragility of the meteoroids. More fragile meteoroids will disintegrate at larger heights, increasing in brightness above the detection level (Koten *et al.* 2004). We can thus confirm that the Leonids are more fragile than the Perseids. This is confirmed by comparing the F-parameter: The Leonids on average peak earlier than the Perseids (Note, however, that the flat-topped shape of the average light curve of the Leonids makes it more difficult to define the F-parameter).

Plotting the F-parameter as a function of magnitude shows that brighter meteors normally peak later. This has been shown for the Leonid 2001 dataset before by a different data interpretation method (Koschny *et al.* 2002) and can be explained by assuming that a single large grain which is not disintegrating is part of the meteoroid. The observed 'empty' region in the lower left area of Figure 2 can be interpreted such that brighter meteors always must contain one or several large grains, which do not disintegrate easily. This shifts the peak of the light cuve to the back.

Current meteoroid ablation models typically assume that meteoroids disintegrate when entering the Earth's atmosphere. The individual fragments ablate and generate light. The shape of the light curve depends on the size distribution of these fragments. Campbell-Brown and Koschny (2004) use *e.g.* Gaussian and power-law distributions to fit different light curves. Typically the Gaussian distribution will better fit the flat-topped light curve; power laws will better fit peaked light curves. Alternatively one can use a Poisson distribution derived from fracture mechanics to describe both shapes (Koschny *et al.* 2002) which is proposed here as it would allow to use only one number (the Poisson coefficient) to describe all light curve shapes.

Flat-topped light curves have been observed before for the Leonids, see Murray *et al.* (2000). They only see the flat-topped light curves for their 1998 data; the 1999 data is more consistent with a peaked light curve. They argue that the 1998 has been ejected from the comet several revolutions before the material encountered in 1999. The longer flight time could imply that the meteoroids had been more fragmented over time. Our observation of flat-topped light curves for the 2001 Leonids is consistent with that proposal - most of the particles recorded by us in Australia are expected to be from the 1699 perihelion (Asher 2000).

4 Conclusion

We have used the Virtual Meteor Observatory (VMO) to retrieve data of double-station observations of the Leonids 2001 and the Perseids 1997, 2007, and 2009. The 2009 data camera from cameras initially operated as single stations; the VMO functionality was used to identify the used camera systems as providing data of the same meteors.

We analyzed the height profiles and light curves of all meteors having more than 8 data points (typically this means having more than 8 video frames). Our height data is consistent with other author's results. The peak of the light curves of our Perseid meteors is somewhat later than in other publications, but showing the same trend.

From the shape of the average light curves we can confirm that the Leonids are more fragile than the Perseid meteors. The Leonids show a more flat-topped light curve. Murray et al. (2000) explain flat-topped versus peaked light curves by assuming a much higher meteor stream age. Our measurements confirm this proposal.

In addition to the scientific results, we conclude that the concept of the VMO is good and the VMO can be used for doing extensive data mining once it has been moved from its current beta-version state to the final version. Additional data on meteors which would go beyond the scope of this paper is easily available and can be retrieved with a simple SQL query, *e.g.* plotting the end height versus photometric mass can be done in one line. However, the current data quality for orbital data still has to be improved. It is recommended that the VMO implement clearly defined data quality criteria. An important additional routine which would be needed in the VMO to allow further studies in the direction shown here would be to add an automated stream association mechanism. Currently, the meteor streams are simply assigned the shower code given by MetRec to the single-station data, which turned out to be not always correct after manually checking the orbital elements.

The VMO will contain single-station data from the IMO video camera network and dedicated double-station data. Discussions are ongoing to include the SonotaCo network (SonotaCo et al. 2010) data. This will make the VMO the largest database for meteor data so far. While the data used here is not yet more numerous than previous studies, all these large datasets will be accessible using exactly the same scripts once the archive is fully operational.

Acknowledgements

The VMO is hosted by the European Space Agency's Research and Scientific Support Department. Thanks to Günter Thörner and his team for providing the necessary support. All data providers of the IMO video network are acknowledged for making their data available through the VMO.

References

D. Asher, Proc. International Meteor Conference, Frasso Sabino 1999, ed. R. Arlt, International Meteor Organization, 5-21 (2000)

G. Barentsen, D. Koschny, J. Mc Auliffe, S. Molau, R. Arlt, EPSC Abstracts, Vol. 3, EPSC2008-A-00293 (2008a)

G. Barentsen, R. Arlt, D. Koschny, P. Artreya, J. Flohrer, T. Jopek, A. Knöfel, P. Koten, J. Mc Auliffe, J. Oberst, J. Toth, J. Vaubaillon, R. Weryk, M. Wisniewski, P. Zoladek, WGN, the Journal of the IMO, 38:1, 10-24 (2008b)

M. Campbell-Brown, D. Koschny, Astronomy & Astrophysics 418, 751-758.

D. Fleming, R. Hawkes, J. Jones, Meteoroids and their parent bodies, Proc. Int. Ast. Symp., Smolenice, Slovakia, 6-12 July 1992, Bratislava: Astronomical Institute, Slovak Academy of Sciences, 1993, edited by J. Stohl and I.P. Williams, 261-264 (1993)

X. Jiang, J. Hu, Planet. Space Sci. 49, 1281-1283 (2001)

D. Koschny, J. Mc Auliffe, G. Barentsen, Earth, Moon, and Planets 102, 247-252 (2008)

D. Koschny, J. Diaz del Rio, WGN 30:4, 87-101 (2002)

D. Koschny, P. Reissaus, A. Knöfel, R. Trautner, J. Zender, Proc. Asteroids, Comets, Meteors (ACM 2002), 29 July – 092 August 2002, TU Berlin, Berlin, Germany, ESA-SP-500 (2002)

D. Koschny, R. Trautner, J. Zender, A. Knöfel, O. Witasse, Proc. Asteroids, Comets, Meteors (ACM 2002), 29 July – 092 August 2002, TU Berlin, Berlin, Germany, ESA-SP-500 (2002)

I. Murray, M. Beach, M. Taylor, P. Jenniskens, R. Hawkes, Earth, Moon and Planets 82-83, 351-367 (2000)

SonotaCo, S. Molau, D. Koschny, EPSC Abstracts, Vol. 5, EPSC2010-798 (2010)

An Investigation of How a Meteor Light Curve is Modified by Meteor Shape and Atmospheric Density Perturbations

E. Stokan • M. D. Campbell-Brown

Abstract This is a preliminary investigation of how perturbations to meteoroid shape or atmospheric density affect a meteor light curve. A simple equation of motion and ablation are simultaneously solved numerically to give emitted light intensity as a function of height. It is found that changing the meteoroid shape, by changing the relationship between the cross-section area and the mass, changes the curvature and symmetry of the light curve, while making a periodic oscillation in atmospheric density gives a small periodic oscillation in the light curve.

Keywords meteor · meteoroid ablation modeling

1 Introduction

The ablation of small objects, meteoroids, in the atmosphere produces light that may be observed on the ground. As the meteoroids enter the atmosphere, particles are removed from the rapidly heating body and excited or ionized. Atmospheric particles may also be excited and ionized in smaller numbers. These excited atoms and ions emit photons in narrow bands that may be analysed for meteoroid chemical composition using a spectrometer, or examined in their time-dependence to suggest velocity, structure, or other physical properties of the meteoroid. Thus, meteors, the streaks of light that occur as meteoroids burn up in the atmosphere, reveal information about the composition and properties of meteoroids when observations are combined with ablation models. Since the meteoroids originate from parent bodies throughout the Solar System, one is able to learn about the structure and history of the Solar System without sending exploration or sample return missions.

When examining the light curve, the graph of meteor magnitude versus time, properties such as the shape and symmetry of the curve can reveal whether the meteoroid is fragmenting, or what sort of cross sectional area it is presenting to the atmosphere, as examined in Beech (2009). In some cases, light curves with varying symmetry may be observed for particles belonging to a single shower, such as the Leonid particles modeled by Campbell-Brown and Koschny in 2004. Periodic oscillations in the light curve, such as those examined by Beech and Brown (2000), or Beech, Illingworth, and Murray (2003), may indicate meteoroid rotation that is as rapid in frequency as 10^2 Hz, but is not rapid enough to make the meteor appear like an evenly-heated sphere. These oscillations occur as local maxima in the light curve, flares, which are distinct from noise.

The purpose of this brief investigation is to qualitatively comment on how a meteor light curve is influenced by two phenomena: variation in meteoroid shape and ablation, and periodic oscillations in atmospheric density. Specifically, we examine whether either of these perturbations can result in flares

E. Stokan (✉) • M. D. Campbell-Brown
Department of Physics and Astronomy, The University of Western Ontario, London, Ontario, Canada, N6A 3K7. Phone: 1 (519) 661-2111 ext. 87985; Fax: 1 (519) 661-2033, E-mail: estokan@uwo.ca

in the meteor light curve. To model meteoroid shape variation, the equation relating an object's cross-section area to mass employed by Beech (2009) is utilized. Periodic atmospheric density oscillation is modelled by introducing an oscillation to the isothermal atmosphere profile. Meteoroid motion and ablation is modeled using the standard equations. Solutions for velocity, mass, and intensity as a function of height are obtained numerically.

2 Method

The fundamental equations of motion and ablation are as follows:

$$\frac{\Lambda(\rho_{atm}SV)V^2}{2} = \frac{\Lambda\rho_{atm}SV^3}{2} = -Q\frac{dm}{dt} \tag{1}$$

$$m\frac{dV}{dt} = -\Gamma S\rho_{atm}V^2 \tag{2}$$

where Λ is the dimensionless heat transfer coefficient, Γ is the dimensionless drag coefficient, ρ_{atm} is the density of the atmosphere, S is the cross-sectional area of the object, V is the object's velocity, and m is the mass of the object.

The cross-section area is made a function of the mass of the object with power α, following Beech, 2009:

$$S = \pi\left(\frac{3m_i}{4\pi\rho_i}\right)^{2/3}\left(\frac{m}{m_i}\right)^{\alpha} \tag{3}$$

Here, m_i is the initial mass of the object, and ρ_i is the initial density (assumed constant throughout the trajectory). α may take any value, with larger positive values of α indicating that the cross-sectional area is a more sensitive function of the mass of the object. $\alpha = 0$ gives an object with a constant cross-section, possibly representing a cylinder that ablates along the height axis, while $\alpha = 2/3$ gives a spherical object that ablates radially, or self-similarly. Negative α gives an object that experiences a larger cross-section area as the mass depletes, which may represent an object that fragments as it ablates.

The atmospheric density profile is represented by an isothermal atmosphere with a small relative oscillation:

$$\rho(h) = \rho_0[1 + A\cos(kh)]exp\left(-\frac{h}{h_0}\right) \tag{4}$$

Oscillation amplitudes between 2% and 10% are employed, as well as wavelengths between 1 and 10 km. Oscillation in atmospheric density may originate from two main sources: physical phenomena such gravity waves or transient oscillations in the atmosphere (small amplitude and large vertical wavelength), or other physical phenomena observed in radiosonde data, which is usually smoothed out (amplitude of about 10%, and possible wavelength of 1 km), as noted in Hedin (1991).

The equations of motion and ablation are recast as the following:

$$\frac{dV}{dh} = \pi\Gamma\left(\frac{3m_i}{4\pi\rho_i}\right)^{2/3}\frac{m^{\alpha-1}(h)}{{m_i}^{\alpha}}\frac{V(h)}{\cos Z}\rho(h) = \xi m^{\alpha-1}(h)V(h)\rho(h) \quad (5)$$

$$\frac{dm}{dh} = \frac{\pi\Lambda}{2Q}\left(\frac{3m_i}{4\pi\rho_i}\right)^{2/3}\left[\frac{m(h)}{m_i}\right]^{\alpha}\frac{V^2(h)}{\cos Z}\rho(h) = \eta m^{\alpha}(h)V^2(h)\rho(h) \quad (6)$$

These equations are solved numerically using simple Euler integration. This gives the velocity and mass of the object as a function of height. The light curve is then produced assuming that the luminous intensity I is proportional to the loss of kinetic energy:

$$I = \tau\frac{dE_k}{dt} = \tau\left(\frac{1}{2}\frac{dm}{dt}V^2 + mV\frac{dV}{dt}\right) = -\tau V^2\cos Z\left(\frac{V}{2}\frac{dm}{dh} + m\frac{dV}{dh}\right) \quad (7)$$

The natural logarithm of the intensity gives a scale that approximates the magnitude for the light curve. Appendix 1 gives meteoroid properties and parameters used in the numerical simulation.

3 Results and Discussion

Varying the shape of the object by varying α produced light curves with different properties. Table 1 summarizes the properties, while Figure 1 shows the light curves and mass loss graphically.

Table 1. Summary of qualitative observations of light curves for objects ablating with different α

	Concavity, symmetry of light curve	Maximum brightness	Height of maximum brightness	Ending height
$\alpha < 0$	Upward, highly asymmetric	Highest	Highest	Same as height of maximum brightness
α small ($0 < \alpha < 1$)	Downward, highly asymmetric	Moderate	Moderate	Above $h = 0$
α large ($\alpha > 1$)	Lowest, more symmetric	Lowest	Lowest	$h = 0$

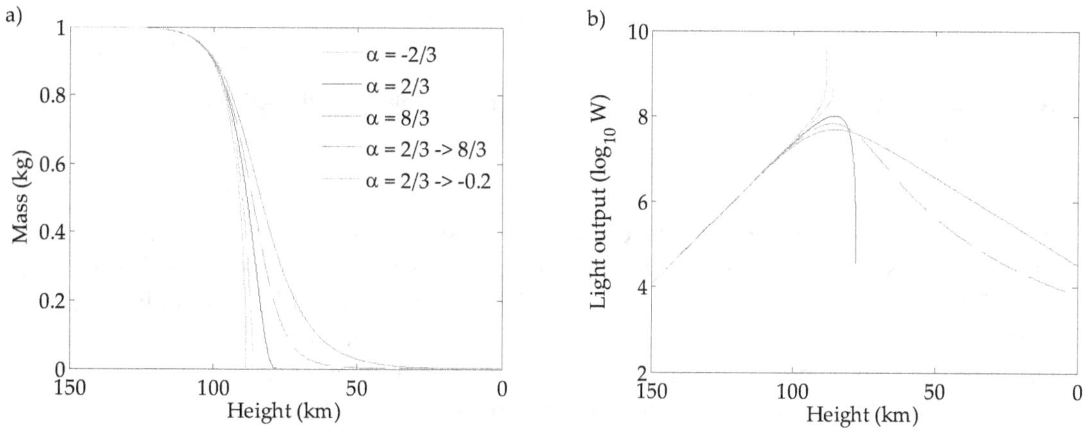

Figure 1. a) Mass and b) rough light curve for ablation with modified shape (α-parameter). The same legend applies to both figures.

For $\alpha < 0$, the light curve is concave upward, with the meteoroid burning up at the highest altitude compared to other choices for α. This represents an object that reveals more cross-section area as the mass decreases, perhaps a fragmenting, pancaking object:

$$S \propto \frac{1}{m^{|\alpha|}} \tag{8}$$

For $\alpha = 2/3$, the light curve is concave downwards and is asymmetrical with a slow rise to a peak brightness, and a rapid drop. This is the standard single-body light curve, that of a self-similar spherical object. As α increases, the object's maximum brightness decreases and is moved to lower heights, making the light curve more symmetric. In the limit of large $\alpha > 1$, the object survives to the ground. This may represent an object that becomes more aerodynamic or resistant to ablation as the mass decreases. In any case, *no* flares, or local maxima in the light curve, are created if α has a constant value through the trajectory of the object. Even varying α from one value to another during object ablation, representing a quickly rotating object that becomes oriented, gives a light curve that initially resembles the curve of first α value, then slowly merges towards that of the second α value, with no flares being observed.

The light curve associated with the oscillatory atmospheric density profile displays oscillations about the light curve with the smooth atmospheric density, as shown in Figure 2.

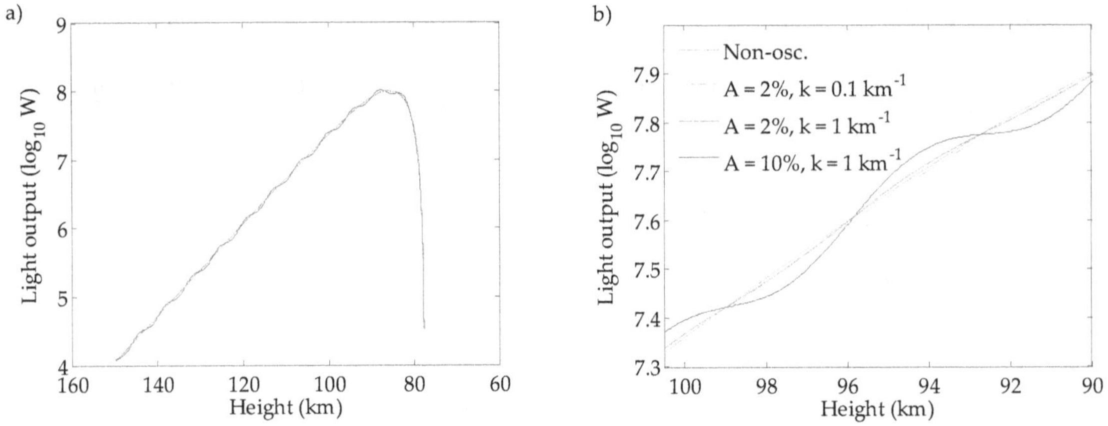

Figure 2. a) Rough light curve and b) enlarged rough light curve for ablation with oscillating atmosphere density, $\alpha = 2/3$. The same legend applies to both figures.

In this case, small flares are observed, but even the largest oscillations with the smallest wavelengths, corresponding to transient oscillations in the atmospheric density data, produce small oscillations in the light curve. Such small oscillations in a measured light curve would likely be indistinguishable from noise. This suggests that periodic flares in a light curve are not likely to be caused by oscillations in atmospheric density. Perhaps some other mechanism, such as meteoroid rotation or periodic charge separation is responsible for oscillatory flares observed in some light curves. This will be investigated in more detail in the future.

References

Beech, M., On the shape of meteor light curves, and a fully analytic solution to the equations of meteoroid ablation, Mon. Not. R. Astron. Soc. **397**, 2081-2086 (2009)

Beech, M., Brown, P., Fireball flickering: the case for indirect measurement of meteoroid rotation rates, Plan. Space Sci. **48**, 925-932 (2000)

Beech, M., Hargrove, M., Classical meteor light curve morphology, Earth, Moon, and Planets **95**, 389-394 (2004)

Beech, M, Illingworth, A., Murray, I. S., Analysis of a "flickering" Geminid fireball, Meteoritics & Plan. Sci. **38** (7), 1045-1051 (2003)

Campbell-Brown, M. D, and Koschny, D., Model of the ablation of faint meteors, Astronomy and Astrophysics **418**, 751-758 (2004)

Ceplecha, Z., Borovicka, J., Elford, W. G., Hawkes, R. L., Porubcan, V., Simek, M., Meteor phenomena and bodies, Space Science Reviews **84**, 327-471 (1998)

Hedin, A. E., Extension of the MSIS thermosphere model into the middle and lower atmosphere, J. Geophys. Research **96**, 1159-1172 (1991)

Holton, J. R., Beres, J. H., Zhou, X., On the vertical scale of gravity waves excited by localized thermal forcing, J. Atm. Sciences **59**, 2019-2023 (2002)

Vargas, F., Gobbi, D., Takahashi, H., Lima, L. M., Gravity wave amplitudes and momentum fluxes inferred from OH airglow intensities and meteor radar windows during SpreadFEx, Ann. Geophys. **27**, 2361-2369 (2009)

Appendix: Values used for numerical simulation

Λ	1	h_0	6.5 km
Q	$6{\cdot}10^6$ J/kg	m_i	1 kg
Γ	1	ρ_i	3.5 kg/m^3
τ	0.1	Z	45°
ρ_0	1.01 kg/m^3	V_i	35 km/s

Dependences of Ratio of the Luminosity to Ionization on Velocity and Chemical Composition of Meteors

M. Narziev

Abstract On the bases of results simultaneous photographic and radio echo observations, the results complex radar and television observations of meteors and also results of laboratory modeling of processes of a luminescence and ionization, correlation between of luminous intensity I_p to linear electronic density q from of velocities and chemical structure are investigated. It is received that by increasing value of velocities of meteors and decrease of nuclear weight of substance of particles, lg I_p / q decreased more than one order.

Keywords meteors · meteor luminosity · ionization

1 Introduction

Studying the interaction of processes of luminescence and ionization and investigating their dependence on the velocity of meteors belongs to the actual questions of meteor physics. Knowledge of these dependences need to address such important and yet unresolved until the end of questions, as a refinement of the scale radio magnitudes, as well as the mass scale as the photo and radar meteors. Attempts to study the interaction of processes of luminescence and ionization of meteors, as well as finding the dependence of the ratio coefficient of luminous to the ionization on the velocity in the range $32 < V < 62$ km /s were made earlier than on the basis of data parallel visual-radar (Greenhow and Hawkins 1952), as well as photographic and radar observations (Davies and Hall 1963; Babadjanov 1969).

However, because of the low accuracy in the first method, and because of statistical heterogeneity and lack of observational data in the second, the results obtained by different authors were significantly different. The dependence of the relationship of light intensity to the linear electron density on the velocities in the range 11 - 31 km/s generally has not been investigated.

2 Dependences of Ratio of the Luminosity to Ionization on Velocity and Chemical Composition of Meteors

In this paper, on the bases of results of simultaneous optical and radio echo observations and the results of laboratory simulation of the luminescence and ionization, the correlation between the intensity of luminescence I_p to linear electron density q from the velocity and chemical composition of meteors are investigated.

M. Narziev (✉)
Institute of Astrophysics of Academy of Sciences Tajikistan, Bukhoro str. 22, Dushanbe 734042, Tajikistan. E-mail: mirhusseyn_narzi@mail.ru

According to the physical theory of meteors, the ratio of luminous intensity I_f to the initial electron line density q is related with the parameters of the meteor body equation:

$$I_p / q = \tau V^3 \mu / 2 \beta \qquad (1)$$

where τ is the luminous efficiency, β - the ionizing probability, V- velocity of the meteor and μ - the mean mass of a meteor atom. According to the equation (1), the ratio I_p /q depends not only on the coefficients of luminous efficiency and ionization, but also on the velocity and chemical composition of meteor bodies.

To investigate the I_p /q from velocity and other factors, we used the results of parallel television and radar observations conducted during periods of maximum activity of meteor showers from 1978 - 1980 in Dushanbe (Narziev and Malyshev 2006, 2009), as well as the data of similar observations of the fainter ($4 < M < 8$) and low-velocity meteors ($10 < V < 36$ km/s) at Cambridge (Massachusetts) (Cook et al., 1973), the results of parallel photo - radar in Dushanbe (Babadjanov 1969), and the Jodrell Bank (Davies and Hall 1963). The basic equipment used for the observations, the method of processing the observational data and initial data on the individual meteors in the aforesaid sources are given in Davies and Hall (1963); Babadjanov (1969); Narziev and Malyshev (2006, 2009); and Cook et al. (1973).

Table 1 confirmed the following dates: N - number of the meteor, V - velocity, H - the height of the point of specular reflection, M and q - the absolute magnitude and the linear electron density at the point of specular reflection, I_p - luminous intensity, calculated from the known formula:

$$\lg I_f = 9.72 - 0.4\, M \qquad (2)$$

The linear electron density for our joint meteors and meteor joint given in [2, 3], was determined from the measured duration of the radar echo. The value of lg I_p /q, calculated for each meteor is given in the sixth column, and in the seventh column source is indicated, which undertook the initial data. For meteors, given in Cook et al. (1973), the table gives the values of lg I_p /q calculated by n - Settlements.

According to the results given in Table 1, the calculated values of lg I_p /q are in the range -5.2 to -2.7. Figure 1 illustrates the distributions lg I_p /q and shows that the values lg I_p /q change in a fairly wide range from -5.5 to -2.5, with a maximum range of -5 to - 4.5. A large spread of values lg I_p /q, as already noted, possibly related to the dependence of the relationship lg I_p /q on the velocity and the difference in the chemical composition of meteors.

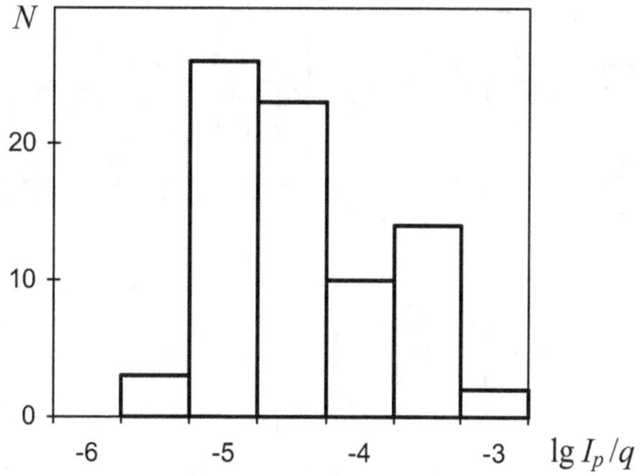

Figure 1. Observed distributions of ratio lg I_p /q.

Table 1. Ratio of luminous intensity I_p to the initial electron line density q by the results combined optical and radio observations of meteors.
(Sources: (A) Narziev and Malyshev 2006, 2009; (B) Davies and Hall 1963; (C) Babadjanov 1969; (D) Cook et al. 1973)

N	V км/с	$H_{км}$	M	lg q	lg I_p/q	Source
1	56.70	104.4	1.50	13.70	- 4.62	(A)
2	36.50	92.0	1.10	13.60	- 4.28	--”--
3	59.99	102.9	- 0.30	13.96	- 4.12	--”--
4	-	98.0	- 0.90	14.73	- 4.55	--”--
5	43.90	101.0	- 1.50	14.60	- 4.28	--”--
6	40.80	99.0	1.63	13.16	- 4.22	--”--
7	57.90	102.3	1.90	13.67	- 4.71	--”--
8	46.20	105.0	1.10	13.98	- 4.70	--”--
9	40.10	92.0	2.50	13.98	- 3.67	--”--
10	21.90	86.0	0.20	13.07	- 3.43	--”--
11	41.40	96.3	1.60	13.60	- 4.52	--”--
12	29.10	90.0	- 1.60	15.08	- 4.72	--”--
13	39.20	95.1	2.20	13.56	- 4.72	--”--
14	29.90	92.3	2.16	13.11	- 4.25	--”--
15	43.40	92.0	0.50	13.77	- 4.25	--”--
16	41.80	91.6	1.90	12.60	- 3.64	--”--
17	41.80	93.0	1.70	13.44	- 4.40	--”--
18	40.60	92.6	1.80	13.30	- 4.30	--”--
19	40.20	92.0	0.20	13.97	- 4.33	--”--
20	30.80	90.9	1.80	13.28	- 4.28	--”--
21	47.70	102.2	2.20	13.49	- 4.65	--”--
22	44.97	93.5	1.60	13.50	- 4.42	--”--
23	42.90	92.7	1.20	13.45	- 4.21	--”--
24	40.80	98.9	2.50	12.88	- 4.16	--”--
25	41.40	94.2	1.00	13.83	- 4.51	--”--
26	37.50	98.5	2.50	12.84	- 4.12	--”--
27	31.00	88.0	-	13.11	-	--”--
28	22.60	86.5	0.0	13.73	- 4.01	--”--
29	43.90	95.0	1.40	13.70	- 4.54	--”--
30	14.30	88.5	- 1.00	14.21	- 4.09	--”--
31	69.40	110.5	- 0.15	14.59	- 4.81	--”--
32	60.90	102.4	1.50	13.55	- 4.43	--”--
33	60.90	106.4	-	14.36	-	--”--
34	62.20	107.5	0.80	14.20	- 4.80	--”--
35	38.50	91.4	- 1.00	14.75	- 4.63	--”--
36	38.00	84.2	0.0	13.34	- 3.62	--”--
37	40.90	87.0	2.05	12.33	- 3.43	--”--
38	52.90	106.6	0.22	14.50	- 4.86	--”--
39	55.50	106.1	2.05	13.92	- 5.02	--”--
40	57.80	105.8	2.28	13.88	- 5.07	--”--
41	59.90	104.1	1.00	13.76	- 4.44	--”--
42	60.90	106.7	1.10	14.04	- 4.76	--”--
43	55.70	102.5	2.00	13.64	- 4.72	(A)
44	63.60	99.0	- 1.75	14.77	- 4.35	--”--
45	65.80	102.0	- 1.20	14.84	- 4.64	--”--
46	63.90	102.7	1.82	13.62	- 4.63	--”--
47	58.10	106.2	1.50	13.99	- 4.72	--”--
48	62.20	106.0	0.80	14.20	- 4.75	--”--
49	65.70	101.1	1.53	14.09	- 4.98	--”--
50	62.30	103.5	1.05	13.63	- 4.33	--”--
51	59.80	105.5	0.20	14.29	- 4.65	--”--
52	60.50	102.5	1.50	13.56	- 4.44	--”--
53	60.40	100.8	2.60	13.27	- 4.59	--”--
54	65.70	99.0	- 3.20	15.71	- 4.71	--”--
55	55.00	105.2	1.40	13.85	- 4.69	--”--
56	61.00	108.0	1.70	14.27	- 5.23	--”--
57	56.10	106.3	0.40	14.12	- 4.56	--”--
1	37.00	94.9	- 1.7	14.99	- 4.59	(B)
2	40.00	110.3	- 1.80	14.88	- 4.44	--”--
3	29.00	92.8	2.00	12.19	- 3.27	--”--
4	27.50	81.5	2.70	12.54	- 3.90	--”--
5a	33.00	101.9	3.30	12.39	- 3.99	--”--
5в	33.00	96.5	1.60	13.75	- 4.67	--”--
6	34.00	96.4	1.60	12.39	- 3.31	--”--
7	26.00	89.0	1.50	12.88	- 3.52	--”--
661345a	71.50	97.8	- 6.30	15.93	- 3.69	(C)
6613456	71.60	97.3	- 4.40	15.19	- 3.71	--”--
670805	60.10	98.2	- 2.10	14.25	- 3.87	--”--
670821	60.50	99.0	- 2.90	15.28	- 4.40	--”--
670866	61.70	107.5	- 3.00	15.50	- 4.58	--”--
670931	61.00	95.0	- 5.80	16.73	- 4.69	--”--
670954	63.70	93.6	- 4.80	14.96	- 3.32	--”--
1	31.20	83.7	5.50	10.16	- 3.08	(D)
2	14.70	97.3	6.87	10.12	- 2.97	--”--
7	17.90	91.9	4.85	10.80	- 3.02	--”--
9	28.80	99.3	4.95	11.10	- 3.36	--”--
12	16.20	90.1	7.75	9.65	- 3.03	--”--
14	36.00	84.0	6.13	10.77	- 3.51	--”--
15	30.10	92.9	6.30	10.72	- 3.52	--”--
19	30.40	100.7	6.20	10.40	- 3.28	--”--
21	32.00	92.3	6.78	10.22	- 3.21	--”--
23	35.70	84.3	7.15	10.24	- 3.38	--”--
24	27.10	91.0	6.50	10.08	- 2.96	--”--
25	20.20	88.9	5.53	10.13	- 2.79	--”--

Dependence of lg I_p/q on the velocity are investigated by observations of 66 meteors that have absolute magnitudes, the prisoners in the interval $-1 < M < +8$. Meteors brighter than magnitude -1^m are excluded for the following reasons: a) In most of the observed cases, these meteors are registered on turning trails. The number of such meteors in our case was 7. b) In addition, bright meteors features with multicenter radio echo duration and displacement of the mirror reflection along the trail. These factors tend to lead to an underestimation of the values of the radio echo duration and the line electron density.

The rest of the meteors were divided into groups according to velocity intervals of 10 km/s and for each group the average value of V and lg I_p/q was calculated. The results are shown in Figure 2 (red circles), where the values of lg I_p/q are on the axis of ordinates and the X-axis shows meteor velocity. From the data presented in the figure, the ratio of lg I_f/q in the range 14 – 25 km/s does not change significantly, and it is shown that in further increasing the velocity to 62 km/s, this ratio decreases more than an order of magnitude.

According to the equation (1), the ratio lg I_p/q can be determined if we know the value of τ and β considering the given value of velocity and chemical composition. Such data for the velocity range of 11 - 53 km/s were obtained from laboratory simulation of the emission and ionization for particles consisting of Fe, Ca, Si, Mg, etc. (Becker and Friichtenicht 1971; Boitnott and Savage 1970; Boitnott and Savage 1971; Friichtenicht and Becker 1973; Slattery and Friichtenicht 1967).

These elements are the parts of stony meteoroids and are often observed in the spectra of meteors. The results of these experiments confirm the dependence of V on τ for model B (Lebedinets 1980). The dependence of β on V for the case of iron particles is obtained in the form (Slattery and Friichtenicht 1967):

$$\beta(Fe) = 1.5 \cdot 10^{-21}\, V^{3.12} \qquad (3)$$

By specifying the chemical composition of dust particles and the numerical values of τ and β according to these experiments, using equation (1), we can calculate the ratio of lg I_p/q for different values of velocity. The calculation results are shown in Figure 2 (white circles on the - Fe). Similar calculations are carried out for copper particles in Figure 2 (triangle Δ - Cu). As from observational data and the results of laboratory simulation it is shown that changing the value of lg I_p/q on the velocity of this change τ from V in model B. The differences between the curves is likely due to difference of chemical composition, partly to measurement errors that occur in the case of observations and data in the laboratory simulation, as well as conditions of the laboratory experiments, which correspond to heights of 70 km. On the basis of the results of simultaneous observations of meteors, lg I_p/q is found with velocity dependence:

$$\lg I_p/q = (6.66 \pm 0.73) - (1.63 \pm 0.35) \lg V$$

where V expressed in cm/s.

We can estimate the influence of chemical composition of meteoroids in the scatter in the value of lg I_p/q, using the results of laboratory simulations. To do this, from (Lebedinets 1980; Becker and Friichtenicht 1971; Boitnott and Savage 1970; Boitnott and Savage 1971; Friichtenicht and Becker 1973; Slattery and Friichtenicht 1967) we had taken numerical values of lg τ and lg β for the velocity V = 40 km/s. Data of lg τ and lg β are calculated values of lg I_p/q for micron-sized dust particles,

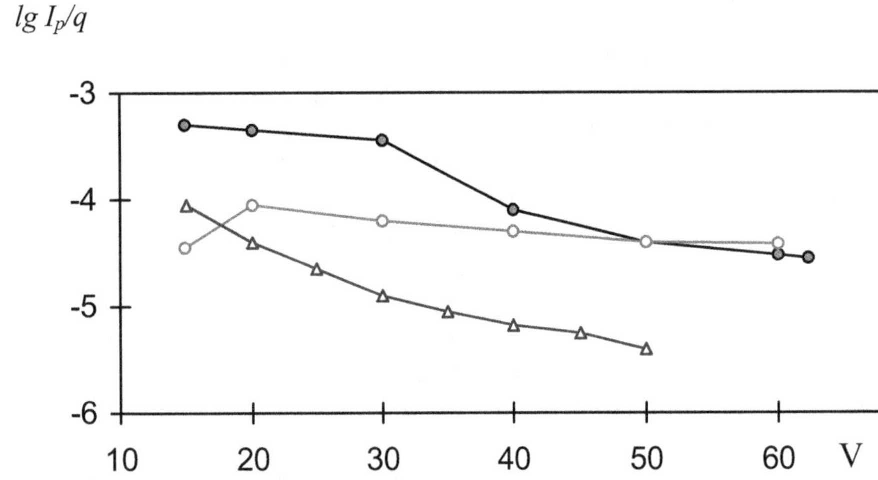

Figure 2. Variation of mean values of lg I_p /q as a function of velocity V.

containing in its composition Mg, Si, Ca and Fe are presented in Table 2. According to the results given in the table, the value of lg I_p /q is not constant, but in all probability is a function of the atomic weight of the substance. For a given value of the velocity, value of lg I_p /q depending on the chemical composition of matter varies from -5.46 to -4.33. If the observed values of lg I_p / q, according to the results of parallel observations at 40 to 42 km/s, vary in the range -4.52 to -3.43. The average observed value lg I_p / q at a velocity V = 41 km/s is -4.2. Thus, based on how the results of parallel optical and radar observations and data from laboratory simulation of the emission and ionization, it follows that the ratio of light intensity to a linear electron density is a function of velocity and chemical composition of meteors.

Table 2. Ratio of lg I_p /q as a functions of chemical composition of the substance.

Elements	lg τ	lg β	lg I_p /q
Mg	- 3.40	- 0.821	- 5.46
Si	- 2.97	- 0.523	- 5.27
Ca	- 2.88	- 0.208	- 5.33
Fe	- 2.03	- 0.225	- 4.33

3 Conclusions

1. For the range of meteor velocities from 14 to 71 km/s and a brightness of up to 7^m – -7^m meteors obtained as a result of parallel optical and radar observations, we calculated the ratio of the logarithm of light intensity to a linear electron density. It was found that the calculated values of the ratio of light intensity to the linear electron density in the range -5.1 to -2.7. The average value of lg I_p /q is -4.5.

2. According to the results of parallel optical and radar observations and the data of laboratory modeling of the phenomenon of a meteor, we studied the relation between the logarithm of the ratio of light intensity to the linear electron density lg I_f /q on the velocity and chemical composition of the meteors. It is received from simultaneous results observations of meteors, and results of laboratory

modeling follows that by increasing value of velocities of meteors lg I_p /q decreased more than one order.

References

1. Babadjanov P.B *Report of Academy of Sciences USSR*, 4, 800, 800-802 (1969)
2. Becker D.G., Friichtenicht J.F. Astrophys. J., 166, N 3, pt 1, 699 - 716 (1971)
3. Boitnott C.A., Savage H.F. Astrophys. J., 161, N 1, pt 1, 351 - 358 (1970)
4. Boitnott C.A., Savage H.F. Astrophys. J., 167, N 2, pt 1, 349 - 355 (1971)
5. Cook A.F, Forti G., McCrosky RE, Posen A., Southworth R., Williams J.T. Evolutionary and physical properties of meteoroids. IAU - Colloquium.Washington. 23-44 (1973)
6. Davies J.G., Hall J.E. Proc. Roy. Soc., A271, N 1344, p. 120-128 (1963)
7. Friichtenicht J.F., Becker D.G. In: Evolutionary and physical properties of meteoroids. Ed. C.L. Hemenwey, P. M. Millman, A. F. Cook. N.Y.: NASA SP – 319, 53 - 82 (1973)
8. Greenhow J.S., Hawkins G.S. Nature. 170, N 4322, p. 355-357 (1952)
9. Lebedinets V. N. Dust in the upper atmosphere and space. The Meteors. Leningrad, Gidrometeoizdat, 248 (1980)
10. Narziev M., Malyshev, I. F. Bulletin of the Institute of Astrophysics Academy of Sciences of the Republic of Tajikistan. 85, 35-45 (2006)
11. Narziev M., Malyshev, I. F. Proceedings of the Academy of Sciences of the Republic of Tajikistan. 4(137), 36-45 (2009)
12. Slattery J.C., Friichtenicht J.F. Astrophys. J. 147, N 1, 235-244 (1967)

CHAPTER 6:

CHEMICAL AND PHYSICAL PROCESSES RESULTING FROM METEOROID INTERACTIONS WITH THE ATMOSPHERE

Atmospheric Chemistry of Micrometeoritic Organic Compounds

M. E. Kress • C. L. Belle • G. D. Cody • A. R. Pevyhouse • L. T. Iraci

Abstract Micrometeorites ~100 μm in diameter deliver most of the Earth's annual accumulation of extraterrestrial material. These small particles are so strongly heated upon atmospheric entry that most of their volatile content is vaporized. Here we present preliminary results from two sets of experiments to investigate the fate of the organic fraction of micrometeorites. In the first set of experiments, 300 μm particles of a CM carbonaceous chondrite were subject to flash pyrolysis, simulating atmospheric entry. In addition to CO and CO_2, many organic compounds were released, including functionalized benzenes, hydrocarbons, and small polycyclic aromatic hydrocarbons. In the second set of experiments, we subjected two of these compounds to conditions that simulate the heterogeneous chemistry of Earth's upper atmosphere. We find evidence that meteor-derived compounds can follow reaction pathways leading to the formation of more complex organic compounds.

Keywords micrometeorite · organic chemistry · atmosphere

1 Introduction

Micrometeorites ~100 μm in diameter carry most of the extraterrestrial material striking the top of the atmosphere, approximately 40 million kg annually [3]. The majority of these particles are most closely related to CM chondrites, and thus should carry a few percent organic material by weight, initially. These particles experience severe heating upon atmospheric entry, reaching their peak temperatures at altitudes of >85 km [2] (see also [4] in this volume for more details on atmospheric entry temperatures). Most micrometeorites are melted either partially or completely, indicating that they reached temperatures sufficient to melt silicate, >1600 K [2] [3]. Such strong heating had been assumed to cause complete destruction of the organic content of the particles in this size range.

In recent years, the new field of astrobiology has generated much interest in the relationship of extraterrestrial organic compounds and the prebioitic environment of early Earth. The process of delivering material to habitable planets generates tremendous heat whether it is via micrometeorites or km-sized objects; thus, this step seems to be a potential dealbreaker for a relationship between interstellar or meteoritic organic compounds and the origin of life. However, in recent years the

M. E. Kress (✉) • A. R. Pevyhouse
Department of Physics & Astronomy, San José State University, CA 95192-0106 USA. Phone: +1-408-924-5255; E-mail: mkress@science.sjsu.edu

C. L. Belle
Department of Biology, The Colorado College, Colorado Springs, CO 80903 USA

G. D. Cody
Geophysical Laboratory, Carnegie Institution of Washington, 5251 Broad Branch Rd NW, Washington, DC 20015 USA

L. T. Iraci
Earth Science Division, NASA Ames Research Center, Moffett Field, CA 94035 USA

questions have been further refined to investigate how infalling material is modified during the delivery process, as opposed to whether this or that molecule can 'survive' delivery. For instance, Court and Sephton [1] found that methane evolves from the pyrolysis of carbonaceous chondrite particles.

Here, we report preliminary results on two sets of experiments: 1) atmospheric entry was simulated by flash-pyrolyzing micrometeorite analogs, producing methane and a variety of organic compounds, and 2) heterogeneous chemistry in Earth's upper atmosphere was simulated with sulfuric acid-catalyzed reactions among two of the pyrolysis products, resulting in the formation of more complex organic compounds.

2 Atmospheric Entry

A fresh fragment from the interior of the Murchison CM 2 carbonaceous chondrite was crushed and sieved to yield 300 μm diameter particles. To reproduce the effects of atmospheric entry encountered by micrometeorites, these particles were flash-heated at 500 K/second to temperatures in excess of 1300 K in a CDS 1000 pyroprobe with heated injector interface. This instrument has been used in pyrolytic analysis of ancient biomacromolecules and extraterrestrial organic solids. Upon release from the solid particle, the pyrolysis products were entrained in a helium stream and deposited on a cold finger (a loop of the GC column immersed in liquid nitrogen). Upon liquid N_2 boil off, the molecular products (pyrolysate) are chromatographically separated on the GC column (a Supleco SPB 50, 50% phenyl-50% dimethyl silicone) employing an Agilent 6890 series GC and analyzed with a HP5972 mass spectrometer.

3.5 wt % of the Murchison meteorite is composed of organic material; of this approximately 30 wt % of these organics are converted into volatiles during flash pyrolysis, the remaining 70 % is a char. The resulting mass spectrum is shown in Figure 1. The majority of the organics were evolved in a temperature range of 500 to 1000 K. The volatile organics appeared to have been completed removed from the particle by a temperature of 1000 K.

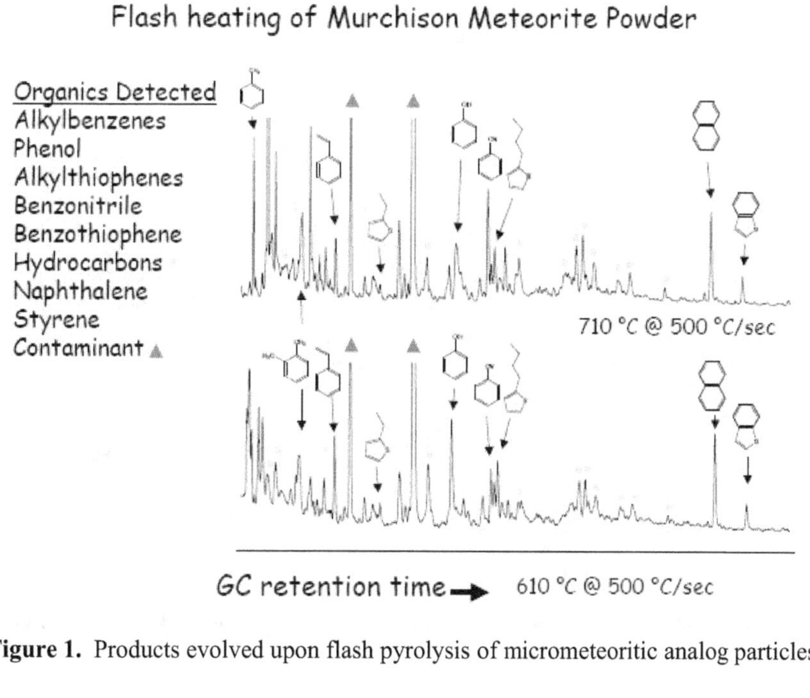

Figure 1. Products evolved upon flash pyrolysis of micrometeoritic analog particles

The compounds that were identified as pyrolysis products included relatively simple compounds including CO, CO_2, H_2O, CH_4, and H_2S. Also evolved from the meteorite during pyrolysis were complex organics, including alkylbenzenes, phenol and alkyl phenols, alkylthiophenes, benzonitrile, benzothiophene, a variety of light hydrocarbons, naphthalene and alkyl-naphthalenes, styrene, and a minor amount of larger polycyclic aromatics including anthracene and phenanthrene. The absolute and relative abundances of these compounds have not yet been quantified.

3 Heterogenous Chemistry in the Upper Atmosphere

Sulfuric acid particles exist in Earth's upper atmosphere, and organic compounds often react strongly with this acid. We have studied the reaction of phenol and styrene, two of the compounds identified in the pyrolysis experiments that are known to independently undergo reactions with sulfuric acid. The sulfuric acid solution was used as a surrogate matrix to mimic upper atmospheric particles.

Theory predicts an acid-catalyzed reaction between phenol and styrene to produce 4-(1-phenylethyl) phenol (shown in Figure 2), and our experiments showed spectral evidence consistent with this pathway (Figure 3). The reaction mixture is compared with 4-cumylphenol which serves as an analog for 4-(1-phenylethyl) phenol, which was not commercially available but has a very similar infrared spectrum. The only difference between these two structures is that 4-cumylphenol has an additional methyl group on the α carbon atom in place of the hydrogen atom. H_2SO_4 concentrations higher than 30 wt% are required to obtain reaction at all temperatures and in a short amount of time. In general, reaction occurs more readily at colder temperatures (5°C compared to 65°C).

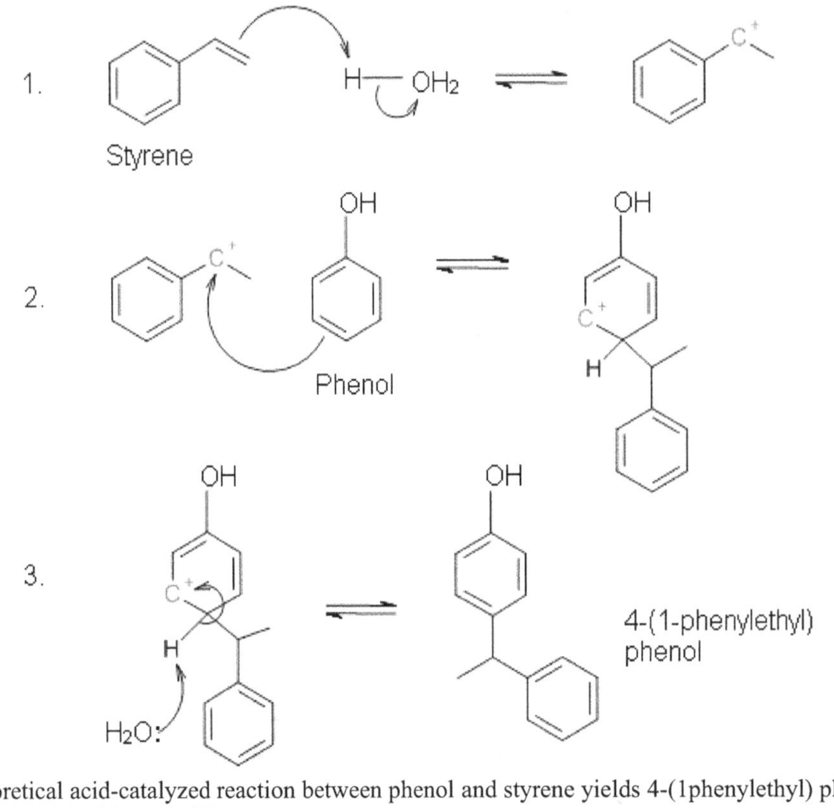

Figure 2. Theoretical acid-catalyzed reaction between phenol and styrene yields 4-(1phenylethyl) phenol. Note loss of $=CH_2$ in step 1 and addition of $-CH_3$ group.

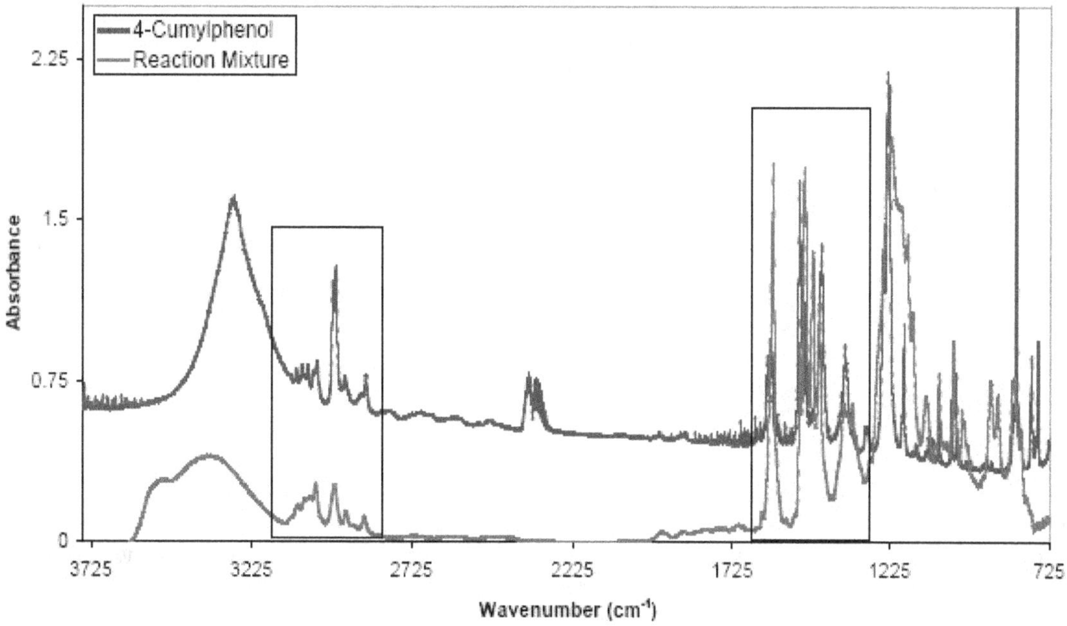

Figure 3. Comparison of 4-cumylphenol IR spectrum (upper curve) with that of the of reaction mixture (lower curve). This reaction mixture was 70wt% sulfuric acid heated to 40°C for 5 minutes and then remained at 20°C for one day. 4-cumylphenol is an analog for the predicted product, 4-(1-phenylethyl) phenol, shown in Figure 2.

4 Summary and Future Work

The fate of organic material entering Earth's atmosphere from space is not well understood. The preliminary results from our experiments show that 1) a wide variety of organic compounds may be released from micrometeorites during atmospheric entry, and 2) these compounds may then go on to react with each other under conditions in the Earth's upper atmosphere. In particular, we found that phenol and styrene are released from flash-pyrolyzed CM chondrite micrometeorite-analogs. We also found that, under conditions analogous to those of the upper atmosphere, phenol and styrene react to produce a compound with a para-disubstituted aromatic ring.

Meteor-derived organic compounds are susceptible to destruction by solar UV, which has a higher flux at altitudes where most of the organic compounds will be released (>85 km). Organic compounds will be destroyed by prolonged exposure to solar UV; this issue is discussed in more detail in Pevyhouse & Kress ([4], this volume). If organic compounds are to persist in the atmosphere, they must be readily mixed to lower altitudes over timescales that are short compared to their photochemical lifetimes. Aromatic compounds are generally more stable to photolysis than are aliphatic hydrocarbons and thus are more likely to participate in heterogeneous chemical reactions leading to greater chemical complexity in the Earth's modern atmosphere.

Future work will entail quantifying the compounds released during entry conditions. Once the abundances these species are measured, they can be incorporated into atmospheric chemical models. The questions of astrobiological interest include investigating the roles that aromatics and light hydrocarbons play in planetary atmospheres. These compounds are strong greenhouse gases, and they also drive smog production in low-O_2 environments. Aromatic compounds also may be important in organic haze production, and they are excellent absorbers of ultraviolet radiation. On the early Earth, high levels of aromatic compounds from infalling debris may have shielded the prebiotic planetary surface from stellar UV. An understanding of these chemical processes may also be critical to pre-empting false positives that masquerade as biomarkers in the atmospheres of exoplanets.

Acknowledgements

MEK and ARP acknowledge research support from the NASA Astrobiology Institute's Virtual Planetary Laboratory (PI: V. Meadows). CLB was supported via the Undergraduate Student Research Program / Universities Space Research Association.

References

1. R.W. Court and M.A. Sephton, Investigating the contribution of methane produced by ablating micrometeorites to the atmosphere of Mars, Earth and Planetary Science Letters, 288, 382-385 (2009)
2. S.G. Love and D.E. Brownlee, Heating and thermal transformation of micrometeoroids entering the earth's atmosphere, Icarus, 89, 26-43 (1991)
3. S.G. Love and D.E. Brownlee, A direct measurement of the terrestrial mass accretion rate of cosmic dust, Science, 262, 550-552 (1993)
4. A. R. Pevyhouse and M.E. Kress, Modeling the entry of micrometeoroids into the atmospheres of Earth-like planets, in Meteoroids 2010, NASA Conference Proceedings, held in Breckenridge, CO, May 2010. Edited by D. Janches (in preparation)

Formation of the Aerosol of Space Origin in Earth's Atmosphere

P.M. Kozak • V.G. Kruchynenko

Abstract The problem of formation of the aerosol of space origin in Earth's atmosphere is examined. Meteoroids of the mass range of 10^{-18}-10^{-8} g are considered as a source of its origin. The lower bound of the mass range is chosen according to the data presented in literature, the upper bound is determined in accordance with the theory of Whipple's micrometeorites. Basing on the classical equations of deceleration and heating for small meteor bodies we have determined the maximal temperatures of the particles, and altitudes at which they reach critically low velocities, which can be called as "velocities of stopping". As a condition for the transformation of a space particle into an aerosol one we have used the condition of non-reaching melting temperature of the meteoroid. The simplified equation of deceleration without earth gravity and barometric formula for the atmosphere density are used. In the equation of heat balance the energy loss for heating is neglected. The analytical solution of the simplified equations is used for the analysis.

As an input parameter we have used the cumulative distribution of space matter influx onto earth on masses in large mass range. Basing on this distribution we have plotted three-dimensional probability density distribution of influx of particles as a function of parameters, which determine the heating and stop altitude of a meteoroid: initial mass m_0, velocity of entry into the atmosphere υ_0 and radiant zenith angles z_{R0}. The obtained three-dimensional distribution had been presented first as a product of three independent distributions on the mentioned parameters, then it was transformed using the equation of deceleration into the distribution on the following parameters: m_0, υ_0 and "altitude of stopping" H_S. The final 2-dimensional distribution on parameters υ_0 and H_S of the aerosols of space origin in the atmosphere was obtained by means of integration of the previous distribution over υ_0.

Keywords meteoroids · meteors · atmosphere aerosol · aerosol formation · space origin

1 Introduction

There are aerosols of both ground and space origin in Earth's atmosphere. Aerosols of the ground origin are presented basically in the lower atmosphere: in the troposphere. The most powerful aerosol layer of the ground-based origin, known also as Junge Layer, is placed at altitudes of 10-25 km. It originated from the condensation of some components of the atmosphere appearing from the photo-chemical transformations of some products of volcano eruptions, for example sulphuric acid vapors. The second confidently established aerosol layer in the atmosphere is placed at altitudes of 80-85 km, corresponding to the minimal atmospheric temperature, in the mesopause. The origin of this aerosol layer in not finally established. Most of scientists, and the authors as well, hold an opinion that all the particles there to be of space origin. Under some special conditions the condensation of water vapors on these particles becomes possible, and we can see, probably, the high-latitudinal silvery clouds.

P.M. Kozak (✉) • V.G. Kruchynenko
Astronomical Observatory of Kyiv Taras Shevchenko National University, Kyiv, Ukraine. Phone: 044-4862762; Fax: 044-4862630; E-mail: kozak@observ.univ.kiev.ua

According to meteor physics investigations (Whipple 1950; Whipple 1951; Levin 1956; Öpik 1956; Lebedinets 1980; Lebedinets 1981; Voloshchuk et al. 1989) most of low-mass particles coming into the atmosphere with initial velocities ~11.2-72.5 km/s lose their energy at altitudes of 140-80 km. Small fragments detaching from already heated bigger particles in the atmosphere cannot be decelerated without almost entire loss of their masses due to evaporation. The deeper penetration of a particle to the atmosphere the lower the probability to save its macro size. This task of motion, deceleration and destruction of a separated particle in "abnormal environment" according to the terminology of Öpik (1956) we were considering in Voloshchuk et al. (1989). Such a conclusion is also given from the experimental investigations of chemical analysis of particles, caught in the atmosphere with the help of high airplanes and balloons. Such particles are similar to coaly chondritics (Nady 1975), having a big amount of helium in their surfaces, which penetrated there from the solar wind. Therefore, these are the primary interplanetary particles, which have come though the atmosphere without intensive heating and are not the products of fragmentation of larger bodies (Brownlee and Hodge 1973).

The amount and distribution of the aerosol of space origin in the atmosphere is connected by some authors with planetary global warming. In this work we will try to examine the problem of formation of the aerosol of space origin in Earth's atmosphere basing on the initial meteoroid distributions on the Earth's heliocentric orbit and the equations of classic meteor physics.

2 Meteor Physics Equations to be Used

In this chapter we consider the basic equations of meteor physics to be used in the work, namely the equation of heating of the meteoroid, and the equation of its deceleration. In addition, the simplification of the equations in order to realize the final investigation analytically is substantiated.

2.1 Complete Equations of Meteoroid Deceleration and Heating

The base assumption for the transformation of a small meteoroid into an aerosol particle, not into a meteor, consists in non-reaching by the meteoroid its melting temperature. Therefore, we have to determine the mass interval, and other parameters of meteoroids, which coming into the Earth's atmosphere, do not reach the melting temperature because of their deceleration and heat radiation.

2.1.1 Heating Balance Equation

The theory of heating of low-mass meteoroids with their deceleration, which plays an important role in this case, were developed by Whipple (1950), Whipple (1951) and later by Fecenkov (1955). They have obtained the name of Whipple's micro-meteorites. It is known (Levin 1956) that the particles having the size less than x_0 warm up to the same temperature (x_0 is the warming up depth at which the temperature of the body is less to e times relatively the surface). According to Öpik (1937) and Levin (1956) such particles have radius $r_0 \leq 10^{-3}$ cm. The change of temperature of such a particle with taking into account the energy loss for heating and radiation can be written as:

$$S_{M0}Edt = m_0 cdT + \beta\sigma\left(T^4 - T_0^4\right)S_{F0}dt, \quad (1)$$

where S_{M0} = *const* and S_{F0} = *const* are the middle section and entire surface area of the particle accordingly, m_0 = *const* is its initial mass, c is the specific heat capacity and σ is Stefan's constant, T and

T_0 are the current temperature for time t and initial temperature of the particle in the field of solar radiation at the distance of 1 a.u., $\beta \leq 1$ is a coefficient of thermal radiation of the meteoroid characterizing the digression from black body radiation, $E = \Lambda \rho_A \upsilon^3/2$ is the energy incoming to unity of the meteoroid surface due to its collision with atmosphere molecules, Λ is the dimensionless coefficient of heat conductivity, ρ_A is the atmosphere density.

2.1.2 Deceleration Equation

If the space particle is not warmed up to the melting temperature it becomes the aerosol particle. So, the next question we should answer: at which altitude will it stop? In order to solve this problem we consider the equation of deceleration, which can be written in the most common vector view as

$$m_0 \frac{d\boldsymbol{\upsilon}}{dt} = -c_R S_{M0} \rho_A \upsilon \boldsymbol{\upsilon} + m_0 \mathbf{g}, \tag{2}$$

or separated into constituent parts:

$$m_0 \frac{d\upsilon}{dt} = -c_R S_{M0} \rho_A \upsilon^2 + m_0 g \cos z_R \tag{3}$$

$$\upsilon \frac{dz_R}{dt} = -g \sin z_R, \tag{4}$$

where υ is the meteoroid velocity, z_R is the zenith angle of its radiant, c_R is the resistance coefficient, g is the free fall deceleration constant.

2.2 Accepted Assumptions

In our calculations we use some assumptions and simplifications. First, we suppose the particles of space origin producing the aerosol are of meteor mass range. Lower bound of meteoroid initial mass is 10^{-18} g according to meteoroid mass distributions presented in literature (Ceplecha et al. 1992), higher bound corresponds to the r_0, and is approximately equal to 10^{-8} g (Öpik 1937). The second assumption is that we consider just warmed up and evaporated particles and neglect the mass loss due to blowing meteoroid molecules away in its "cold" state. The next, the most doubtful assumption consists in the fact we use the barometric formula for the atmosphere density:

$$\rho_A(H) = \rho_A(0) \exp(-\frac{H}{H^*}). \tag{5}$$

Here $\rho_A(0)$, H^* are the atmosphere density at the sea level and altitude of the homogeneous atmosphere accordingly. For precise calculations one should use the numerical solution of the equations (1) and (2) and take the real atmosphere density distribution from modern models of atmosphere, especially for altitudes over approximately 120 km. We use the formula (5) here just for the purpose of obtaining the analytical solution of (3) and (4) in order to understand the physics of the aerosol layer formation. Then, we consider the sporadic meteoroids as the main source of aerosol particles, i.e. the particles which are supposed to be of the stone composition. Finally, we calculate the mean aerosol influx during a year.

2.3 Simplification of the Equations

According to Öpik (1937), the small meteoroid spends almost all its energy for the thermal radiation if its radius $r \leq 10^{-3}$ cm (corresponds to $m_0 \approx 10^{-8}$ g for spherical particles), so we can neglect the first term in the equation (1):

$$T^4 - T_0^4 = \frac{S_{M0}\Lambda\rho_A\upsilon^3}{2\beta\sigma S_{F0}}. \tag{6}$$

Since we deal with low-mass particles we can suppose they are decelerated relatively fast, so we can neglect the gravity term in the equations (2). The equations (3) and (4) in this case transform into the equation

$$m_0\frac{d\upsilon}{dt} = -c_R S_M \rho_A \upsilon^2, \tag{7}$$

and $z_R = z_{R0} = const$.

Also we use the relation between the time t and altitude H of the particle

$$dH = -\upsilon\cos z_R dt \tag{8}$$

and express the middle section and surface area of the particle through the shape parameter A: $A = S_M/V^{2/3}$, where V is the meteoroid volume. Supposing the particle is spherical $S_{F0} = 4S_{M0} = 4A(m_0/\rho_M)^{2/3}$, the shape parameter for spherical particles to be $A = \pi(3/4\pi)^{2/3}$.

2.4 Variation Parameters, Constants and Final Equations

Using (6), (7), (8), the shape parameter and barometric formula (5) we obtain

$$T^4 - T_0^4 = \frac{\Lambda\rho_A\upsilon_0^3}{8\beta\sigma}\exp\left(-\frac{3c_R AH^*}{m_0^{1/3}\rho_M^{2/3}\cos z_{R0}}\rho_A\right) \tag{9}$$

$$\upsilon = \upsilon_0\exp\left(-\frac{c_R AH^*}{m_0^{1/3}\rho_M^{2/3}\cos z_{R0}}\rho_A\right). \tag{10}$$

Reaching by the particle of maximal temperature along its trajectory can be derived from $dT/d\rho_A = 0$, and so from (9):

$$\rho_{AT\max} = \frac{m_0^{1/3}\rho_M^{2/3}\cos z_{R0}}{3c_R AH^*}.$$

Putting it back into (9) we obtain

$$T_{\max}^4 - T_0^4 = \frac{\Lambda m_0^{1/3} \rho_M^{2/3} \cos z_{R0} \upsilon_0^3}{24 \beta \sigma c_R A H^* \exp(1)}. \tag{11}$$

Thus, the condition of transformation of space particle into the aerosol can be expressed now as

$$T_{\max} \le T_{melt}, \tag{12}$$

where T_{max} has to be expressed from (11), T_{melt} is the melting temperature of the particle.

Looking at the equations (10) and (11) we can note that there are three parameters of a meteoroid (under the assumptions made above) having an influence onto its belonging to the class of aerosols or meteors, and to the altitude of stopping in the case of the aerosols. These are initial mass of the particle m_0, velocity υ_0, and zenith radiant angle z_{R0}. The ranges of their variations are: $m_0 = 10^{-18} - 10^{-8}$ g according to Ceplecha et al. (1992) for the lower limit and Öpik (1937) for the higher limit (see above), $\upsilon_0 = 11.2 - 72.5$ km/s, i.e. the particles belonging to the solar system are considered, $z_{R0} = 0° - 90°$, all possible entrance angles are taken into account.

Expressing (12) through the variation parameters we obtain the final inequality of separation of the meteoroids onto aerosols and meteors

$$m_0^{1/3} \upsilon_0^3 \cos z_{R0} \le C_T, \tag{13}$$

where $C_T = 24 \beta \sigma c_R A H^* \exp(1)(T_{melt}^4 - T_0^4)/\Lambda \rho_M^{2/3}$.

If the condition (13) is realized we can find the altitude of stopping H_S of the aerosol particle from (10), supposing the velocity of stopping υ_S is a small enough value. Here we continue to use the equation (10) except the Stokes formula for low velocities, where the deceleration is proportional to the first power of the velocity, so H_S can be found from the expression

$$\upsilon_0 = \upsilon_S \exp\left(\frac{C_V \rho_A(0) \exp(-\frac{H_S}{H^*})}{m_0^{1/3} \cos z_{R0}} \right), \tag{14}$$

where $C_V = c_R A H^* \rho_M^{-2/3}$.

During the calculations the following values of constants are taken (Levin, 1956): $\Lambda = 1$, $\sigma = 5.67032 \times 10^{-5}$ erg·cm^{-2}·K^{-2}·s^{-1}, $\beta = 1$, $c_R = 1$, $H^* = 7 \times 10^5$ cm, $\rho_A(0) = 1.6 \times 10^{-3}$ g/cm^3, $\rho_M = 3$ g/cm^3, $T_0 = 276$ K, $T_{melt} = 1600$ K, $\upsilon_S = 0.5$ km/s. For an iron particle $\Lambda = 0.75$; $\rho_M = 7.6$ g·cm^{-3}; $c_R = 1.25$; $T_{melt} = 1800$ K.

3 The Statistical Approach to the Process of Space Origin Aerosol Formation in the Atmosphere

Here we propose the statistical model for the description of atmospheric aerosol formation from meteoroids. We will construct the 3-dimensional distribution on variation parameters having an influence onto the probability of the aerosol formation and onto the altitude of the aerosol layer. Let's represent the distribution as a multiplication of three *independent* single-parameter probability density distributions:

$$p_{m\upsilon z}(m_0, \upsilon_0, z_{R0}) = p_m(m_0) p_\upsilon(\upsilon_0) p_z(z_{R0}) . \qquad (15)$$

It is obvious that this distribution should be normalized to unity over all three parameters, i.e. there must be

$$\int\int\int p_{m\upsilon z}(m_0, \upsilon_0, z_{R0}) dm_0 d\upsilon_0 dz_{R0} = 1 ,$$

where the integration is carried out inside all possible ranges of parameter values.

3.1 Primary Distribution of Meteoroids

Let us find all three 1-dimensional primary probability density distributions, and start from the distribution on mass.

3.1.1 Probability Density Distribution on Initial Mass

There can be found in the literature distributions of space matter onto Earth as cumulative distributions of number of particles on their masses, for example Ceplecha (1992), Kruchynenko (2002), Kruchynenko (2004). We will use the linear dependence (Kruchynenko 2002, Kruchynenko 2004) for the further calculations:

$$\log_{10} N(m_0^{'} \geq m_0) = C_0 - k \log_{10} m_0 , \qquad (16)$$

where $N(m_0' \geq m_0)$ is a number of particles with masses not less than m_0 coming into all Earth atmosphere during a year, $C_0 = 7.86$, $k = 0.892$.

The probability density distribution on mass $p_m(m_0)$ according to cumulative distribution (16) can be described by Pareto distribution:

$$\begin{aligned} p_m(m_0 < m_{0l}) &= 0 \\ p_m(m_0 \geq m_{0l}) &= \frac{k m_{0l}^k}{m_0^{k+1}} \end{aligned} \qquad (17)$$

where m_{0l} is chosen freely. The probability density function is normalized to unity in the value range 0 – +∞. There are the following obvious consequences:

$$F(m_0) = \int_{m_{0l}}^{m_0} p_m(m_0) dm_0 = 1 - \int_{m_0}^{+\infty} p_m(m_0) dm_0 = 1 - \frac{m_{0l}^k}{m_0^k} , \qquad (18)$$

$$\frac{dN(m_0)}{N_l(m_{0l} \le m_0 \le +\infty)} = dF(m_0) = p_m(m_0)dm_0, \qquad (19)$$

where F(m_0) is the cumulative probability, $N_l(m_0 \ge m_{0l})$ is a sample of all particles in the chosen range, which can be found from (16) as $N_l(m_{0l}) = 10^{C_0}/m_{0l}^k$. We suppose $m_{0l} = 10^{-18}$ g, so $N_l = 8.24\times10^{23}$.

3.1.2 Probability Density Distribution on Initial Velocity

The probability density distribution on velocity $\rho_\upsilon(\upsilon_0)$ will be chosen according to radar meteor observations, for example (Voloshchuk et al. 1989):

$$p_\upsilon(\upsilon_0) = PG(\overline{\upsilon_1}, \sigma_{\upsilon1}) + (1-P)G(\overline{\upsilon_2}, \sigma_{\upsilon2}), \qquad (20)$$

where

$$G(\upsilon) = \frac{1}{\sigma_\upsilon\sqrt{2\pi}}\exp(-\frac{(\upsilon-\overline{\upsilon})^2}{2\sigma_\upsilon^2})$$

are Gaussians with the following parameters: $\overline{\upsilon_1}$ = 32.32 km/s, $\sigma_{\upsilon1}$ = 6.51 km/s, $\overline{\upsilon_2}$ = 54.26 km/s, $\sigma_{\upsilon2}$ = 5.15 km/s. The value P is changing during a year. For the mean value we choose $P \approx 0.33$ (Voloshchuk et al. 1989). It is obvious that the probability density function is normalized to unity in the range $0 - +\infty$.

3.1.3 Probability Density Distribution on Initial Radiant Zenith Angle

The probability density distribution on radiant zenith angle $\rho_z(z_{R0})$ will be derived from the following thoughts: let suppose that the number of particles $dN(r, r + dr)$ entering into earth atmosphere from some direction in the range dr in some spatial angle $d\Omega$ (see Figure 1) per time unity can be expressed as $dN(r, r + dr) \sim 2n_0\pi r dr d\Omega dt$, where n_0 is a spatial concentration of meteoroids. Since $r = R_\oplus \sin z_R$, we have $dN(z_R, z_R + dz_R) \sim 2n_0\pi R_\oplus^2 \sin z_R \cos z_R dz_R d\Omega dt$. So we have to use the sine-cosine distribution $\sin z_{R0} \cos z_{R0}$.

After normalization to unity we obtain the final distribution on zenith radiant angle

$$p_Z(z_{R0}) = 2\sin z_{R0}\cos z_{R0}. \qquad (21)$$

Strictly saying, this distribution will be distorted by the Earth gravity, but we use it due to its simplicity.

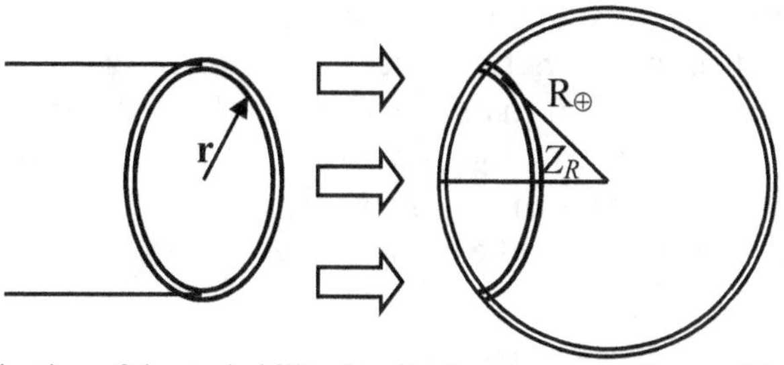

Figure 1. To the derivation of the probability density function on radiant zenith angle. $R_\oplus$ is Earth's radius.

3.2 Separation of the Distribution into Aerosols and Meteors

The primary distribution of meteoroids can be conceived as a geometrical 3-dimensional cube, where the three considered parameters m_0, υ_0 and z_{R0} determine the dimensions along its three ribs-axes, limited by the permissible parameter values. The "intensity" in each point inside such a cube is expressed by the value of $p_{m\upsilon z}(m_0, \upsilon_0, z_{R0})$. The real number of particles dN in the range of dm_0, $d\upsilon_0$, and dz_{R0} can be found from (19). If we make a few sections perpendicularly to the cube rib describing the mass m_0 we will obtain the 2-dimensional pictures in coordinates $\upsilon_0 \leftrightarrow z_{R0}$ for the fixed mass values, where the relative value of $p_{m\upsilon z}$ can be expressed with the help of lines of the similar values, for instance. In the Figure 2 we show only two maximums of the $p_{m\upsilon z}$ corresponding to modal values of bimodal distribution of velocity and the maximum of zenith radiant angle value $z_{R0} = 45°$. Figure 2a corresponds to $m_0 = 10^{-12}$ g, Figure 2b to $m_0 = 10^{-9}$ g. The regions of aerosols and meteors are separated by solid line according to inequality (13).

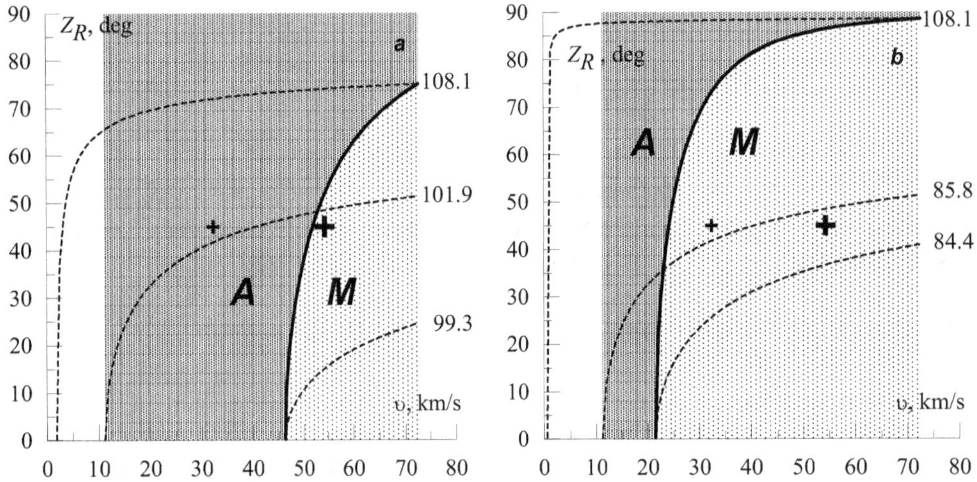

Figure 2. The separation of meteoroids onto aerosols and meteors. Letter ***A*** and dark gray color corresponds to aerosols, otherwise to meteors (letter ***M*** and light gray color). The picture ***a*** corresponds to mass $m_0 = 10^{-12}$ g, ***b*** to $m_0 = 10^{-9}$ g. Dashed lines describe the equal altitudes of stopping. Signs "+" show positions of modes of the distribution.

We can see in Figure 2 that the region of aerosols (denoted in the picture with the letter *A* and dark gray color) is increasing while the mass is decreasing (the region of meteors denoted as *M* with lighter gray color is decreasing accordingly). Therefore, there must be a mass value lower of which all particles remain aerosols. The probability for a meteoroid to become a meteor is proportional to its velocity and cosine of zenith radiant angle. Setting according parameters to their maximal values $\upsilon_0 = 72.5$ km/s and $\cos z_{R0} = 1$ we get the critical mass value $m_{0cr} \approx 1.7\times10^{-14}$ g. Finally, all space particles entering into the Earth atmosphere remain aerosols if their masses are lower than the critical value, then the rate of aerosols is decreasing almost down to zero while the mass is increasing up to the value of approximately $m_0 = 10^{-8}$ g. This rate $g_A(m_0)$ can be easily calculated with the help of the formula

$$g_A(m_0) = \int_0^{\pi/2} p_z(z_{R0})dz_{R0} \int_{\upsilon_{0\,MIN}}^{\upsilon_0(z_{R0})} p_\upsilon(\upsilon_0)d\upsilon_0 \,.$$

The power coefficient in cumulative mass distribution k from (16) is running a range of values: k = 0.892 for $m_0 \le 1.7\times10^{-14}$ g, then k = 1.087 for $m_0 \approx 10^{-13} / 10^{-12}$ g, k = 1.189 for $m_0 \approx 10^{-11} / 10^{-10}$ g, k = 1.438 for $m_0 \approx 10^{-9} / 10^{-8}$ g (the average value for all the mass range $m_0 \approx 10^{-13} / 10^{-8}$ g is k = 1.232).

3.3 Transformation of the Primary Distribution to New Variables

The relation (14) connects four variation parameters: three ones included into the primary distribution, and the fourth one, the altitude of stopping H_S. This parameter can be called conditionally the "free" parameter. The dotted curves corresponding to some of its values (minimal altitude of stopping, minimal and maximal velocities altitudes) to be expressed in kilometers are shown in Figure 2.

Since the main aim of our investigations is to plot the two-dimensional distribution $p_{mH}(m_0, H_S)$ of the aerosol formation into the atmosphere, we will solve it in two steps. The first one is to change the variable z_{R0} in the primary distribution $p_{m\upsilon H}(m_0, \upsilon_0, z_{R0})$ to the altitude of stopping H_S of the aerosol particle. The second step will be consisting in the reducing of 3-dimensional distribution $p_{m\upsilon H}(m_0, \upsilon_0, H_S)$ to $p_{mH}(m_0, H_S)$ by means of integration of $p_{m\upsilon H}(m_0, \upsilon_0, H_S)$ over all range of υ_0.

According to statistical probability density distribution transformations and taking into account that only one variable is changing ($z_{R0} \to H_S$) we can write

$$p_H(H_S) = p_{z(H_S)}(z_{R0}(H_S))\left|\frac{\partial z_{R0}(H_S)}{\partial H_S}\right|, \qquad (22)$$

where $z_{R0}(H_S)$ and determinant of transition $\frac{\partial z_{R0}(H_S)}{\partial H_S}$ can be found from (14). Let us denote

$$C_Z(m_0, \upsilon_0, H_S) \equiv \cos z_{R0} = C_V \rho_A(H_S)/m_0^{1/3} \ln \upsilon_0/\upsilon_S .$$

Then we can write

$$p_Z(z_{R0}(H_S)) = 2C_Z\sqrt{(1-C_Z^2)}, \quad \frac{\partial z_{R0}}{\partial H_S} = \frac{1}{H^*}\frac{C_Z}{\sqrt{1-C_Z^2}},$$

and put them into (22):

$$p_H(m_0, \upsilon_0, H_S) = \frac{2}{H^*}C_Z^2(m_0, \upsilon_0, H_S) \qquad (23).$$

The final view of the obtained distribution $p_{m\upsilon H}(m_0, \upsilon_0, H_S)$ while taking into account (23) is shown in Figure 3 for the same masses as in Figure 2. The "free" parameter now is the cosine of the zenith radiant angle, and the dashed curves correspond to different values of z_{R0} expressed in degrees. The value $z_{R0} = 0$ is placed lower than others in Figure 3 and shown with a solid curve. The region to be placed lower than value $z_{R0} = 0$ is forbidden for both aerosol and meteor particles.

An interesting fact is that the inequality (13) is now transformed in the "stable" state in new coordinates and does not depend on the mass:

$$H_S \ge H^* \ln\left(\frac{C_V \rho_A(0)\upsilon_0^3}{C_T \ln(\upsilon_0/\upsilon_S)}\right)$$

It is shown in Figure 3 with a solid diagonal line.

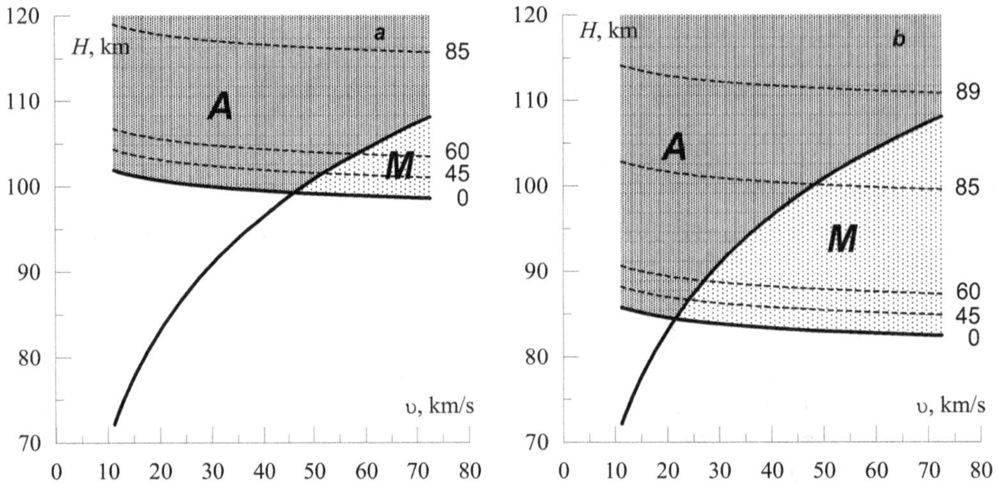

Figure 3. The same distributions as in Fig. 2 but in new variables. Dashed lines describe the equal radiant zenith angles of meteoroids.

3.4 Resultant Distribution Reducing

The final transformation of the distribution (23) consists in reducing it to the two-dimensional state by means of integration over υ_0. The limits of the integration can be easily determined from Figure 3 and according formulae. We obtain the following

$$p_{mH}(m_0, H_S) = p_m(m_0) \int\limits_{\upsilon_1(m_0,H_S)}^{\upsilon_2(m_0,H_S)} p_\upsilon(\upsilon_0) p_H(m_0, \upsilon_0, H_S) d\upsilon_0 . \qquad (24)$$

If we denote the integral, which has to be taken numerically, as

$$I_\upsilon(m_0, H_S) = \int\limits_{\upsilon_{01}(m_0,H_S)}^{\upsilon_{02}(m_0,H_S)} \frac{p_\upsilon(\upsilon_0)}{\ln^2(\upsilon_0 / \upsilon_S)} d\upsilon_0 ,$$

the final formula for the formation of the aerosol of space origin in the atmosphere can be written as

$$p_{mH}(m_0, H_S) = p_m(m_0) \frac{2}{H^*} \left(\frac{C_V \rho_0 \exp(-H_S / H^*)}{m_0^{1/3}} \right)^2 I_\upsilon(m_0, H_S) . \qquad (25)$$

The function $p_{mH}(m_0, H_S)$ is shown in Figure 4.

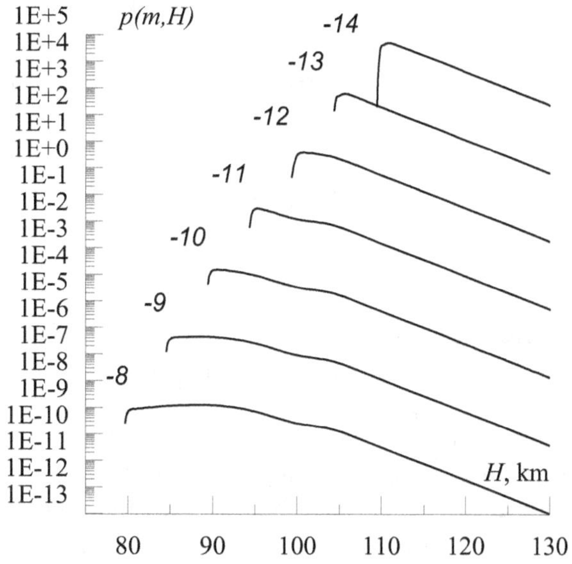

Figure 4. Final two-dimensional distribution of influx of aerosol of space origin into Earth's atmosphere. 10-logarithm scale

4 Conclusion

As it can be seen from the Figure 4 the minimal altitude which can be reachable by the aerosol stone particle of space origin is approximately ~79.6 km. This value corresponds to the meteoroid which is moving vertically with the velocity ~16.6 km/s. A particle of the same mass and with lower velocity will stop higher, with higher velocity will transform into a meteor.

The meteoroids with the mass less or equal to ~1.7×10^{-14} g remain the aerosols always. For masses 10^{-14} / 10^{-8} g cumulative distribution coefficient k increases from 0.892 to 1.438 while the mass increases.

The Figure 4 also demonstrates that aerosols of mass range 10^{-14} / 10^{-8} g stop in relatively thin altitude range 80-120 km. Evidently, the aerosols do not stay at these altitudes forever but immediately start to move downwards under gravitational force and the resistance force of air, which can be described by Stokes formula. How it occurs is the goal for the future work.

References

D.E. Brownlee, P.W. Hodge Space Res., 13/2, 1139-1151 (1973).
Ceplecha Z. Astron. Astrophys., 263, 361-366 (1992).
Fecenkov V.G. Meteoritics, 12, 3 – 14 (1955).
Kruchynenko V.G. Kin. And Phys. Nebesn. Tel, 18, 2, 114 – 127 (2002).
Kruchynenko V.G. Kin. And Phys. Nebesn. Tel, 20, 3, 269 – 282 (2004).
B.Y. Levin Moscow: AN SSSR, 296 pp. (1956).
V.N. Lebedinets Leningrad: Hidrometeoizdat, 250 pp. (1980).
V.N. Lebedinets Leningrad: Hidrometeoizdat, 272 pp. (1981).
B. Nady Amsterdam - New York, 747 pp. (1975).
E.J. Öpik Publ. Obs. Astr. Tartu, 29, 5, 67 (1937).
E.J. Öpik Irish Astron. J., 4, 3/4, 84-135 (1956).
Y.I. Voloshchuk, B.L. Kashcheev, V.G. Kruchynenko Kyiv: Naukova dumka, 294 pp. (1989).
F. L. Whipple Proc. Nat. Acad. Sci. Amer., 36, 12, 686-695 (1950).
F. L. Whipple Proc. Nat. Acad. Sci. Amer., 37, 1, 19-29 (1951).

Composition of LHB Comets and Their Influence on the Early Earth Atmosphere Composition

C. Tornow • S. Kupper • M. Ilgner • E. Kührt • U. Motschmann

Abstract Two main processes were responsible for the composition of this atmosphere: chemical evolution of the volatile fraction of the accretion material forming the planet and the delivery of gasses to the planetary surface by impactors during the late heavy bombardment (LHB). The amount and composition of the volatile fraction influences the outgassing of the Earth mantle during the last planetary formation period. A very weakened form of outgassing activity can still be observed today by examining the composition of volcanic gasses. An enlightenment of the second process is based on the sparse records of the LHB impactors resulting from the composition of meteorites, observed cometary comas, and the impact material found on the Moon. However, for an assessment of the influence of the outgassing on the one hand and the LHB event on the other, one has to supplement the observations with numerical simulations of the formation of volatiles and their incorporation into the accretion material which is the precursors of planetary matter, comets and asteroids. These simulations are performed with a combined hydrodynamic-chemical model of the solar nebula (SN). We calculate the chemical composition of the gas and dust phase of the SN. From these data, we draw conclusions on the upper limits of the water content and the amount of carbon and nitrogen rich volatiles incorporated later into the accretion material. Knowing these limits we determine the portion of major gas compounds delivered during the LHB and compare it with the related quantities of the outgassed species.

Keywords impacts · solar nebula · hydrodynamic · chemistry

1 Fate of Volatiles During Planet Formation

Table 1 shows that the major gasses (CO_2, H_2O, N_2, O_2) making 98-100% of the atmospheres of the three large rocky planets clearly vary in their concentrations. However, a completely different situation is observed for Mercury. Its atmosphere is incredible thin, contains relatively large hydrogen and helium concentrations, and, in addition to oxygen, one finds a high fraction of sodium (29%). Both, the amount of hydrogen and helium and the existence of a large Na fraction indicate a strong interaction between the planet and the solar wind. This strong interaction is supported by the small distance to the Sun which causes a high radiation intensity (see Table 1) as well. Compared to the small radius and mass of the planet, it has an outsized iron core (note, its high density in Table 1) which could have been the result of a large mantle-stripping impact (Benz et al., 1988). Since the pressure and chemical composition of

C. Tornow (✉) • S. Kupper • E. Kührt • U. Motschmann
Inst. of Planetary Research, German Aerospace Research Center (DLR), Berlin. Phone: +493067689427; Fax: +493066055340: E-mail: carmen.tornow@dlr.de

M. Ilgner
Astrophysical Institute and Observatory, Friedrich Schiller University, Jena

U. Motschmann
Institute of Theoretical Physics, Technical University Braunschweig

Mercury's atmosphere differ so largely from the corresponding values of the other planets, we concentrate our study to Earth and partially Mars and Venus.

Table 1. Bulk, orbital and atmospheric parameter of the four rocky planets as observed today (http://nssdc.gsfc.nasa.gov/planetary/factsheet, Prinn & Fegley (1987), and). Note, that Mercury atmosphere additionally contains a large fraction of Na (29 %). The normalisation values used in column 1 are: $R_{\oplus} = 6.37 \times 10^6$ m, $M_{\oplus} = 5.97 \times 10^{24}$ kg, $\rho_{\oplus} = 5.515$ g/cm^3, $L_{\oplus} = 1.37 \times 10^3$ W/m^2, 1 AU = 1.496×10^{11} m, $B_{\oplus} = 5 \times 10^{-5}$ T, and 1 bar = 10^5 Pa.

Parameter		**Mercury**	**Venus**	**Earth**	**Mars**
mean radius / $R_{\oplus}$		0.383	0.950	1.00	0.532
mass / $M_{\oplus}$		0.0553	0.815	1.00	0.107
mean density / $\rho_{\oplus}$		0.984	0.951	1.00	0.713
solar irradiance / $L_{\oplus}$		6.67	1.91	1.00	0.431
semi-major axis / AU		0.387	0.723	1.00	1.52
magnetic field / $B_{\oplus}$		$\sim 10^{-2}$	$< 10^{-5}$	1.00	-
surface pressure / bar		10^{-15}	92	1.014	6.36×10^{-3}
atmospheric composition with respect to major gasses in %	CO_2	-	96.5	0.038	95.3
	H_2O	-	2×10^{-4}	~ 1	3×10^{-4}
	N_2	-	3.5	78.08	2.7
	O_2	42	-	20.95	0.13
	H_2	22	10^{-3}	5.5×10^{-5}	-
atmospheric composition with respect to rare gasses in ppm	^4He	6×10^4	12	5.24	1.4
	^{20}Ne	-	7	18.2	2.5
	^{36}Ar	-	31	9.34×10^4	1.6×10^4
	^{84}Kr	-	0.025	1.14	0.3
	^{130}Xe	-	< 0.009	0.09	0.08

1.1 Planet Formation

Two aspects influence the chemical composition of a planetary atmosphere, the formation process of the planet and the planetary evolution due to internal forces (e.g. magnetic fields, volcanism, plate motion, erosion, evolution of life) and external phenomenons (e.g. solar wind, impacts). The formation process needs to be considered since it has influenced the amount and composition of the volatile fraction of the accretion material. This fraction was produced by hydrides and oxides of N and C bearing molecules in the SN. Its amount and composition depend on the formation time of the planet and the distance to the protosun. The evolution effect is characterised on the one hand by relatively short and powerful events (e.g. impacts or volcanism) and on the other hand by continuous processes with a low immediate influence (e.g. magnetic fields or solar wind).

The influence of Earth evolution on the fractional abundances of the major gasses in the atmosphere is shown in Figure 1. Due to the sparse records not much is known about the Hadean eon (4.6-3.8×10^{-9} years) which also comprises planet formation. However, in order to understand where the carbon dioxide, water, and nitrogen content of the early atmosphere was coming from, one has to consider the scenario of inner planet formation in detail. It is based on core accretion and can be divided into four periods:

- pebble formation (> 1 mm) by dust coagulation and settlement into disk midplane with a ~ 10^4 y timescale,
- planetesimal formation (> 10^2 km) due to gravitational collapse of pebble clusters formed in various turbulence producing instability regions with a 10^3 - 10^4 y timescale,
- protoplanet formation (10^2-10^3 km) by gravitational cleaning of related feeding zones with a 10^5 - 10^6 y timescale and in two phases, which are
    - a runaway accretion phase with a relative growth rate given by $dM/Mdt \sim M^{1/3}$, and
    - an oligarchic accretion phase with a relative growth rate given by $dM/Mdt \sim M^{-1/3}$
- planet formation (10^4 km) by chaotic accretion due to giant impact events causing mergers of protoplanets (e.g. Moon forming impact) with a time-scale between 10^7 and 10^8 years.

Concerning the first two phases, it was shown by Johansen et al. (2007), Lyra et al. (2008), and Brauer et al. (2008) that the planetary embryos with a radius larger than 10^3 km could have been formed after a period of coagulation and settling. Planetesimal formation causes a mainly a physical modification of the accretion material. If one compares porosity values observed for cometary dust P_{CD} ~ 0.85 (Greenberg & Li, 1999) with porosities of the C-and D-type asteroids (0.5 - 0.6) (Trigo-Rodriguez & Blum, 2009) one realises the increased compactification due to collisions. This fits perfectly to observations of enstatite chondrites (Macke et al., 2009) coming from large solid bodies which are highly compactified (porosity ≤ 0.06). In addition to compactification protoplanet and planet formation leads to chemical modification resulting in an increase of insoluble organic matter and a decrease of the soluble fraction. This modification results in an increase of carbonaceous matter and a decrease of H and N containing molecules.

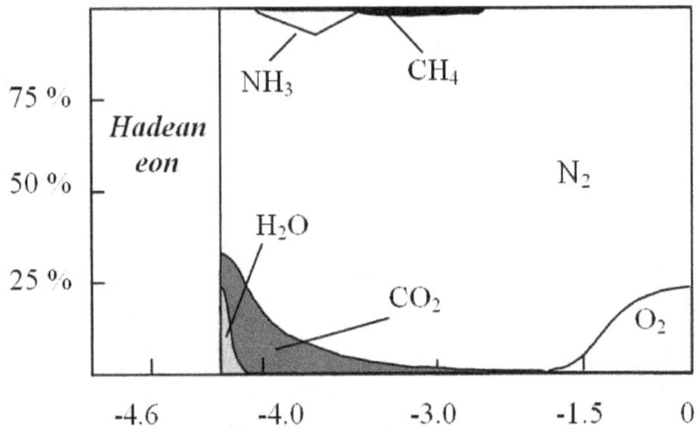

Figure 1. Concentration in percentage, C, shown for the major atmospheric gasses of the Earth versus time in Gyr (1 Gyr = 10^8 years) whereby today is set to 0 Gyr (data except for NH_3 are from Kasting, 2004 and Kaltenegger et al., 2007). Note, that the time is logarithmically scaled and the concentrations of the reducing molecules CH_4 and NH_3 are given in 100% – $C(CH_4)$ and 100% – $C(NH_3)$, respectively.

In the simulations of O'Brien et al. (2006) a mixture of protoplanets with Mars-like masses and many large planetesimals is assumed to be the initial population of the accretion of rocky planets. This assumption agrees with the products of runaway and oligarchic accretion and describes the final, chaotic period of accretion. The chaotic period explains why the volatile concentration of the Earth does not agree with an equilibrium condensate formed at the pressure and temperature in the SN at 1 AU (Prinn & Fegley, 1987). In this period one has to take into account an outgassing of the planetary mantle of the three planets.

There is much evidence that water and CO_2 are typical substances outgassed from the mantle. According to Matsui (1993), Zahnle (1998) and references therein during the chaotic accretion period a magma ocean (depth: ~ 2000 km) with a steam atmosphere of ≥100 bar and a surface temperature of ~ 1500 K has been formed on Earth. In the course of 5×10^7 years (Elkins-Tanton, 2008) the surface has cooled enough to allow the formation of a proto-ocean. According to model results (Kuramoto & Matsui, 1993, Elkins-Tanton, 2008) a local magma ocean could have been formed for Mars as well, but the ocean must have been more shallow in order to form a wet mantle and allow water outgassing. In contrast, due to the more intensive solar radiation on Venus (see Table 1) a hydrosphere was probably not formed on this planet (Abe, 1988).

1.2 Water

Now, we have to ask for the sources of water, carbon dioxide, and nitrogen which are contained in the early atmosphere (Figure 1). At first, there are indications that the planetesimals contained water gathered by physisorption and chemisorption (Stimpfl et al., 2006). The high adsorption energy of chemisorption found for forsterite ensures that water is held by the mineral surface at environmental temperatures of 700 K -1000 K. These values are typical for the inner region of the SN. Consequently water could have contained already in protoplanets formed in the inner SN. According to Morbidelli et al. (2000) during chaotic accretion a further reservoirs results from the outer asteroid belt. The parent bodies of carbonaceous chondrites and, if their number was large, main belt comets (Hsieh and David Jewitt, 2006) could have contributed to a large fraction of water. Observations have shown that D/H ratio of these bodies (~ 1.3×10^{-4}; Kerridge, 1985) is comparable to $D/H_{SMOW} = 1.56\times10^{-4}$, whereby SMOW stands for standard mean ocean water.

1.3 Carbon

In the inner region of the SN carbon is contained in the dust grains since main components are SiC compounds and refractive organic matter (e.g. kerogen-like substances). In addition large amounts of carbon is stored in polycyclic aromatic hydrocarbons (PAHs) which are nano-size particles collected by the larger dust grains during their settling to the midplane (Zubko et al., 2004). In the outer region of the SN, i.e., behind the snow line, carbon bearing molecules were incorporated in the ice mantle of dust grain or later as CH_4 clathrates in pebble clusters (Lunine & Stevenson, 1985).

1.4 Nitrogen

The sources of nitrogen are less known. It is very likely that the SN has contained N_2, but observations (Armitage et al., 2003; Sicilia-Aguilar et al., 2007) suggest that the gas of the nebula was blown away after less than 10 Ma, depending on the frequency range and intensity of the stellar UV radiation in the environment of the SN. A protoplanet, which can be formed in 10^5 to 10^6 years, has gathered enough mass to keep the SN gas as a primary atmosphere. According to the calculations of Genda & Abe (2003) in which the Moon forming impact was considered, it is likely that the Earth was able to keep at least 70 % of its primary atmosphere. In addition, an N-bearing substance (Si_3N_4) was found in ordinary chondrites (Lee et al., 1995). Clément et al. (2005) have detected features in the infrared spectrum of carbon stars which coincide well with the main features of laboratory Si_3N_4 spectra. Consequently, these nitrides are of interstellar origin. Further, N_2 could have been added to the Earth atmosphere during the LHB. We will consider this possibility in more detail in section 3.

2 Atmospheric Composition After Earth Formation

Due to the formation of life on Earth the current atmospheric composition differs clearly from the composition directly after the formation of the planet. In order to understand the influence of the LHB comets on the early Earth atmosphere we need a solidified assumption concerning the composition directly after planet formation as a starting point. According to the reflections in the previous section the atmosphere of the rocky planets contained as major gases CO_2 and N_2. Table 1 shows that for Mars and Venus the carbon dioxide fraction is large (95 - 96%) while the nitrogen fraction is relatively small (3 - 4%). The water fraction disappeared on both planets. Mars has lost its water due to the disappearance of its magnetic field. Thus, in addition to thermal ejection the solar wind could have stripped away its atmosphere. The surface cooling and pressure decreasing have given a situation in which water ice sublimated and due to the solar UV radiation the molecule dissociated. H_2 has left the planet and O has oxidised minerals on the planetary surface. However, a part of the water ice has survived and is probably buried under the dust. Concerning Venus, it was already mentioned that no hydrosphere was formed due to the high temperature. Similar to Mars, Venus has presumably no magnetic field and the water vapour molecules have been dissociated by the strong solar UV radiation. In contrast to Mars, Venus has lost large amounts of hydrogen and oxygen by nonthermal processes such as ion pick-up (Lammer et al., 2006). If one assumes no large differences in the chemical composition of the accretion material and compares the current D/H ratios (Lammer et al., 2008) of Earth (1.5×10^{-4}), Mars (8.1×10^{-4}), and Venus (2×10^{-2}) it follows that the loss of H_2O molecules on Earth was least important.

If one constructs an atmospheric composition of the early Earth we take a CO_2/N_2 ratio as observed for today for Mars and Venus. As a result, 78% N_2 of the Earth atmosphere today correspond to 3-5 % N_2 for the early case. The resulting early pressure varies between 15-26 bar produced by a CO_2 atmosphere. Is the related amount of carbon available on Earth? Table 2 presents the current mixing ratios for the most important volatiles at the time directly after planet formation. We see, that on Venus nearly the complete amount of carbon dioxide, nitrogen, and water are contained in the atmosphere. On Mars and Earth this is true for nitrogen only. A large amount of CO_2 on Earth and Mars is in a condensed phase. On Mars we have CO_2 ice and on Earth the equilibrium reaction

$$Mg_2SiO_4 + 4CO_2 + 4H_2O \rightleftharpoons 2Mg^{2+} + 4HCO_3^- + H_4SiO_4$$

which describes weathering via hydrolysis and carbon dioxide dissolution in water, controls the amount of carbon in the condensed and gaseous phase. The mineral Mg_2SiO_4 symbolises olivine, i.e. forsterite, HCO_3^- denotes a bicarbonate ion, and H_4SiO_4 is silicic acid. Other, more complex, weathering reactions are possible as well, for instance with feldspar ($KAlSi_3O_8$). According to Pidwirny (2006) there are 7-10×10^{22} g carbon dioxide available on Earth and the resulting pressure ~ 20 bar. From Table 2 one realizes a much larger amount of water (Lide, 2001) which is given by 1.4×10^{24} g which would produce a pressure of ~ 290 bar. This is close to the upper limit of the calculations of Zahnle, 1998. However an outflow during the phase of magma ocean and steam atmosphere as well as a large water component stored in the planetary mantle cannot be excluded. As the surface has cooled down sufficiently a shallow oceans formed about 4.4×10^{-9} years ago (compare with Wilde et al., 2001). The CO_2 and water amounts fit well to the data given in Table 2.

Table 2. Global mass fraction of volatile substances stored in the bulk planet and in the atmosphere today. The masses of the planets follow from Table 1. The data are given in Goody & Walker (1972).

Substance	Site	Venus	Earth	Mars
carbon dioxide	bulk planet	1.2×10^{-4}	1.8×10^{-5}	3.1×10^{-6}
	atmosphere	8.7×10^{-5}	4.4×10^{-10}	3.0×10^{-8}
nitrogen	bulk planet	1.5×10^{-6}	1.2×10^{-6}	2.0×10^{-10}
	atmosphere	1.1×10^{-6}	5.8×10^{-7}	1.7×10^{-10}
water	bulk planet	2.0×10^{-9}	2.3×10^{-4}	3.9×10^{-6}
	atmosphere	1.6×10^{-9}	2.5×10^{-9}	2.2×10^{-12}

Finally we have to consider the different types of accretion material. Based on equilibrium calculations and for an atmospheric state derived from an impact atmosphere (Abe & Matsui, 1987) a gas composition is determined by Schaefer & Fegley (2010). The obtained data important to evaluate our assumed atmospheric composition are shown in Table 3 for four different chondrite types (CI, CM are carbonaceous chondrites with very pristine material, L is an ordinary chondrite with a low amount of oxidized iron, and EH is an enstatite chondrite with a high amount of iron and non-oxidized iron). The most pristine material is fond for carbonaceous chondrites of the type CI while the CM chondrites experienced an extensive aqueous alteration. L and EH chondrites contain reducing material and CI and CM produce a neutral composition. For our evaluation we use the CO_2/N_2 ratio, which is given for an early atmosphere by a value ranging between 15-32. A composition of CM and L chondrites produces nearly the same range: 15-33. The same order of agreement was not reached for the ratio H_2O/CO_2 which gives ~ 15 for the early atmosphere and 3-5 for CM and L chondrites. We have not used the EH values since in this case the agreement to early Earth rations becomes worse.

Table 3. Gas compositions of impact generated atmospheres from chondritic planetesimals at 1500 K and 100 bars.

substance	CI	CM	L	EL
H_2O	69.47	73.38	17.43	5.71
CO_2	19.39	18.66	5.08	9.91
N_2	0.82	0.57	0.33	1.85
H_2	4.36	2.72	42.99	14.87
CO	3.15	1.79	32.51	67.00
H_2S	2.47	2.32	0.61	0.18

Now we have determined an early chemical composition and found that the early atmosphere was mainly neutral. However for the formation of life one needs a more reducing environment. Since SN chemistry is hydrogen chemistry the LHB comets could have a more reducing influence. Thus, the retention of the primary atmosphere and the delivery of volatile molecules by LHB comets will increase the reducing character of the Earth atmosphere and improve the chances of life formation. Observations from Schopf (1993) and Brazier et al. (2002) have shown that life on the Earth probably formed somewhere around 3.5×10^9 years or perhaps even earlier (Mojzsis et al., 1996; van Zuilen et al., 2002; Cate & Mojzsis, 2006). Unfortunately, there is not much evidence left from this time to describe the geological state of the planet and the thermodynamic one of its atmosphere.

3 Calculation of Nitrogen Bearing Molecules in SN

We simulate chemical processes in each of the three evolution periods considered in our solar nebula model. We discriminate between

- a quasi-static prestellar core,
- a collapsing protostellar core, and
- an evolving turbulent disk.

Our purpose is to identify chemical species that were incorporated into comets in a sufficiently large number. Especially, we have made great efforts in order to derive a realistic and compact hydrodynamic models to describe the evolutionary periods of the solar nebula.

Table 4. The three phases of the multi-zone solar nebula model.

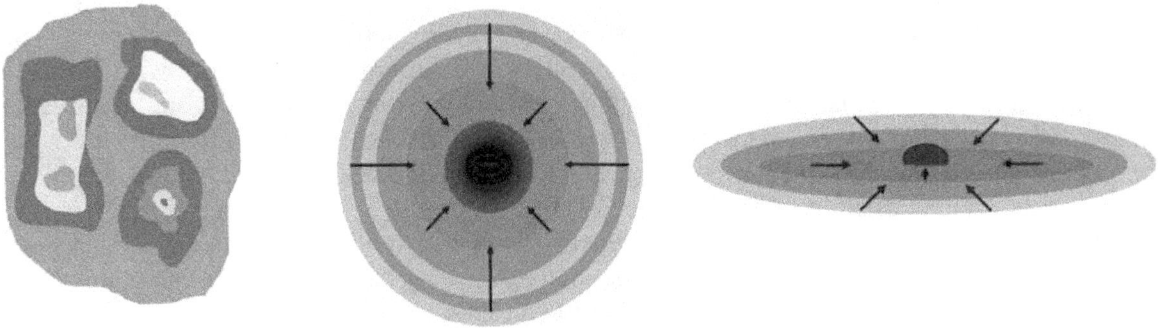

dark cloud with quasi-static prestellar cores

collapsing protostellar core

evolving turbulent disk with protosun

3.1 Quasi-static Prestellar Core

The quasi-static evolution of a prestellar core is modelled with a linear time dependency of the temperature and density. Systematic flow processes are not considered. The negligence of flows and unsteady evolution events such as shock waves or cloud collisions is justified since the temperature and density of the cloud core change over the large time interval of nearly 15 million years. The relative abundances of species i in the gas and ice phase x_i and x_i^*, respectively, are calculated from a set of kinetic equations. The rates for the chemical reactions are computed from data of Woodall et al. (2007) and Aikawa et al. (1997).

Table 5 contains the initial abundances. We have restricted our set of species to compounds having no more than seven atoms. From the calculated abundance evolution we obtain the time dependence of the ratios shown in Figure 2. One recognises an increasing amount of non-polar ice and bounded heavy isotopes in the course of the prestellar core evolution. A large H_2 to H ratio seems to be advantageous for the formation of CO_2 relative to H_2O.

Table 5. Initial abundances relative to hydrogen abundance.

H	H_2	D	He	O	C+	N	Si
0.9	0.1	1.5×10^{-5}	0.14	1.8×10^{-4}	7.3×10^{-5}	2.1×10^{-5}	6.0×10^{-11}

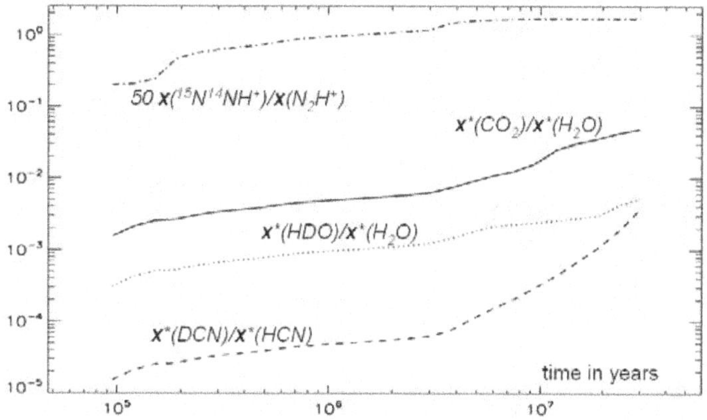

Figure 2. Time dependency of the nitrogen isotope ratio in the gas phase (note the factor of 50 to present all curves in the same figure), the D/H ratios in the ice phase and the CO_2/H_2O ratio for the polar to non-polar ice fraction calculated for the slowly evolving quasi-stationary prestellar core.

3.2 Collapsing Protostellar Core

The gravitational collapse of a cloud core causes the central density to increase over more than 15-16 orders of magnitudes. At the end of this process a stellar core, the T Tauri star, and a young disk have formed in the centre of the solar nebula. Therefore, a numeric simulation of this type of collapse is a complex task.

We have derived an analytical solution to solve the continuity, momentum and Poisson equation for a collapsing cloud core in four radial zones using spherical symmetry. According to Saigo et al. (2008) the spherical symmetry has no serious drawbacks as long as the rotation rate is low 10^{-15} s^{-1}. The mathematics of this solution will be described in a different publication. In Figure 3 we present the calculated radial density, mass, velocity, and temperature profiles at different times. In order to include the influence of the formed protostellar disk we have coupled our collapse solution to the disk model derived by Stahler et al. (1994).

The values of the four radial profiles in Figure 3 are given for an Eulerian grid. However, the computation of the chemical abundance evolution of the gas and ice phase following from the continuity equation of each species can be simplified if one uses a transformation to a Lagrangian grid defined by the initial positions of the gas-ice parcels at the beginning of the collapse. The resulting total time dependencies of the density and temperature are calculated for an inner gas parcel moving from 2.5 to 1.3 AU. In this case the temperatures are high enough to guarantee the loss of the ice phase due to the evaporation of the icy grain mantles. In order to study the temporal progress of depletion of the ice phase species we have computed the ratio of the current to the initial abundance for selected compounds. The obtained values are presented in Figure 4.

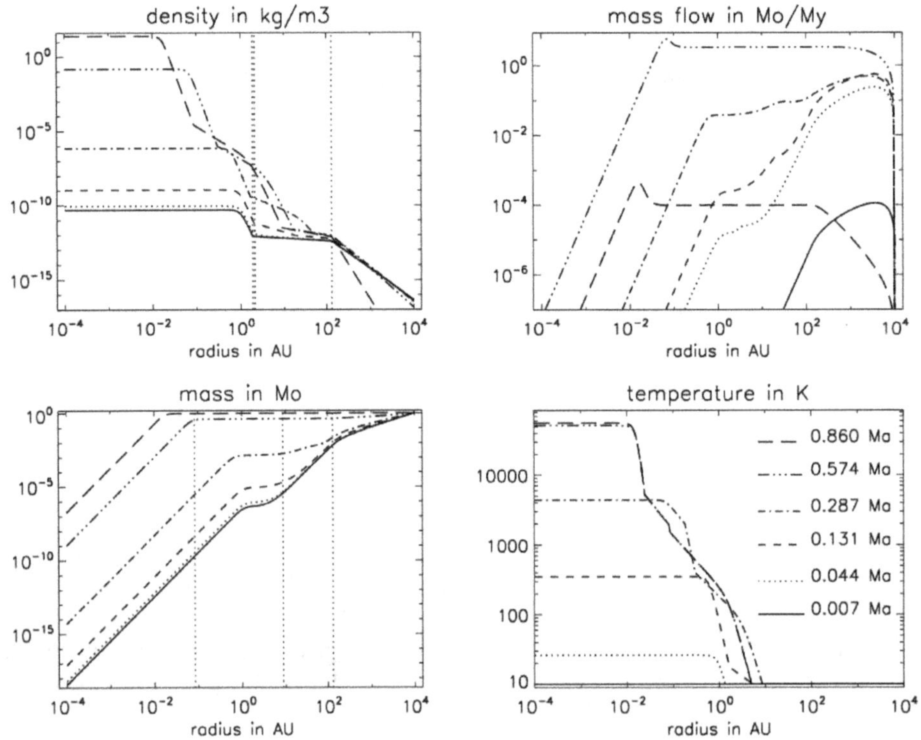

Figure 3. Radial profiles of density, mass, mass flow, and temperature for selected time points calculated with our analytical multi-zone model of the solar nebula. The vertical dotted lines in the left plots show the distribution of the zones at the beginning (upper plot) and at the end of the collapse period (lower plot).

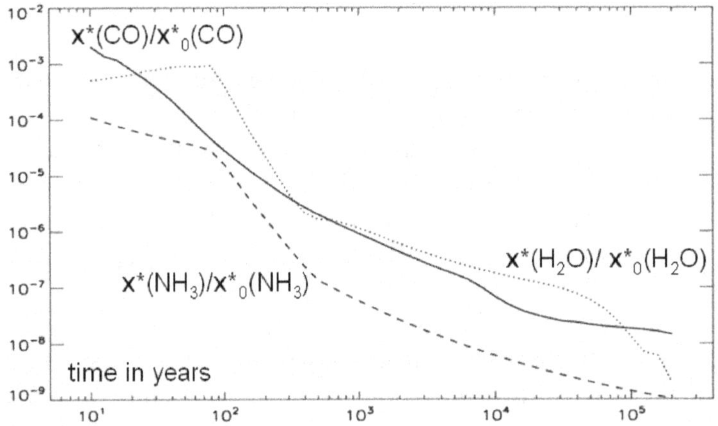

Figure 4. Time dependency of the ratio of the current to the initial abundance for CO, H_2O, and NH_3 calculated for the period of the collapsing protostellar core.

3.3 Evolving Turbulent Disk

The disk model of Stahler et al. (1994), is valid for a young disk only. In order to study the chemical evolution of the gas and ice species in a mature disk we have used the non-stationary model of Davis (2003). This model describes the disk cooling and depletion in the course of its evolution. Due to the gas flow we have to switch to the Lagrangian grid again in order to compute the abundance values. The necessary initial data follow from the final abundance results calculated for the collapse period. In contrast to our collapse model the Davis model is based on axial symmetry. In order to keep a simple radial dependency without angular variations, the relative abundances are derived with respect to the column density. For time intervals much larger than 10^7 the corresponding number density would be less then 0.01 cm^{-3}, i.e. a gas disk is not existent anymore. Therefore, at most 10 million years are of physical interest only. Figure 5 shows the time behaviour of the same ice ratios as seen in Figure 2. However, one recognizes clear differences although in both cases the ice phase abundancies are growing with respect of their initial values. For the evolving disk, there is a superposition of the time dynamics of the disk parameter itself and the time dynamics of the chemical processes. Thus, the shapes of the disk related abundance ratios versus time are less monotonic than the same curves of the prestellar core. Further, disk density of the considered gas parcel decreases whereas core density increases slowly.

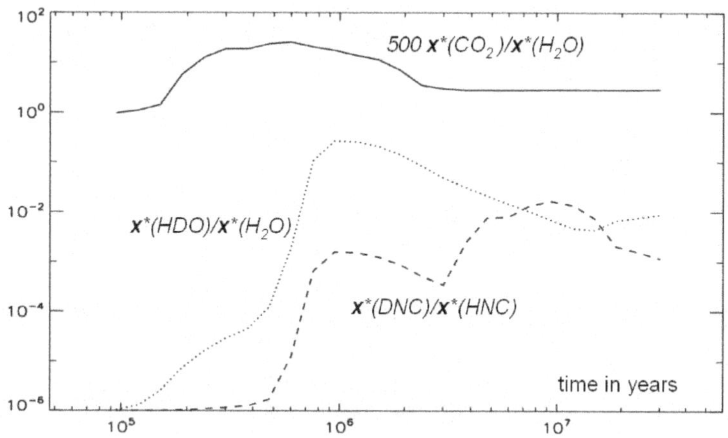

Figure 5. Time dependency of the D/H ratios in the ice phase and the CO_2-H_2O molecular ratio for the polar to non-polar ice fraction calculated for the evolving disk.

4 Conclusion and Outlook

We have motivated the assumption that the ratio of CO_2/N_2 was nearly similar (i.e. ~ 15) for the atmospheres of Earth, Mars, and Venus directly after planet formation. In order to calculate the primarily reducing contribution of LHB comets to the Earth atmosphere we have combined a hydrodynamical model of the SN with a kinetic model to simulate the chemical evolution. Especially we have developed an analytical solution for the collapse period that gives the chance to simulate this process very efficiently. Both models, the hydrodynamic and the chemical, were thoroughly tested to guarantee the consistency of merging the evolution periods of the solar nebula using the transition from an Eulerian to a Lagrangian grid. However, the transition from the spherically collapsing cloud core to the disk is complicated and further research needs to be done for the transition between the different temperature models.

From chemical calculations a distinct difference between disk and prestellar core chemistry becomes conspicuously. It is related to the higher dynamics in the disk on the one hand and to its complex initial chemical state on the other. The effects of both phenomenons are entangled and further research needs to be done to investigate their influence independently. Figure 6 allows to estimate the amount of nitrogen bearing molecules. According to Gomes et al. (2005) nearly 10^{22} g of material from LHB comets have reached the Earth surface. The ice formed by soluble matter amounts 25 - 33%. Thus one gets 2.5×10^{21} g and the corresponding N amount is not more than 1 - 5 % giving $\geq 2.5\times10^{19}$ g (see indications in Figure 6). If we compare this contribution with the current mass of the biosphere (10^{19} g). Consequently, the LHB comets might have delivered an amount of reducing and soluble material important for life formation in a otherwise neutral atmosphere. In a next study we will calculate the amount of reducing gasses from the SN retained by the Earth during its formation process.

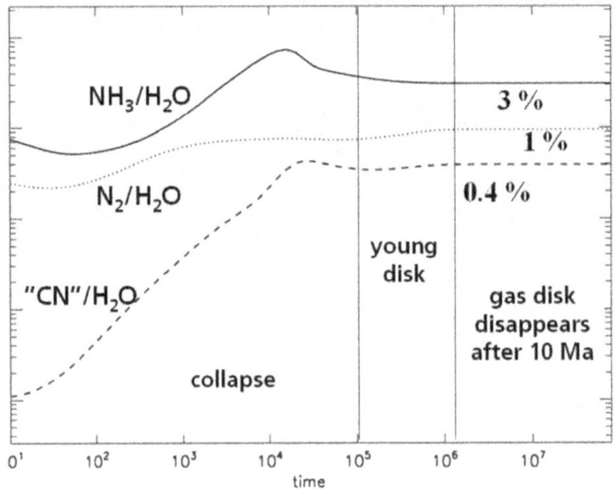

Figure 6. Abundance ratios versus time in years. The evolution of the three major nitrogen bearing molecules in the ice phase of the solar nebula is illustrated, whereby "CN" stands for the abundance of HCN + HC_3N.

References

Abe, Y., 1988, Conditions Required for Formation of Water Ocean on an Earth-Sized Planet. Abstracts of Lunar and Planetary Science Conference, v.19, page 1

Aikawa, Y.; Umebayashi, T.; Nakano, T.; Miyama, S.M. 1997, Evolution of Molecular Abundance in Protoplanetary Disks. Astrophysical Journal Letters v.486, pp.L51-L54.

Armitage, Philip J.; Clarke, Cathie J.; Palla, Francesco, 2003, Dispersion in the lifetime and accretion rate of T Tauri discs. Monthly Notice of the Royal Astronomical Society, v. 342, pp. 1139-1146.

Brasier, M. D., Green, O. R., Jephcoat, A. P., Kleppe, A. K., Kranendonk, J. V., Lindsay, J. F., Steele, A. and Grassineau, N. V., 2002, Questioning the evidence for Earth's oldest fossils, Nature, v. 416, pp. 76-81.

Benz, W., Slattery, W. L., Cameron, A. G. W., 1988, Collisional stripping of Mercury's mantle. Icarus, vol. 74, pp. 516-528.

Brauer, F., Henning, Th., Dullemond, C. P., 2008, Planetesimal formation near the snow line in MRI-driven turbulent protoplanetary disks. Astronomy and Astrophysics, v. 487, Issue 1, 2008, pp.L1-L4

Cates, N. L.; Mojzsis, S. J., 2006, Geochronology and Geochemistry of a Newly Identified Pre-3760 Ma Supracrustal Sequence in the Nuvvuagittuq Belt, Québec, Canada. 37th Annual Lunar and Planetary Science Conference, March 13-17, 2006, League City, Texas, abstract no.1948.

Clément, D., Mutschke, H., Klein, R., Jäger, C., Dorschner, J., Sturm, E., Henning, Th., 2005, Detection of Silicon Nitride Particles in Extreme Carbon Stars. The Astrophysical Journal, v. 621, pp. 985-990.

Davis, S.S. 2003, A Simplified Model for an Evolving Protoplanetary Nebula. Astrophysical Journal, v.592, pp. 1193-1200.

Drake, Michael J.; Righter, Kevin, 2002. Determining the composition of the Earth. Nature, v. 416, pp. 39-44.

Genda, Hidenori and Abe, Yutaka, 2003, Survival of a proto-atmosphere through the stage of giant impacts: the mechanical aspects. Icarus, v. 164, pp. 149-162

Goody & Walker, 1972, Planetary Atmospheres. Englewood Cliffs, NJ (USA): Prentice-Hall,

Gomes, R.; Levison, H. F.; Tsiganis, K.; Morbidelli, A., 2005, Origin of the cataclysmic Late Heavy Bombardment period of the terrestrial planets. Nature, v. 435, pp. 466-469.

Greenberg, J. Mayo and Li, Aigen, 1999, Morphological Structure and Chemical Composition of Cometary Nuclei and Dust. Space Science Reviews, v. 90, pp. 149-161

Hsieh, Henry H. and Jewitt, David, 2006, A Population of Comets in the Main Asteroid Belt. Science, v. 312, pp. 561-563

Johansen, Anders, Oishi, Jeffrey S., Mac Low, Mordecai-Mark, Klahr, Hubert, Henning, Thomas, Youdin, Andrew, 2007, Rapid planetesimal formation in turbulent circumstellar disks. Nature, v. 448, pp. 1022-1025.

Kaltenegger, L.; Traub, Wesley A.; Jucks, Kenneth W., 2007, Spectral Evolution of an Earth-like Planet, The Astrophysical Journal, v. 658, pp. 598-616.

Kasting, J. F., 2004, When Methane Made Climate; July 2004; Scientific American Magazine; 8 Page(s)

Kerridge, J. F., 1985, Carbon, hydrogen and nitrogen in carbonaceous chondrites Abundances and isotopic compositions in bulk samples. Geochimica et Cosmochimica Acta, v. 49

Kuramoto & Matsui, 1993, Core formation, wet early mantle, and H2O degassing on early Mars. In Lunar and Planetary Inst. Workshop on Early Mars: How Warm and How Wet?, Part 1 pp. 15-17

Lammer, Helmut; Kasting, James F.; Chassefière, Eric; Johnson, Robert E.; Kulikov, Yuri N.; Tian, Feng, 2008, Atmospheric Escape and Evolution of Terrestrial Planets and Satellites. Space Science Reviews, v. 139, pp. 399-436

Lammer, H.; Lichtenegger, H. I. M.; Biernat, H. K.; Erkaev, N. V.; Arshukova, I. L.; Kolb, C.; Gunell, H.; Lukyanov, A.; Holmstrom, M.; Barabash, S, Zhang, T. L., Baumjohann, W., 2006, Loss of hydrogen and oxygen from the upper atmosphere of Venus. Planetary and Space Science, v. 54, pp. 1445-1456.

Lee, M. R., Russell, S. S., Arden, J. W., Pillinger, C. T., 1995, Nierite (Si3N4), a new mineral from ordinary and enstatite chondrites. Meteoritics, v. 30, pp. 387

Lide, David R., 2009, CRC Handbook of Chemistry and Physics. Taylor & Francis,

Lunine, J. I. and Stevenson, D. J., 1985, Thermodynamics of clathrate hydrate at low and high pressures with application to the outer solar system. Astrophysical Journal Supplement Series, v. 58, July 1985, pp. 493-531

Lyra, W., Johansen, A., Klahr, H., Piskunov, N., 2008, Embryos grown in the dead zone. Assembling the first protoplanetary cores in low mass self-gravitating circumstellar disks of gas and solids. Astronomy and Astrophysics, v. 491, pp.L41-L44

Macke et al., 2009, EH and EL Enstatite Chondrite Physical Properties: No Difference in Iron Content. 72nd Annual Meeting of the Meteoritical Society, held July 13-18, 2009 in Nancy, France. Published in Meteoritics and Planetary Science Supplement., p.5047

Matsui, T. and Abe, Y., 1987, Evolutionary tracks of the terrestrial planets. Earth, Moon, and Planets, v. 39, pp. 207-214.

Matsui, T., 1993, Early evolution of the terrestrial planets: accretion, atmosphere formation, and thermal history. Primitive solar nebula and origin of planets, p. 545 - 559

Mojzsis, S.J., Arrhenius, G., McKeegan, K.D., Harrison, T.M., Nutman, A.P. and Friend, C.R.L., 1996, Evidence for life on Earth by 3800 million years ago, Nature v. 384, pp. 55-59.

Morbidelli, A., Chambers, J., Lunine, J. I., Petit, J. M., Robert, F., Valsecchi, G. B., Cyr, K. E., 2000, Source regions and time scales for the delivery of water to Earth. Meteoritics & Planetary Science, v. 35, pp. 1309-1320

O'Brien, David P., Morbidelli, Alessandro, Levison, Harold F., 2006, Terrestrial planet formation with strong dynamical friction. Icarus, v. 184, pp. 39-58

Pidwirny, M., 2006, Greenhouse effect. Encyclopaedia of Earth, Editor Cutler J. Cleveland. Topic Editor, Hanson, H.. Environmental Information Coalition (EIC) of the National Council for Science and the Environment (NCSE), Washington D.C., USA.

Prinn, R. G. and Fegley, B., 1987, The atmospheres of Venus, earth, and Mars - A critical comparison. Annual review of earth and planetary sciences. Volume 15 Palo Alto, CA, Annual Reviews, Inc., 1987, p. 171-212

Saigo, K., Tomisaka, K., and Matsumoto, T., 2008, Evolution of First Cores and Formation of Stellar Cores in Rotating Molecular Cloud Cores. The Astrophysical Journal, v. 674, pp. 997-1014.

Schaefer, L. and Fegley, B., 2010, Chemistry of atmospheres formed during accretion of the Earth and other terrestrial planets, Icarus, v. 208, pp. 438-448.

Schopf, J. W., 1993, Microfossils of the Early Archean Apex Chert: New Evidence of the Antiquity of Life, Science, v. 260, pp. 640-646.

Sicilia-Aguilar, A., Hartmann, L. W., Watson, D., Bohac, C., Henning, T., Bouwman, J., 2007, Silicate Dust in Evolved Protoplanetary Disks: Growth, Sedimentation, and Accretion. The Astrophysical Journal, v. 659, pp. 1637-1660

Stahler, S.W., Korycansky, D. G., Brothers, M.J., and Touma, J., 1994, The early evolution of protostellar disks. The Astrophysical Journal, v.431, pp. 341-358.

Stimpfl, M., Drake, M. J., de Leeuw, N. H., Deymier, P., Walker, A. M., 2006, Effect of composition on adsorption of water on perfect olivine surfaces. Geochimica et Cosmochimica Acta, v. 70, pp. A615-A615.

Trigo-Rodriguez, J. M. and Blum, J., 2009, Tensile strength as an indicator of the degree of primitiveness of undifferentiated bodies. Planetary and Space Science, v. 57, pp. 243-249

Wilde, S. A., Valley, J. W., Peck, W. H. and Graham, C. M., 2001, Evidence from detrital zircons for the existence of continental crust and oceans on the Earth 4.4 Gyr ago, Nature 409, pp. 175-178.

Woodall, J., Agúndez, M., Markwick-Kemper, A. J., and Millar, T. J., 2007, The UMIST database for astrochemistry 2006. Astronomy and Astrophysics, v.466, pp.1197-1204.

van Zuilen, M.; Lepland, A.; Arrhenius, G., 2002, Reassessing the evidence for the earliest traces of life. Nature, v. 418, pp. 627-630.

Zahnle, K., 1998, Origins of Atmospheres. Origins, ASP Conference Series, v. 148, p.364

Zubko, V., Dwek, E., and Arendt, R.G., 2004, Interstellar Dust Models Consistent with Extinction, Emission, and Abundance Constraints. The Astrophysical Journal Supplement Series, v.152, pp.211-249.

Modeling the Entry of Micrometeoroids into the Atmospheres of Earth-like Planets

A. R. Pevyhouse • M. E. Kress

Abstract The temperature profiles of micrometeors entering the atmospheres of Earth-like planets are calculated to determine the altitude at which exogenous organic compounds may be released. Previous experiments have shown that flash-heated micrometeorite analogs release organic compounds at temperatures from roughly 500 to 1000 K [1]. The altitude of release is of great importance because it determines the fate of the compound. Organic compounds that are released deeper in the atmosphere are more likely to rapidly mix to lower altitudes where they can accumulate to higher abundances or form more complex molecules and/or aerosols. Variables that are explored here are particle size, entry angle, atmospheric density profiles, spectral type of the parent star, and planet mass. The problem reduces to these questions: (1) How much atmosphere does the particle pass through by the time it is heated to 500 K? (2) Is the atmosphere above sufficient to attenuate stellar UV such that the mixing timescale is shorter than the photochemical timescale for a particular compound? We present preliminary results that the effect of the planetary and particle parameters have on the altitude of organic release.

Keywords atmospheric entry · micrometeor · modeling · organic chemistry

1 Introduction

Micrometeorites ~200 μm in diameter carry most of the incoming mass to the modern Earth, approximately 30 million kg annually [2]. Love and Brownlee (1991) [3] found that micrometeors in this size range experience severe heating upon atmospheric entry. Peak heating occurs at an altitude of > 85 km within seconds of atmospheric entry, typically to temperatures in excess of 1600 K, sufficient to melt silicate and metals [3].

Recent experiments have simulated the flash-heating experienced by micrometeors upon atmospheric entry [1], [4]. Both of these groups found that methane is released, and Kress et al. [1] also found that other light hydrocarbons and a variety of more complex organics are released at temperatures of ~ 500 to 1000 K. In the current study, we identify the altitudes at which these temperatures are reached, which is an essential first step to determining the ultimate fate of these compounds.

The influence that PAHs and methane could have on a planetary atmosphere depends on the altitude at which they are released from an incoming particle. The altitude at which a molecule is released determines its fate. Vertical mixing will bring a molecule deeper down into the atmosphere, where its photochemical lifetime is longer. The photochemical lifetime of a substance is the time it takes for destruction mechanisms to reduce its concentration to 1/e its original amount. The deeper in the atmosphere an organic compound is released, the greater the probability of it being vertically mixed. Methane, CH_4, for example, will be broken into residual compounds by photolysis if released above Earth's stratopause due to Lyman-alpha radiation (λ = 121.6 nm). Methane also is destroyed at this

A. R. Pevyhouse (✉) • M. E. Kress
Department of Physics and Astronomy, San Jose State University, San Jose, CA 95192. Phone: 650-219-9502; E-mail: apevyhouse@santarosa.edu

altitude by reactions with O(^{1}D) and OH [5]. After several steps, these reactions will convert methane to CO_2. However, if released at an altitude of 70 km or lower, methane has a long enough photochemical lifetime to allow mixing [5].

We first apply the atmospheric entry model to the Earth. We then extend this study to plausible Earth-like planets of different masses and atmospheric densities to identify the parameter space in which micrometeors release organics close to a planet's surface. Varying planet mass was found to not result in organics being ablated under a greater portion of atmosphere.

We find that, for the modern Earth, organics are typically released at an altitude such that the timescale for methane to mix lower into the atmosphere is very long compared to its photochemical destruction timescale at that altitude [5].

2 Modeling the Atmospheric Entry of Micrometeorites

In this study, infalling micrometeorites were simulated numerically to generate temperature profiles for a variety of particle sizes and entry parameters. Entry parameters of interest were initial velocity and entry angle. The physics of atmospheric entry is that of Love and Brownlee (1991) [3].

Numerical modeling using an Euler algorithm was done to simulate atmospheric entry of micrometeorites. This model takes a continuous evaporation approach while the particle is treated as an isothermal sphere of density ρ_{met} = 3 g/cm^3. Incoming micrometeorites are heated due to collisions with atmospheric molecules. Particle temperature is determined by balancing the power imparted to it from atmospheric molecules, P_{in}, to the rate at which thermal energy is being dissipated by radiative and evaporative mechanisms, such that

$$P_{in} = 0.5\rho_{atm} s v^3 \tag{1}$$

and

$$T = \left(\frac{P_{in}}{4\pi r^2 \sigma \varepsilon}\right)^{1/4} \tag{2}$$

where ε is the emissivity of the particle, T is the particle's temperature, σ is the Stefan-Boltzmann constant, s is the particle's geometric cross section, v is the particle's velocity with respect to the atmosphere and ρ_{atm} is the density of the atmosphere (a function of altitude). The change in velocity due to atmospheric drag and gravity is

$$d\mathbf{v} = \left(-0.75\frac{\rho_{atm} v^2}{\rho_{met} r}\hat{v} + \mathbf{g}\right) dt \quad . \tag{3}$$

We first reproduced the Love and Brownlee (1991) [3] results using the United States Standard Atmosphere of 1976 [6] as the atmospheric model. These calculations served as a benchmark for those for hypothetical earth-like planets, whose mass and surface atmospheric density were treated as free parameters, and whose atmospheric density was assigned a simple exponential decay law.

To estimate the altitude at which organic compounds may be released in the atmospheres of hypothetical Earth-like planets, atmospheric pressure and planetary mass were treated as free parameters. The atmospheric density profiles for these worlds were approximated by assigning a simple exponential decay function. This exponential decay model treated the atmosphere as isothermal at a temperature of 288.15 K with a constant molecular weight of 28.97 g/mol. Using this atmospheric profile, the effect of varying a planet's mass and atmospheric pressure on the altitude at which a micrometeorite first reaches 500 K was investigated. Results obtained for a world with 1 Earth mass and 1 atm surface pressure were compared against the results obtained using the U.S. 1976 Standard Atmosphere [6].

3 Results

The altitude at which volatile organic compounds are released is defined as the altitude range for which an incoming particle would be between 500 and 1000 K.

Results for a 1 Earth mass planet with 1 atm atmospheric pressure were compared to those of Love and Brownlee (1991) [3] who used the 1976 Standard Atmosphere as the atmospheric model. Use of an exponential decay function to model a planetary atmosphere consistently resulted in a higher calculated altitude of organic release compared to the altitude calculated using the U.S. 1976 Standard Atmosphere [6] (Figure 1). Heating rates also differed between the two atmospheric models. A 100 μm diameter particle entering at 80 deg and 20 km/s experienced a heating rate of 56 K/s under the exponential decay model compared to 43 K/s using the U.S. 1976 Standard Atmosphere [6]. Heating rates were determined between 500 to 1000 K.

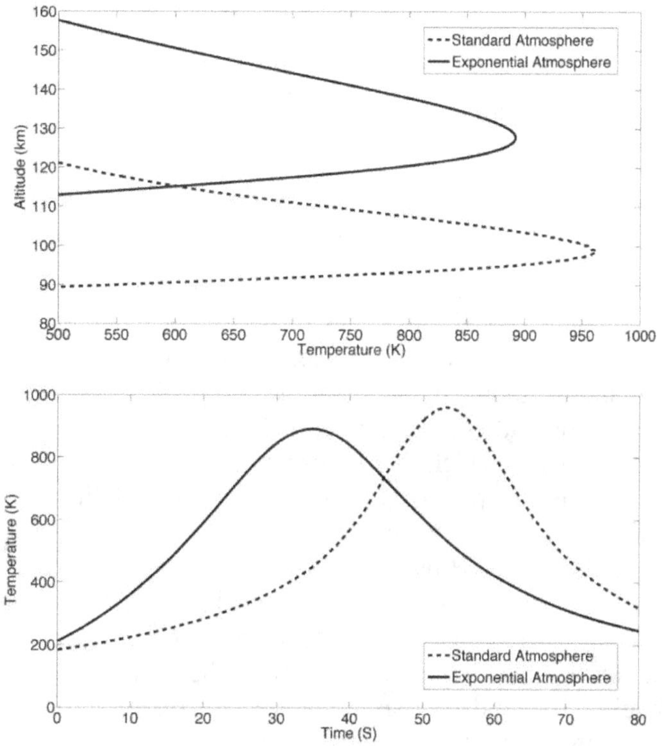

Figure 1. Comparison of results from the U.S. 1976 Standard Atmosphere [6] and exponential decay model for a 50 μm diameter particle entering at 80 deg and 12 km/s. Top: Particle temperature as a function of altitude. Bottom: Particle temperature as a function of time. Note that the particle reached its peak temperature later when in the standard atmosphere compared to an atmosphere whose pressure is exponentially decaying.

Figure 2 shows the dependence of organic release altitude on planetary mass and atmospheric pressure. Planetary mass was shown to have a greater effect on the altitude of organic release compared to planetary surface pressure. A micrometeorite falling through the atmosphere of a planet with a mass of 0.1 Earth mass and 0.1 atm surface pressure first reached 500 K at an altitude of 345 km. Increasing atmospheric pressure to 10 atm increased this altitude to 494 km. The same difference in atmospheric surface pressure resulted in only an 11 km difference for a planet of 10 Earth masses.

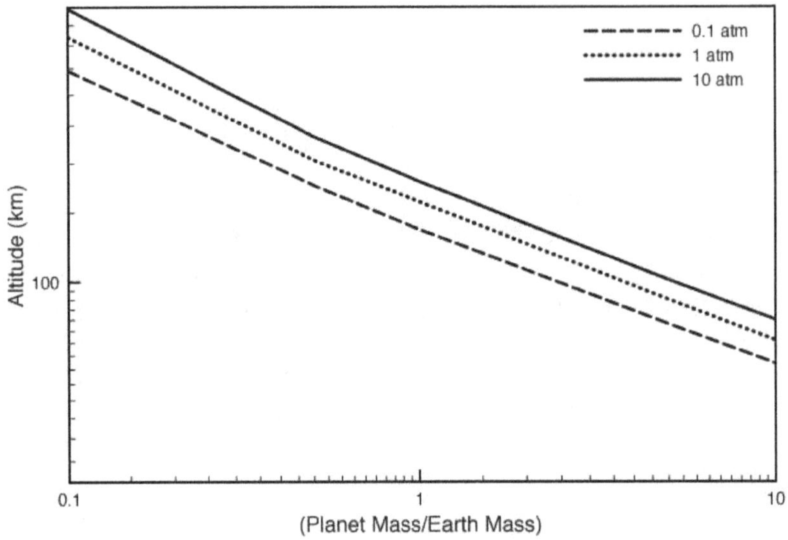

Figure 2. The altitude at which a 100 μm diameter particle first reaches 500 K as a function of planetary mass. The initial velocity of the particle is 12 km/s with an entry angle of 45°. Note the effect of increasing planetary mass on lowering the altitude at which a particle first reaches 500 K.

4 Discussion

Approximately 2×10^7 kg/yr of extraterrestrial material is deposited into Earth's atmosphere each year. The amount of organic carbon deposited can be estimated to be 10% of this total [2]. The level of ablated micrometeoritic organic compounds in a planetary atmosphere is determined by the competing rates of material deposition and degradation. Degradation of organic molecules occurs by photolysis due to exposure to solar UV radiation and chemical reactions with atmospheric molecules. The rate of this degradation depends on the altitude at which these organic compounds are released.

Uncertainties in the determination of the altitude range volatile organics are released from incoming micrometeorites originate from four factors.

The first factor is the Love and Brownlee (1991) [3] model. It does not take into account that meteorites have more than one phase. Micrometeorites in this model are treated as generic silicates. Therefore, an organic phase that evaporates at lower temperatures compared to silicates is not taken into consideration. The limitation of this model comes from the physics of the micrometeorite being determined to the 90% level by the silicates that are present. In reality, the loss of organics and ice will keep the particle cooler for longer due to the energy used for the phase change of these components. This lower temperature will allow for the particle to reach a lower altitude before reaching 500 K. The use of the Love and Brownlee (1991) [3] should therefore be considered as providing a conservative estimate on the altitude at which organics are released from micrometeorites.

The atmospheric model used is the second factor. The size of micrometeorites makes them very sensitive to any changes in atmospheric density. The difference in results between the U.S. 1976 Standard Atmosphere [6] and the Exponential Decay model makes the point that better representations of exoplanet atmospheres should be used.

The other two factors are the estimation of a heat transfer coefficient and lack of a rate equation for the release of volatile organics. A heat transfer coefficient will determine the percentage of input power that goes into heating the particle. A rate equation for the evaporation of organics determines were in the temperature range 500 to 1000 K organics are released. This is critical in determining the lower altitude boundary a particle will release organics. The need for an organic evaporation rate equation is discussed below.

The rate at which a particle is heated will determine the range of time organic compounds are released. Slow heating rates will give more time for volatile organics to be outgassed from a particle compared to quicker rates. Heating too quickly can result in organic compounds in the particle to be transformed to char before they are able to diffuse out of the particle.

A particle entering at 12 km/s and an angle of 0 deg was found to have a heating rate ~ 300 K/s. The same 12 km/s particle entering at 80 deg had a reduced heating rate of ~ 40 K/s. The heating rate of 500 K/s used by Cody was higher than any rate found for particles with an initial velocity less than 20 km/s. Such a high heating rate should be considered a worse case scenario of heating. Under such a high rate of heating, it is unknown if the volatile organics contained in a particle are all outgassed before being charred.

Further experiments on Murchison samples are needed to determine a rate equation for the evaporation of organic compounds. A rate equation will give insight into where in the temperature range of organic release organics are ablated. It is unknown if the majority of organics are released when a particle reaches 500 K, volatilization of organics is a continuous process over the entire temperature range of organic release, or if the majority of organics are ablated as the particle approaches 1000 K.

4.1 Effect of Stellar Class on Lyman-alpha Exposure

The further into a planetary atmosphere micrometeorites release organics, the greater the protection from lyman-alpha radiation. Lyman-alpha will degrade organic molecules on a time scale less than atmospheric vertical mixing times if released under too little atmosphere. The intensity of planetary exposure to Lyman-alpha depends on the temperature of a planet's home star and its distance to it. The liquid water habitable zone (LW-HZ) is defined as the region in space around a star in which a planet would be able to maintain liquid water on its surface [7]. Figure 3 shows the continuum flux of Lyman-alpha through the LW-HZ of F0, G2, and M0 stars as defined by Kasting et al. [7].

M-stars comprise about 75% of all main-sequence stars. Their hydrogen burning lifetimes are much longer than G2V stars like our Sun. Comparison of the intensity of Lyman-alpha radiation between a G star and an inactive M dwarf indicates ~ 10^{-7} reduction in Lyman-alpha intensity. This reduction in Lyman-alpha could slow the rate of rate of organic degradation in the atmosphere on an M-star planet. However, too low a level of UV radiation has been thought to inhibit the biogenesis of complex macromolecules. The volatile UV output from M-star flares have been hypothesized to be needed for the synthesis of large complex macromolecules [8].

The spectral distribution of radiation incident on an M-star planet has been theorized to result in a thicker ozone layer compared to the Earth [9]. A broader ozone layer could increase the photochemical lifetime of ablated molecules.

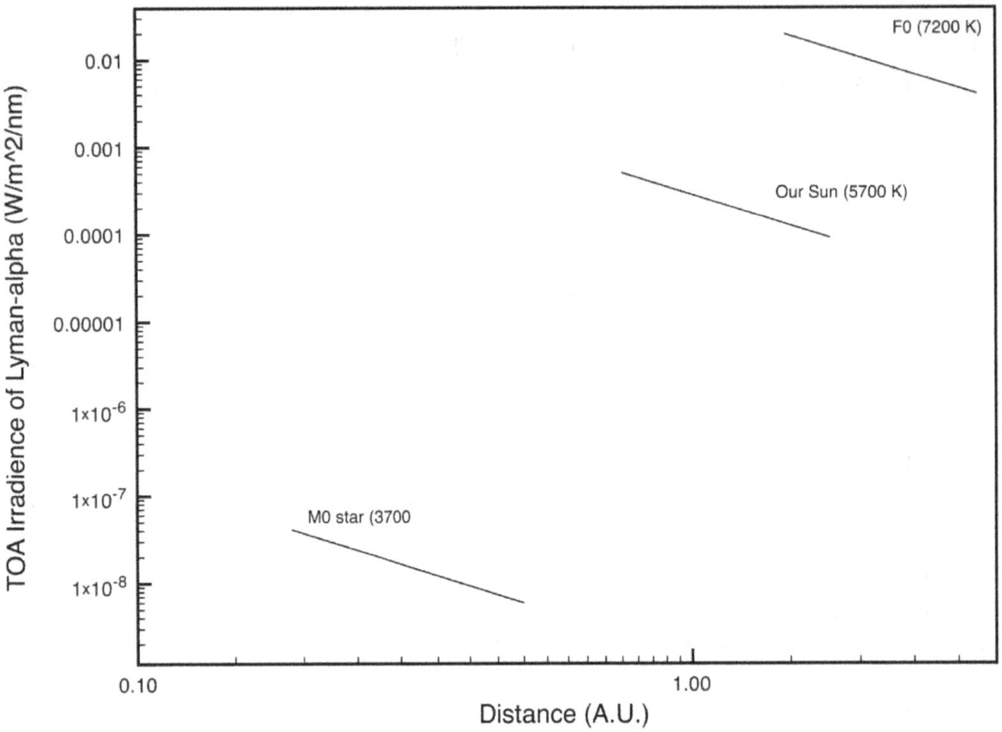

Figure 3. Irradiance at the top of the atmosphere (TOA) of Lyman-alpha for M0, G2, and F0 stars. Irradiance values were calculated from Planck's function. The range of habitable zone for each stellar class follows those published by Kasting et al. [7]

Another source of protection could come from the high probability of planets with the LW-HZ being tidally locked. Synchronous rotation does not necessarily mean atmospheric freeze out [10]. Therefore, the side of the planet always facing away from the star could provide a protected environment for ablated volatile organics. Future work should include modeling atmospheric mixing on tidally locked planets to investigate further the micrometeoritic contribution of volatile organics to these worlds.

5 Conclusion

For organic compounds to reach altitudes were exposure to UV radiation is low enough that it will not degrade, the compounds need to be either photochemically stable (e.g. PAHs) or the parent micrometeorite reaches 500-1000 K at lower altitudes. Although survival of methane in our modern atmosphere looks grim, that does not mean the release of organics in other atmospheres is not important. Smaller stars radiating less UV than our Sun may provide a longer time frame for ablated material to be vertically mixed into the atmosphere. At constant planetary density, increasing planet mass lowers the altitude 500 K and is first reached by an incoming particle but does not necessarily result in organic ablation occurring under a greater percentage of a planets atmosphere.

The need for atmospheric models of exoplanets was demonstrated in this study. Results differed by 35 km in altitude between the U.S. 1976 Standard Atmosphere [6] and exponential decay model atmosphere. This is due to the exponential decay model calculating a denser atmosphere compared to the U.S 1976 Standard Atmosphere [6] for altitudes above 100 km. This result was independent of entry angle for a 50 μm diameter particle entering at 12 km/s.

Progress into the micrometeoritic contribution of volatile organics to the atmosphere of planets and moons has been made in this study. Heating rates for further lab experiments have been clarified as well as the need to determine a rate equation for the release of volatile organics. Determination of an upper altitude for when a particle first reaches 500 K under a worst case scenario of heating has been made. Although progress has been made, further work needs to be done in three main areas: (1) determine a rate equation for the evaporation of organics under different rates of heating. (2) investigate the altitude range a particle first reaches 500 K while varying the heat transfer coefficient. (3) use the exponential model to simulate atmospheres with various combinations of atmospheric temperature and pressure that are favorable for liquid water to be present on a planetary surface. This will allow the study to be extended to a broader variety of exoplanets.

Acknowledgements

MEK and ARP acknowledge research support from the NASA Astrobiology Institute's Virtual Planetary Laboratory (PI: V. Meadows).

References

1. M.E. Kress, C.L. Belle, G.C. Cody, A.R. Pevyhouse and L.T.Iraci, Atmospheric chemistry of micrometeoritic organic compounds, NASA Conference Proceedings from Meteoroids 2010, D. Janches, editor (2010)
2. S.G. Love and D.E. Brownlee, A direct measurement of the terrestrial mass accretion rate of cosmic dust, Science, 262, 550-552 (1993)
3. S.G. Love and D.E. Brownlee, Heating and thermal transformation of micrometeoroids entering the earth's atmosphere, Icarus, 89, 26-43 (1991)
4. R.W. Court and M.A. Sephton, Investigating the contribution of methane produced by ablating micrometeorites to the atmosphere of Mars, Earth and Planetary Science Letters, 288, 382-385 (2009)
5. G.P. Brasseur and S.Solomon, Aeronomy of the Middle Atmosphere, 3rd edn. D.Reidel Publishing Company (1984)
6. R.A. Minzner, The 1976 Standard Atmosphere Above 86-km Altitude. NASA SP-398, NASA Special Publication, 398 (1976)
7. J.F. Kasting and D.P. Whitmire and R.T. Reynolds, Habitable Zones around Main Sequence Stars, Icarus, 101, 108-128 (1993)
8. A.P. Buccino and G.A. Lemarchand and P.J.D Mauas, UV habitable zones around M stars, Icarus, 192, 582-587 (2007)
9. A. Segura and J.F. Kasting and V. Meadows and M. Cohen and J. Scalo and D.Crisp and R.A.H Butler and G. Tinetti, Biosignatures from Earth-Like Planets Around M Dwarfs, Astrobiology, 5, 706-725 (2005)
10. M. Joshi, Climate Model Studies of Synchronously Rotating Planets, Astrobiology, 3, 415-427 (2003)

A Numerical Study of Micrometeoroids Entering Titan's Atmosphere

M. Templeton • M. E. Kress

Abstract A study using numerical integration techniques has been performed to analyze the temperature profiles of micrometeors entering the atmosphere of Saturn's moon Titan. Due to Titan's low gravity and dense atmosphere, arriving meteoroids experience a significant "cushioning" effect compared to those entering the Earth's atmosphere. Temperature profiles are presented as a function of time and altitude for a number of different meteoroid sizes and entry velocities, at an entry angle of 45°. Titan's micrometeoroids require several minutes to reach peak heating (ranging from 200 to 1200 K), which occurs at an altitude of about 600 km. Gentle heating may allow for gradual evaporation of volatile components over a wide range of altitudes. Computer simulations have been performed using the Cassini/Huygens atmospheric data for Titan.

Keywords micrometeoroid · Titan · atmosphere

1 Introduction

On Earth, incoming micrometeoroids (~100 μm diameter) are slowed by collisions with air molecules in a relatively compact atmosphere, resulting in extremely rapid deceleration and a short heating pulse, often accompanied by brilliant meteor displays. On Titan, lower gravity leads to an atmospheric scale height that is much larger than on Earth. Thus, deceleration of meteors is less rapid and these particles undergo more gradual heating. This study uses techniques similar to those used for Earth meteoroid studies [1], exchanging Earth's planetary characteristics (e.g., mass and atmospheric profile) for those of Titan. Cassini/Huygens atmospheric data for Titan were obtained from the NASA Planetary Atmospheres Data Node [4].

The objectives of this study were 1) to model atmospheric heating of meteoroids for a range of micrometeor entry velocities for Titan, 2) to determine peak heating temperatures and rates for micrometeoroids entering Titan's atmosphere, and 3) to create a general simulation environment that can be extended to incorporate additional parameters and variables, including different atmospheric, meteoroid and planetary data.

The micrometeoroid entry simulations made using Titan atmospheric data assume that, as on Earth, micrometeors are heated by collision with molecules in the atmosphere. Unlike on Earth where heating pulses last a few seconds and reach temperatures sufficient to melt silicates (> 1600 K [1]), micrometeors on Titan experience a more gradual thermal exchange lasting several minutes and the particles do not reach such high temperatures. The long duration of this gradual heating and cooling may allow ices and volatile organic species (such as small PAHs) to be evaporated throughout Titan's upper atmosphere.

M. Templeton (✉) • M. E. Kress
Dept. of Physics & Astronomy, San José State University, San José, CA 95192-0106. Phone: +1-408-924-5255; E-mail: templeton100@gmail.com

2 Atmospheric Entry Model

The method used in these simulations is that of Love & Brownlee [1] for micrometeoroids entering Earth's atmosphere. Meteoroids are assumed to be spherical and of uniform composition and density, $\rho_{met} = 3$ g/cm^3, with a starting radius r of 100 μm and an entry angle of 45°. g is the acceleration due to gravity for Titan, 1.352 m/s^2. A full two-dimensional simulation is performed to correctly account for Titan's curvature.

The change in velocity due to atmospheric drag and gravity is

$$d\mathbf{v} = \left(-0.75 \frac{\rho_{atm} v^2}{\rho_{met} r} \hat{v} + \mathbf{g} \right) dt \tag{1}$$

where ρ_{atm} is the local density of Titan's atmosphere calculated from the Huygens probe's pressure and temperature data and v represents the velocity of the meteoroid with respect to the atmosphere. Heating of meteoroids is due to the impacts with atmospheric molecules, in this case primarily nitrogen and methane. The rate of energy transfer, P_{in}, to the meteoroid is described by:

$$P_{in} = 0.5 \rho_{atm} s v^3 \tag{2}$$

where s is the geometric cross section of the meteoroid under study. The temperature T of the particle is determined by a balance of frictional heating and radiative cooling:

$$T = \left(\frac{P_{in}}{4\pi r^2 \sigma \varepsilon} \right)^{1/4} \tag{3}$$

where σ is the Stefan-Boltzman constant and ε is the meteoroid's emissivity.

Atmospheric data were obtained from NASA's Planetary Data System Atmospheres Node website. The data set id is HP-SSA-HASI-2-3-4-MISSION-V1.1 [4]. Figure 1 shows a plot of atmospheric temperature versus altitude for the combined Huygens data set.

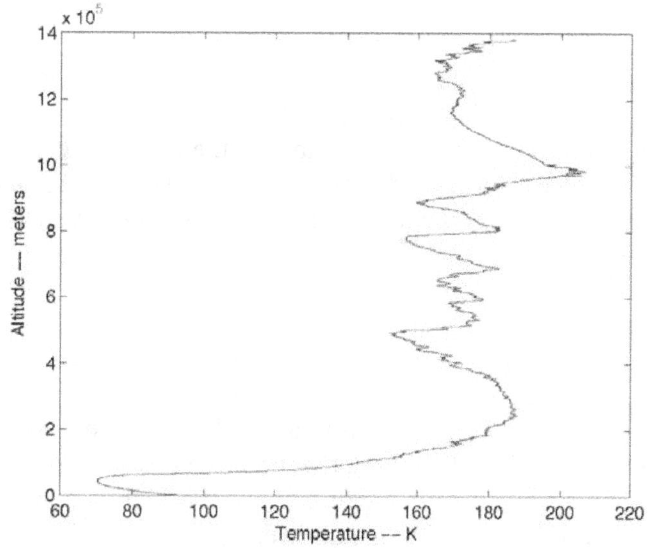

Figure 1. Temperature profile of Titan's atmosphere from the Cassini Huygens mission [4]

3 Results

In this analysis, the only parameter that is varied is the entry velocity of the micrometeoroid. Figure 2 shows altitude versus temperature for meteoroid entry velocities from 1 to 15 km/s, chosen to span the range from Titan's escape velocity (2.6 km/s) and orbital velocity (5.6 km/s) to Saturn's orbital velocity (9.7 km/s).

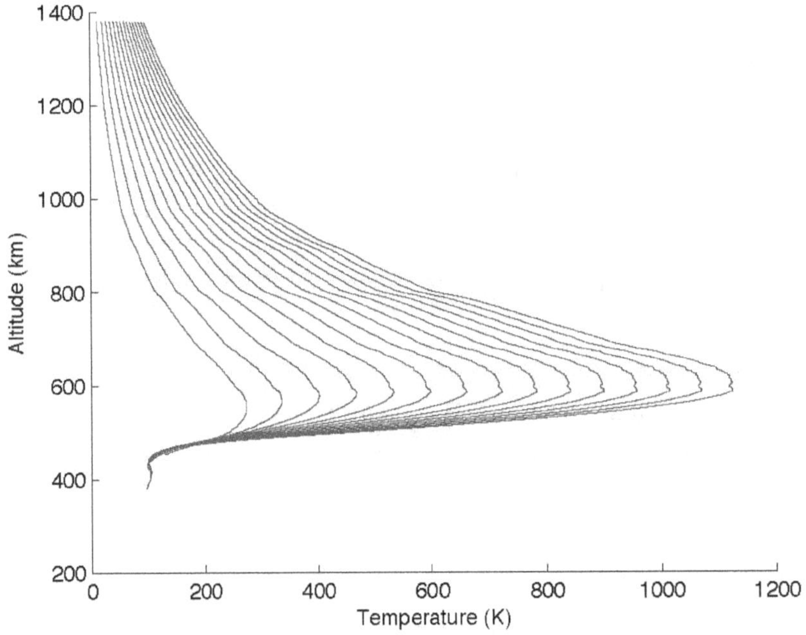

Figure 2. Micrometeoroid temperature as a function of altitude for entry velocities of 1 to 15 km/s. The curve that peaks at the highest temperature is 15 km/s.

The evaporation of meteoroid material due to heating was modeled by the Langmuir formula using a variety of values for the vapor pressure as have been used in other meteor evaporation studies [3]. Varying the vapor pressure value over this range did not significantly alter these results.

This result agrees well with previous studies [2] in that the micrometeors reach peak heating at approximately 600 km, and are heated over a timescale of minutes. The slowest particles (1 km/s) only reach a temperature of about 200 K, whereas the fastest particles (15 km/s) are heated to 1200 K.

4 Discussion

Meteors decelerate once they have encountered roughly their own mass of atmospheric molecules. Compared to Earth [1], micrometeors entering Titan's atmosphere will experience significantly less severe heating, because Titan's gravity is only ~ 14% that of Earth. Titan's atmospheric scale height is thus larger than Earth's, making it a more diffuse medium through which to decelerate and allowing for more time to radiate away the frictional heat of atmospheric entry.

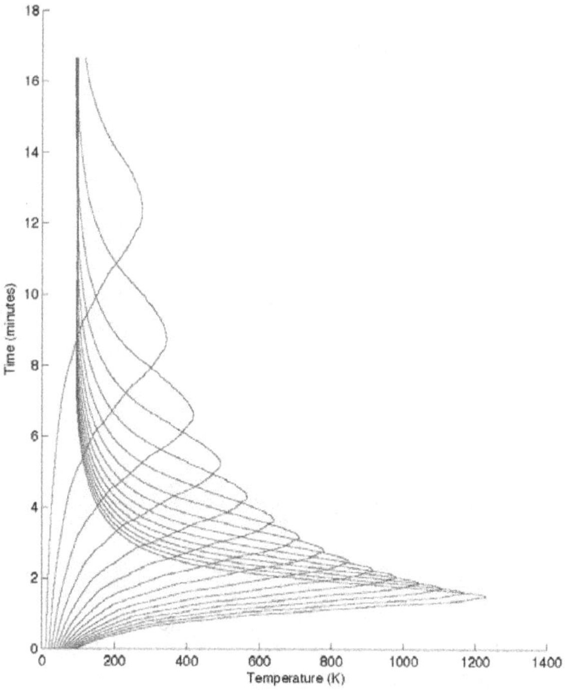

Figure 3. Micrometeoroid temperature as a function of time for entry velocities of 1 to 15 km/s. The curve that peaks at 200 K and ~ 12 minutes is that for 1 km/s and the curve that peaks at ~ 1200 K at 2 minutes is for 15 km/s.

A 12 km/s micrometeor of 100μm diameter and 45° entry angle will reach a peak temperature of 1800 K after 13 seconds [1]. By comparison, the same micrometeor entering Titan's atmosphere will not exceed 1000 K and will require about two minutes to reach its peak temperature.

Detailed knowledge regarding ranges of input velocities, size distribution, average composition, etc. is incomplete for meteor sources in the neighborhood of the outer planets. The assumptions made here assume similarity to the situation observed in our part of the solar system. If meteoritic material in the area of the outer planets is more cometary in origin with a higher percentage of water ice, then a lower meteoroid density and a modified entry velocity range may be more appropriate. The specific heats of vaporization and melting are very different for water ice compared to that used in Earth-based meteor studies [1]. This difference will keep the particle's temperature lower since energy is more efficiently partitioned into melting and evaporation. The slowest micrometeors may possibly retain some water ice, while the fastest will likely lose all of the ices and most of their organic compounds.

5 Conclusions

Titan's low gravity and large scale height means that micrometeors undergo relatively slow heating and cooling compared to those entering Earth's atmosphere. Molecules liberated from meteoroids during their descent will likely be able to participate in photochemical and heteorgeneous reactions

Recent experiments have shown that flash-heated CM chondrite micrometeorites will evolve organic compounds, including PAHs and light hydrocarbons, at temperatures from 500 to 1000 K [5]. Similar experiments should be conducted at slower heating rates to observe what organic compounds may be released under the more gentle heating expected in Titan's atmosphere. These compounds can be incorporated into chemical models for Titan's atmosphere. In particular, micrometeorites may be involved in the presence of oxygen-bearing compounds and also small polycyclic aromatic hydrocarbons in Titan's atmosphere.

Acknowledgements

MEK acknowledges research support from the NASA Astrobiology Institute's Virtual Planetary Laboratory (PI: V. Meadows).

References

1. S.G. Love and D.E. Brownlee, Heating and thermal transformations of micrometeoroids entering Earth's atmosphere. Icarus 89, 26-43 (1991)
2. S.G. Love and D.E. Brownlee, Heating and thermal transformations of micrometeoroids entering Earth's atmosphere. Icarus 89, 26-43 (1991)
3. E.J. Öpik, Physics of Meteor Flight in the Atmosphere. Wiley-Interscience, Hoboken, NJ
4. NASA Planetary Atmospheres Data Node, Data Set ID: HP-SSA-HASI-2-3-4-MISSION¬V1.1 (2006)
5. M.E. Kress, C.L. Belle, G.C. Cody, A.R. Pevyhouse and L.T.Iraci, Atmospheric chemistry of micrometeoritic organic compounds, NASA Conference Proceedings from Meteoroids 2010, D. Janches, editor (2010)

Global Variation of Meteor Trail Plasma Turbulence

L. P. Dyrud • J. Hinrichs • J. Urbina

Abstract We present the first global simulations on the occurrence of meteor trail plasma irregularities. These results seek to answer the following questions: when a meteoroid disintegrates in the atmosphere will the resulting trail become plasma turbulent, what are the factors influencing the development of turbulence, and how do they vary on a global scale. Understanding meteor trail plasma turbulence is important because turbulent meteor trails are visible as non-specular trails to coherent radars, and turbulence influences the evolution of specular radar meteor trails, particularly regarding the inference of mesospheric temperatures from trail diffusion rates, and their usage for meteor burst communication. We provide evidence of the significant effect that neutral atmospheric winds and density, and ionospheric plasma density have on the variability of meteor trail evolution and the observation of non-specular meteor trails, and demonstrate that trails are far less likely to become and remain turbulent in daylight, explaining several observational trends using non-specular and specular meteor trails.

Keywords meteor trail · plasma· turbulence · simulation

1 Introduction

The daily occurrence of billions of meteor trails in the Earth's upper atmosphere presents a powerful opportunity to use remote sensing tools to better understand the meteoroids that produced them, and the atmosphere and ionosphere in which their trails occur. One of the most promising tools employed in this endeavor are high-power-large-aperture (HPLA) radars. Such radars routinely observe two distinct types of meteor echoes, head echoes and non-specular meteor trails. Head echoes are the radar reflection from targets with short durations, usually less than 1 millisecond at a given range, and moving at apparent meteoroid velocities [Close et al., 2002; Janches et al., 2000; Mathews et al., 2001, Janches et al. 2008, Chau and Galindo, 2008, Dyrud et al.. 2008]. When radars are pointed perpendicular to the magnetic field, head echoes are often, but not always, followed by echoes lasting seconds to minutes [Dyrud et al., 2005; Zhou et al., 2001, Malhotra et al., 2007]. Because these echoes occur simultaneously over multiple radar range gates, the term non-specular echoes has been adopted by many authors in order to differentiate them from the meteor echoes from specular meteor radars, which require a trail to align perpendicular to the radar beam [Ceplecha et al., 1998; Cervera and Elford, 2004]. It is now understood that non-specular trails are reflections from plasma instability generated field aligned irregularities (FAI) [Chapin and Kudeki, 1994a, Oppenheim et al., 2000, Zhou et al., 2001, Dyrud et al., 2001, Dyrud et al., 2002, Dyrud et al., 2007, Close et al., 2008]. However, the influence that turbulent trails has on specular observations of meteor trails has only been briefly studied [Hocking, 2004,

L. P. Dyrud (✉)
Communications and Space, Sciences Laboratory, Pennsylvania State University, University Park, PA, USA. E-mail: Lars.Dyrud@jhuapl.edu

J. Hinrichs • J. Urbina
Applied Physics Laboratory, John Hopkins University, Columbia, MD, USA

Galigan et al., 2004], and we do not yet understand the degree to which meteor trails are inherently plasma unstable. This paper seeks to address some of these unknowns.

We focus on the role that neutral atmospheric wind and density, and ionospheric plasma density has on the development of meteor trail turbulence and evolution. Our goal is to understand how regional, diurnal and seasonal variability in these background parameters will influence the role that plasma turbulent meteor trails has on various applications and scientific studies. Most prominently, turbulent trails are thought to have a diffusion rate that can exceed the nominal cross-field ambipolar diffusion rate by up to an order of magnitude, significantly altering trail evolution, duration and reflectability [Dyrud et al., 2001]. The effects of this turbulent evolution are important for specular radar derivations of diffusion rate and therefore neutral temperature (T_n) [Hocking et al., 1999, Kumar, 2007], meteor burst communication [Fukuda et al., 2003], and scientific studies involving non-specular trail observations in general [Dyrud et al., 2005, 2007, Malhotra et al., 2007].

In order to understand the global variation of meteor trail turbulence, we expanded a model of the evolution of an individual meteor from atmospheric entry to trail instability and diffusion (See Dyrud et al. [2005, 2007] for a detailed description of the model) by incorporating climatological models for the relevant ionospheric and atmospheric parameters. For readers interested in the global modeling of the incoming meteor flux see Janches et al., [2006] and Fentzke and Janches [2008].

Our model was originally used to simulate artificial radar Range-Time-Intensity (RTI) images for comparison with facilities like the 50 MHz Jicamarca Radar and other coherent radars [Chau et al., 2008, Oppenheim, 2007, Dyrud et al. 2004, Dyrud et al., 2007, Hinrichs, 2008]. This program simulates head echoes and non-specular trails for meteoroids of a chosen velocity, mass, and composition, entering the Earth's atmosphere. Our new program runs this individual meteor model, and then measures several key parameters pertaining to trail plasma instability, with this paper focusing on the duration of trail plasma instability. Instability duration is closely associated with the duration of an individual non-specular trail observation. Further, duration also acts as a guide for researchers interested in specular meteor trail observations, and meteor burst communication, by indicating when, where and to what degree they can expect turbulent versus laminar meteor trail evolution. By analyzing trail variation on a global scale, we show that properties of the atmosphere and ionosphere play a critical role in the observation and interpretation of meteor trails observations, and that as a result, the characteristics of meteor trail evolution are considerably more variable then previously expected.

2 Model Description

The model used here simulates meteoroid entry into the atmosphere, including ablation, ionization, thermal expansion and plasma stability based upon the meteor Farley-Buneman Gradient-Drift (FBGD) instability [Dyrud, 2001, 2002, Oppenheim, 2000, 2003a,b]. We have now enhanced the capability of this program by automating location and time specific ionospheric and atmospheric data from three main climatological models: Cospar international reference atmosphere (CIRA) [CIRA, 2005], the International Reference Ionosphere (IRI2000) [Bltiza, 2001], and the Horizontal Wind Model (HWM) [Hedin et al., 1996]. The parameters required from these models include electron density, atmospheric mass density, neutral temperature and wind speed, and from these we also derive ion and electron collision frequencies based upon the formulas from Banks and Kockarts, [1972] for a given location and time. This information is used to make location specific meteor simulations, which are then called multiple times to build up global maps of the meteor trail characteristics. While we recognize that these climatological models do not capture the full variability of aeronomical and ionospheric parameters, they do allow for an examination of the resulting climatological and global variation expected for meteor trail evolution. An aspect which previously has never been explored in light of the known influence of plasma turbulence.

An example meteoroid comparison to model results from an individual meteoroid simulation, for a single location and local time are displayed in Figure 1a. We display the results of this model as a simulated Range-Time-Intensity image of a head echo and non-specular trail, similar to those produced by a coherent radar observations [Chapin et al. 1994, Close et al. 2002, Oppenheim et al. 2007, Dyrud al. 2005, Malhotra et al. 2007]. This figure displays the head echo trace as diagonal colored line, with color corresponding to electron line density per meter divided by 10^6, such that it may appear on the same color bar as FBGD growth rate. We have worked on numerous head echo models (See Dyrud and Janches, [2008]) but have opted to plot a parameter which is related to head echo strength, but is also of direct physical relevance to the development of plasma instability. To the right of the head echo, this plot displays the calculated, non-negative, FBGD growth rate as a function of time and altitude for a diffusing meteor column. Examination of Figure 1a reveals that only a limited altitude portion of the trail is immediately plasma unstable, with the width of this unstable portion decreasing in time. The total duration of plasma instability for this example is approximately 15 seconds, which is defined as the time from trail generation at a given altitude to the time the growth rate becomes negative at that same altitude. In order to test our model, data comparison was done with trail observations from Fort Macon North Carolina with simulated meteors. This comparison is shown in Figure 1b. We continue with a presentation of the duration of trail turbulence if such a meteor where to occur simultaneously across the globe.

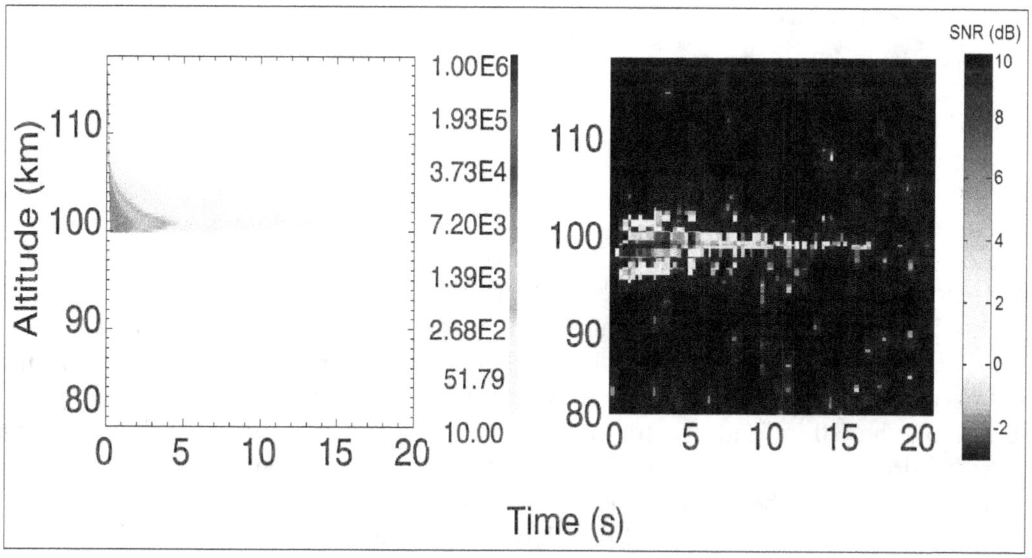

Figure 1a. A simulated RTI simulation for a meteoroid near North Carolina compared to a observed meteoroid in North Carolina (Latitude = -35° and Longitude = -55°) conditions for 00:00 UT, on June 27th . The simulated meteoroid mass is 10 μg, traveling at a velocity of 70 km/s, composed of an atomic mass of 30 AMU. The duration of the trail turbulence is approximately 15 seconds. The color bar shows instability growth rate in s^{-1} for the trail, and the simulated head echo displays electron line density per meter divided by 10^6 (units chosen such that they appear on the same scale).

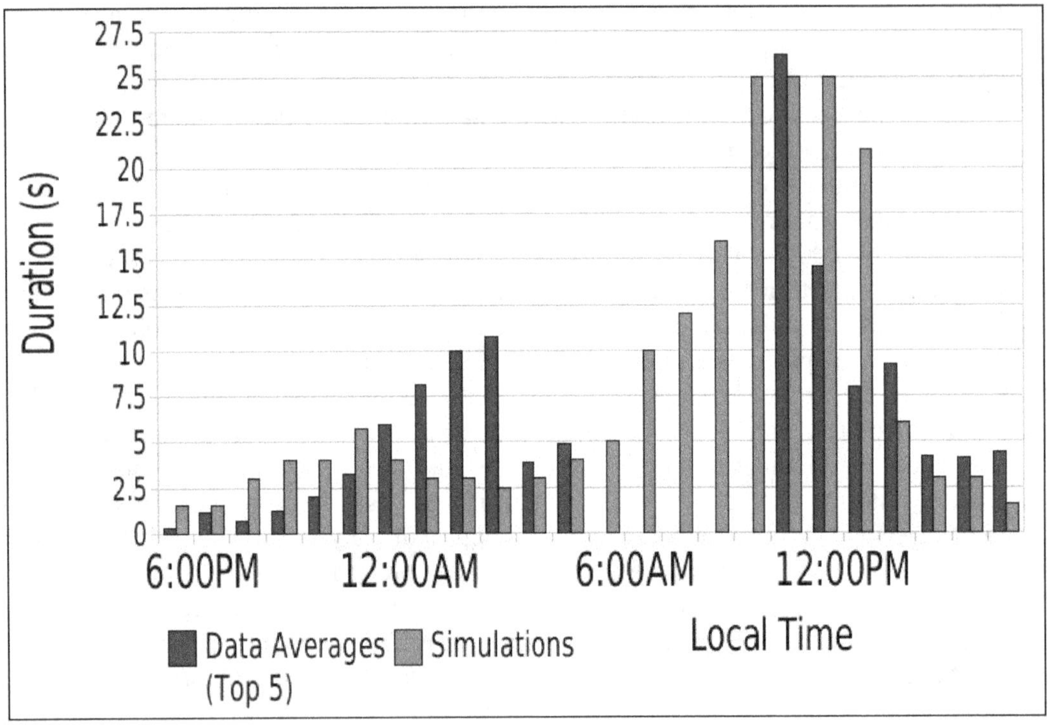

Figure 1b. A bar graph comparing meteor trail simulations from North Carolina with data observed at Fort Macon North Carolina. The data was obtained by taking the average of the top 5 largest meteors at each hour and comparing it with meteor simulations from the model previously described. Data was not taken for 6 hours, starting at 6am, because the radar was not on. Notice how much of the data points compare well with the simulated data, yet some outliers occur in the data.

3 Global Model Results

Here we examine the duration of meteor trail turbulence as function of location, for a trail produced by a 10 microgram meteor traveling at 70 km/s on June 27th at 00:00 UT, with a zenith angle of 45°. These characteristic meteoroid parameters were chosen because this is a commonly measured size class of meteoroids among the billions of daily meteors [Mathews, 2001, Chau, Dyrud and Janches, 2008]. The meteor simulation of the type shown in Figure 1a, is repeated several hundred times across a 2° latitude and longitude grid, with the analyzed results displayed in Figure 2a. This plot shows, in color, the duration of plasma instabilities within the meteor trail for each location, which are seen to vary between 9 seconds and 0, where 0 indicates no trail turbulence is generated. The first striking observation from this figure is the dramatic global variation of meteor trail evolution, even for trails produced by the very same meteoroid. Some of the features shown are: a clear day to night variation, i.e. that duration is significantly longer in the dark regions of the globe, and since we show a January day one can see that more of the Northern hemisphere contains longer turbulent durations than the more sunlit southern hemisphere. We also see that that duration is in general longer near equatorial regions with enhancements that appear in the Northern Atlantic Ocean, over South America and Africa. The variation in this figure is caused by variation in the main drivers for instability, which are primarily background ionization levels, and the magnitude of the neutral wind blowing both perpendicular to the trail and the geomagnetic field.

We point out here that only the HWM was used for determination of meteor zone winds, and therefore these results will likely not to apply near the equator and at the highest latitudes where ionospheric drifts due to electrojets dominate the neutral winds. Hinrichs et al. [2008], has specifically analyzed a 24 hour meteor simulation for the Jicamarca radar location with inclusion of an electrojet drift model to show that the magnitude of the electrojet drift strongly modulates trail duration. We expect to incorporate climatological models for the high and low latitude electrojets into this global simulator in the future. We continue with a presentation of both the meteoric and atmospheric parameters responsible for variability in meteor trail evolution.

4 Meteor Properties

We now investigate meteoric properties and their influence on meteor trail duration. Two meteoric properties that have the greatest effect on meteor trail duration are mass and velocity. Figure 2b is a global simulation with identical parameters as 2a except the mass of the meteor has been decreased from 1.0 μg to 0.1 μg in order to investigate the effect of mass. As seen by comparing Figures 2a and 2b, a meteor with a mass of 1.0 μg will produce longer duration meteor trails compared to a smaller massed meteor. A more massive meteor produces steeper plasma density gradients, and penetrates to lower altitudes where polarization fields are the strongest. Not only do meteors of larger mass produce longer duration meteor trails at night, but during the daytime meteors of higher masses are now turbulent in regions where 0.1 μg meteors were not.

Unlike mass, an increased meteoric velocity doesn't always have a complementary effect to meteor trail duration. Yet, velocity has a significant impact on meteor trail duration. Figure 3 shows the drastic effect velocity has on meteor trails duration. A slow meteor traveling at 15 km/s has less ablation and ionization, which produces a relatively short lived meteor trail, if any trail at all. A very fast meteor traveling at 75 km/s has so much energy that all of its mass becomes ionized at such high altitudes that short trails are produced, due to weak polarization fields above ~100 km. The longest meteor trails observed are created by meteors of speeds in between both extremes. A velocity ranging from 35-40 km/s allows the meteor to reach lower altitudes where polarization electric fields become stronger, yet still generates steep density gradients. The impact that meteor velocity has on global meteor trail variability is shown by comparing Figure 2b and Figure 4. Figure 4 shows a global simulation of a 0.1 μg meteoroid, identical to that in Figure 2b, but with a velocity of 35 km/s. One may see that the slower velocity results in longer duration trails and more daytime trails.

5 Atmospheric Properties

In order to understand the atmospheres role in meteor trail evolution we investigate the parameters which have profound effects on trail evolution. We find that these parameters are electron density present in the ionosphere and the horizontal winds that a meteor experiences. Small changes in atmospheric properties result in dramatic global variability. Since electron density and winds effect meteor trail duration we must further investigate the variability seen in these parameters to understand a global meteor trail outlook.

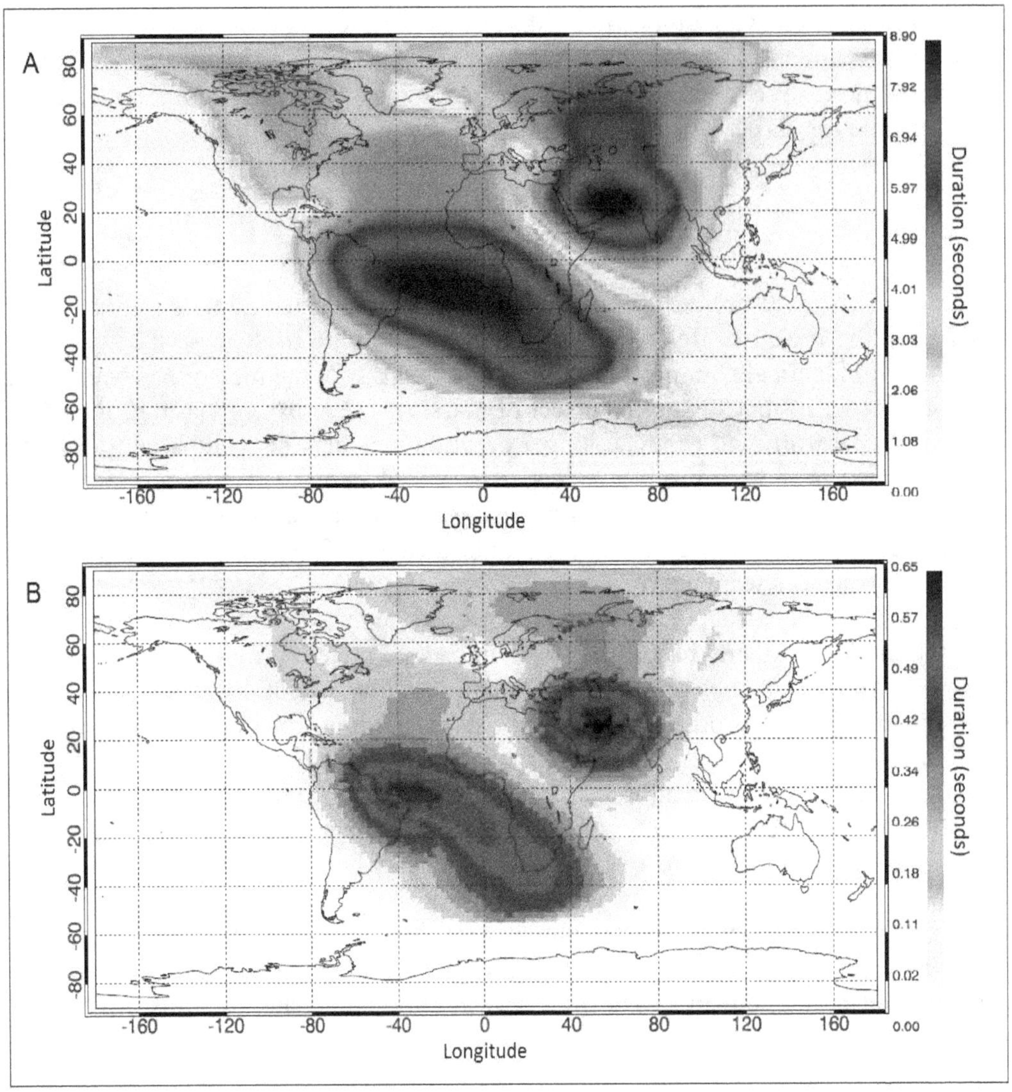

Figures 2. (a) A global simulation of the duration of meteor trail turbulence of a single 1 μg meter simulated across the world traveling a 55 km/s at 00:00 UT on January 1st. The units of the color bar are in seconds after meteor trail creation. Each pixel results from the measured duration of a simulation of the type shown in Figure 1a. The location of the meteor presented in Figure 1 is denoted by a (*) near North Carolina. Figure 2a illustrates the effect that the atmospheric properties electron density and horizontal wind speed have on a meteor's trail duration. (b) A global simulation of the duration of meteor trail turbulence of a single 0.1 μg meter simulated across the world traveling a 55 km/s at 00:00 UT on January 1st. The units of the color bar are in seconds after meteor trail creation. Each pixel results from the measured duration of a simulation of the type shown in Figure 1a. Notice that a meteor of lesser mass meteor experiencing the identical atmosphere as Figure 2a will produce meteor trails of shorter duration or no meteor trail at all depending on the location.

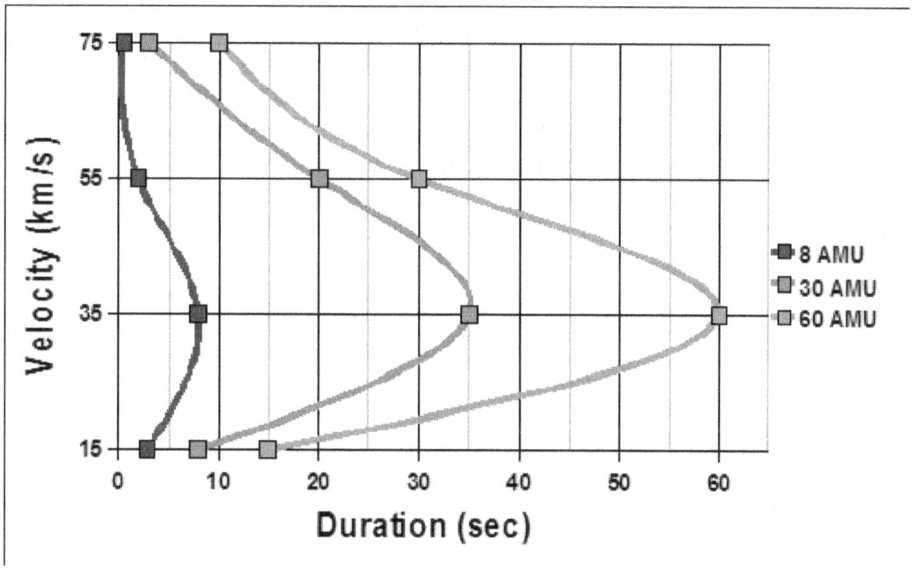

Figure 3. The duration of 3 meteors of different atomic masses (8, 30 and 60) and how its velocity effects meteor trail duration. The duration of a meteor trail is plotted in seconds and the meteors velocity is in km/s. Notice how slow and fast traveling meteor will produce short duration meteor trails in comparison with mid range meteor velocities. A velocity of 35-40 km/s has the right amount of energy to create a long duration trail.

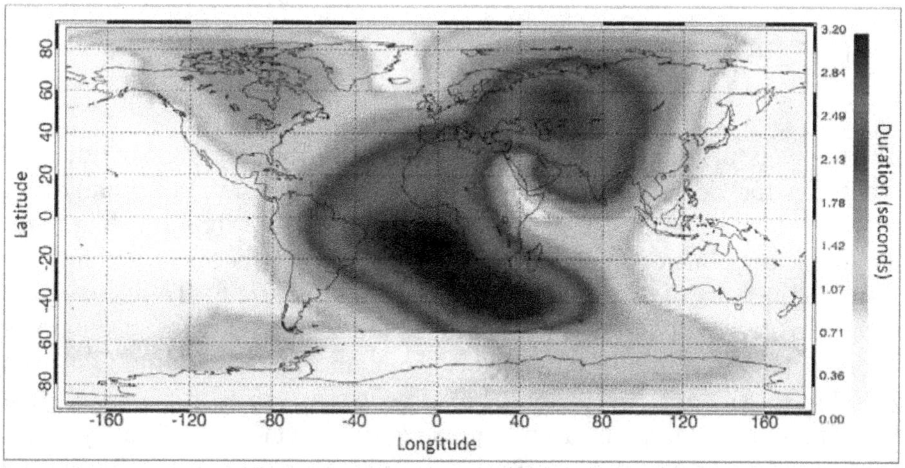

Figure 4. A global simulation of the duration of meteor trail turbulence of a single 0.1 μg meter simulated across the world traveling a 35 km/s at 00:00 UT on January 1st. The units of the color bar are in seconds after meteor trail creation. Each pixel results from the measured duration of a simulation of the type shown in Figure 1a. Notice the impact that a meteors velocity has on meteor trail evolution and trail duration. In comparison with Figure 2b which has a speed of 55 km/s, a meteor traveling at 35 km/s has a preferred velocity for producing a long duration meteor trail. Not only are longer duration trails produced but also trails in areas where they were absent in Figure 2b.

Electron density is the first atmospheric property we look at since its effect on trail duration is quite evident. Electron density is important because densities in the day differ from nighttime densities by a factor of two and electron density directly effects meteor trail evolution and duration. A meteor subject to high electron densities will produce a shorter plasma turbulent trail than if lower densities were present. Figure 5 shows the diurnal cycle of electron density at 0:00 UT on January 1st. The global structure of present electron density determines the area across the globe in which conditions favor meteor trail evolution.

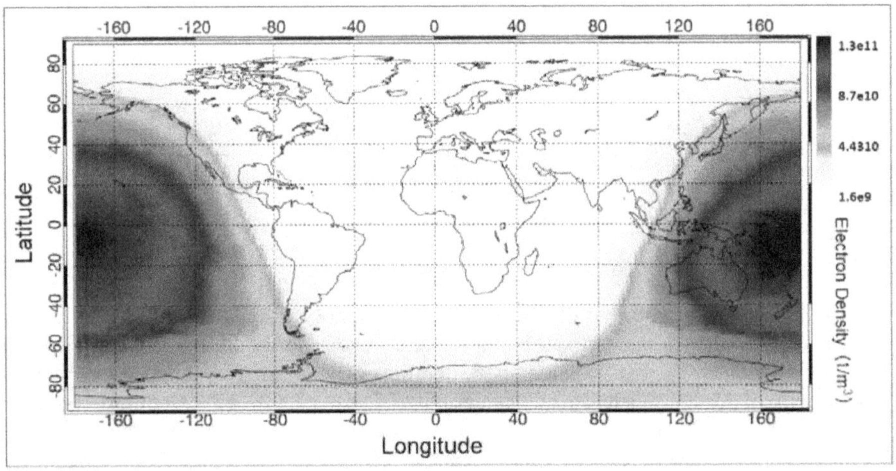

Figure 5. The electron density of the Earth's ionosphere that is present in Figure 2a and b. Density is measured in $1/m^3$ at 0:00 UT on January 1st. The color bar shows densities that range from high density present during daytime hours and low densities present at night. Notice the distinct diurnal cycle of high density daytime located at the east and west and the nighttime low density area located at the center of the world. Higher electron densities inhibits meteor trail evolution.

Electron density is an important factor in meteor trail evolution, but identical meteors that encounter constant electron density still have variability in trail duration around the globe. This is attributed to the winds that a meteor encounters. A meteor that is exposed to high winds will have a longer duration than the same meteor that is exposed to lower wind speeds. The impact that wind speed has on trail duration of meteors of different speeds is shown in Figure 6. To demonstrate the effect that global winds have on meteor trail duration we examined the horizontal wind speed at the altitude where maximum duration of the meteor trail occurs, this is shown in Figure 7. In the figure wind speeds vary from 0.5 m/s to 107 m/s. Although this is not the total wind that a meteor encounters, the wind speed at the altitude of maximum duration gives a good picture of the winds that directly influence a meteors plasma trail. Both the distinct global structure of winds at the altitude of maximum duration and the diurnal cycle of electron density are essential in understanding the global variability presented in this paper.

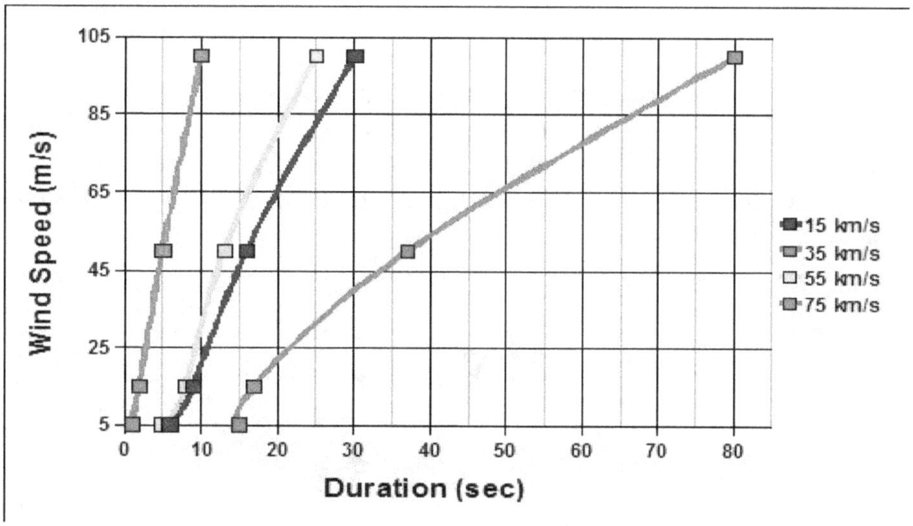

Figure 6. A plot of trail duration versus wind speed for 4 different velocities of 1.0μg meteors. The influence that the increased magnitude of horizontal wind speeds have on a meteor's plasma trail and its duration. The plot used data from simulations of the meteors trail durations of 4 identical meteors traveling at different speeds (15, 35, 55 and 75km/s). These meteors are simulated at different wind speeds. Notice that meteor trail duration is directly linked to wind speeds.

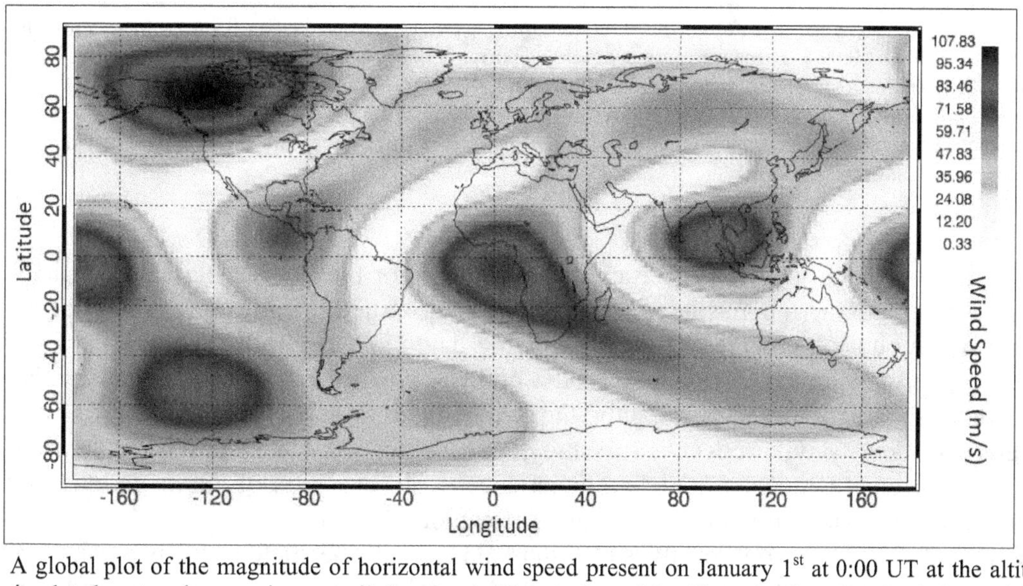

Figure 7. A global plot of the magnitude of horizontal wind speed present on January 1st at 0:00 UT at the altitude at which the simulated meteor has maximum trail duration in Figures 2a and b. The color bar shows wind speed in m/s. The wind pattern shown here is present throughout the day and is fixed in local time. Notice the large variations in the magnitude of horizontal wind speed which effects trail evolution.

6 Seasonal Variability

The effects of the atmosphere on meteor trail evolution can be seen not only globally, but seasonally as well. We now look as seasonal variability of meteor trails since the atmospheric properties we presented vary seasonally. We investigate seasonal variation by inspecting simulations near the equinoxes and solstices. Seasonal variability in meteor trail duration can be seen by comparing Figure 2b and 8. Differences in the global plots of meteor trail duration of identical meteors on January 1st and March 20th are the result of subtle yet key changes in the atmosphere throughout the seasons.

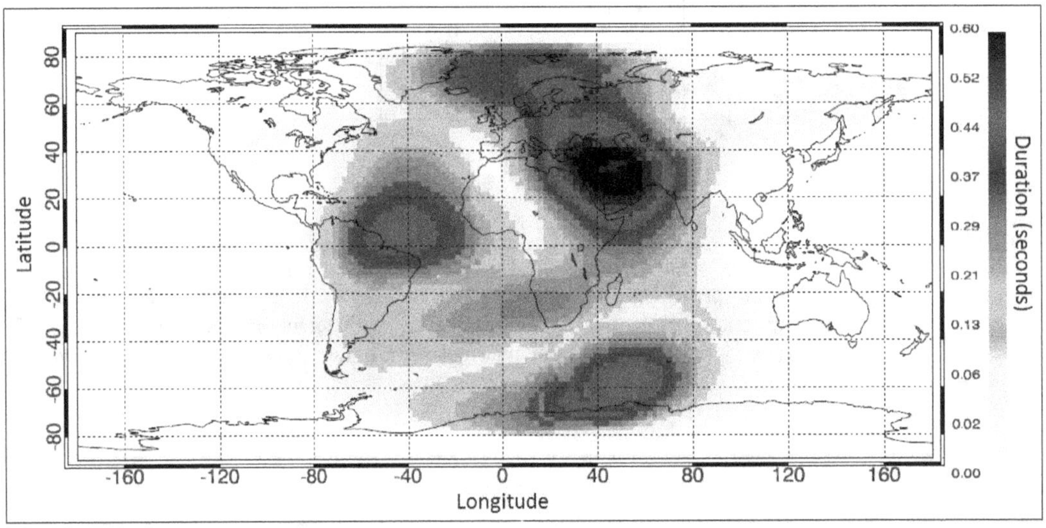

Figure 8. A global view of the duration of a meteor's trail. This simulation is of a 0.1 μg meteor traveling at 55 km/s on March 20th at 0:00 UT, measured in seconds. The units of the color bar are in seconds after meteor trail creation. Each pixel is a simulated meteor as seen in Figure 1a. Figure 8 takes place on March 20th, otherwise it is the identical conditions that is simulated in Figure 2b. Notice the differences in the global structure of trail duration compared to Figure 2b, which is simulated on January 1st. The differences are caused by the changes in the atmosphere throughout the seasons.

One difference in atmospheric properties is the structure of the present electron density. Figure 5 showed winter in the northern hemisphere. The shape of this structure varies throughout the seasons and is based on the amount of sun present throughout the day. For example June's electron density is a horizontally flipped version of Figure 5 since the night is present longer in the southern hemisphere. Since both hemispheres experience roughly the same amount of sunlight during the months near equinox, electron density will reflect accordingly. The other change in atmosphere throughout the seasons is the horizontal winds that a meteor experiences. The structure of the magnitude of horizontal wind at the altitude of maximum duration changes throughout the year.

The combination of electron density and winds along with the meteors own parameters help the understanding of the great differences seen in both day/night observations and seasonal variability of meteor trials. With this understanding we now have a better idea of the scope that small variations in atmosphere has on worldwide variability of meteor trail evolution.

7 Discussion

This paper presents a drastic global and seasonal variability in plasma turbulent meteor trail duration. We find that variations in trail duration are caused by two atmospheric properties, electron density of the ionosphere and the magnitude of horizontal winds. While the observational studies of meteor trail turbulence and non-specular meteor trails remain sparse in terms of geographical and local time coverage, several observational trends have been reported in the literature. The model we constructed is critical for placing data from individual sites in the context of a global meteor flux into a local atmosphere and ionosphere. Here we review some observations on diurnal trends.

[Chapin and Kudeki, 1994a] and [Chapin and Kudeki, 1994b] published some of the first observations of non-specular trails from the Jicamarca radar. While it was not the focus of their paper, the difference in trail occurrence and duration before and after sunrise can was clearly shown in their Figure 4. The figure shows two distinct periods of meteor observations; the first half contains over 125 meteor echoes before sunrise near 6:20 LT, followed by an abrupt decrease in the number trails observed. After 6:20 only 20 meteor echoes are seen throughout the second half of observation.

Recently, Oppenheim et al. 2008 drew specific attention to the diurnal variability of non-specular echoes at Jicamarca. Before dawn, 341 non-specular trails were observed for 1288 head echoes and only 81 trails for 1240 head echoes after dawn. They suggested that this was evidence of a previously published theory by Dimant and Oppenheim [2006a, b] that predicted stronger zeroth order ambipolar fields at night, and therefore an enhanced driver for instabilities. In contrast, we provide an alternative explanation for this day night variability, which involves not just background electron density but the presence of background electric fields or winds that drive polarization fields within the meteor trail [See Dyrud et al. 2007]. The results presented here show that day/night variability is a global phenomenon, and not limited to electrojet regions.

Zhou et al. [2001] presented observations of head echoes and non-specular trials from the MU 50 MHz radar in Japan. This experiment was conducted with the radar pointing both perpendicular, and off -perpendicular to the geomagnetic field. They noted that essentially all head echoes had a corresponding non-specular trail in the perpendicular to **B** geometry, while the off- perpendicular had essentially no trails, but similar counts of head echoes. Their data were collected from 00:00 to 08:30 LT over 4 nights, but made no comment on pre and post sunrise differences. These results cemented the view that non-specular echoes result from plasma instability induced FAI. While not the primary focus Close et al. [2008] recently demonstrated that larger meteoroids are more likely to produce non-specular echoes than smaller.

Simek [2005] examined the seasonal and diurnal variability of specular meteor trail durations to show that mean sunlit durations were 2.27 - 0.11 seconds, but that night durations were 1.95 - 0.06 seconds. These general trends fit what we expect and report here, that enhanced diffusion as a result of trail turbulence during predominantly night-time meteors will reduce trail duration. However, the values reported here include a number of influencing factors such as changing echo altitude as a function of local time. However, specular echo duration as a function of altitude, which helps isolate the effects of trail turbulence, has been examined by Singer et al. [2008]. They showed that low altitude decay times decreased at high latitude in summer, and that strong echo trails had longer decay times than weaker echo trails (stronger echoes likely typify higher electron line densities produced by larger meteoroids). However, examination of this author's Figure 2 shows that these trends are reversed at the highest altitude of observation (94 km). The results reported here explain this seasonal and meteor size trend reversal at higher altitudes. In the summer hemisphere trails are more likely to be produced in a sunlit ionosphere, and therefore remain turbulent for shorter periods of time or not at all. If turbulent decay rates are faster than laminar decay rates as reported by Dyrud et al. [2001] we expect summer trails to possess, on average, longer decay times. Since larger meteoroids produce larger plasma density gradients we also expect larger trails (or stronger echoes) to possess faster decay times. Further, these effects of turbulent diffusion are more pronounced at higher altitude as also discussed by Dyrud et al. [2001].

In a study of specular trail diffusion as a function of radar pointing to **B,** Hocking [2004] suggested that " … future theoretical analysis need to include externally imposed electric fields in order to produce accurate simulations of diffusion rates..". This is what we have included in this study. Hocking [2004] examined decay times as a function of radar azimuth angle and time of day, and found that there was far stronger anisotropic diffusion at greater altitudes above 93 km, and that winds and electric fields appear to influence the diffusion rate in general, and the overall anisotropy.

As the above summary of studies show, the existing non-specular and specular trial observations do support a day to night variation in the occurrence of meteor trail plasma turbulence. The studies also show that larger meteoroids are more like to produce turbulent trails. Further, our simulations here indicate that this day/night occurrence variation is one that is predicted to be global. However, the detailed variability is a result of the altitudinal wind profiles and magnitude. Understanding this variability will require substantially increased observations, both in terms of geographical and local time coverage, and comparison with data from other instruments. We conclude by noting that the driving factors accounting for meteor trail turbulence are many and complexly intertwined, thus it is not the focus of this short letter to describe all the competing forces but to publicize the predicted dramatic variability to researchers in various meteor related fields. We are working on a detailed analysis of the various contributions and expect to report them in an upcoming publication, but can summarize the general trends here. The primary drivers for turbulence duration are background ionization: turbulence lasts longer at night, wind or drift velocity: higher winds or drifts produce longer turbulent durations, meteoroid mass: larger meteoroids produce longer turbulent durations, velocity: velocities near 35 km/s (with some modification with entry angle and a particular mass) longer lasting turbulence because they deposit their mass at preferred altitudes for turbulence, between 90-105 km altitude.

We expect that a complete understanding and characterization of all the driving forces behind meteor trail turbulence will improve our understanding of non-specular trails, but also dramatically improve our ability to use specular trail observations to derive atmospheric temperature and other parameters, by isolating decay rates from the influence of turbulence.

Acknowledgments

Lars Dyrud and Jason Hinrichs' work was supported by NSF grants ATM-0613706 and ATM-0638912.

References

Banks, P. M., and G. Kockarts (1973), Aeronomy: Part A, Chapter 9, *Academic Press*.
Bilitza, D., International Reference Ionosphere 2000, Radio Science 36, #2, 261-275, 2001.
Ceplecha, Z., et al. (1998), Meteor Phenomena and Bodies, *Spa. Sci. Rev.*, *84*, 327-471.
Cervera, M. A., and W. G. Elford (2004), The meteor radar response function: Theory and application to narrow beam MST radar, *Planetary and Space Science*, *52*, 591-602.
Chapin, E., and E. Kudeki (1994a), Radar interferometric imaging studies of long duration meteor echo observed at Jicamarca, *Journal of Geophysical Research*, *99*, 8937-8949.

Chapin, E., and E. Kudeki (1994b), Plasma-wave excitation on meteor trails in the equatorial electrojet, *Geophysical Research Letters*, *21*, 2433-2436.

Close, S., et al. (2002), Scattering characteristics of high-resolution meteor head echoes detected at multiple frequencies, *Journal of Geophysical Research (Space Physics)*, *107*, 9-1.

Close, S., T. Hamlin, M. Oppenheim, L. Cox, and P. Colestock (2008), Dependence of radar signal strength on frequency and aspect angle of nonspecular meteor trails, J. Geophys. Res., 113, A06203, doi:10.1029/2007JA012647.

Committee on Space Research (COSPAR). The COSPAR International Reference Atmosphere (CIRA-86), [Internet]. British Atmospheric Data Centre, 2006-, *Date of citation*. Available from http://badc.nerc.ac.uk/data/cira/.

Chau, J. L., Galindo, F., First definitive observations of meteor shower particles using a high-power large-aperture radar, Icarus, Volume 194, Issue 1, p. 23-29, 2008

Dimant, Y. S., and M. M. Oppenheim (2006), Meteor trail diffusion and fields: 2. Analytical theory, *J. Geophys. Res.*, 111, A12313, doi:10.1029/2006JA011798, 2008a

Dimant, Y. S., and M. M. Oppenheim (2006), Meteor trail diffusion and fields: 1. Simulations, *J. Geophys. Res.*, 111, A12312, doi:10.1029/2006JA011797., 2008b

Dyrud, L. P., M. M. Oppenheim, and A. F. vom Endt (2001), The Anomalous Diffusion of Meteor Trails, *Geophys. Res. Lett.*, 28(14), 2775–2778.

Dyrud, L. P., Meers M. Oppenheim, Sigrid Close and Stephen Hunt, Interpretation of Non-Specular Radar Meteor Trails, Geophys. Res. Lett., 2002GL015953, 2002

Dyrud, L., et al. (2005), The meteor flux: it depends how you look, *Earth, Moon \& Planets*, *95*, 89-100, 2005a

Dyrud, L. P., E. Kudeki, and M. M. Oppenheim, Modeling long duration meteor trails, J. Geophys. Res., doi:10.1029/2007JA012692, Vol. 112, No. A12, A12307, 2005b

Dyrud, L. P., E. Kudeki, and M. M. Oppenheim, Modeling long duration meteor trails, J. Geophys. Res., doi:10.1029/2007JA012692, Vol. 112, No. A12, A12307, 2007

Dyrud, L. P., and D. Janches, Modeling the meteor head echo using Arecibo radar observations, *J. of Atmos. And Solar-Terr. Phys.*, 70, 2008a, 1621-1632

Fentzke, J. T., and D. Janches (2008), A semi-empirical model of the contribution from sporadic meteoroid sources on the meteor input function in the MLT observed at Arecibo, *J. Geophys. Res.*, 113, A03304, doi:10.1029/2007JA012531.

Fukuda A., K. Mukumoto, Y.Yoshihiro, M. Nagasawa, Y. Yamagishi-, N. Sato-, H. Yang., M. W. Yao, and L. J. Jin, Adv. Polar Upper Atmos. Res., 17, 120-136, 2003

Galligan, D. P., G. E. Thomas, W. J. Baggaley, On the relationship between meteor height and ambipolar diffusion, Journal of Atmospheric and Solar-Terrestrial Physics Volume 66, Issue 11, , July 2004, Pages 899-906.

Hedin, A.E., Fleming, E.L., Manson, A.H., Schmidlin, F.J., Avery, S.K., Clark, R.R., Franke, S.J., Fraser, G.J., Tsuda, T., Vial, F., Vincent, R.A. Empirical wind model for the middle and lower atmosphere. J. Atmos. Terr. Phys., 58, 1421–1447, 1996

J. Hinrichs, Dyrud, L. P., and, J. Urbina, Annales Geophysicae, Diurnal Variation of Non-Specular Meteor Trails, 2008.

Hocking, W. K. (1999), Temperatures Using Radar-Meteor Decay Times., *Geophys. Res. Lett.*, 26(21), 3297–3300.

Hocking, W. K. Experimental Radar Studies Of Anisotropic Diffusion Of High Altitude Meteor Trails., Earth, Moon, and Planets (2004) 95: 671–679

Janches, D., et al. (2000), Micrometeor Observations Using the Arecibo 430 MHz Radar, *Icarus*, *145*, 53--63.

Janches, D., C. J. Heinselman, J. L. Chau, A. Chandran, and R. Woodman (2006), Modeling the global micrometeor input function in the upper atmosphere observed by high power and large aperture radars, *J. Geophys. Res.*, 111, A07317, doi:10.1029/2006JA011628.

Malhotra, A., J. D. Mathews, and J. Urbina (2007), Multi-static, common volume radar observations of meteors at Jicamarca, *Geophys. Res. Lett.*, 34, L24103, doi:10.1029/2007GL032104.

Mathews, J. D., et al. (2001), The micrometeoroid mass flux into the upper atmosphere: Arecibo results and a comparison with prior estimates, *Geophysical Research Letters*, *28*, 1929.

Oppenheim, M. M., A. F. vom Endt, and L. P. Dyrud (2000), Electrodynamics of Meteor Trail Evolution in the Equatorial E-Region Ionosphere, *Geophys. Res. Lett.*, 27(19), 3173–3176.

Oppenheim, M. M., G. Sugar, E. Bass, Y. S. Dimant, and J. Chau (2008), Day to night variation in meteor trail measurements: Evidence for a new theory of plasma trail evolution, *Geophys. Res. Lett.*, 35, L03102, doi:10.1029/2007GL032347.

Oppenheim, M. M., L. P. Dyrud, and L. Ray (2003a), Plasma instabilities in meteor trails: Linear theory, *J. Geophys. Res.*, 108(A2), 1063, doi:10.1029/2002JA009548.

Oppenheim, M. M., L. P. Dyrud, and A. F. vom Endt (2003b), Plasma instabilities in meteor trails: 2-D simulation studies, *J. Geophys. Res.*, 108(A2), 1064, doi:10.1029/2002JA009549.

Singer, W. R. Latteck, L. F. Millan, N. J. Mitchell, J. Fiedler, Radar Backscatter from Underdense Meteors and Diffusion Rates,Earth Moon Planet (2008) 102:403–409, DOI 10.1007/s11038-007-9220-0
Urbina, J., E. Kudeki, S. J. Franke, S. Gonzales, Q. Zhou, and S. C. Collins, Geophys. Res. Lett., 27, 2853-2856, 2000.
Zhou, Q. H., et al. (2001), Implications of Meteor Observations by the MU Radar, *Geophysical Research Letters*, *28*(7), 1399.

CHAPTER 7:

BOLIDE OBSERVATIONS AND FLIGHT DYNAMICS

Passage of Bolides through the Atmosphere

O. Popova

Abstract Different fragmentation models are applied to a number of events, including the entry of TC_3 2008 asteroid in order to reproduce existing observational data.

Keywords meteoroid entry · fragmentation · modeling

1 Introduction

Fragmentation is a very important phenomenon which occurs during the meteoroid entry into the atmosphere and adds more drastic effects than mere deceleration and ablation. Modeling of bolide fragmentation (100 – 10^6 kg in mass) may be divided into several approaches. Detail fitting of observational data (deceleration and/or light curves) allows the determination of some meteoroid parameters (ablation and shape-density coefficients, fragmentation points, amount of mass loss) (Ceplecha et al. 1993; Ceplecha and ReVelle 2005). Observational data with high accuracy are needed for the gross-fragmentation model (Ceplecha et al. 1993), which is used for the analysis of European and Desert bolide networks data. Hydrodynamical models, which describe the entry of the meteoroid including evolution of its material, are applied mainly for large bodies (>10^6 kg) (Boslough et al. 1994; Svetsov et al. 1995; Shuvalov and Artemieva 2002, and others). Numerous papers were devoted to the application of standard equations for large meteoroid entry in the attempts to reproduce dynamics and/or radiation for different bolides and to predict meteorite falls. These modeling efforts are often supplemented by different fragmentation models (Baldwin and Sheaffer, 1971; Borovička et al. 1998; Artemieva and Shuvalov, 2001; Bland and Artemieva, 2006, and others).

The fragmentation may occur in different ways. For example, few large fragments are formed. These pieces initially interact through their shock waves and then continue their flight independently. The progressive fragmentation model suggests that meteoroids are disrupted into fragments, which continue their flight as independent bodies and may be disrupted further. Similar models were suggested in numerous papers, beginning with Levin (1956) and initial interaction of fragments started to be taken into account after the paper by Passey and Melosh (1980). The progressive fragmentation model with lateral spreading of formed fragments is widely used (Artemieva and Shuvalov, 1996; Nemtchinov and Popova, 1997; Borovička et al. 1998; Bland and Artemieva, 2006).

The second mode of fragmentation is the disruption into a cloud of small fragments and vapor, which are united by the common shock wave (Svetsov et al. 1995). This fragmentation occurs during the disruption of relatively large bodies. If the time between fragmentations is smaller than the time for fragment separation, all the fragments move as a unit, and a swarm of fragments and vapor penetrates deeper, being deformed by the aerodynamical loading like a drop of liquid (Hills and Goda 1993 and others). This liquid-like or "pancake" model assumes that the meteoroid breaks up into a swarm of small

O. Popova (✉)
Institute for Dynamics of Geospheres Russian Academy of Sciences, Leninsky prospect 38, bldg.1, Moscow 119334, Russia. Phone: +7 495 939 70 00; Fax: +7 499 137 65 11; E-mail: olga@idg.chph.ras.ru

bodies, which continue their flight as a single mass with increasing pancake-like cross-section. The smallest fragments can be evaporated easily and fill the volume between larger pieces. Initially formed fragments penetrate together deeper into the atmosphere and the fragmentation proceeds further. But large fragments may escape the cloud and continue the flight as independent bodies.

The formation of a fragment–vapor cloud was observed in the breakup of a meteoroid on 1 February 1994 (McCord et al. 1995), in the fragmentation of the Benesov bolide (Borovička et al. 1998), and in other cases. The total picture of fragmented-body motion is comparatively complicated. Both scenarios are realized in the real events (Borovička et al. 1998).

2 The Entry of TC_3 2008

2.1 Observational Data

The entry of asteroid TC_3 2008 over Sudan was observed by numerous eyewitnesses and a few detecting systems, including Meteosat satellites (Borovička and Charvat, 2009), infrasonic array and US Government satellites (Jenniskens et al. 2009). Meteorites named Almahata Sitta were recovered in December 2008. Meteorite searches allowed collectors to find about 300 fragments with total mass up to 3.95 kg (Jenniskens et al. 2009). Masses (from 1.5 g to 283 g) were found along a 29 km path.

Almahata Sitta was classified as an anomalous polymict ureilite (Jenniskens et al. 2009). Different lithologies including a number of non-ureilite fragments (enstatite and ordinary chondrites) were found among retrieved samples. All pieces are fresh and unweathered, so they probably had been incorporated into asteroid TC_3 2008 and did not originate from an earlier meteor event (Bischoff et al. 2010). This indicates that the asteroid was probably a collection of different lithologies, which were included as distinct stones within the asteroid body. The measured bulk density of Almahata Sitta varies from fragment to fragment (2.9 – 3.1 g/cm^3) and porosity is about 15-20% (Kohout et al. 2010). These values are close to the typical ureilite values (3.05 g/cm^3 and 9%; Britt and Consolmagno, 2003). Welten et al. (2010) estimated the macroporosity of the asteroid as high as about 50%. One small piece of Almahata Sitta was disrupted in the laboratory and its measured tensile strength was about 56 ± 25 MPa (Jenniskens et al. 2009).

The initial diameter of this meteoroid was estimated as 4.1 ± 0.3 m based on asteroid visual magnitude (Jenniskens et al. 2009). Corresponding pre-atmospheric mass (ρ = 2.3 g/cm^3) is about 83 ± 2 5 t (Jenniskens et al. 2009). This estimate correlates well with the mass obtained based on infrasound signal (87 ± 27 t). The irradiated energy recorded by US DoD satellites allows an estimated initial mass of 56 t, assuming an integral luminous efficiency of 9.3% based on optical events calibrated by infrasound registration (Brown et al. 2002a). According to theoretical estimates (Nemtchinov et al. 1997) the integral luminous efficiency is slightly lower for this low velocity entry - 6.8-8.2%; these values result in initial mass of about 63-77 t.

The lower mass estimate (~20 t) is suggested by Kohout et al. (2010) and is based on assumptions of higher albedo and essential macroporosity of the asteroid. But this low mass estimate corresponds to very high value of integral luminous efficiency (~26%), which seems not probable.

The light curve recorded by US DoD satellites wasn't published, but it was released that the signal consisted of three peaks, while the most energy was radiated in the middle of a 1-s pulse at 37 km altitude and a final pulse 1 s later (at about 33 km altitude) (Jenniskens et al. 2009). Analysis of Meteosat 8 images allows the estimation of bolide brightness at two random heights, 45 and 37.5 km, where it reached −18.8 and −19.7 magnitude, respectively (Borovička and Charvat, 2009). The peak brightness was probably brighter than -20^{mag} (Borovička and Charvat, 2009). A schematic version of possible light curve is shown on Figure 1. Minimal detectable intensity is assumed to be about $2\ 10^9$ W/sr. Shapes of light peaks are arbitrary, but the total irradiated energy corresponds to reported value of $4\ 10^{11}$ J (or 0.096 kt; USAF press release).

Analysis of Meteosat 8 images allows the conclusion that dust release due to meteoroid breakup occurred at altitudes 44, 37 and possibly at 53 km. The broken pressures were estimated as 0.2-0.3 MPa (at 46-42 km altitude) and 1MPa (at 33 km) (Jenniskens et al. 2009).

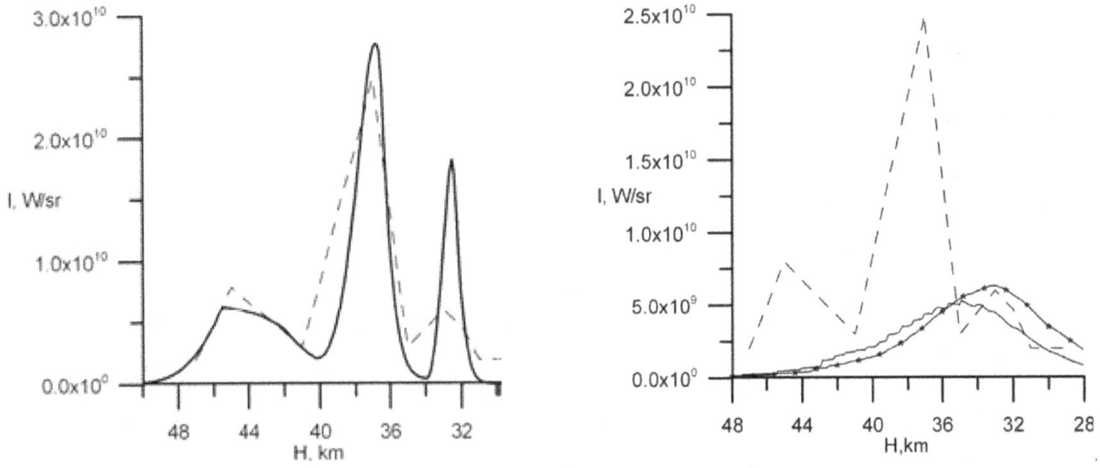

(a) (b)

Figure 1. (a) Schematic light curve of Almahata Sitta (dashed line) and an example of model light curves of Almahata Sitta in the frame of pan-cake model. (b) Schematic light curve of Almahata Sitta (dashed line) and model curves obtained in the frame of disruption onto two fragments (pointed line) and disruption into a number of fragments.

2.2 Modeling Efforts

2.2.1 Pan-cake Model

The presence of three peaks in the Almahata Sitta light curve indicates that there were three main stages of fragmentation. Similar light curves for a number of satellite observed bolides were successfully reproduced in the frame of pan-cake (or liquid-like) models (Svetsov et al. 1995; Nemtchinov et al. 1997; Popova and Nemtchinov 2008). Although liquid like models mentioned above are applicable mainly for large impactors, which are destroyed so intensively that fragments couldn't be separated (>~4-10 m in size) (Svetsov et al. 1995; Bland and Artemieva 2006), its modifications provide reasonable energy release. These models may be suitable for catastrophic disruption, when a huge number of fragments are formed.

The shape of light curve depends on chosen model parameters (rate of dust cloud spreading, mass fraction fragmented in every break up, assumed strength at the breakup). One possible light curve of Almahata Sitta is shown on Figure1a. The meteoroid with initial mass of 83 tons and bulk density 2.5 g/cm^3 initially disintegrated at the altitude of about 50.4 km under the aerodynamical loading about 0.15

MPa on two-three big pieces and a cloud of small fragments and dust, which may be described in the frame of pan-cake model. Formation of this cloud is accompanied by the first flare in the light curve. The next fragment (or few fragments) is broken up by aerodynamical loading of about 0.6 MPa at 40.2 km altitude. And the last fraction of meteoroid was disrupted at 33.9 km altitude under loading of about 1.5 MPa. The fractions of initial mass fragmented at different altitudes are roughly 33, 47 and 20% (i.e. ~27.6, 38.4 and 16 t). The integral luminous efficiency was about 6.5%, corresponding to a total irradiated energy of about 0.099 kt. Slightly different values of mass fractions (25, 65 and 10 %) and strengths (0.15, 0.4 and 1.5 MPa) also permits reproduction of the triple peaked light curve. About 70% of the initial mass is evaporated, and about of 30% of it (~25 t) remains in the atmosphere as a decelerated cloud of dust.

According to the statistical strength theory (Weibull, 1951) and direct observations on natural rocks (e.g., Hartmann, 1969) the strength of a body in nature tends to decrease as body size increases. The effective strength is usually expressed as $\sigma = \sigma_s(m_s/m)^\alpha$, where σ and m are the effective strength and mass of the larger body, σ_s and m_s are those of small specimen, and α is a scaling factor. There are no precisely determined values of scaling factor α, but for stony bodies the exponent is estimated to be in the range of 0.1–0.5 (Svetsov et al. 1995). It has not been proven that theses values hold for meteorite strength, though that is commonly assumed in meteoroid fragmentation theories (e.g., Baldwin and Sheaffer, 1971; Tsvetkov and Skripnik, 1991; Nemtchinov and Popova, 1997; Borovička et al. 1998; Artemieva and Shuvalov, 2001; Bland and Artemieva, 2006).

The inferred strength at breakups depart from the values, which are predicted by the strength scaling law with exponent $\alpha \sim 0.25$. Bland and Artemieva (2006) suggest using a small variation in strength (about 10% around predicted values), but there is much more significant deviation. Even application of larger variations in strength (up to 50% of predicted value) reproduces only double peak curves, and the altitude difference between peaks is smaller than observed one.

The pan-cake model is not capable of providing a mass-velocity distribution of meteoroid fragments; it cannot predict the meteorite strewn field. Besides, the same luminous efficiency is used for the solid fragment and for the cloud of vapor if their sizes are equal.

2.2.2 Progressive Fragmentation Models

The possibility to describe the fate of individual fragments, to determine meteorite strewn or crater fields is the main and extremely important advantage of the progressive fragmentation type models. The number of fragments changes in the process of the disruption from 1 (a parent body) to an arbitrarily large value, depending on the assumed properties of the meteoroid. These types of models usually incorporate the strength scaling law mentioned above and different assumptions about distribution of formed fragments on mass.

Bland and Artemieva (2006) suggested that each fragmentation of a single body results in two fragments with smaller mass and usually higher strength (although a small (<10%) variation in strength was considered). Each fragment is subjected to additional fragmentations later if the dynamic loading exceeds the updated fragment strength.

Disruption into two fragments supplemented with the strength scaling law leads to a single peak light curve. Corresponding modeling efforts are shown on Figure 1b. An initial meteoroid mass $m_s \sim 83$ tons, initial strength $\sigma_s \sim 0.15$-0.2 MPa and $\alpha \sim 0.25$ are assumed. The observed value of strength at initial breakup is used as sample strength. The usage of stony meteorite sample strength (~30MPa for 0.01 kg) results in higher initial strength and lower altitude of fragmentation beginning (~40 km). Masses of daughter fragments are chosen randomly in every breakup. Heat transfer coefficient $C_h \sim 0.1$

corresponds to ablation parameter of about 0.016 s^2/km^2 similar to the characteristic value for stony bodies (0.014 s^2/km^2; Ceplecha et al. 1998) and to the value used by Bland and Artemieva (2006). The used luminous efficiencies in the satellite detectors passband were obtained in the course of radiative hydrodynamic numerical simulations (Golub' et al. 1996; Nemtchinov et al. 1997).

Fragmentation starts at 47 km altitude and proceeds down to about 29 km altitude. A large fraction of initial mass lands on the ground ($M_{fall} \sim 24$ t) in more than 5000 pieces. The largest fragment reaches 10-20 kg (size of largest fragment is mainly determined by suggested strength scaling law and entry velocity). Most of the fallen mass is contained in the largest fragments. The integral luminous efficiency is about 2.5 - 3% and total light energy in the satellite detectors passband is about 0.03 – 0.05 kt TNT. Obtained values vary slightly from one set of calculations to another due to random choice of fragment size at breakup, but they are close to each other on average. The light pulse starts and ends at lower altitudes than the schematic one for the real event, the model light intensity and total irradiated energy are lower.

The single disruption event may result in a number of fragments. The mass distribution of fragmented rocks is often described by a power law (Hartmann 1969; Fujiwara et al. 1989). The power law distribution was also used in the description of meteorites (Jenniskens et al. 1994; Hildebrand et al. 2006) and in modeling of meteoroid entry (Nemtchinov and Popova, 1997; Borovička et al. 1998). Following Hartmann (1969), the cumulative fragment distribution in the breakup is assumed $N \sim m^{-b}$, where N is the cumulative number of fragments of mass > m; b is the negative slope in a log(N)-log(m) plot. The slope holds the same if a logarithmic-incremental plot is used ($F \sim m^{-b}$, where F = number of fragment within a logarithmic increment, $d\log m$). The value $b \sim 0.6$ is accepted (Hartmann 1969). Disruption into several groups of fragments is considered. The average mass in neighboring groups changes in $\sqrt{2}$ times. The size of the largest daughter fragment is chosen randomly in every breakup, the number of groups and number of fragments in a group are determined based on parent fragment mass and fragment distribution mentioned above. All other parameters are the same as in the previous case.

The formation of a number of fragments causes the appearance of flashes in the light curve and slightly shifts the light curve to higher altitude (Figure 1b). The total fallen mass is still close to previous case ($M_{fall} \sim$ 20-24 t), but the fallen mass has wider distribution. The total number of fragments increases essentially up to 10^4 – 10^5. Nevertheless, the Almahata Sitta entry is poorly described by this and previous approaches. The fallen mass is too huge and irradiated energy is small.

2.2.3 Luminous Efficiency

The model light curves and total irradiated energy are dependent on assumed values of luminous efficiency. In general, luminous efficiencies vary with meteoroid size, velocity, altitude of flight and meteoroid composition. The dependence of luminous efficiencies f in the satellite detectors passband on altitude is given on Figure 2 for H-chondrite meteoroids (Golub et al. 1996; Nemtchinov et al. 1997). Luminous efficiencies mainly increase with meteoroid size and velocity and become higher at lower altitudes (Figure 2). The values of luminous efficiencies for achondrite bodies probably differ from H-chondrite ones due to the different composition of vapor in the radiative volume. In the entry modeling, the same luminous efficiency is used for the solid fragment and for the cloud of vapor if their sizes are equal. The model, which allows the determination of these coefficients, also has some limitations (Golub' et al. 1996), but currently it provides the best known estimates of luminous efficiency f for satellite observed light curves.

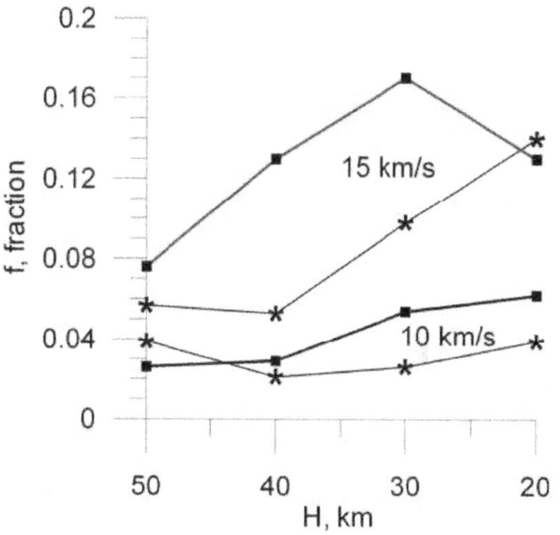

Figure 2. Luminous efficiencies for H-chondrite bodies versus altitude for two velocities (10 and 15 km/s) and two different sizes ($R \sim 0.14$ m (stars) and 1.4 m (squares))

Moreover, these luminous efficiencies were a basis for the determination of integral luminous efficiency η, i.e. the relation between total irradiated energy and initial kinetic energy for satellite observed bolides (Nemtchinov et al. 1997). An independent estimate of integral luminous efficiency η was obtained by Brown et al. (2002a) based mainly on infrasound registrations of 13 events. There were 3 meteorite falls among these events, compositions of other meteoroids were unknown (Brown et al 2002a). These estimates of η agree well with each other (Popova and Nemchinov 2008).

Roughly, it may be estimated that the light intensity has the precision of about two times. It should be also noted here that the light curve on Figure 1 is not really observed, it is only a sketch.

2.2.4 Hybrid Model

A large number of fragments may be formed simultaneously, but the progressive fragmentation model considers their flight and radiation independently. This type model deals better with few fragments, which are well separated. The progressive fragmentation model does not well describe the case of production of a large number of poor separated fragments. Different configurations of fragments may occur during the disruption process and influence further motion and radiation of fragments (Artemieva and Shuvalov, 1996, 2001). A number of separated fragments may be formed whereas smaller fragments and dust probably have no time to be separated and form a spreading cloud. The suggestion that randomly chosen part of mass in the break up forms an expanding cloud of dust causes the appearance of the flares on the light curve (Figure 3a) and the increase of radiated energy up to 0.04-0.07 kt TNT, but these values are still lower than observed ones. Shape of light pulse varies from one numerical run to another. The total fallen mass decreases down to 6-14 t in 10^3-10^5 fragments. Fallen mass is essentially overestimated. Larger fraction of mass should be converted into the dust in breakups.

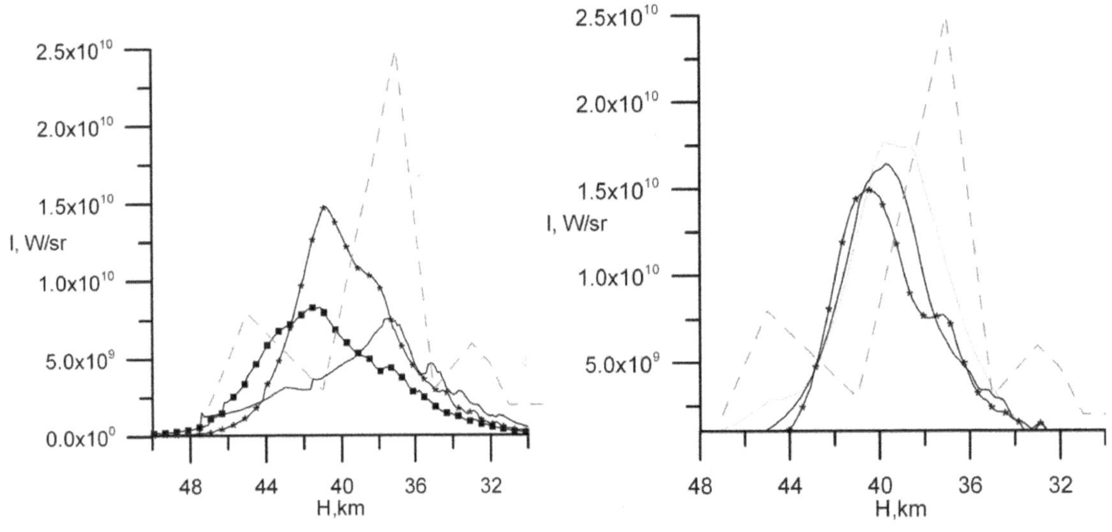

Figure 3. (a) Schematic light curve of Almahata Sitta (dashed line) and three model curves obtained under assumption that in every breakup some part of mass formed spreading cloud of vapor and dust. (b) Schematic light curve of Almahata Sitta (dashed line), two model curves obtained under assumption that in every breakup only few fragments are formed, some part of mass (~30% in average) formed spreading cloud (black and pointed curves); fragmentation onto two parts, one of which is converted into dust spreading cloud (gray curve).

During the progressive fragmentation of Moravka meteoroid (initial mass estimate ~1.2 ton) at the altitudes 30-40 km (Borovička and Kalenda, 2003) every break up of parent fragment resulted in formation of 1-3 relatively large fragments and dust (invisible on videorecord). Dust mass reached 10-90% of the parent fragment mass. Light curves obtained under the assumption that the number of fragments in the breakup is relatively small are shown on Figure 3b. Number of fragments in breakup is about 1-10 (2-3 in average) and some part of mass is converted into spreading dust cloud (~30% in average). Fallen mass in these cases is about 1.5-2.5 t, number of fragments is about 1000, integral luminous efficiency is about 4 – 5% and E_r ~ 0.07-0.08 kt, but light pulse becomes more narrow (Figure 3b) even if the deviation of fragment strength from strength scaling law is allowed.

In the limiting case of the disruption into two parts – one fragment and dust cloud – the fallen mass decreases to about 5-15 kg in one piece. Integral luminous efficiency increases up to 5-6% and E_r ~ 0.08 – 0.09 kt (Figure 3b). In order to get few peaks on the light curve the strength of fragments should essentially deviate from assumed strength scaling law, but the pulse is still narrow even if the strength may change on about 50% (Figure 3b).

It is possible to increase the mass fraction converted into dust clouds artificially and to fit observed light energy and shape of light pulse, but it is done above in the frame of pan-cake model. Light curve may be fitted if fallen mass is smaller about 100-400 kg, which seems to be an upper estimate.

3 Comparison With Other Events

3.1 Dust Formed in the Breakups

As it was mentioned above (Section 2.1), formation of dust clouds were directly observed during the entry of TC_3 2008 (Borovička and Charvat 2009). The amount of warm decelerated dust was estimated as at least 10 t, that is in the same order as our estimates (~25 t).

The Almahata Sitta entry confirmed that a large part of stony meteoroid mass and energy may be deposited in the atmosphere during the entry. Dust clouds are often observed at breakup events during observations of meter-sized meteoroids. These clouds are formed typically at 30-60 km altitude, but the data on particle size and on the mass fraction of the parent body, which was dispersed into dust, are scarce.

Attempts to collect dust from meteoroid disruption were done for two separate events, Revelstoke and Allende. The air through which a fireball had been observed to pass was sampled for meteoritic debris. Particulate matter was collected on special filters, which was mounted on aircraft and flown downwind from the site of the meteorite fall at 10-12 km altitude (Carr, 1970). According to Carr (1970), Revelstoke and Allende represented two different types of events. In the case of Revelstoke (type I carbonaceouse chondrite, corresponding sound wave energy is estimated as 10^{12}-10^{13} J, i.e.~1 kt) only a small 1 g of material was found in the fall area (possibly the result of rough terrain in the fall area, but may be the result of essential breakup in the atmosphere), large amount of debris still present in the atmosphere three days after event. Air samples contained a substantial excess over background of magnetite and transparent glass spherules and in addition contained a substantial number of irregular opaque particles high in Ni. Sizes of collected particles were mainly 2-4 μm (<10-25 μm). The Allende event was quite different (type III carbonaceouse chondrite, initial mass estimate >2000 kg, intial energy ~10^{12} J) >500 kg was found on the ground. Allende filters were clean – only a small number of particles were collected. The difference between sample and background is less than a factor of four, although some amount of opaque and glass spherules (<10 μm) were collected. Carr (1970) suggested that the difference in collected air samples and on fall sites resembles two different types of meteoroid breaks in the atmosphere.

A dust cloud formed due to fragmentation of large meteoroid (initial mass estimate 600 - 1900 t) was recorded during routine lidar observation of the atmosphere (Klekociuk et al. 2005). The meteoroid was fragmented at 32 km altitude and a dust cloud was recorded 7.5 hours later. The total mass of dust in this cloud was estimated as about 1000 t (that is lower estimates, because according to satellite observations, second fragmentation of meteoroid occurred at 25 km altitude). Dust size and concentration were estimated as 0.4-0.98 μm and 2-6 10^6 м$^{-3}$. Data on this event suggests that a large fraction of initial meteoroid mass may be released in the atmosphere as dust. Micron-sized particles may exist in the atmosphere during weeks-months and may play an important role both in climate processes and ozone layer dynamics (Klekociuk et al. 2005). No material was collected or found.

3.2 Tagish Lake and Carancas

The fraction of initial mass recovered as meteorites is mainly about f_m ~ 0.1-3 % for 11 meteorite falls with detailed tracking data on atmospheric passage (Popova et al. 2010). The recovered mass is smaller than estimated total fallen mass partially due to incomplete finding. The highest fractions are obtained for two smallest and slowest meteorites (~10%, Lost City and Innisfree), the smallest fractions (<10^{-4})

are found for Tagish Lake and Almahata Sitta meteorites, probably due to their specific structure and composition.

The Tagish Lake material is classified as ungrouped carbonaceous chondrite with a very high porosity of 25-49%, and the largest meteorite fragment constitutes only about 10^{-5} of its initial meteoroid mass (Hildebrand et al. 2006). Modeling efforts done to describe the entry of the Tagish Lake meteoroid demonstrated that about of 80-90% of instantaneous mass of the body was lost in the main breakup at the altitude of about 34-35 km (Brown et al. 2002b; Ceplecha 2007). Hildebrand et al.(2006) concluded that most of the initial meteoroid mass (~60-90 t) was deposited at 30-40 km altitude as the dust. Attempts to apply pan-cake and progressive fragmentation models to the Tagish Lake case also confirmed that essential amount of its mass was deposited as the dust in the atmosphere similar to Almahata Sitta case. About 1000 kg of initial mass may be converted into meteorites. This estimate is of the same order as the mass estimates at the end of luminous trajectory (~1300 – 2700 kg) obtained by Brown et al.(2002) and Ceplecha (2007). Hildebrand et al.(2006) estimated the total fallen mass as about 100-1000 kg, and only 16.3 kg was collected.

Borovička and Charvat (2009) compared the apparent strengths at fragmentation for a few bolides and suggested that the Tagish Lake meteoroid is the best analog to asteroid TC_3 2008. Presence of non-ureilite fragments among retrieved samples of Almahata Sitta shows inhomogeneous structure of the asteroid body (Bischoff et al. 2010). Besides, some authors suggest high macroporosity of TC_3 2008 (Borovička and Charvat 2009; Welten et al. 2010; Kohout et al. 2010). The parent bodies of Tagish Lake and Almahata Sitta meteorites were probably very fragile and inhomogeneous. They were catastrophically disrupted during the atmospheric passage producing dust clouds, and their stronger parts became meteorites.

The opposite case was observed in the Carancas event, where the fall of a stony meteorite caused the formation of a 13-m wide impact crater. This ordinary chondrite meteoroid probably did not experienced significant atmospheric fragmentation (Borovička and Spurny 2008), although there was no detailed observational data. The meteoroid mass was estimated as about M~1300-10000 kg (Borovička and Spurny 2008) or even as 10000-50000 kg. (Kenkmann et al. 2009). The Carancas event confirms that meteoroid strength and fragmentation scenario can vary significantly from case to case. But it should be noted here that small crater formation on the Earth is an extremely rare event due to disruption of meteoroids in the atmosphere, whereas 10-30 similar sized bodies enter the atmosphere every year (Nemtchinov et al. 1997; Brown et al. 2002a).

3.3 Mbale

If the fraction of initial mass recovered as meteorites f_m exceeds about 1-5%, probably there is no large dust deposition during the passage. Entry of these meteoroids is reasonably described in the frame of progressive fragmentation models. Comparison of model predictions with strewn fields permits better understanding of the details of meteoroid breakups, although in many cases the incomplete recovery adds uncertainties in strewn field data.

About 150 kg of material in more than 850 pieces were collected on a strewn field of a size 3x7 km after the fall of L5/6 ordinary chondrite Mbale in 1992 (Jenniskens et al. 1994). Its pre-entry mass was estimated as 400-1000 kg (more probably ~1000 kg) based on cosmogenic radionuclide data. Entry velocity was roughly estimated as 13.5 km/s. It was assumed that small fragmentation started probably above 25 km, but the main catastrophic breakup occurred at 10-14 km altitude.

Application of progressive fragmentation model to the Mbale entry allowed estimation of fallen mass as 200-250 kg in 100-3000 fragments (in dependence on assumed breakup model) covering a strewn field of about 1x7-9 km. Wind drift, which is essential for gram-sized fragments, wasn't taken into account. Multiple breakups occurred at the altitudes 22-35 km under the loading of about 0.8-1.3 MPa. Strength scaling law (with allowed random strength deviations) was used. The values of breakup loading are in the same range as for other observed meteoroid fragmentations (Popova et al. 2010), but it should be noted here that observed strength of meteoroids at breakup substantially deviates from this scaling law (Popova et al. 2010). The Mbale meteorite fragment distribution is better reproduced if a number of pieces following power law distribution are formed in every breakup. Model results satisfactorily describe the observed strewn field.

More strewn fields should be modeled in the future in order to better understand the details of the breakup process (strength at breakup; fragment distribution at breakup, etc).

4 Summary

Different fragmentation scenarios occur during the passage of meteoroids 100 – 10^6 kg through the atmosphere. There are a number of events which deposited essential fraction of their masses as dust in the atmosphere. Observational data are still incomplete to make definite conclusion, what fraction of incoming bodies is fragile enough to deposit this dust and how it is related with their structure/composition etc. But even bodies, which deposited much of mass as a dust/vapor, are able to produce meteorites. The total picture of fragmented-body motion is comparatively complicated. Better statistics are needed to estimate parameters of incoming cosmic material and to predict its behavior in the atmosphere. A full set of data, including detailed light curves, photographic trajectories, spectra, acoustic and seismic signals, and data on the composition of found meteorites are highly desirable.

References

Artem'eva N.A., Shuvalov V.V. 1996. Interaction of shock waves during passage of disrupted meteoroid through atmosphere. Shock Waves 5, 359–367.

Artemieva, N.A., Shuvalov, V.V., 2001. Motion of a fragmented meteoroid through the planetary atmosphere. J. Geophys. Res. 106, 3297-3310.

Baldwin, B., Sheaffer, Y., 1971. Ablation and breakup of large meteoroids during atmospheric entry. J. Geophy. Res. 76, 4653-4668.

Bischoff, A., Horstmann, M., Laubenstein M.,Haberer, S. 2010. Asteroid 2008 TC3 – Almahata Sitta: not only a ureilitic meteorite, but a breccias containing many different achondritic and chondritic lithologies. Lunar Planet. Sci. 41. Abstract 1763

Bland, P.A., Artemieva, N.A., 2006. The rate of small impacts on Earth. Meteorit. Planet. Sci. 41, 607-631.

Borovička, J., Popova, O.P., Nemtchinov, I.V., Spurný, P., Ceplecha, Z., 1998. Bolides produced by impacts of large meteoroids into the Earth's atmosphere: comparison of theory with observations. I. Benesov bolide dynamics and fragmentation. Astron. Astrophys. 334, 713-728.

Borovička, J., Kalenda, P., 2003. The Morávka meteorite fall: 4. Meteoroid dynamics and fragmentation in the atmosphere. Meteorit. Planet. Sci. 38, 1023-1043.

Borovička, J., Spurny , P., 2008. The Caranacas meteorite impact – encouter with a monolithic meteoroid. Astron.Astrophys. 485, L1-L4.

Borovička, J., Charvat Z., 2009. Meteosat observation of the atmospheric entry of 2008 TC3 over Sudan and associated dust cloud. Astron.Astrophys. 485,1015-1022.

Boslough, M.B., Crawford, D.A., Robinson, A.C., et al.: Mass and Penetration Depth of Shoemaker-Levy 9 Fragments from Time-Resolved Photometry, GRL, 21, 1555–1558, 1994.

Britt, D.T., Consolmagno, G.J., 2003. Stony meteorite porosities and densities: a review of the data through 2001. Meteorit. Planet. Sci. 38, 1161-1180.

Brown, P., Spalding, R.E., ReVelle, D.O., Tagliaferri, E., Worden, S.P., 2002a. The flux of small near-Earth objects colliding with the Earth. Nature 420, 294-296.

Brown, P.G., ReVelle, D.O., Tagliaferri, E., Hildebrand, A.R., 2002b. An entry model for the Tagish Lake fireball using seismic, satellite and infrasound records. Meteorit. Planet. Sci 37, 661-675.

Carr M.H., 1970, Atmospheric collection of debris from the Revelstoke nd Allende fireballs. Geochimica et Cosmochimica Acta 34,689-700.

Ceplecha, Z., ReVelle, D.O.. 2005. Fragmentation model of meteoroid motion, mass loss, and radiation in the atmosphere. Meteorit. Planet. Sci. 40, 35-54.

Ceplecha Z., Spurný P., Borovička J., Keclíková J. 1993. Atmospheric fragmentation of meteoroids, Astron. Astrophys. 279, 615-626.

Ceplecha, Z., Borovička, J., Elford, W.G., ReVelle, D.O., Hawkes, R.L., Porubčan, V., Šimek, M., 1998. Meteor phenomena and bodies. Space Sci. Reviews 84, 327-471.

Fujiwara A., Cerromi P., Ryan E. et al. 1989. Experiments and scaling laws for catastrophic collisions. In: Asteroids II, eds.R.Binzel, T.Gehrels and M.Matthews, Univ.Arizona Press, Tucson, Arizona, 240-265.

Golub', A.P., Kosarev, I.B., Nemtchinov, I.V., Shuvalov, V.V., 1996. Emission and ablation of a large meteoroid in the course of its motion through the Earth's atmosphere. Solar System Res. 30, 183-197.

Hartmann, W.K., 1969. Terrestrial, Lunar, and Interplanetary Rock Fragmentation. Icarus 10, 201.

Hildebrand, A., McCausland, P.J.A. , Brown, P.G. , Longstaffe, F.J. , Russell, S.D.J. , Tagliaferri, E., Wacker, J.F., Mazur, M.J. , 2006. The fall and recovery of the Tagish Lake meteorite. Meteorit. Planet. Sci. 41, 407-431.

Hills J.G., Goda M.P. 1993. The fragmentation of small asteroids in the atmosphere. Astron. J.,105, 1114-1144.

Jenniskens P, Betlem H, Betlem J et al. 1994. The Mbale meteorite shower. Meteoritics 29,246–254.

Jenniskens P., and 34 colleagues, 2009. The impact and recovery of asteroid 2008 TC3. Nature. 458, 485-488.

Kenkmann T., Artemieva N.A., Wunnemann K., Poelchau M.H., Elbeshausen D., Nunes del Prado H., 2009. The Carancas meteorite impact crater, Peru: Geologic surveying and modeling of crater formation and atmospheric passage. Meteorit. Planet. Sci. 44, 985-1000.

Klekociuk A.R., Brown P.G., Pack D.W., ReVelle D.O., Edwards W.N., Spalding R.E., Tagliaferri E., Yoo B.B., Zagari J. 2005. Meteoritic dust from the atmospheric disintegration of a large meteoroid . Nature, 436,1132-1135.

Kohout T., Kiuru R., Montonen M., Scheirich P., Britt B., Macke R. Consolmagno G., 2010. 2008 TC3 asteroid internal structure and physicalproperties inferred from study of the Almahata Sitta meteorites. Icarus, under review.

Levin, B. Yu (1956) Fizicheskaya teoriya meteorov i meteornoe veshchestro v Solnechnoi sisteme (Physical Theory of Meteors and Meteoric Matter in the Solar System) Nauka, Moscow, 294 (in Russian)

McCord T.B., Morris J. Persing D. et al. Detection of a meteoroid entry into the Earth's atmosphere on February 1, 1994. 1995. JGR, 100, 3245-3249.

Nemtchinov, I.V., Popova, O.P., 1997. An analysis of the 1947 Sikhote-Alin event and a comparison with the phenomenon of February 1, 1994. Sol. Sys. Res. 31, 408-420.

Nemtchinov I.V., Svetsov V.V., Kosarev I.B. et al, 1997. Assessment of kinetic energy of meteoroids detected by satellite-based light sensors. Icarus, 130, 259–274.

Passey Q.R., Melosh H.J., 1980. Effects of atmospheric breakup on crater field formation. Icarus 42,211–233.

Popova, O., Nemchinov, I. 2008. Bolides in the Earth Atmosphere. In: Adushkin V., Nemchinov I. (Eds.). Catastrophic events caused by cosmic objects, Springer, pp.131-163.

Popova O., Borovička J., Hartmann W., Spurny P. et al. 2010. Very low strength of interplanetary meteoroids and small asteroids. Under review.

Svetsov, V.V., Nemtchinov, I.V., Teterev, A.V., 1995. Disintegration of large meteoroids in Earth's atmosphere: Theoretical models. Icarus 116, 131-153.

Shuvalov, V.V., Artemieva, N.N., 2002. Numerical modeling of Tunguska-like impacts. Planet. Space Sci. 50, 181-192.

Tsvetkov, V.I., Skripnik, A.Ya., 1991. Atmospheric fragmentation of meteorites according to strength theory. Astronom. Vestnik 25, 364-371 (in Russian).

Weibull, W.A., 1951. A statistical distribution function of wide applicability. J. Applied Mechanics 10, 140-147

Welten K.C., Meier M. M., Caffee M. W., Nishiizumi K., Wieler R., Jenniskens P., Shaddad M. H., 2010. High porosity and cosmic ray exposure age of asteroid 2008 TC3 derived from cosmogenic nuclides. Lunar Planet. Sci. 41. Abstract 2256.

Constraining the Drag Coefficients of Meteors in Dark Flight

R. T. Carter • P. S. Jandir • M. E. Kress

Abstract Based on data in the aeronautics literature, we have derived functions for the drag coefficients of spheres and cubes as a function of Mach number. Experiments have shown that spheres and cubes exhibit an abrupt factor-of-two decrease in the drag coefficient as the object slows through the transonic regime. Irregularly shaped objects such as meteorites likely exhibit a similar trend. These functions are implemented in an otherwise simple projectile motion model, which is applicable to the non-ablative dark flight of meteors (speeds less than ~ 3 km/s). We demonstrate how these functions may be used as upper and lower limits on the drag coefficient of meteors whose shape is unknown. A Mach-dependent drag coefficient is potentially important in other planetary and astrophysical situations, for instance, in the core accretion scenario for giant planet formation.

Keyword meteorite · drag

1 Introduction

The scientific value of meteorites motivates many efforts to recover them quickly. To expedite the collection of meteorites resulting from observed fireballs, one must better constrain where they may have landed. Given an object's instantaneous position and velocity (in three dimensions), its mass (as inferred from deceleration) and its drag coefficient, its landing site can be constrained.

In recent years, detailed observations of meteors have been made by the European Fireball Network [16] [18], the Desert Fireball Network [3], and the Southern Ontario Network [22], resulting in the recovery of many meteorites. These detection networks can measure a meteor's position and velocity with very high precision. The Neuschwanstein Bolide was observed by the European Fireball Network; its angle is constrained to ±0.07°, the altitude to ±50 m, and the speed to ±800 m s^{-1} [17] [19]. Information about the mass of the object can also be derived from observations [17].

Considering the high accuracy to which observables can be measured, the drag coefficient introduces most of the uncertainty in calculating the landing site of these observed meteors. Detailed analysis of the drag coefficients of meteorites is daunting because meteorites are irregularly-shaped objects, and thus have different drag coefficients, not only for each individual object but also for the infinite number of orientations each can assume as they tumble, spin and ablate during flight. Nevertheless, there has been much effort to determine the drag coefficient for meteorite-like shapes (e.g. [23] [10]). Several studies have used or derived constant values for drag coefficients (e.g. [5] [23] [10] [21]).

R. T. Carter
Department of Physics, University of Oregon, Eugene, OR 97403 USA. E-mail: rcarter2@uoregon.edu

P. S. Jandir, Department of Physics, University of California, Riverside, CA 92521 USA

M. E. Kress (✉), Department of Physics & Astronomy, San José State University, CA 95192 USA. E-mail: mkress@science.sjsu.edu

Mach number, M, is an object's speed relative to the local speed of sound. In Earth's atmosphere, the speed of sound depends on the square root of the temperature of air and thus is primarily a function of altitude. At sea level, the speed of sound is approximately 340 km/s. We derived empirical equations for the drag coefficient as a function of Mach number from the data available in the aeronautical engineering literature [9] [11] [12]. In these studies, the drag coefficients for spheres were measured up to Mach 10 and for cubes up to Mach 3. From these data, we derive functions for the drag coefficients $C_D(M)$ for spheres and cubes. We use these two $C_D(M)$ functions as lower and upper limits for the drag on a meteor. We note that drag coefficients dependent on Mach number have been used in meteor physics before, e.g. [6] and often define the drag coefficient as Γ, which is related to C_D via the relation $\Gamma = 0.5C_D$. Here, when we refer to the drag coefficient, we use the C_D convention.

In section 2, we present the $C_D(M)$ functions for spheres and cubes, and explain how the equations of motion are solved. In section 3, we show how Mach-dependent drag affects the landing site for the idealized cases of spherical and cubic 'meteors.' We illustrate how the search area is smaller when using a $C_D(M)$ whose limits are set by a sphere and cube, compared to a search area delimited with constant C_D. In section 4, we discuss how $C_D(M)$ becomes increasingly important for meteors whose mass, velocity, altitude and entry angle cause them to spend more time at lower velocities where the drag coefficient varies the most. In section 5, we summarize our findings and discuss how a Mach-dependent drag coefficient may improve upon other planetary and astrophysical calculations.

2 Model

The drag coefficients for several regular shapes, including a sphere and a cube, have been experimentally measured over a wide range of Mach numbers [9] [11] [12]. In these ballistic range tests, pellets in the shape of spheres and cubes were shot out of high velocity guns. For all speeds, the drag coefficient for a cube is always about a factor of two greater than that for a sphere. However, spheres and cubes exhibit similar trends in $C_D(M)$ (see Figure 1). At low speeds, the drag coefficient of a cube is 1.09 and that of a sphere is 0.46. These both almost double as the speed increases to Mach 1.5, and then they level off to 1.7 and 0.9 respectively, at high Mach numbers. Therefore, the Mach number has as much of an effect on drag coefficient as does the shape, when comparing spheres and cubes.

Using these experimental data, best fit functions were derived to estimate the drag coefficient of spheres and cubes as a function of Mach number (Figure 1). Both of these drag coefficient functions are piece-wise. At lower speeds, C_D depends on M^2, and the best fit functions are quadratic. In the supersonic ($M > 1$) regime, the best fit functions are both a sum of two exponentials. The drag coefficient of a cube as a function of Mach number is

$$\begin{aligned} C_D(M) &= 0.60M^2 + 1.04, & 0 \le M \le 1.150 \\ C_D(M) &= 2.1e^{-1.16(M+0.35)} - 6.5e^{-2.23(M+0.35)} + 1.67, & M > 1.150 \end{aligned} \quad (1)$$

The drag coefficient of a sphere as a function of Mach number is

$$\begin{aligned} C_D(M) &= 0.45M^2 + 0.424, & 0 \le M \le 0.722 \\ C_D(M) &= 2.1e^{-1.2(M+0.35)} - 8.9e^{-2.2(M+0.35)} + 0.92, & M > 0.722 \end{aligned} \quad (2)$$

In the hypersonic regime, ($M > 5$), the drag coefficient for a cube levels off to a constant value of 1.67. That of a sphere approaches 0.92. In general, the drag coefficients of simple shapes tend to approach a constant value as Mach number increases. This tendency, known as Mach number independence, is well documented in aeronautics [2]. However, Mach number independence does not mean that models of meteors traveling at arbitrarily high Mach numbers should use a constant drag coefficient. At sufficiently high speeds, enough heat is generated to dissociate the air molecules surrounding a meteor. This changes the fluid properties of the air which, in turn, affects the drag coefficient [1]. Thus, our Mach-dependent drag coefficient is most applicable to a meteor's 'dark flight', which commences when a meteor's speed drops below Mach 10 [4] [20] [14]. At speeds greater than this, one must consider energy losses via radiation and ablation in the equations of motion. [7] [8]

We implement the Mach-dependent drag coefficients in a two-dimensional projectile motion calculation. Input parameters include the meteor's altitude, its velocity (speed and angle), its final (post-ablation) mass and the drag coefficient $C_D(M)$ of the object. Our model uses the fourth order Runge-Kutta method to solve the differential equations of motion,

$$\frac{\partial^2 x}{\partial t^2} = -\frac{F_{D,x}}{2m}\left(\frac{\partial x}{\partial t}\right) \quad (3)$$

$$\frac{\partial^2 y}{\partial t^2} = -\frac{mg - F_{D,y}}{2m}\left(\frac{\partial y}{\partial t}\right) \quad (4)$$

where x is the horizontal position along the direction of flight, y is the vertical position of the bolide, m is the mass of the bolide, $F_{D,x}$ is the force due to drag in the x-direction (= $\frac{1}{2}C_D(M)\rho A\cos\theta v^2$) and likewise for the y-direction, v is the speed of the object, A is the cross-sectional area of the meteoroid, θ is the angle of the velocity vector with respect to horizontal, and ρ is the density of air. Atmospheric density, temperature, and pressure are calculated as a function of altitude using the U.S. 1976 Standard Atmosphere [13]. We emphasize that this expression for the drag force is appropriate for an object that is in dark flight (velocity less than about Mach 10).

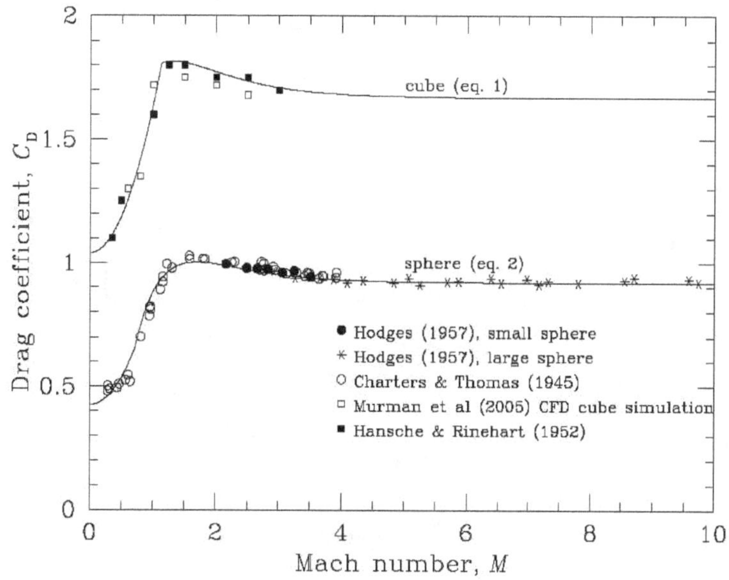

Figure 1. Drag coefficients as a function of Mach number for a sphere and a cube. Our best fit functions for a sphere and a cube (equations 1 and 2) are plotted along with experimental data [9] [11] [12]. Also shown are more current numerical results for a cube [15].

3 Results

We begin by applying the model to the idealized case of a sphere. This allows us to calculate the exact landing site, using our Mach-dependent drag coefficient for a sphere. We compare this to the landing position calculated using a constant low-Mach drag coefficient for a sphere of 0.464.

For the initial angle, speed, and altitude of this bolide, we use the values measured for the Neuschwanstein main body fragment as observed just before dark flight. We chose this meteor because its properties were very well measured by the European Fireball Network: studies of this event determined an angle of 49.23° ± 0.07° measured from the horizontal, a speed of 3.1 ± 0.8 km s^{-1}, and an altitude of 16.06 ± 0.05 km [19] [17]. The mass of the Neuschwanstein main body fragment is estimated to be 15 kg [17]. We will refer to these as the 'Neuschwanstein parameters' in the discussion that follows.

When the drag coefficient of this spherical meteor is kept constant at its low-velocity value of 0.464, the drag force is greatly underestimated. The sphere is predicted to land approximately 1880 m further downrange compared to the result obtained with the Mach-dependent drag coefficient given in Equation 1.

We did the same analysis for a cube-shaped meteor. The landing site computed with the Mach-dependent drag coefficient given in Equation 2 lands about 806 m further uprange than that calculated with a constant drag coefficient for a cube of 1.094. The cube is less affected because its greater drag causes it to spend more time with a C_D closer to its low-Mach value. The sphere, on the other hand, stays at high speeds longer, and thus spends more time with a C_D that is much greater than its low-Mach value. The more aerodynamically-shaped the object, the more important is the use of a Mach-dependent drag coefficient.

In Figure 2, we show the landing ellipses for a meteor coming in from the left. The landing ellipse is the area in which the resulting meteorite is likely to fall, and is determined by uncertainties in the meteor's motion, shape and mass. This meteor has Neuschwanstein parameters as described above.

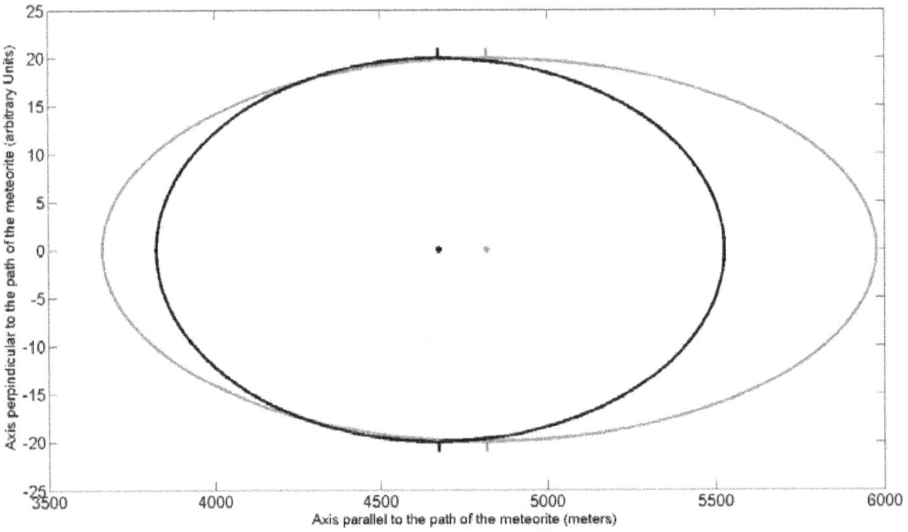

Figure 2. Landing ellipses for a meteor with Neuschwanstein parameters (see text). The length of the black ellipse is bound by the Mach-dependent drag coefficients $C_D(M)$ for a sphere and cube (equations 1 and 2). The gray ellipse is bound by using a constant drag coefficient of 0.7 and 1.6. The centroid of the Mach-dependent calculation is shifted uprange by approximately 150 m; the area of the constant-drag ellipse is 1.36 times that of the $C_D(M)$ ellipse.

The vertical axis of the ellipse is perpendicular to the meteor's velocity. Its units are arbitrary, determined in practice by crosswinds and uncertainty in the meteor's position and velocity. The horizontal axis is parallel to the meteor's flight. The gray ellipse is calculated with constant values for C_D and the black ellipse is calculated using our $C_D(M)$ method. The points indicate the positions of the centroids of the ellipses.

The major axis of the constant C_D ellipse is determined by the low and high drag estimates for a meteor, 0.7 and 1.6, respectively. The value of 1.6 is that of a cube at high Mach numbers [23], which we assign to be the upper limit. A C_D of 0.7 is the low-Mach value from Ceplecha [6], which we use here as a lower limit on the drag coefficient.

The length of $C_D(M)$ ellipse is bound by the $C_D(M)$ functions for a sphere and cube (Equations 1 and 2). The centroid of the $C_D(M)$ ellipse is shifted approximately 150 m uprange from the centroid of the constant-drag ellipse. The area of an ellipse is πab, where a and b are the semi-major and semi-minor axes, and thus the ratio of the two areas is the ratio of the semi-major axes (given the same uncertainty in semi-minor axes). The area of the landing ellipse calculated using constant values for drag is about 1.36 times greater than the area of the one bound by our Mach-dependent drag coefficients for spheres and cubes.

4 Discussion

When using our Mach-dependent drag model to constrain the search ellipse for real meteors, the assumption is that the object will be less aerodynamic than a sphere and more aerodynamic than a cube. Our discussion here also implicitly assumes that the irregularly-shaped object will have a Mach-dependent drag coefficient that follows the same trend as that for spheres and cubes (i.e. a decrease in drag coefficient as the object slows through the transonic regime).

The Neuschwanstein bolide was a deeply-penetrating fireball that entered dark flight at an altitude of just over 16 km. Dark flight often begins at much higher altitudes. A recent example is the Bunburra Rockhole meteorite, which began dark flight at an altitude of 30 km [20]. In these types of falls, the altitude introduces more time for drag to operate, which means a substantially larger search area, regardless of what method is used to calculate it. Thus, constraining the search area even by 27% (from the example of the Neuschwanstein fireball) can be helpful to meteorite hunters.

We conducted an exploration of parameter space to determine what meteors would be most affected by our Mach-dependent drag coefficient. Not surprising, the Mach-dependent drag was more important for bolides that entered dark flight at higher altitudes and at shallower angles. The mass had a less dramatic effect: only for masses smaller than 1 kg did the landing ellipse change substantially. The speed also had an important influence on the landing ellipse. We show a plausible scenario in Figure 3, in which we constrain the landing site of a (non-ablating) meteor with a mass of 15 kg, at an altitude of 36 km, a speed of 7.1 km s^{-1}, and an angle of 29° measured from the horizontal. (We chose a velocity higher than the 3 km/s upper limit on the non-ablative phase of dark flight to force the meteor to spend more time in the hypersonic regime where C_D does not vary as much. Meteors with lower initial speeds will have more of their flight take place in the $M < 5$ regime where the C_D varies significantly.) In this case, the areas of the landing ellipses calculated with the constant-drag method and our method have the same ratio (about 1.36). However, the centroid of the ellipse found with our method is about 1.4 km uprange, and this ellipse is no longer entirely contained within the constant-drag ellipse (see Figure 3).

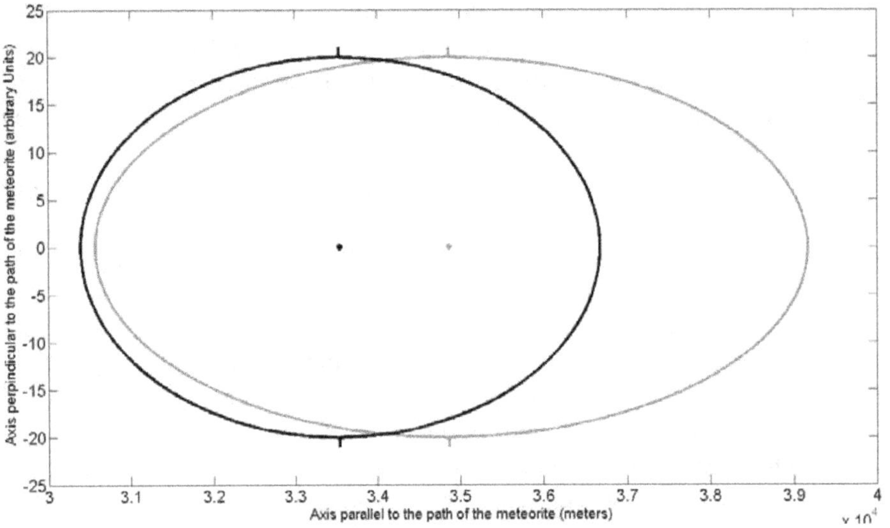

Figure 3. Landing ellipses for a hypothetical but plausible meteor. The length of the black ellipse is bound by the Mach-dependent drag coefficients $C_D(M)$ for a sphere and cube (equations 1 and 2). The gray ellipse is bound by using a constant drag coefficient of 0.7 and 1.6. The centroid of the Mach-dependent calculation is shifted uprange by approximately 1400 m; the area of the constant-drag ellipse is 1.36 times that of the $C_D(M)$ ellipse.

When a meteor has been observed and successfully recovered, one can empirically derive a constant drag coefficient that reproduces the known landing site. However, this technique has no predictive power. It is unlikely that the same constant C_D will work in a model of any other meteorite with different initial conditions, even if it has similar mass and shape. A difference in angle, speed, and starting altitude will affect the amount of time it spends in the high vs. low Mach/drag regimes. Instead, a Mach-dependent drag coefficient, similar to equations 1 and 2, can derived for meteorites that have been recovered after observed falls.

5 Conclusions

The drag coefficient, C_D, is an important parameter in determining the landing site of observed meteors, but the value for C_D is notoriously elusive. Aerodynamic studies of regular shapes (spheres and cubes) reveal that an object's speed affects its drag coefficient as much as its shape does. This trend likely holds true for irregularly-shaped objects such as meteorites. All but the most oddly-shaped meteoroids likely have shapes such that their aerodynamic properties cause them to suffer less drag than cubes and more drag than spheres of the same mass traveling at the same speed.

We have derived functions for drag coefficients as a function of Mach number from experimental data for spheres and cubes. These functions can be readily implemented into any model for meteor trajectories that currently uses a constant value for the drag coefficient. Dark flight trajectories and landing ellipses were then computed using these drag coefficient functions, and compared to those computed using constant drag coefficients. The landing ellipses bound by spheres and cubes is smaller compared to those calculated using constant drag coefficients from 0.7 to 1.6. We also find that the centroids of the ellipses are shifted uprange, sometimes by more than 1 km.

More work must be done to understand the changes in drag coefficients due to the dissociation of air at the high temperatures generated around the incoming bolide. Accounting for this phenomenon is necessary in order to build a complete model of meteor flight. This issue has been explored by scientists in the field of aerothermodynamics, but has not yet been applied to meteor physics. Current studies in aeronautics may yield useful results for meteor scientists, for instance studies of irregular solids undergoing atmospheric entry (satellite wreckage and other space debris, e.g. [15].

Finally, we anticipate other astrophysical and planetary applications of a Mach-dependent drag coefficient. For instance, the drag coefficient is an important factor in calculating the motion of planetesimals in gas-rich astrophysical disks such as the solar nebula, and also in the giant planet subnebulae. In these scenarios, the drag coefficient is assumed to be constant, usually 1 (e.g. [21]). Variable drag coefficients may have important consequences for the timescales for planet formation and for the coagulation of solids in planet-forming disks around other stars.

Acknowledgements

The authors thank the US National Science Foundation for supporting RTC and PSJ via a Research Experience for Undergraduates grant to SJSU. MEK acknowledges support from the NASA Astrobiology Institute. We would also like to thank Gary Allen, Jeff Brown, Stuart Rogers, Jeff Hollingsworth, Laura Iraci, Prestin Martin, and Jeff Johnson for their suggestions for the manuscript and James Schombert, Scott Murman, Wendy Sullivan, Alejandro Garcia, Michael Kaufman, and Joshua Johnson for additional help.

References

1. Anderson, J.D., Fundamentals of aerodynamics, McGraw-Hill, New York (2001)
2. Anderson, J.D., Hypersonic and high-temperature gas dynamics, American Institute of Aeronautics and Astronautics, Reston, Virginia (2006)
3. Bland, P.A., Astron. Geophys., 45, 520 (2004)
4. Borovicka, J., Kalenda P., Meteorit. Planet. Sci., 38, 1023 (2003)
5. Brown, P., Hildebrand, A.R., Green, D.W.E., Page, D., Jacobs, C., ReVelle, D., Tagliaferri, E., Wacker, J., Wetmiller, B., Meteorit. Planet. Sci., 31, 502 (1996)
6. Ceplecha, Z., Bull. Astron. Inst. Czechosl., 38, 222 (1987)
7. Ceplecha, Z., Borovicka, J., Elford, W.G., Revelle, D.O., Hawkes, R.L., Porubcan, V., Simek, M., Space Sci. Rev., 84, 327(1998)
8. Ceplecha, Z., ReVelle, O., Meteorit. Planet. Sci., 40, 35 (2005)
9. Charters, A.C., Thomas, R.N., J. Aeronautical Sci., 12, 468 (1945)
10. Gritsevich, M.I., Stulov, V.P., Solar System Res., 42, 118 (2008)
11. Hansche, G.E., Rinehart, J.S., J. Aeronautical Sci., 19, 83 (1952)
12. Hodges, A.J., J. Aeronautical Sci., 24, 755 (1957)
13. Lewis, B., The Complete 1976 Standard Atmosphere, http://www.mathworks.com/matlabcentral/fileexchange/13635-complete-1976-standard-atmosphere (2007)
14. McCrosky, R.E., Posen, A., Schwartz, G., Shao, C.-Y., SAO Special Report, 336, 1 (1971)
15. Murman, S.M., Aftosmis, M.J., Rogers, S.E., American Institute of Aeronautics and Astronautics, 43, 1223 (2005)
16. Oberst, J., Molau, S., Heinlein, D., Gritzner, C., Schindler, M., Spurný, P., Ceplecha, Z., Rendtel, J., Betlem, H., Meteorit. Planet. Sci., 33, 49 (1998)
17. Oberst, J., Heinlein, D., Kohler, U., Spurny, P., Meteorit. Planet. Sci., 39, 1627 (2004)
18. Porubcan, V., Svoren, J., Husárik, M., Kanuchová, Z., Contrib. Astron. Obs. Skalnat´e Pleso, 39, 101 (2009)
19. Spurný, P., Heinlein, D., Oberst, J., Asteroids, Comets, Meteors, ed: B. Warmbein, ESA Publications Division, Noordwijk, 137 (2002)

20. Spurný, P., Bland, P.A., Borovicka, J., Shrbeny, L., McClafferty, T., Singleton, A., Brevan, W. R., Vaughan, D., Towner, M.C., Deacon, G., 40th Lunar and Planetary Sciences Conf.,1498, 1 (2009)
21. Tanigawa T., Ohtsuki, K., Icarus, 205, 658 (2010)
22. Weryk, R.J., Brown, P.G., Domokos, A., Edwards, W.N., Krzeminski, Z., Nudds, S.H., Welch, D.L., Earth, Moon, Planets, 102, 241 (2008)
23. Zhdan, I.A., Stulov, V.P., Stulov, P.V., Turchak, L.I., Solar System Research, 41, 505 (2007)

The Trajectory, Orbit and Preliminary Fall Data of the JUNE BOOTID Superbolide of July 23, 2008

N. A. Konovalova • J. M. Madiedo • J. M. Trigo-Rodriguez

Abstract The results of the atmospheric trajectory, radiant, orbit and preliminary fall data calculations of an extremely bright slow-moving fireball are presented. The fireball had a -20.7 maximum absolute magnitude and the spectacular long-persistence dust trail (Fig 1 and 2) was observed in a widespread region of Tajikistan twenty eight minutes after sunset, precisely at $14^h 45^m 25^s$ UT on July 23, 2008. The bolide was first recorded at a height of 38.2 km, and attained its maximum brightness at a height of 35.0 km and finished at a height of 19.6 km. These values are very much in line with other well-known fireballs producing meteorites. The first break-up must have occurred under an aerodynamic pressure P_{dyn} of about 1.5 MPa, similar to those derived from the study of atmospheric break-ups of previously reported meteorite-dropping bolides. Our trajectory, and dynamic results suggest that one might well expect to find meteorites on the ground in this case. The heliocentric orbit of the meteoroid determined from the observations is very similar to the mean orbit of the June Bootid meteor shower, whose parental comet is 7P/Pons-Winnecke (Lindblad et al. 2003). If the parent was indeed a comet, this has implications for the internal structure of comets, and for the survivability of cometary meteorites.

Keywords fireball · meteoroid · atmospheric trajectory · radiant · orbit · fall data

1 Introduction

On July 23, 2008 at $14^h 45^m 25^s$ UT, two minutes after sunset, an extremely bright slow-moving fireball and the spectacular dust trail that it left behind were witnessed by numerous casual witnesses in a widespread region of Tajikistan. The area from which it was seen was within a radius of about 300 km. The sky in the area of the fireball was completely clear. The event was bright enough to be recorded by video and photo cameras, and, therefore, the time of the fireball passage is reliable derived from these observations. The intensity of the flash was so great that at about 100 km from the epicenter it was possible to be noticed by persons through the windows inside rooms. The majority of the eyewitnesses agree that suddenly the fireball became very bright. Some observers reported intense sounds, such as "cracking" and "thunder" just after the fireball appearance. A sonic boom probably associated with the bolide fragmentation at the height of the brightest flare was heard as far as 100 km away from the burst location. Many witnesses watched the resulting dust trail, which was perceptible for about 20 minutes

N. A. Konovalova (✉)
Institute of Astrophysics of the Academy of Sciences of the Republic of Tajikistan, Bukhoro, str. 22, Dushanbe 734042, Tajikistan. E-mail: nakonovalova@mail.ru.

J. M. Madiedo
Facultad de Ciencias Experimentales. Universidad de Huelva. 21071 Huelva, Spain

J. M. Trigo-Rodriguez
Institute of Space Sciences (CSIC-IEEC). Campus UAB, Facultat de Ciències, Torre C5-p2. 08193 Bellaterra, Spain

Figure 1. The trail of the July 23, 2008 fireball from HisAO (Tajikistan). The photograph by U. Hamroev was taken at 14:45 UT

Figure 2. The trail of the July 23, 2008 fireball from Vose (Tajikistan). The photograph by M. Ahmetzyanov was taken at 14:58:58 UT

and was expressive distorted by the atmospheric winds. The trail of the fireball was very bright exhibiting a blue-white and on the end orange-red colors. The fireball was observed by a visible-light satellite system which detected a brightest path of fireball light. The total radiated energy was 2×10^{11} J, which is equivalent to a total released energy of about 0.05 kT (Brown 2008). The fireball's flare reached an absolute magnitude of -20.7 putting it in the superbolide category. One day after the event we interrogated inhabitants and a few witnesses have furnished numerous photographs of the dust trail of which two double-station photographs (baseline of 11.3 km) fortunately were taken immediately after the flight of fireball, showing clear references for being calibrated. The trajectory of the fireball was in fact photographed only during the later stages of its path. The terminal point of the fireball can be seen close to the western horizon.

2 Data Obtained

On the basis of two available double-station records (baseline of 11.3 km) the astrometric calibration of the fireball apparent trajectory in reference to the stars was made following the standard procedure and method described in (Katasev 1966). The procedure to obtain the astrometric measurements was based on the use of a Zeiss Ascorecord device. Measuring the rectangular coordinates of the positional stars and any feature point (beginning, terminal, and all flares and depressions) on the fireball trail, such measurements were converted to equatorial coordinates by using the astrometric method of the METEOR software package developed by the Meteor department (Institute of Astrophysics, Tajikistan). As a result of the astrometric measurements we were able to determine the fireball atmospheric trajectory, radiant, velocity, and orbit. The exact duration of the fireball was known from the data of a visible-light satellite system (Brown 2008). The fireball was first recorded at a height, H_b of 38.2 ± 0.5 km when the velocity, v_b was 14.3 ± 0.5 km/s. The fireball traveled a 19-km observed luminous trajectory and terminated its light at a low altitude H_e of 19.6 ± 0.5 km when the fireball decelerated to 5.8 ± 0.5 km/s. The slope of the trajectory was extremely steep - the zenith distance of the radiant was only of about 10° and the difference between the beginning and the terminal height was 18.6 km. The brightest flare was near the beginning of the trajectory at the height H_{max} = 35.0 ± 0.5 km when the first break-up must have occurred under an aerodynamic pressure P_{dyn} of about 1.5 MPa. At the heights of other two small flares the aerodynamic pressure was 2.9 MPa and 3.1 MPa respectively. The apparent radiant was in Bootes, which suggest that the bolide belongs to the J.Bootid meteor shower. The resulting data on the atmospheric trajectory and coordinate of radiant, calculated by software METEOR are given in Table 1 and Table 2.

Table 1. Atmospheric trajectory data

	Beginning	Maximum light	Terminal
Velocity (km/s)	14.3 ± 0.5	13.1 ± 0.5	5.8 ± 0.5
Height (km)	38.2 ± 0.5	35.0 ± 0.5	19.6 ± 0.5
Abs. magnitude	-	- 20.7	-

Table 2. Radiant data

Radiant (J2000.0)	Observed	Geocentric	Heliocentric
α_R (deg)	221.83 ± 2.1	219.52 ± 2.1	-
δ_R (deg)	+32.40 ± 2.1	+30.95 ± 2.1	-
initial velocity v_∞ (km/s)	16.0	11.6	38.5

In order to obtain an accurate orbit, it is necessary for the fireball to be observed from multiple stations. In spite of the presence of only two records of the 23 July, 2008 fireball, the resulting data have a good accuracy. The initial (pre-atmospheric) velocity, v_∞ = 16 km/s was used for the meteoroid orbit computations. The resulting elements of the heliocentric orbit for the equinox 1950.0, calculated by software METEOR are given in Table 3.

Table 3. Orbital data

Orbit (J2000.0)	
Semimajor axis (AU)	3.32
Eccentricity	0.694
Perihelion distance (AU)	1.015
Aphelion distance (AU)	5.624
Argument of perihelion (deg)	176.76.
Ascending node (deg)	119.709
Inclination (deg)	11.95°

These orbital results were also tested by using the software of the Spanish Fireball Network (SPMN). It is remarkable that the computed orbit is very similar to the mean orbit of the June Bootid meteor shower (Lindblad et al. 2003). Both the *D*-criterion of Southworth and Hawkes (Southworth and Hawkes 1963) and Drummond (Drummond 1981) were applied to the obtained orbit (D_{SH} = 0.114 and D_{Dr} = 0.082). The meteoroid, with a mass of about 24 tons and a kinetic energy, possible in the order of 0.5 – 0.6 kt (personal communication Dr. O. Popova) entered the Earth's atmosphere with velocity of about 16 km/s. Penetrating deeply into the atmosphere, the aerodynamic pressure increased progressively on the front part of the body. At a height of 35 km the pressure on the surface reached 1.4 MPa causing the fracture of the body, so its mass started to break and crumble. Among the relatively few documented cases of collisions with bodies of such a great mass, this is a remarkable event because the body, probably of cometary origin, penetrated deeply into atmosphere and probably dropped meteorite. We hope to have the chance of recovering meteorites from this event occurred over an inhabited region.

3 Data of Dark Flight

The dark flight was simulated by using the standard procedure described in (Ceplecha 1987). The deceleration at the terminal point of the trajectory obtained from the estimated values of fireball velocity vs. height was -7.6 km/s^2, with a trajectory inclination of about 80º. The simulation was performed by taking into account the modeled atmospheric conditions (provided by the British Atmospheric Data Center) with a software package developed by the Spanish Meteor Network (SPMN) which uses a standard Runge-Kutta calculation procedure. Spherical shape was assumed and a value of the drag factor at the terminal point of 0.58 was used. Under these conditions, the terminal mass of the meteoroid, which depends on its density, would vary from 1.5 ± 0.3 kg (d = 3.7 g/cm^3) to 4.2 ± 0.3 kg (d = 2.2 g/cm^3). Most of the surroundings near of the impact area are covered by agricultural fields, without stones or rocks that would complicate meteorite searches, as the soil consists basically of clay. Thus the favorable circumstances of an almost vertical fireball trajectory and favorable countryside give a hope of successful search of the meteorite. At present, the search of additional and more detailed data is in progress and expeditions to the impact area will be organized in short in order to try to find the meteorite.

Acknowledgements

We are grateful to Dr. O. Popova for the presentation our work at the conference METEOROIDS 2010. We are grateful to all the witnesses of the fireball of July 23, 2008 who volunteered their observations, and especially to those who offered photographs of the fireball.

References

P. Brown, htpp://astroalert.su/files/bolide, (2008)
Z. Ceplecha, Astronomical Institutes of Czechoslovakia, Bulletin 38, (1987)
J.D. Drummond, Icarus, 45, 545, (1981).
L.A. Katasev, Investigations of the meteors in the Earth's atmosphere. L.:Gidrometeoizdat ,336, (1966)
B.A. Lindblad; L. Neshisan; V. Porubcan; J. Svoren, Earth, Moon and Planets, 93, 4, 249, (2003)
R.B. Southworth, G.S. Hawkes, Smith. Contrib. Astrophys., 7, (1963)

Infrasonic Detection of a Large Bolide over South Sulawesi, Indonesia on October 8, 2009: Preliminary Results

E. A. Silber • A. Le Pichon • P. G. Brown

Abstract In the morning hours of October 8, 2009, a bright object entered Earth's atmosphere over South Sulawesi, Indonesia. This bolide disintegrated above the ground, generating stratospheric infrasound returns that were detected by infrasonic stations of the global International Monitoring System (IMS) Network of the Comprehensive Nuclear-Test-Ban Treaty Organization (CTBTO) at distances up to 17 500 km. Here we present instrumental recordings and preliminary results of this extraordinary event. Using the infrasonic period-yield relations, originally derived for atmospheric nuclear detonations, we find the most probable source energy for this bolide to be 70 ± 20 kt TNT equivalent explosive yield. A unique aspect of this event is the fact that it was apparently detected by infrasound only. Global events of such magnitude are expected only once per decade and can be utilized to calibrate infrasonic location and propagation tools on a global scale, and to evaluate energy yield formula, and event timing.

Keywords infrasound · bolide · fireball · airburst · impact

1 Introduction

Medium sized Near Earth Objects (NEOs) (>10 m diameter) may penetrate deep into the atmosphere, though rarely, and cause significant damage on the ground (Chapman, 2008) and could potentially perturb regional climate trends (Toon et al. 1997). However, currently available models cannot accurately define the critical impactor size at which the regional climate is affected (Bland and Artemieva, 2003). A part of the problem is limited observational data, as records of significant NEOs are scarce. Therefore, various observational methods, including infrasound, are critical to re-evaluate airburst models and determine with more accuracy the size at which an object can influence the local climate.

Records of significant NEO impacts are rare. Klekociuk et al. (2005) and Arrowsmith et al. (2008) report multi-instrumental observations of two different impactors with energies of 20-30 kilotons of TNT (1 kT = 4.185×10^{12} J) occurring in the fall of 2004, while Brown et al. (2002) present infrasound data for two somewhat less energetic events over the Pacific in 2001. In all cases these events occurred over open ocean and much of the energetics information was compiled from records of the associated airwaves detected by infrasonic stations.

E. A. Silber (✉) • P. G. Brown
Department of Physics and Astronomy, University of Western Ontario, London, ON N6A 3K7, Canada. Phone:1-519-661-2111 x82385; Fax:1-519-661-2033; E-mail: elizabeth.silber@uwo.ca

A. Le Pinchon
CEA/DAM/DIF, F-91297 Arpajon, France

Infrasound is low frequency sound extending below the 20 Hz hearing threshold of the human ear and just above the natural oscillation frequency of the atmosphere (>0.01 Hz, Brunt-Väisälä frequency). It has the ability to propagate over long distances with very little attenuation, thus enabling the study of remote explosive sources (Hedlin et al., 2002). The International Monitoring System (IMS), operated by the Comprehensive Nuclear-Test-Ban Treaty Organization (CTBTO), features as one of its monitoring technologies, a global network of 42 fully certified infrasonic stations designed to detect nuclear explosions (CTBTO web: http://www.ctbto.org).

Bright meteors (also known as fireballs) fall into the category of events that can be detected and consequently studied using infrasound (ReVelle, 1976, 1997; Brown et al., 2002a). Fireballs are produced by large meteoroids which may penetrate deep into the atmosphere and generate a cylindrical blast wave during their hypersonic passage, which decays to low frequency infrasonic waves that propagate over great distances (ReVelle, 1976; Edwards, 2010; Le Pichon et al., 2002a, Brown et al., 2002; Brown et al., 2003). Global impacts detected infrasonically can provide a valuable tool in the estimation and validation of the influx rate of meter sized and larger meteoroids (Brown et al. 2002; Silber et al. 2009). Very often, infrasound offers the only available record when it comes to major impacts over open ocean. Infrasound observations can provide crucial trajectory and energetics information for interesting events which otherwise lack such information (e.g. the Carancas crater forming impact in Peru in 2007 (Brown et al., 2008; Le Pichon et al., 2008)). Here we present evidence that a significant NEO impact occurred on October 8, 2009 over South Sulawesi, Indonesia based primarily on infrasonic recordings of the blast wave detected across the globe; this may have been one of the most energetic impactors to collide with the Earth in recent history.

2 The Indonesian Bolide

At 2:57 UTC (10:57 a.m. local time) on October 8, 2009 a loud rumbling sound and ground shaking startled the people of the town of Bone, South Sulawesi, Indonesia (4.5°S, 120°E). Eyewitnesses who ran out their homes in fright saw a very bright object flying across the sky, subsequently disintegrating in the mid air, leaving a thick dusty smoke trail behind (Surya news report, in Indonesian: http://www.surya.co.id/2009/10/09/ledakan-misterius-guncang-sulsel.html). A news article stated that there are reports from local residents that the surviving remnants of the object may have crashed into the sea (Surya news report, in Indonesian: http://www.surya.co.id/2009/10/09/leda kan-misterius-guncang-sulsel.html).

Shortly thereafter, the national media, including Metro TV of Jakarta and two news agencies, The Jakarta Globe and The Jakarta Post, released a number of reports, including an amateur video of the smoke trail (Figure 1). Features and the appearance of the smoke trail are consistent with dust trails of other fireballs observed in a similar manner (e.g. the Tagish Lake fireball (Hilderbrand et al, 2006)), indicating a probable meteoritic origin of the event. As per The Jakarta Globe, the airburst caused damage to several houses in Panyula village (The Jakarta Globe, available at: http:// www.thejakartaglobe.com/home/astronomer-sulawesi-blast-bigger-than-atom-bomb-and-caused-by-met eorite/338073) and the police department in Bone was flooded with reports of audible sounds extending as far as 11 km from Latteko, Bone district, South Sulawesi (The Jakarta Globe: available at: http://thejakartaglobe.com/home/mysterious-explosion-panics-locals-in-south-sulawesi-police-still-investigating/334246). Unfortunately, there was one casualty, a 9 year old girl with an underlying heart condition who went into cardiac arrest upon hearing the thunderous sounds (The Jakarta Globe, available at: http://www.thejakartaglobe.com/home/astronomer-sulawesi-blast-bigger-than-atom-bomb-

and-caused-by-meteorite/338073). Initially, local people speculated that the event was caused by a falling airplane; however, South Sulawesi Police spokesman Sr. Comr. Hery Subiansauri confirmed that no aircraft was involved nor any other air incident had occurred. The extraterrestrial nature of the event was confirmed by Thomas Djamaluddin, head of the Lapan Center for Climate and Atmosphere Science (The Jakarta Post, available at: http://www.thejakartapost.com/news/2009/10/08/blast-may-be-result-falling-space-waste-or-meteorite-lapan.html).

Upon scrutinizing scrutinizing these reports, we undertook a thorough investigation of infrasonic records of all IMS infrasound stations to search for possible signals from the air explosion.

Figure 1. A screenshot from Metro TV news report showing an amateur video of the smoke trail, twisted by the wind (You Tube, available at: http://www.youtube.com/watch?v=yeQBzTkJNhs&videos=jkRJgbXY-90).

3 Data Processing and Analysis

We were able to examine a total 31 infrasound stations in the IMS network which were providing data at the time of the event. Probable signals originating from 4.5°S, 120°E were detected at 17 IMS stations (Figure 2), which we correlated with the event. Table 1 summarizes data from all stations which detected the signal, sorted by distance. The signal was extraordinary in two aspects: first, it was detected by many infrasound stations, some of which are at extreme distances (>17,000 km), and second, that most of the signal energy is contained in very low frequencies, indicative of a source yielding very high energy. Infrasonic signals were analyzed using two independent methods, Matseis 1.7 (Harris and Young, 1997; Young et al., 2002) and Progressive Multi-Channel Correlation Method (PMCC) (Cansi, 1995).

First, infrasound data across each station have been array processed in windows (typically of 30-60 second length) to search for coherent signals with consistent back-azimuth measurements for several adjacent windows using the analysis package Matseis 1.7 (Harris and Young, 1997; Young et al., 2002). To determine the arrival azimuth for a coherent signal, we used the standard method of cross-correlating the output between each sensor of an array and performed beamforming of the signals across the array

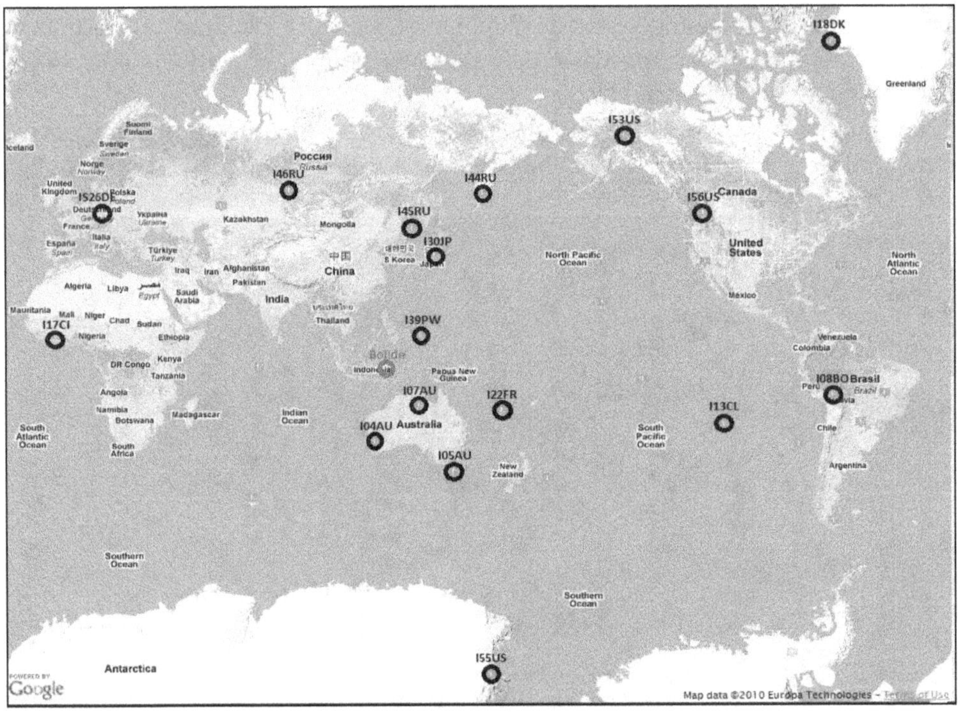

Figure 2. A global map (courtesy of CTBTO, web: http://www.ctbto.org) showing all stations (black circles) that detected the Indonesian bolide event circled in red.

Table 1. Summary of all detections, sorted by distance. We include results for two methods of signal detection (MatSeis and PMCC).

Distance (km)	Station ID	Latitude (deg)	Longitude (deg)	True Back Azimuth (deg)	Observed Back Azimuth (deg)	Arrival time	Signal Duration (s)	Minimum Celerity (m/s)	Maximum Celerity (m/s)	Peak-to-peak Amplitude via PMCC (Pa)	Peak-to-peak Amplitude via MatSeis (Pa)	Period at max Amplitude via PMCC (s)	Period at max PSD via MatSeis (s)	Period at max Amplitude via MatSeis (s)
2099	I39PW	7.5	134.5	230	264	04 :39 :51	1235	283	340	...	1.57	...	13.65	14.87
2291	I07AU	-19.9	134.3	316	318	04 :55 :46	850	287	320	2.823	3.091	6.96	7.88	5.79
3350	I04AU	-34.6	116.4	7	9	05 :59 :18	1370	271	305	0.471	0.526	5.36	7.31	7.11
4920	I30JP	35.3	140.3	210	211	07 :33 :43	1280	280	302	0.642	0.6077	25.60	7.88	7.89
5009	I05AU	-42.5	147.7	319	319	07 :37 :01	690	280	292	0.542	0.874	10.50	29.26	25.23
5386	I22FR	-22.2	166.8	284	285	07 :45 :08	1340	290	312	0.165	0.127	5.30	20.48	21.07
5543	I45RU	44.2	132.0	196	197	08 :04 :54	1450	278	300	1.192	1.1873	10.70	17.07	19.79
7296	I46RU	53.9	84.8	222	224	09 :46 :19	1490	281	298	0.803	...	15.20	...	...
7323	I44RU	53.1	157.7	141	141	09 :49 :46	2450	268	294	0.363	0.7896	6.99	18.62	18.29
8577	I55US	-77.7	167.6	311	305	10 :55 :07	1060	289	299	0.168	0.145	12.10	17.07	17.62
10573	I53US	64.9	-147.9	270	270	12 :49 :47	830	291	297	0.488	0.418	12.70	12.80	14.66
11594	I26DE	48.8	13.7	80	80	14 :28 :51	185	278	279	0.04	...	5.48	...	...
11900	I18DK	6.7	-4.9	350	340	14 :15 :26	1100	284	292	0.693	0.645	18.10	25.60	21.81
12767	I56US	48.3	-117.1	293	322	14 :54 :45	1520	286	292	0.765	0.764	14.70	13.65	11.83
13636	I13CL	15.3	-23.2	244	240	16 :26 :53	1310	273	281	0.618	0.606	12.10	11.38	11.31
13926	I17CI	-33.7	-78.8	91	87	17 :05 :34	615	270	274	0.128	0.1347	12.10	9.31	8.64
17509	I08BO	-16.2	-68.5	203	218	18 :54 :45	30	...	305	...	0.933	...	17.07	16.34

(Evers and Haak, 2001). A sample output is shown in Figure 3. In total 15 positive detections were identified in this way, using the approximate location and timing from media reports and expected typical stratospheric propagation speeds as a guide to isolate the period of most probable signal arrival on each array. This procedure was repeated for multiple bandpasses to try and isolate any coherent signal from the station noise.

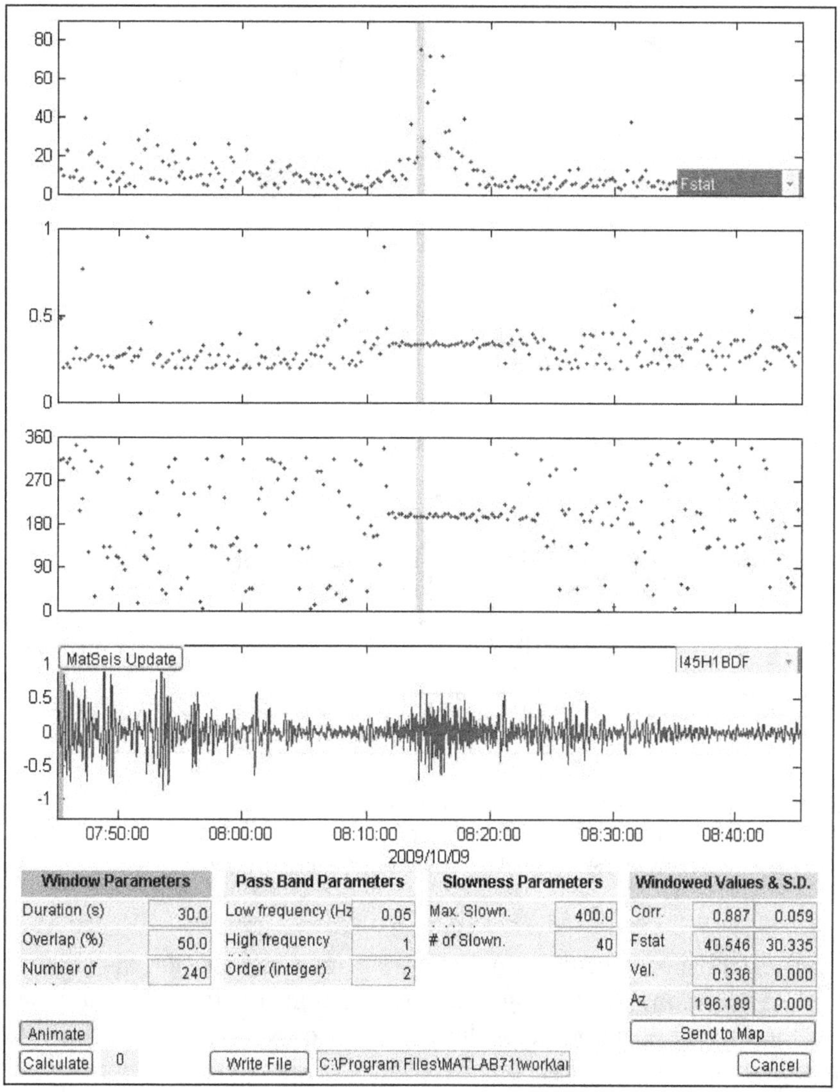

Figure 3. An example of signal observed at I45RU, located 5543 km from the source. The top window is the F-statistic, a measure of the relative coherency of the signal across the array elements in any particular window, the second window represents the apparent trace velocity of the acoustic signal across the array in the direction of the peak F-stat, while the third window shows the best estimate for the signal back-azimuth in the direction of maximum F-stat for each window. The fourth window shows the bandpassed raw pressure signal for one array element.

The second method, PMCC, for analysing the data, sensitive to coherent signals with very low signal-to-noise ratio (SNR), yielded positive detections at a total of 16 IMS stations. This technique has been successfully implemented in detections of other bolides (cf. Arrowsmith et al., 2008), as it searches for coherent signals in both frequency and time windows, selecting detections of similar parameters to identify coherent signals (Brachet et al., 2010) (Figure 4).

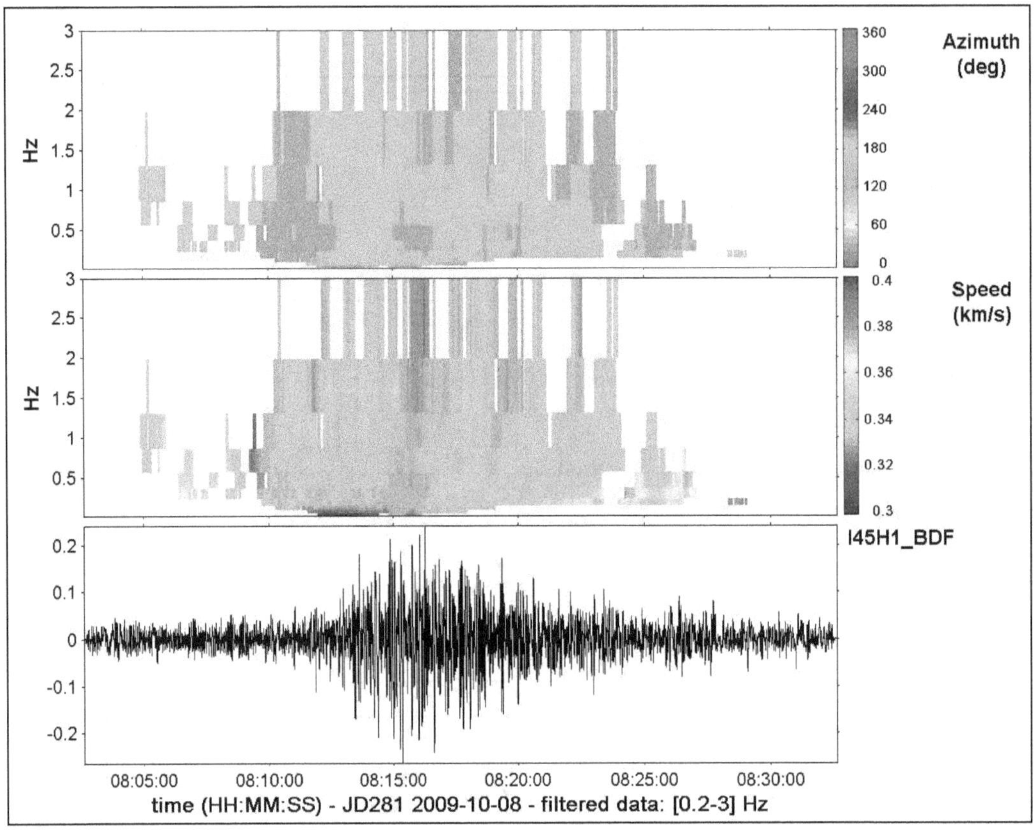

Figure 4. Results from array processing using the PMCC algorithm for the IMS station I45RU. The top window gives the observed azimuth, while the middle window represents the trace velocity of the signal. The bottom window shows the bandpassed raw pressure signal for one array element.

We have also established a geolocation using the nine closest stations (Figure 5) by utilizing a non-linear system of equations describing the propagation of the detection waves through the atmosphere, where the inverse location algorithm is based on Geiger's approach (1910). The location results are obtained assuming a homogeneous half-space with a typical celerity value of 290 m/s for each individual phase without azimuthal correction (Brown et al., 2002). In order to determine the location errors, the 95% confidence ellipses are estimated by repeatedly running the linearized least-squares inversion with arbitrary sub-sets of the input data within ±10° and ±30 m/s ranges of uncertainties for the azimuths and celerity, respectively.

The maximum peak-to-peak amplitude was determined by bandpassing the stacked, raw waveform using a second-order Butterworth filter and then applying the Hilbert Transform (Dziewonski and Hales, 1972) to obtain the peak of the envelope. The filter cutoff frequencies were typically 0.05 Hz for the low frequency and up to 2.1 Hz for the high frequency (with few exceptions) and were determined using a power spectral density (PSD) method where the signal segment of the waveform was superimposed over the average of the prior and post background noise (of equal length), all being divided into equal windows (50-170 seconds in length, depending on station), establishing a frequency band which lies above the noise. Therefore, the low and high frequency cutoffs would be selected where the signal rises above the noise on the low end or descends into the noise on the high end of the spectrum, respectively.

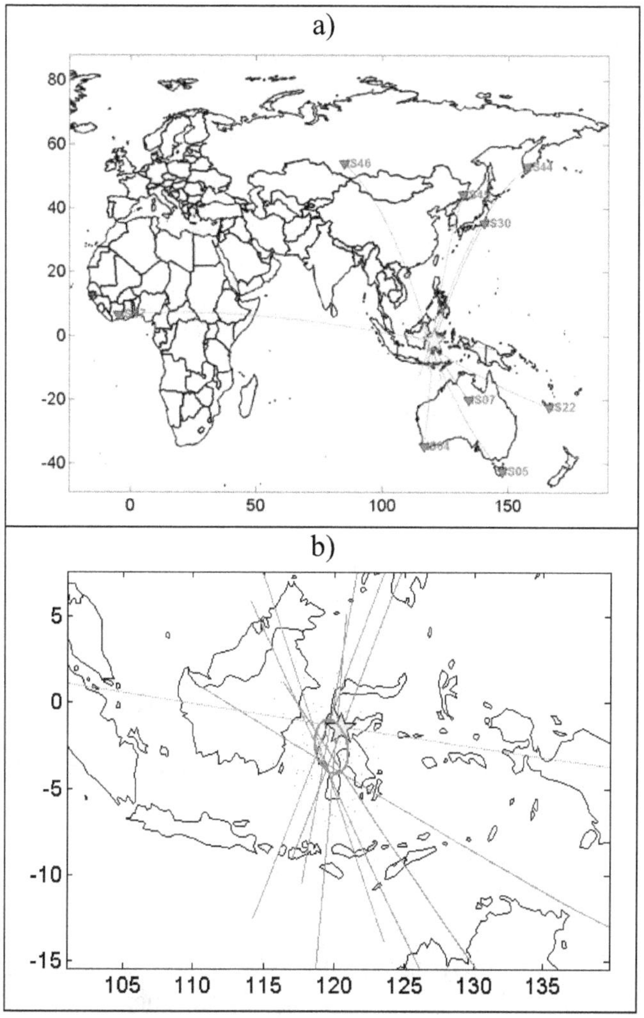

Figure 5. Map showing the geolocation. The best fit solution was obtained using nine stations closest to the Indonesian bolide event.

To measure the dominant period at maximum peak-to-peak amplitude, two independent techniques were employed. First, the dominant period at maximum frequency was acquired from the residual power spectral density (PSD) obtained using the method described above, except the noise PSD was subtracted from the signal PSD. The inverse of the frequency at maximum residual PSD was used to obtain the dominant period. Second, the period at maximum peak-to-peak amplitude was tabulated by measuring the zero crossings of the stacked waveform at each station (cf. ReVelle, 1997) in the same bandpass. The periods obtained using these two techniques show a very strong 1:1 correlation (Figure 6), indicating that this methodology is robust in itself.

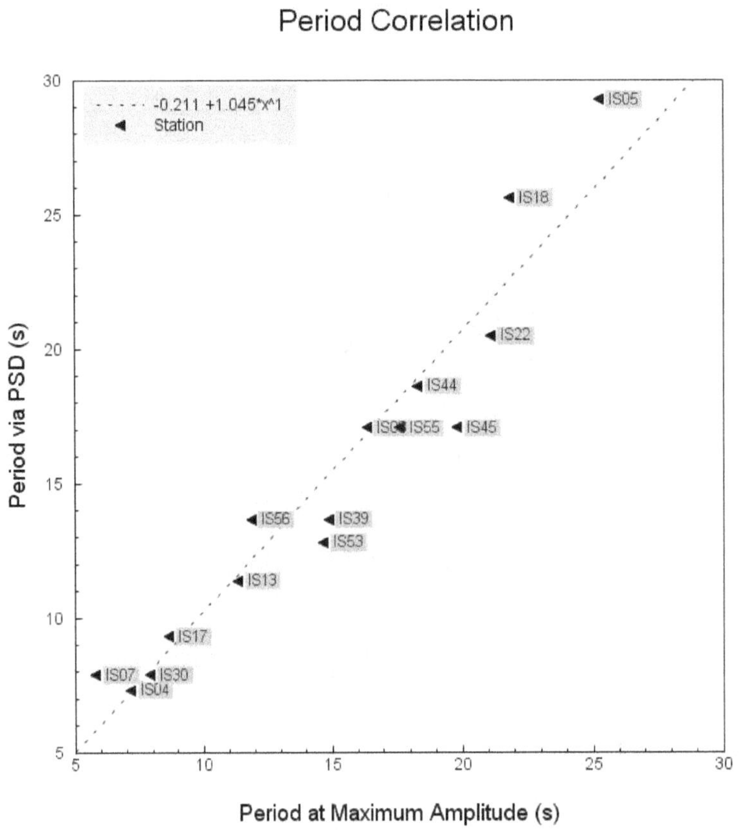

Figure 6. The dominant period correlation using two methods: PSD (vertical axis) and zero-crossings (horizontal axis).

4 Estimating the Source Energy

There are several empirical relations, relying on either the period at maximum amplitude or range and signal amplitude, which can be utilized in estimating source energy from infrasound measurements (Edwards et al., 2006). The yield estimates based on infrasonic amplitude are very uncertain in this instance as the propagation distances are much larger than is typical and outside the range limits where such relations have been developed (Edwards et al, 2006). In general, infrasonic period is less modified during propagation than amplitude (cf. Mutschlecner et al., 1999; ReVelle 1997; ReVelle 1974) and thus the period relationship is expected to be more robust. The Air Force Technical Application Centre (AFTAC) period-yield relations which are commonly used for large atmospheric explosions, are given by ReVelle [1997], as:

$$\log(E/2) = 3.34\log(P) - 2.58 \qquad E/2 \leq 100kt \qquad (1)$$

$$\log(E/2) = 4.14\log(P) - 3.61 \qquad E/2 \geq 40kt \qquad (2)$$

Here, E is the total energy of the event (in kilotons of TNT), P is the period (in seconds) at maximum amplitude of the waveform. Since these relations were originally derived from nuclear explosions, the factor ½ must be incorporated in order to account for energy loss due to radiation for low altitude nuclear airbursts (Glasstone and Dolan, 1977). Even though there are a number of effects that may adversely influence and change the period at maximum amplitude during long range propagation of infrasound, this approach remains more robust than the maximum amplitude based relations, since it shows better agreement with energy estimates for bolide events which had their energies estimated by other methods (Silber et al, 2009; Brown et al., 2002).

5 Results and Discussion

There are total of 17 detections, 16 obtained with PMCC and 15 obtained with MatSeis (Table 1). These detections overlap, except for the signal detected via MatSeis at the Bolivian station (I08BO), 17 509 km from the source. This signal, though very weak and short in duration (~30 seconds) compared to other signals (>185 seconds), shows a strong correlation to the bolide. The correlation indicators are the arrival time, the signal velocity, the dominant period and the apparent agreement between the observed and expected azimuth. The first arrival was detected almost two hours after the event at the closest IMS station, I39PW, at 04:39:51 UTC, while it took nearly 15 hours for the last bits of the signal to arrive to I08BO. Duration of the signal at each station (not including I08BO) was quite significant, ranging from 3 minutes up to 41 minutes. All infrasound signals from the event show similar characteristics, such as long period and very low frequency content, consistent with a large blast radius and consequently a large energy source (ReVelle, 1976). Furthermore, average signal celerities are between 270 m/s and 320 m/s, indicative of stratospheric duct signal returns.

The presence of high altitude winds affects the propagation of the signal in such way that it amplifies the downwind propagation, while it attenuates upwind propagation (c.f. Mutschlecner and Whitaker, 2010; Davidson and Whitaker, 1992; Reed 1969a). Most of the detecting stations are located east from the source and in October the stratospheric winds are predominantly westerly in the northern hemisphere (Webb, 1966). Average signal celerities (defined by the ratio between the horizontal propagation range and the travel time) are between 0.27 and 0.32 km/s, which is consistent with stratospheric duct signal returns. We also searched for possible antipodal signals, but found none.

The geolocation ellipse (Figure 5), computed using azimuths and arrival times, points to 4.9°S and 122.0°E with mean residuals of 2.9°. The source time estimated from this location is 02:52:22 with a residual of 1320 s. The accuracy of the source location strongly depends on the atmospheric wind and temperature profiles at the place and time of the event.

To establish the best possible energy estimate of the Indonesian bolide, the average global period as well as individual periods, using both previously described zero-crossings and PSD methods, for each station were utilized. Table 2 shows the summary of energy estimates. The combined average periods of all phase-aligned stacked waveforms at each station produce a global average of 14.8 seconds (zero crossings method) and 15.3 seconds (PSD method), corresponding to a mean source energy of 42.7 kt of TNT and 47.3 kt of TNT, respectively. Using the measurements from nine stations with the highest signal-to-noise ratio energy yield is 66.1 kt of TNT (zero crossings method) and 78.1 kt of TNT (PSD method). The standard deviation of energy measurements across all stations is approaching the measurement itself, but this is expected because the signal usually emanates from different portions of the bolide trail as observed at different stations. Our best source energy estimate is 70 ± 20 kt TNT, with the error bounds representing the spread in the average from the different approaches (Table 2).

Table 2. List of all detecting stations and their periods measured via two methods (zero-crossings at maximum amplitude in time domain and frequency at maximum PSD in frequency domain), as well as energy measurements for each station, where appropriate AFTAC relations were used (equation (1) or equation (2)).

Energy estimate as a function of period				
Station ID	Period via zero crossings (s)	Energy (kt of TNT)	Period via PSD (s)	Energy (kt of TNT)
IS04	7.11	3.68	7.31	4.05
IS05	25.23	312.64	29.26	577.07
IS07	5.79	1.85	7.88	5.19
IS08	16.34	59.33	17.07	68.61
IS13	11.31	17.37	11.38	17.71
IS17	8.64	7.06	9.31	9.06
IS18	21.81	155.69	25.60	332.00
IS22	21.07	138.75	20.48	126.15
IS30	7.89	5.22	7.88	5.19
IS39	14.87	43.30	13.65	32.56
IS44	18.29	86.42	18.62	91.75
IS45	19.79	112.45	17.07	68.61
IS53	14.66	41.25	12.80	26.25
IS55	17.62	76.29	17.07	68.61
IS56	11.83	20.19	13.65	32.56
Average E (kt of TNT)		**72.10**		**97.69**
Energy estimate based on averaged global period				
Total	14.81	**42.73**	15.27	**47.30**

Energy estimate as a function of SNR				
Station ID	Period via zero crossings (s)	Energy (kt of TNT)	Period via PSD (s)	Energy (kt of TNT)
IS04	7.11	3.68	7.31	4.05
IS05	25.23	312.58	29.26	577.07
IS07	5.79	1.86	7.88	5.19
IS18	21.81	155.65	25.60	332.00
IS44	18.29	86.46	18.62	91.75
IS45	19.79	112.50	17.07	68.61
IS53	14.66	41.30	12.80	26.25
IS55	17.62	76.33	17.07	68.61
IS56	11.83	20.17	13.65	32.56
Average E (kt of TNT)		**90.06**		**134.01**
Energy estimate as a function of SNR (period average)				
Total	16.88	**66.10**	14.46	**78.11**

6 Conclusions

The Indonesian bolide of 8 October, 2009, detected infrasonically on a global scale, was perhaps the most energetic event since the bolide of 1 February, 1994 (McCord et al., 1995) and may have exceeded it in total energy. We have no other instrumental records of this event other than casual video records of the dust trail emphasizing again the value of infrasonic monitoring of atmospheric explosive sources. Low frequency waves were observed at 17 IMS stations of the CTBTO network, making it one of the best infrasonically documented events (DTRA Verification Database, available at: http://www.rdss.info).

Using an average impact velocity for Near Earth Objects (NEO) of 20.3 km/s, the energy limits (50-90 kt of TNT) suggested by this analysis correspond to an object 8-10 m in diameter. Given our upper limit in energy and a lowest possible entry velocity of 11.2 km/s, the upper limit to the mass for this meteoroid is < 6000 tonnes. Based on the flux rate from Silber et al. (2009), such objects are expected to impact the Earth on average every 10-22 years. Additional instrumental records of this unique event would prove valuable in understanding in more detail its interaction with the atmosphere and documenting possible local atmospheric perturbations.

Additional instrumental records of this exceptional event, such as seismic, ground video recordings, satellite and possible meteorites, would prove valuable in understanding such occurrences and documenting possible local atmospheric perturbations. Since events like this one are rather rare, it is essential to maximize all aspects of such observations in order to validate propagation models at global scale, implement and better understand the spatial and temporal influences of atmospheric dynamics over propagation times, especially over long distances, and to evaluate energy yield formula and establish what information, not available via other techniques, can be derived from infrasonic measurements.

Acknowledgements

EAS and PGB thank the Natural Sciences and Engineering Research Council of Canada and Natural Resources Canada, and express their gratitude to the Meteoroid Environment Office of the National Aeronautics and Space Administration for funding the Meteoroids 2010 conference.

References

S.J. Arrowsmith, D.O. Revelle, W. Edwards, P. Brown, Earth Moon Planets (2008) doi: 10.1007/s11038-007-9205-z

N.A. Artemieva, P.A. Bland, P.A. Meteorit. Planet. Sci. 38 (2003)

N. Brachet, D. Brown, R. Le Bras, P. Mialle, J. Coyne, in Infrasound Monitoring for Atmospheric Studies (Springer Netherlands, 2010) pp. 77-118

P. Brown, D.O. ReVelle, E.A. Silber, W.N. Edwards, S. Arrowsmith, L.E. Jackson, G. Tancredi, D. Eaton, J. Geophys. Res.-Planet (2008) doi:10.1029/2008JE003105

D. Brown, C.N. Katz, R. Le Bras, M.P. Flanagan, J. Wang A.K. Gault, Pure. Appl. Geophys. (2002) doi: 10.1007/s00024-002-8674-2

P. Brown, R.E. Spalding, D.O. ReVelle, E. Tagliaferri, S.P. Worden, Nature. (2002a) doi: 10.1038/nature01238

P.G. Brown, P. Kalenda, D.O. ReVelle, J. Borovicka, Meteorit. Planet. Sci. (2003) doi: 10.1111/j.1945-5100.2003.tb00296.x

P.G. Brown, R.W. Whitaker, D.O. ReVelle, E. Tagliaferri, Geophys. Res. Lett. (2002) doi: 10.1029/2001GL013778

Y. Cansi, Geophys. Res. Lett. (1995) doi: 10.1029/95GL00468

C.R. Chapman, Earth Moon Planet (2008) doi:10.1007/s11038-007-9219-6

M. Davidson, R.W. Whitaker, Los Alamos National Laboratory Report LA-12074-MS (1992)

A. Dziewonski, A. Hales, in Methods in Computational Physics (Academic Press, New York, 1972), pp. 39–84

W. Edwards, in Infrasound Monitoring for Atmospheric Studies (Springer Netherlands, 2010), pp. 361-414

W. N. Edwards, P.G. Brown, D. O. ReVelle Estimates of meteoroid kinetic energies from observations of infrasonic airwaves, Journal of Atmospheric and Solar-Terrestrial Physics (2006) doi: 10.1016/j.jastp.2006.02.010

L.G. Evers, H.W. Haak, Geophys. Res. Lett. (2001) doi: 10.1029/2000GL011859

L.Geiger, K. Ges. Wiss. Gött. 4, 331-349 (1910)

S. Glasstone, P.J. Dolan, in The Effects of Nuclear Weapons (United States Department of Defence and Department of Energy, Washington, DC, USA, 1977)

J.M. Harris, C.J. Young, Seismol. Res. Lett. 68, 307–308 (1997)

M. Hedlin, M. Garćes, H. Bass, C. Hayward, E. Herrin, J. Olson, C. Wilson, EOS (2002) doi: 10.1029/2002EO000383

A.R. Hildebrand, P.J.A. McCausland, P.G. Brown, F.J. Longstaffe, S.D.J. Russell, E. Tagliaferri, J.F. Wacker, M.J. Mazur, Meteorit. Planet. Sci. (2006) doi: 10.1111/j.1945-5100.2006.tb00471.x

A. Le Pichon, J.M. Guérin, E. Blanc, D.J. Raymond, Geophys. Res. Lett. (2002a) doi:1029/2001JD001283

A. Le Pichon, K. Antier, Y. Cansi, B. Hernandez, E. Minaya, B. Burgoa, D. Drob, L.G. Evers, J. Vaubaillon, Meteorit. Planet. Sci. (2008) doi: 10.1111/j.1945-5100.2008.tb00644.x

A.R. Klekociuk, P.G. Brown, D.W. Pack, D.O. ReVelle, W.N. Edwards, R.E. Spalding, E. Tagliaferri, B.B. Yoo, J. Zagari Nature (2005) doi: 10.1038/nature03881

T.B. McCord, J. Morris, D. Persing, E. Tagliaferri, C. Jacobs, R. Spalding, L. Grady, R. Schmidt, J. Geophys. Res. (1995) doi: 0148-0227/95/94JE-0280250

J.P. Mutschlecner, R.W. Whitaker, in Infrasound Monitoring for Atmospheric Studies (Springer Netherlands, 2010), p. 455-474

J.P. Mutschlecner, R.W. Whitaker, L.H. Auer, Los Alamos National Laboratory Technical Report, LA-13620-MS (1999)

J.W. Reed, Sandia Laboratories report SC-RR-69-572 (1969a)

D.O. ReVelle, in Annals of the New York Academy of Sciences, ed. by J.L. Remo (New York Academy of Sciences, 1997), p. 822

D. O. ReVelle, J. Geophys. Res. (1976) doi: 10.1029/JA081i007p01217

D.O. ReVelle, Acoustics of Meteors, PhD Dissertation, University of Michigan, 1974

E.A. Silber, D.O. ReVelle, P.G. Brown, W.N. Edwards, J. Geophys. Res.-Planet (2009) doi: 10.1029/2009JE003334

O.B. Toon, K. Zahnle, D. Morrison, R.P. Turco, C. Covey, Rev. Geophys. (1997) doi: 10.1029/96RG03038

W.L. Webb, Structure of the Stratosphere and Mesosphere (Academic Press, New York, 1966)

C.J. Young, E.P. Chael, B.J. Merchant, Proceedings of the 24th Seismic Research Review (2002)

CHAPTER 8:

RADAR OBSERVATIONS

Analysis of ALTAIR 1998 Meteor Radar Data

J. Zinn • S. Close • P.L. Colestock • A. MacDonell • R. Loveland

Abstract We describe a new analysis of a set of 32 UHF meteor radar traces recorded with the 422 MHz ALTAIR radar facility in November 1998. Emphasis is on the velocity measurements, and on inferences that can be drawn from them regarding the meteor masses and mass densities. We find that the velocity vs altitude data can be fitted as quadratic functions of the path integrals of the atmospheric densities vs distance, and deceleration rates derived from those fits all show the expected behavior of increasing with decreasing altitude. We also describe a computer model of the coupled processes of collisional heating, radiative cooling, evaporative cooling and ablation, and deceleration – for meteors composed of defined mixtures of mineral constituents. For each of the cases in the data set we ran the model starting with the measured initial velocity and trajectory inclination, and with various trial values of the quantity $m\rho_s^2$ (the initial mass times the mass density squared), and then compared the computed deceleration vs altitude curves vs the measured ones. In this way we arrived at the best-fit values of the $m\rho_s^2$ for each of the measured meteor traces. Then further, assuming various trial values of the density ρ_s, we compared the computed mass vs altitude curves with similar curves for the same set of meteors determined previously from the measured radar cross sections and an electrostatic scattering model. In this way we arrived at estimates of the best-fit mass densities ρ_s for each of the cases.

Keywords meteor · ALTAIR · radar analysis

1 Introduction

This paper describes a new analysis of a set of 422 MHz meteor scatter radar data recorded with the ALTAIR High-Power-Large-Aperture radar facility at Kwajalein Atoll on 18 November 1998. The exceptional accuracy/precision of the ALTAIR tracking data allow us to determine quite accurate meteor trajectories, velocities and deceleration rates. The measurements and velocity/deceleration data analysis are described in Sections II and III. The main point of this paper is to use these deceleration rate data, together with results from a computer model, to determine values of the quantities $m\rho_s^2$ (the meteor mass times its material density squared); and further, by combining these $m\rho_s^2$ values with meteor mass estimates for the same set of meteors determined separately from measured radar scattering cross sections, to arrive at estimates of the mass densities ρ_s.

The computer model, described in Section IV and Appendix A, treats the simultaneous processes of meteor heating through air molecule collisions, blackbody radiation emission, evaporation, sputtering,

J. Zinn (✉) • P. L. Colestock • R. Loveland
Los Alamos National Laboratory, Los Alamos, NM. E-mail: jzinn@lanl.gov

S. Close
Stanford University, Stanford, CA

A. MacDonell
Boston University, Boston, MA

and deceleration – for meteors of specified assumed initial mixtures of mineral constituents. The model assumes in each case that the meteors are spherical, and remain so without fragmenting. It includes an imbedded table of atmospheric mass densities vs altitude, and data on (1) vapor pressure vs temperature, (2) heat of sublimation, (3) vapor molecular weight, and (4) melting point – for each of the assumed constituent species. Other inputs to the model include, for each individual case, (1) the initial meteor velocity and trajectory inclination (i.e. at the top of the atmosphere), (2) trial values of the initial $m\rho_s^2$ (i.e. values before entering the atmosphere).

The data include 32 individual meteor traces, where the meteors all appear to be in the mass range 10^{-6} to 10^{-4} grams, and the altitudes are such that air molecule collision mean free paths are much larger than the meteor dimensions. Thus air molecule collisions with the meteor can be regarded as isolated events, and fluid-dynamic effects do not apply (large Knudsen number). In our data analysis we fit the reduced data on velocities vs altitude and trajectory inclination as least-squares quadratic functions of the path-integrated air column densities, using tabular data on air densities vs altitude. We then compute the corresponding deceleration rates. We find, as expected, that for all the traces the deceleration rates increase with decreasing altitude.

The model equations and variables are listed in Appendix A. Appendix B describes a quasi-analytic solution of the ablation equations for a 1-component meteor, using the steady-state approximation. It shows that at the lowest altitudes the meteor temperatures are determined mainly by an equilibrium between collisional heating and evaporative cooling. And the ablation coefficients tend to approach a common value equal to the vapor molecular weight divided by twice the heat of vaporization, and independent of the initial meteor velocity.

2 Experimental

The ALTAIR High-Power-Large-Aperture radar facility is located on the Kwajalein Atoll (9° N, 167° E) in the Republic of the Marshall Islands. ALTAIR has a 43-m diameter mechanically-steered parabolic dish, and simultaneously transmits a peak power of 6 MW at two frequencies (VHF-160 MHz, and UHF-422 MHz). (Close et al 2000, Close et al 2004). The radar characteristics are described in detail in those references. It is particularly suited for precise measurements of small targets at long ranges. Extensive measurements going back to 1983 show stable rms tracking accuracies of ±15 milli-degrees in angle and ±6 m in range. In the present paper we discuss a UHF data set consisting of 32 meteor traces obtained on November 18, 1998. The radar sample window encompassed slant ranges corresponding to heights mostly between 90 to 110 km. 150 μs pulsed waveforms were used, with a range sample spacing corresponding to about 7.5 meters . The instantaneous meteor 3-dimensional positions were determined from the monopulse range and angular measurements, and the velocities were determined by direct numerical differencing of the positions vs time (Close et al, 2002).

In this paper we do not yet make use of a much larger set of ALTAIR meteor data obtained in 2007 and 2008, or results of an ongoing analysis of these data where line-of-sight velocities are determined from measured Doppler frequency shifts of the reflected radar signals (Loveland et al, 2010). We expect that the velocities thus determined will be of higher accuracy than those derived from the 1998 data described in this paper. We will report analyses of the newer results in a later paper.

3 Data Analysis

For each of the 32 meteor traces (using the tabulated altitudes, velocities and vertical velocity components vs time) we begin by performing a quadratic least-squares fit to the velocities vs the air path traversed (Q), where

$$Q \equiv \int_z^\infty \rho \, ds, \qquad (1)$$

ρ is the local air density, and ds is the element of distance along the meteor path to the altitude z. The ρ's were taken from the CIRA '61 tabulations (COSPAR International Reference Atmosphere 1961), and the Q integrals were evaluated for each point along each trace using the measured trajectory inclination angles. (We will regard these atmospheric density data as given, and note that they are probably more accurately determined than are the meteor masses or mass densities that we will derive from the radar data). Then from the Q derivative of this fitted quadratic velocity vs Q function we compute the corresponding deceleration rates as functions of z. Figure 1 is a composite plot of the fitted velocities vs altitude for the 32 traces. It will be noted that they all show velocities decreasing with decreasing altitude, and all of them show some downward curvature. Likewise, the deceleration rates increase with decreasing altitude, as they should. Figure 2 is a composite plot of the decelerations (negative accelerations) vs altitude for the 32 traces, derived from the velocity fits. (Note that one meteor streak appears to be interstellar in origin, with a velocity exceeding 72.8 km/s. We will perform orbital analysis on this streak in the future to confirm this result.) We note also that in five of the cases the initial value of dv/dt has come out to be positive, due presumably to inaccuracies in the velocity data. These cases will be discarded as flawed.

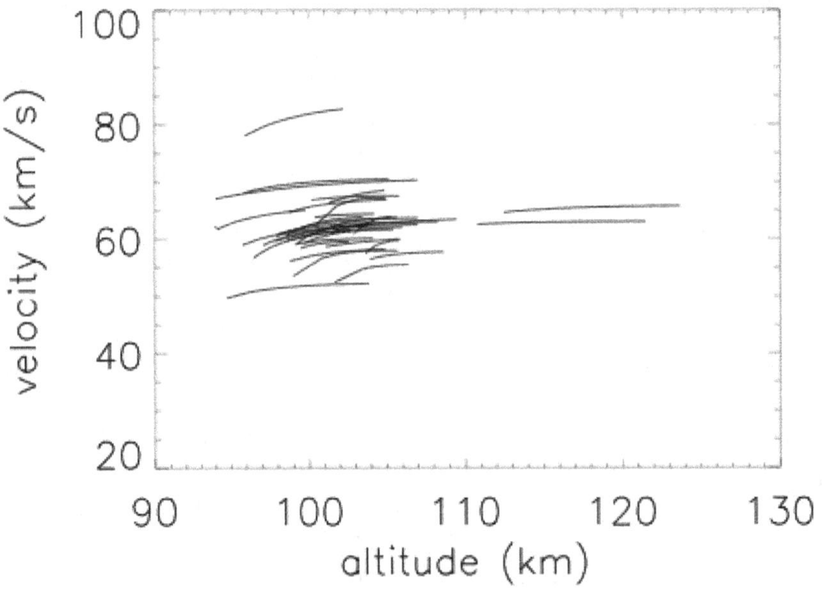

Figure 1. Composite plot of the fitted velocities vs altitude for the 32 cases – from the least-squares quadratic fits of the measured velocities vs Q .

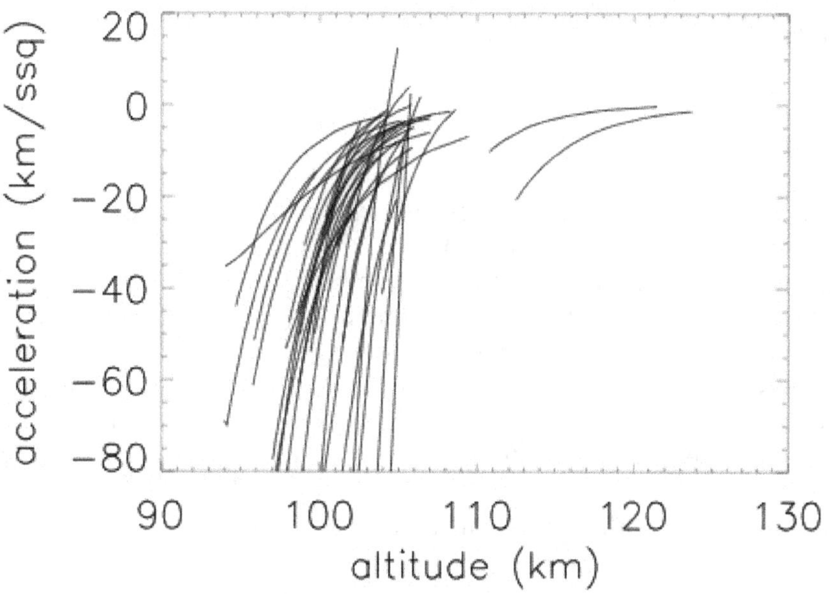

Figure 2. Composite plot of the accelerations vs altitude for the 32 cases – as derived from the quadratic fits to the velocities

In our analysis we will assume that the meteors are spherical. Then the energy flux on the meteor surface due to air molecule collisions is $\pi r^2 \rho v^3/2$, where r and v are the meteor radius and velocity, and ρ is the local air density. Consistent with other authors (e.g. Pecina & Ceplecha, 1982; Opik, 1958; Bronshten, 1983) we will write the meteor mass loss rate as

$$dm/dt \;=\; - \pi r^2 \rho v^3 \sigma \,, \qquad (2)$$

where m is the meteor mass and σ is the "ablation coefficient". The rate of deceleration of the meteor is

$$dv/dt \;=\; - \pi r^2 \rho v^2/m \;. \qquad (3)$$

Combining equations 1 and 2 we obtain

$$dm/m \;=\; \sigma \, v \, dv \;. \qquad (4)$$

If we make the convenient (but not necessarily valid) assumption that σ is constant, then Eq 3 can be integrated, giving

$$\ln(m/m_1) \;=\; (\sigma/2)(v^2 - v_1^2) \,, \qquad (5)$$

where m_1 and v_1 are the initial values of m and v along a given meteor radar trace. The constant-σ assumption would be appropriate if, for instance, the meteor mass loss was dominated by "sputtering".

In an alternative model (e.g. Vondrak et al, 2008; Janches et al, 2009; Lebedinets, 1973), the mass loss is dominated by thermal evaporation of the meteor constituents. The instantaneous evaporation rate is determined by the instantaneous temperature. In the next section we describe our own numerical model of these coupled processes.

4 Numerical Model

The consensus of most current theoretical studies of the ablation and slowing down of small meteors in the atmosphere (Lebedinets, 1973; Janches et al, 2009; Vondrak et al, 2008) is that: (1) Very rapid heating occurs due to collisions with air molecules, moderated by energy losses due to blackbody emission from the meteor surface and due to evaporation. (2) The heating leads to vaporization of meteor (generally preceded by melting). (3) Some sputtering occurs, in addition to the vaporization. (4) The air molecule collisions also lead to deceleration of the meteor. (5) With very small meteors the rate of internal heat conduction is sufficient to maintain a uniform temperature distribution within the meteor. (6) Meteors are composed of mixtures of chemical constituents, and each will vaporize at its own rate.

It the present model we further assume that the meteor is spherical, and that after melting it does not disintegrate.

The rate of heating of the meteor through air molecule collisions is $(dH/dt)_{coll} = \pi r^2 \rho v^3/2$, where r is the instantaneous meteor radius, v its instantaneous velocity, and ρ the local air density, and it is assumed that all the energy of a collision is transferred to the meteor. The rate of loss of energy by blackbody emission is $(dH/dt)_{rad} = -4\pi r^2 \sigma_{SB} T^4$, where σ_{SB} is the Stephan-Boltzmann constant and T is the instantaneous temperature. The vapor pressure of the ith chemical constituent of the meteor is given by the Claussius-Clapeyron equation

$$P_{vap(i)} = A_i \exp(-C_i/T) , \qquad (6)$$

where A_i and C_i are constants characteristic of the particular constituent. The evaporative flux of each constituent from the surface is given by the Langmuir relation

$$F_{evap(i)} = C_{flx} P_{vap(i)} / (\mu_{vap(i)} T)^{1/2} \quad \text{(molecules/cm}^2\text{s)} , \qquad (7)$$

(Taylor and Langmuir 1933), where $\mu_{vap(i)}$ is the molecular weight of the vapor. . If $P_{vap(i)}$ is in dynes/cm^2 and $\mu_{vap(i)}$ is in grams, then the constant C_{flx} is equal to 3.40×10^7. Then the rate of energy loss from the meteor surface due to evaporation of each constituent is

$$(dH/dt)_{evap} = -4\pi r^2 \Delta H_{sblm(i)} F_{evap(i)} , \qquad (8)$$

where $\Delta H_{sblm(i)}$ is the heat of sublimation (erg/molecule). Then the rate of change of the meteor temperature is

$$dT/dt = [(dH/dt)_{coll} + (dH/dt)_{rad} + \Sigma_{(i=1,N)}(dH/dt)_{evap(i)}]/C_p , \qquad (9)$$

where C_p is the specific heat. In the above equation it is assumed that each constituent vaporizes at a rate independent of the other constituents, as long as that constituent is still present (i.e. has not totally evaporated). Values of the parameters $\Delta H_{sblm(i)}$, A_i, C_i, $\mu_{vap(i)}$ and melting point for several likely meteor constituents are listed in Table 1, below.

Table 1. Physico-chemical parameters for meteor constituents, and references. The numbered references are: 1. Ferguson et al, 2004; 2. Brewer et al, 1948; 3. Clarke and Fox, 1969; 4. Wickramasinghe and Swamy, 1968; 5. Brewer and Porter, 1954; 6. Akopov, 1999; 7.Fabian,1993; 8. Patnaik, 2002.

Constituent	ΔH_{sblm}	A	C	μ_{vap}	Melting Pt	References
	(erg/molec)	dyne/cm^2	Deg K	grams	Deg K	
Fe (iron metal)	6.62e-12	5.06e+12	4.836e+4	9.30e-23	1811	1
C (graphite)	1.495e-11	9.74e+15	1.006e+5	3.99e-23	---	2, 3
SiO_2	9.64e-12	4.16e+11	6.99e+4	9.97e-23	1923	4
MgO	8.65e-12	9.16e+14	6.27e+4	6.64e-23	3073	5
FeO	1.03e-11	1.01e+16	7.47e+4	1.20e-22	1653	6, 7, 8

Table 1 shows the physico-chemical parameters that we have assumed for several possible meteor constituents, including A, C, ΔH_{sblm}, and some references. In most cases the references do not give the quantities A, C and ΔH_{sblm} directly, and in those cases we have had to calculate those quantities by fitting Eq 6 to data on vapor pressures measured at two or more temperatures, and assuming that ΔH_{sblm} is equal to the Boltzmann constant k_B times C. The references listed in Table 1 are mostly sources of vapor pressure and/or boiling point data.

The rate of loss of mass from the meteor due to evaporation, is

$$(dm/dt)_{evap} = -4\pi r^2 \sum_{(i=1,N)} (\mu_{vap(i)} F_{evap(i)}) . \qquad (10)$$

There will also be some mass loss due to sputtering, given by $(dm/dt)_{sputt} = \pi r^2 \rho v^3 \sigma_{sputt}$, where σ_{sputt} is the ablation coefficient associated with sputtering (units of s^2/cm^2). The meteor radius r is related to the mass m by $r = (3m/4\pi\rho_s)^{1/3}$, where ρ_s is the mass density of the solid meteor.

Finally, the rate of deceleration of the meteor is given by Eq 3.

$$dv/dt = - \pi r^2 \rho v^2/m . \qquad (3)$$

We have developed our own computer model that incorporates the above processes in the form of a set of ordinary differential equations expressing the rates of change of meteor mass, velocity, radius, temperature, etc. as functions of time. The input meteor composition can be either a pure compound or a mixture of compounds. The differential equations are detailed in Appendix A. This model appears to be very similar to the one described by Vondrak et al, 2008.

Some key questions are, of course: (1) What is the meteor composition? (2) What is its mass density? (3) Does the meteor actually remain intact after it melts? and (4) What is the contribution of sputtering to the total ablation coefficient?

Figures 3a-3f show comparisons, for a set of six traces, of computed vs measured decelerations vs altitude. For inputs to the computations for each trace we take (1) the measured initial velocity; (2) the measured trajectory inclination angle; (3) an assumed initial value of $m\rho_s^2$, which is shown on the plot; (4) an assumed initial composition -- namely an equimolar mixture of SiO_2, FeO and MgO (which corresponds roughly to the expected decomposition products of olivine, a mineral that is commonly found in stony meteorites); (5) an assumed energy for sputtering, E* = 15 eV per molecule, giving a constant sputtering contribution of 2.1×10^{-12} s^2/cm^2 to the total ablation coefficients; (6) a mass density

ρ_s of 1.0 g/cm^3. With these assumptions Figures 3a-3f show good agreement between the computed results and the data. In all, we found satisfactory agreement in 20 of the 32 cases.

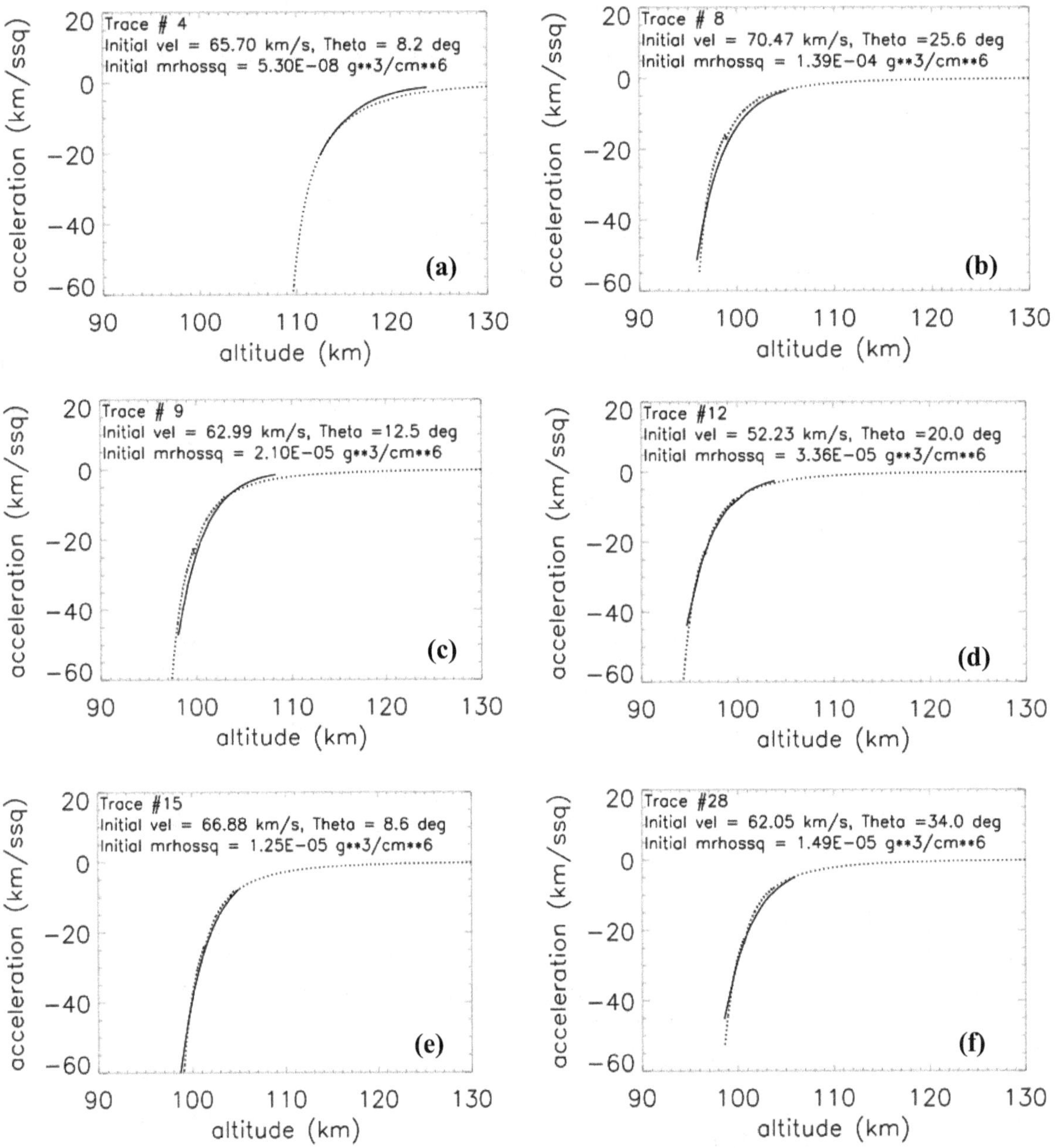

Figures 3a-3f. In each of these plots the solid curve is the acceleration derived from the data fit, and the dotted curve is the one computed with the model (with inputs described in the text).

We also ran computations with other assumed mass densities (ρ_s) and sputtering energies (E*), although the results will not be shown here. From comparisons of the results with Figs 3a-3f we found that the computed deceleration rates did not depend at all on the assumed density ρ_s . This is to be expected, since the initial meteor mass is equal to the input $m\rho_s^2$ divided by ρ_s^2, and the separate dependencies of the acceleration rates on m and ρ_s are always connected through the $m\rho_s^2$.

We also found that without the assumed relatively large sputtering contribution to the ablation coefficients the agreement with the data was less good than the extent of agreement shown in Figs 3a-3f.

Although the computed deceleration values are independent of the separate values of m and ρ_s (once the value of the product $m\rho_s^2$ is prescribed), the mass m is of course equal to $m\rho_s^2/\rho_s^2$. Because of this it is possible to arrive at rough estimates of both m and ρ_s separately, using the measured values of decelerations and radar cross sections in combination, and using the Close et al electrostatic scattering model (Close et al, 2004) together with our present ablation and deceleration model. Figures 4a and 4b show two examples of such attempts to determine both m and ρ_s from the experimentally determined $m\rho_s^2$ and "mass1" (mass from the radar cross sections). In Figure 4a, representing trace #8, the four solid curves are the computed inertial mass values vs altitude derived from the best-fit $m\rho_s^2$ (from Fig 3b) assuming four different values of ρ_s, namely 0.1, 0.316, 1.0 and 3.16 g/cm^3, while the dashed curve is mass1. In this case it appears that the best-fit density ρ_s is about 1 g/cm^3, and the initial pre-ablation mass is about 1×10^{-4} g. Figure 4b is a similar set of plots, but representing trace #9 (from Fig 3c). In this case the best-fit ρ_s is about 0.3 and the initial mass is again about 1×10^{-4} g. We have made similar plots (not shown here) for each of the other measured traces, and we find that the average best-fit ρ_s is about 0.5, but with a spread of values between about 0.1 and 1.

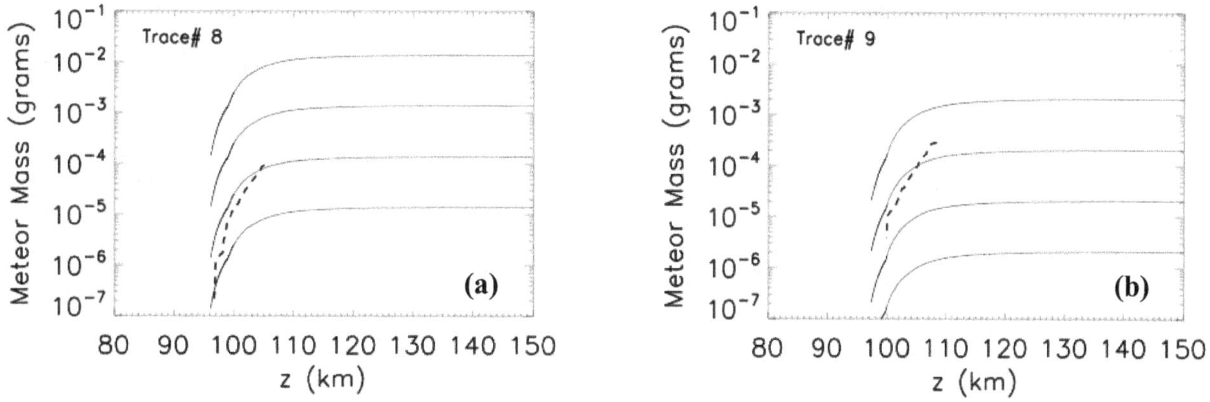

Figures 4. (a) Here the solid curves are the inertial masses (for trace #8) computed with the numerical model, using the best-fit value of $m\rho_s^2$ together with four different assumed values of the meteor density ρ_s, namely (from top to bottom) 0.1, 0.316, 1.0 and 3.16 g/cm^3 . The dashed curve is the "mass1" (mass determined from the measured cross sections together with the electrostatic scattering model. (b) Same as for (a), but representing trace #9.

To elaborate on some further details of the model computations: Figures 5a and 5b show more results from one of the runs, namely the one representing trace #8. Figure 5a shows the computed meteor temperatures vs altitude, showing the successive evaporation of MgO, FeO and SiO_2; and Figure 5b shows the computed variations of the effective ablation coefficient σ with altitude, including the total σ and the separate evaporative contribution.

For meteors composed of mixtures of materials the total vapor pressure at any point is the sum of the vapor pressures of the individual constituents, irrespective of their relative amounts. Then the evaporation rate for each component should be given by the Langmuir equation (Eq 7), irrespective of the fraction of that component in the mixture. Then at each instant all of the constituents will be evaporating simultaneously at rates proportional to their individual vapor pressures – until such times as each successive constituent disappears by evaporation. One result of this is that the meteor temperature rises in a series of discrete steps, where the steps correspond to the disappearances of successive components. This is illustrated in Figure 5a. The ablation coefficients also exhibit a stepwise character, but with sharp decreases between successive steps, as is shown in Figure 5b. It is notable of course that the ablation coefficients are by no means constant, in contradiction to the assumption in Eq 5.

Our assumed constant sputtering contribution to the ablation coefficient produces a substantial difference in the computed ablation and deceleration rates. Figure 5c shows the computed ablation coefficient vs altitude for the same case as that shown in Figures 5a and 5b, where in the computation the sputtering energy E* was raised to 1000 eV per molecule, so that the sputtering contribution to sigma was reduced to 3.1×10^{-14} s^2/cm^2 , which would be in better agreement with the laboratory data. The result was a considerable reduction in the effective average ablation coefficients and a reduction in the meteor deceleration rates.

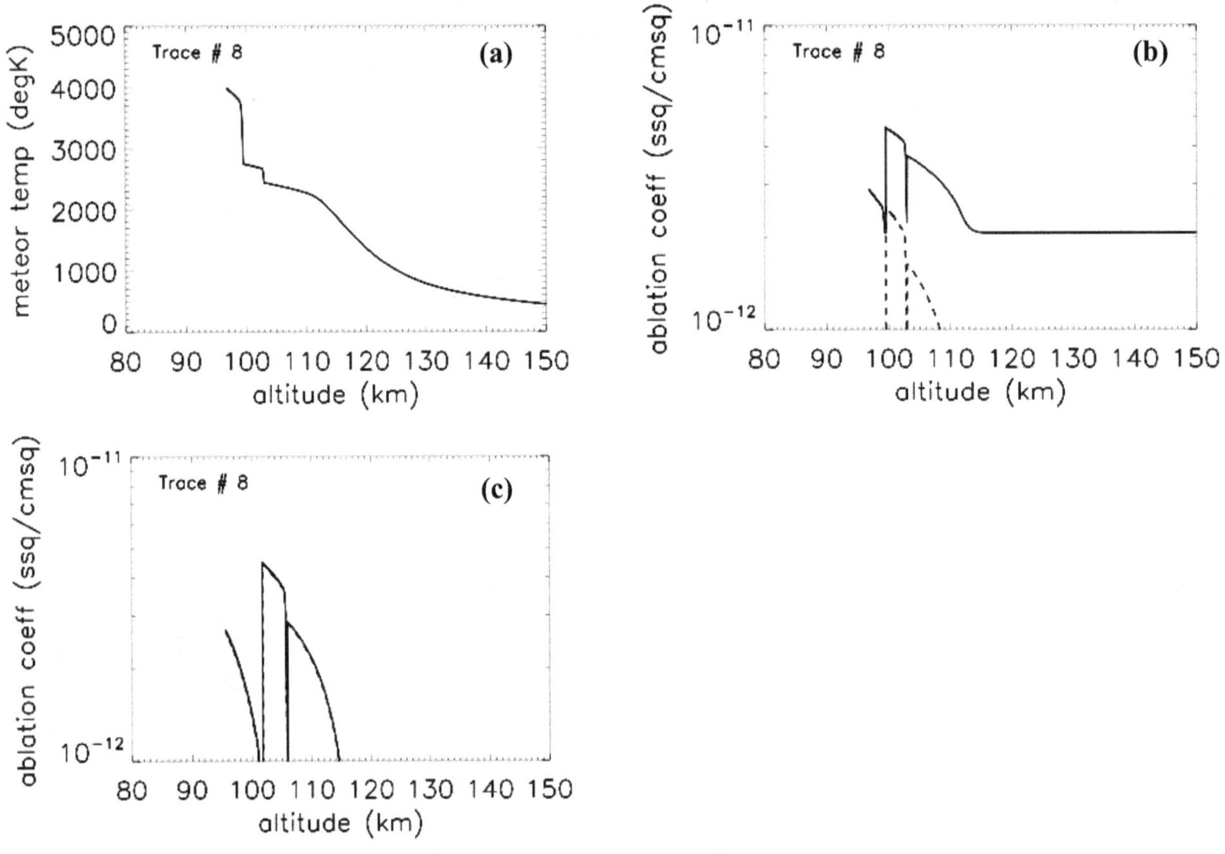

Figures 5. (a) Computed temperature history for the same meteor as in 3a. (b) Computed ablation coefficient vs altitude for the same meteor as in 3a, and 5a. The dashed curve is the evaporative contribution, and the solid curve is the total. (c) Computed ablation coefficient vs altitude for the same case as in Figures 5a,b, when in the computation the assumed sputtering energy E* is raised to 1000 eV.

5 Direct Determination of $m\rho_s^2$ from the Deceleration Data

The quantity $m\rho_s^2$ can be determined directly from the fitted velocity and deceleration rate data without the need to use the computer model, but assuming only that the meteor is a sphere. Then the rate of deceleration is as given by Eq 3. For a sphere of density ρ_s the quantity πr^2 is $\pi r^2 = 1.209\ (m/\rho_s)^{2/3}$. Then, combining these two equations we obtain

$$m\rho_s^2 = [-1.209\ \rho\ v^2/\ (dv/dt)]^3 \tag{11}$$

The values of $m\rho_s^2$ thus determined are of course very sensitive to errors in the measured/fitted deceleration rates. If we nevertheless proceed to evaluate the $m\rho_s^2$ from the data fits for 27 of the measured traces, and plot them as functions of altitude, the result is Figure 6. Only about twenty of these curves seem to be believable, namely those that slope upward to the right and are concave downward. This set of twenty is the same as the twenty for which we found agreement between the computed and measured deceleration rates as described in the previous section.

Despite the expected inaccuracies in these $m\rho_s^2$ values, it is of interest to compare them with the corresponding values that we determined in the previous section from fitting the model-computed decelerations to the data. Table 2 shows, for each of the twenty chosen traces, (1) the initial (uppermost) altitude, (2) the initial value of $m\rho_s^2$ at that altitude, as determined directly from the data using Eq 11, (3) the $m\rho_s^2$ value at the same altitude as computed with the model, and (4) the value extrapolated to the top of the atmosphere using the model. As expected, the agreement between the values in columns 3 and 4 is not very good, but nor is it extremely bad in most cases. The worst disagreement is for traces 4 and 27, which are also exceptional in that their altitudes are more than fifteen kilometers higher than the rest.

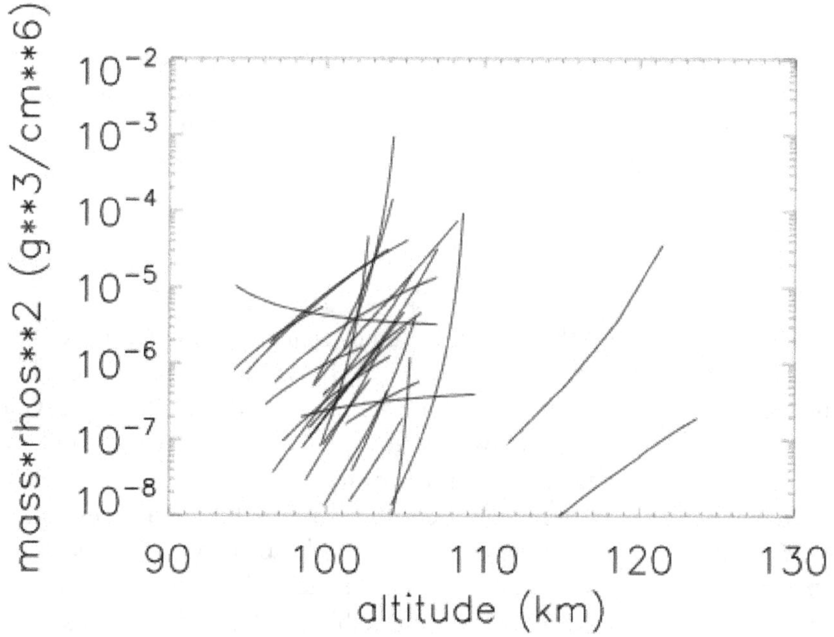

Figure 6. A composite plot of the combined variable $m\rho_s^2$ for 27 traces.

Table 2. Comparisons between values of $m\rho_s^2$ determined directly from the data via Eq 11 and values determined by the method described in Section III.

Trace #	Initial altitude (z1) (kilometers)	$m\rho_s^2$ at z1 from Eq 11	$m\rho_s^2$ at z1 from model	$m\rho_s^2$ ($z = \infty$) from model
1	104	2.e-6	1.6e-6	8.4e-6
4	124	2.e-7	3.0e-8	4.6e-8
8	105	3.e-5	8.5e-5	1.5e-4
9	108	1.e-4	1.4e-5	2.0e-5
10	107	3.e-5	6.e-6	1.0e-5
11	102	1.5e-6	4.5e-6	7.e-5
12	104	2.e-5	2.1e-5	3.2e-5
13	104	1.e-6	4.0e-6	1.2e-5
15	105	3.e-6	4.1e-6	1.2e-5
16	103	4.e-7	1.4e-6	1.6e-5
17	105	2.e-5	6.e-6	1.4e-5
20	104	3.e-7	4.3e-7	4.1e-6
23	105	2.e-7	3.e-7	3.e-6
24	106	3.e-6	1.0e-6	4.2e-6
25	107	1.e-5	2.1e-5	3.4e-5
27	123.5	3.e-5	2.0e-7	2.7e-7
28	106	4.e-6	7.e-6	1.5e-5
29	106	4.e-7	1.1e-6	3.1e-6
30	105	4.e-6	2.7e-6	6.8e-6
32	99.5	4.e-6	1.3e-5	8.2e-5

This procedure (i.e. using Eq 11) has the obvious advantage that it does not use any assumptions about the meteor composition, whereas in using the model a composition must be assumed. In both cases we assume a spherical meteor shape. Using the model has the advantage that it allows us to extrapolate the $m\rho_s^2$ to the top of the atmosphere.

6 Discussion

With the present 32-trace data set the velocities and trajectory inclinations were arrived at by differencing the measured 3D position vs time data. In view of the expected errors inherent to numerical differencing procedures, it has been encouraging to find that these 3D velocity data can be fitted so well as quadratic functions of Q. However, it is also not surprising to find that when we try to infer the $m\rho_s^2$ quantities directly from these data, as in the previous section, that many of the $m\rho_s^2$ vs altitude curves look crazy. We are currently in the process of analyzing a much larger set of ALTAIR meteor data from 2007-2008, where it appears to be possible to obtain more accurate velocities from range Doppler measurements. We are hopeful that when these data are available we can go through these same procedures to obtain a larger set of more reliable $m\rho_s^2$ values. With such a data set we will be able to extract more detailed information about the evaporation rates, ablation coefficients etc.

For purposes of evaluating the initial values of $m\rho_s^2$ we have chosen to use the computer model to find the values that produce the best fits to the deceleration rate data. However, a serious problem with that is that the model results are sensitive to the assumed chemical compositions of the meteors, which are of course not known. Our assumption of the olivine-like composition was convenient because the necessary data on vapor pressures and heats of sublimation of the decomposition products were available in the literature.

In the process of comparing the model results to the deceleration data we found that the fits were improved when we assumed a rather large sputtering contribution to the effective ablation coefficients, namely 2×10^{-12} s^2/cm^2. This value is significantly larger than values that have been determined in laboratory measurements of sputtering from energetic ion bombardment of solid target materials (Behrisch 1981, Bodhansky et al 1980, Lebedinets and Shushkova 1970, Ratcliff et al 1997, Tielens et al 1994)). However, with meteors entering the atmosphere the collision fluxes are much larger than in the laboratory experiments, and for most of the time the meteors are molten. Then the laboratory results may not be directly comparable

The model results show that the meteor temperatures almost invariably exceed the melting points before very much ablation occurs. Nevertheless, in our twenty selected cases the ablation and deceleration rates appear to vary smoothly, without obvious evidence of fragmentation. This seems quite surprising. However, in the remaining twelve cases the failure to fit our model could be an indication of fragmentation.

In our computer model we have assumed that the vapors emitted by the meteors are molecular rather than atomic. This seems to differ from the assumptions in the model described by Vondrak et al 2008, and Janches et al 2009. In view of the fact that the dissociation energies of, for instance, SiO_2, MgO and FeO are very much larger than their sublimation energies, it seems unlikely that the evaporation products would be atomic. On the other hand, subsequent collisions of the evaporated molecules with background air molecules would certainly lead to dissociation and/or ionization.

7 Summary

It appears that with most of these 32 radar traces the range and altitude vs time measurements are of sufficient quality to allow us to extract reliable velocities, trajectory inclinations and deceleration rates. In about 80% of the cases the velocities can be fitted with good accuracy as quadratic functions of the integrals of the air densities along the measured trajectories, and the time derivatives of these functions provide reasonable values of deceleration rates. We have used these fitted velocities and deceleration rates together with a computer model to determine best-fit values of the quantity $m\rho_s^2$, the product of the initial meteor mass times its mass density squared, successfully in 20 of the 32 cases. The model, which we have described, treats the coupled processes of meteor deceleration through air molecule collisions and the associated heating of the meteor, together with cooling by blackbody emission and by evaporation of its constituents, and the rate of loss of mass through evaporation and by sputtering. This procedure does not provide information about the separate quantities m and ρ_s. However, separate estimates of the masses m have been obtained from the measured radar scattering cross sections, using an electrostatic scattering model. By combining these m values with the $m\rho_s^2$ we have obtained values of ρ_s, almost all of which fall in the range between 0.1 to 1 g/cm^3. We have also described a process by which we can obtain $m\rho_s^2$ values directly from the velocity and deceleration data without using the computer model, although the results are very sensitive to errors in the decelerations.

APPENDIX A - The Mathematical Model

Definitions of variables:

t = time (s)

z = altitude (cm)

ρ_s = meteor mass density (g/cm^3)

r = meteor radius (assumed spherical)

M = meteor mass = $(4\pi/3)\rho_s r^3$

m_i = mass of the ith meteor constituent (g)

f_i = mass fraction of the ith constituent

v = meteor velocity (cm/s)

T = meteor temperature (assumed isothermal)

θ = trajectory zenith angle

$\rho(z)$ = local air density (g/cm^3)

H = meteor total enthalpy (ergs)

$\Delta H_{vap(i)}$ = heat of vaporization (erg/g) of the ith constituent

ΔH_{sput} = enthalpy loss by sputtering (ergs)

E^*_{sput} = energy required for sputtering of one gram (erg/g)

$\mu_{m(i)}$ = molecular weight of the ith constituent (g/molec))

$\mu_{vap(i)}$ = molecular weight of the ith vapor constituent (g/molec)

σ_{SB} = Stephan-Boltzmann constant (erg cm^{-2} deg^{-4} s^{-1})

C_p = specific heat of meteor material (erg/g)

$P_{vap(i)}$ = vapor pressure of the ith constituent (d/cm^2)

$A_{vap(i)}$ and $C_{vap(i)}$ = constants for the ith meteor constituent

Index i refers to the ith chemical constituent.

Differential (and other) equations:

$M = \sum_i m_i$

$f_i = m_i / M$

$dz/dt = -v \cos\theta \, dt$

$dv/dt = -\pi r^2 \rho v^2 / m$

$(dH/dt)_{coll} = 0.5 \, \pi r^2 \rho v^3$

$dm_i/dt = (dm_i /dt)_{sput} + (dm_i /dt)_{evap}$

$(dm_i/dt)_{evap} = -4\pi r^2 \mu_{vap(i)} \{3.51\times10^{+19} P_{vap(i)} /(\mu_{vap(i)} T)^{1/2}\}$ (if $m_i > 0$. otherwise zero)

$(dm_i/dt)_{sput} = -f_i \mu_{m(i)} (dH/dt)_{coll} / E^*_{sput}$

$dH/dt = (dH/dt)_{coll} + (dH/dt)_{rad} + (dH/dt)_{evap}$

$(dH/dt)_{rad} = -4\pi r^2 \sigma_{SB} T^4$

$(dH/dt)_{evap} = \sum_i \Delta H_{vap(i)} (dm_i/dt)_{evap}$

$T = H/(C_p M)$

$P_{vap(i)} = A_{vap(i)} \exp(-C_{vap(i)}/T)$

APPENDIX B – Meteor Temperatures and Evaporative Ablation—Quasi-Analytic Solutions of the Steady-State Equations – for a Single-Component Meteor

Simple calculations show that an incoming meteor must be heated by air molecule collisions to quite high temperatures, which are mitigated by the emission of blackbody radiation and by evaporative cooling associated with Langmuir evaporation of the meteor constituents. In the present case the meteors are quite small, so that heat conduction is fast enough to assure that their internal temperature profiles are isothermal. The collisional energy input rate is $\pi r^2 \rho v^3/2$. The radiative energy loss rate is $4\pi r^2 \sigma_{SB} T^4$, where T is the meteor temperature and σ_{SB} is the Stephan-Boltzmann constant. The vapor pressure for a single molecular constituent is approximated by the Clapeyron-Claussius relation

$$P_{vap} = A \exp(-C/T) , \qquad \text{(B-1)}$$

where A and C are constants characteristic of the particular evaporating meteor constituent. And the evaporative flux from the surface is given by the Langmuir relation

$$F_{evap} = C_{flx}\, P_{vap} / (\mu_{vap}\, T)^{1/2} \quad cm^{-2}s^{-1} , \qquad \text{(B-2)}$$

(Taylor and Langmuir 1933), where μ_{vap} is the molecular weight of the vapor. If P_{vap} is in dynes/cm^2 and μ_{vap} is in grams, then the constant C_{flx} is equal to 3.40×10^7. Then the rate of energy loss from the meteor surface due to evaporation is $4\pi r^2\, \Delta H_{sblm}\, F_{evap}$, where ΔH_{sblm} is the heat of sublimation (erg/molecule). We expect that the collisional heating and the radiative and evaporative cooling will balance each other, so that at each point in the meteor trajectory the temperature should be given by the steady-state relation

$$\pi r^2 \rho v^3/2 - 4\pi r^2 \sigma_{SB} T^4 - 4\pi r^2 \Delta H_{sblm} F_{evap} = 0 . \qquad \text{(B-3)}$$

(This equation is equivalent to Eq 3 of Hunt et al, 2004, or Eq 2 of Vondrak et al, 2008, although we assume the steady-state condition dT/dt = 0 (or negligible).)

This equation can be solved for T by Newton-Raphson iteration, and when T is determined we can calculate the evaporative mass loss rate

$$dm/dt = - 4\pi r^2 \mu_{vap} F_{evap} \qquad \text{(B-4)}$$

at each point along the trajectory. Then using this equation together with Eq 2 we can solve for the evaporative contribution to the effective ablation coefficient σ, which is now a function of altitude. From Eqs B-3 and B-4 we can see that this effective σ is a function of the air density ρ, the velocity v and the thermodynamic properties of the meteor material (or individual meteor constituents), but it is not directly dependent on the meteor mass or the mass density or the trajectory inclination angle.

Figures B-1(a,b), B-2(a,b) and B-3(a,b) show computed temperatures and effective ablation coefficients as functions of altitude for meteors composed of pure SiO_2, or MgO, or FeO, respectively. These three compounds are expected to be the decomposition products of the mineral olivine, which is commonly found to be a dominant constituent in stony meteorites (Korotev, 2006).

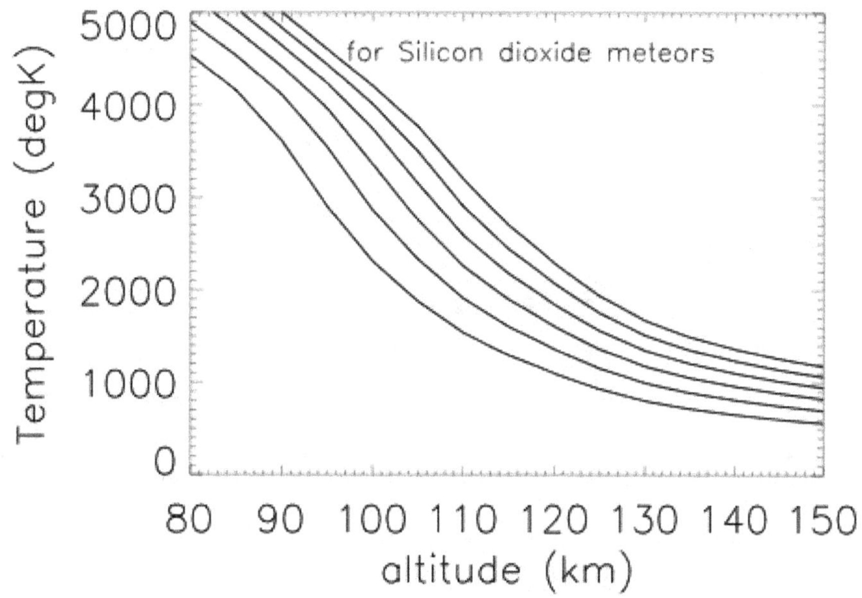

Figure B1a. Composite plots of steady-state temperatures vs altitude for SiO_2 meteors with velocities of 30, 40, 50, 60, 70 and 80 km/s (in that order from bottom to top).

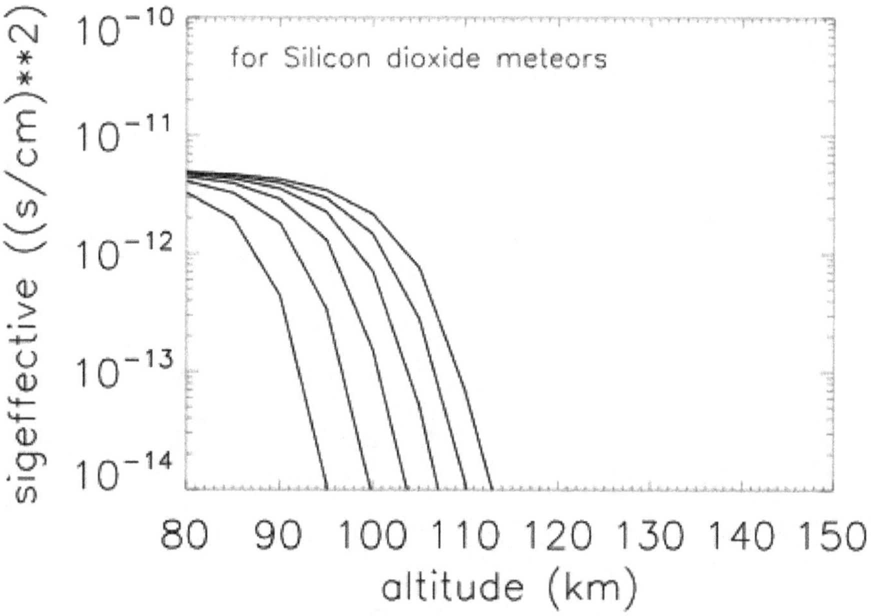

Figure B1b. Composite plots of effective ablation coefficients vs altitude for the same set of cases, and in the same order.

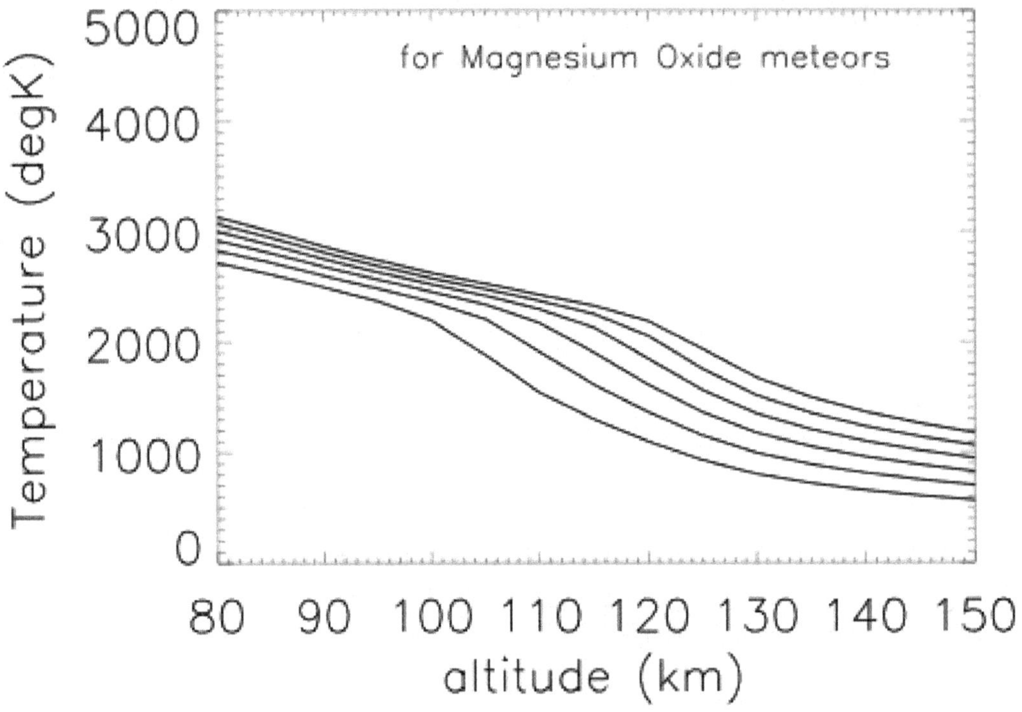

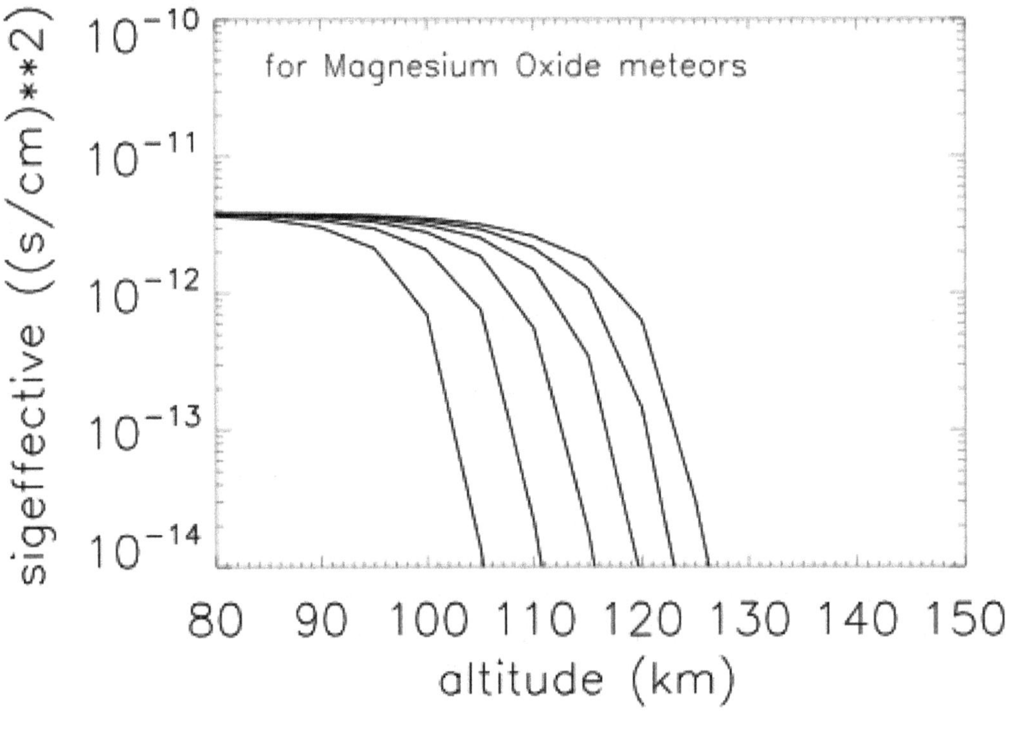

Figures B2a,b. Same as in Figures B1a,b, but for magnesium oxide meteors.

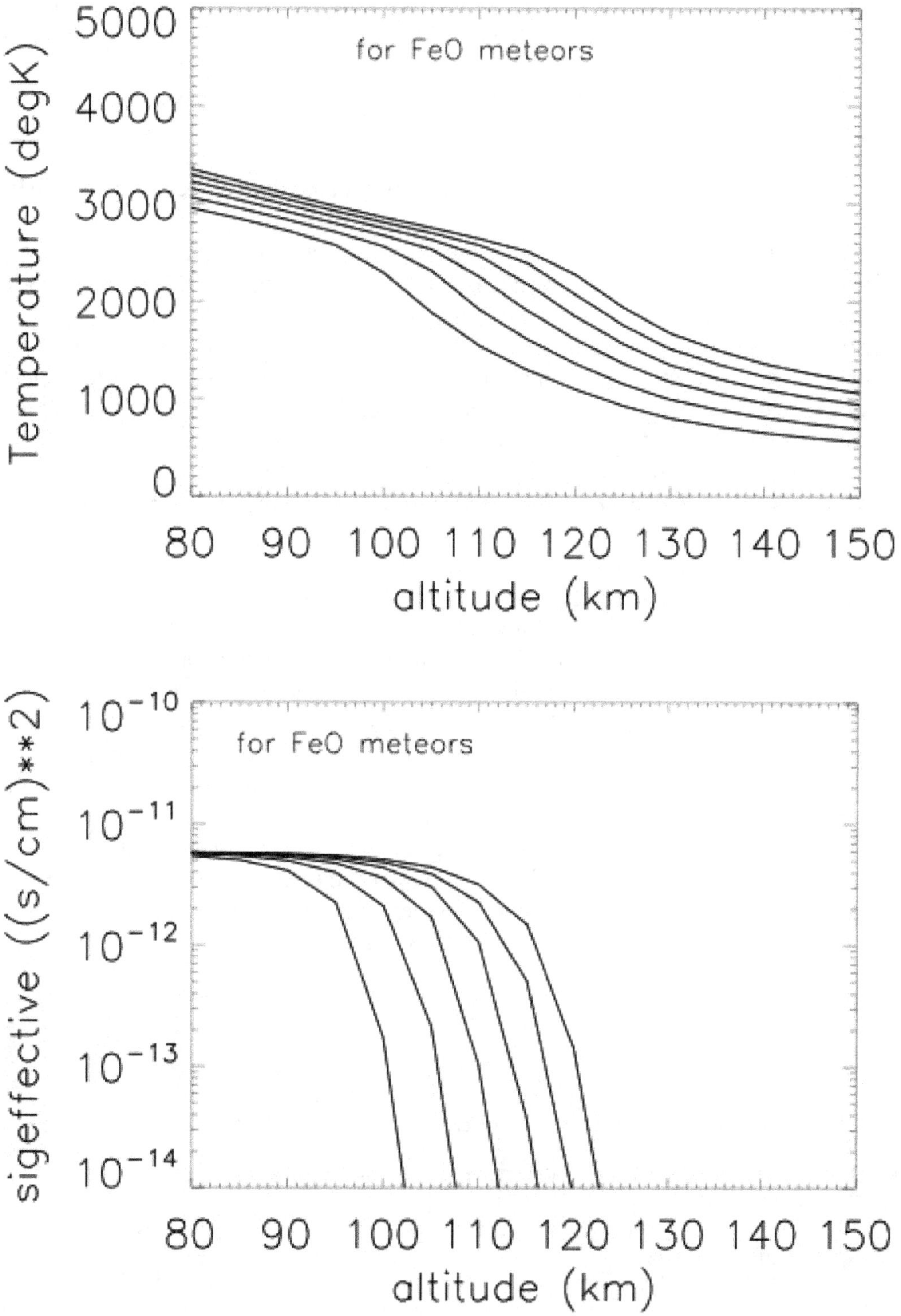

Figures B3a,b. Same as in Figures B1a,b, but for ferrous oxide meteors.

It is interesting to note that in figures B-1b, B-2b and B-3b in each case the effective sigma's tend to converge to a common value at the lowest altitudes. This is due to the fact that at the lowest altitudes the temperatures are so high that the evaporative cooling rate dominates over the radiative cooling rate. Then the second term in Eq B-3 can be ignored in comparison with the third, and combining Eqs B-3, B-4 and 2 gives

$$\sigma = \mu_{vap} / (2\ \Delta H_{sblm}) \quad \text{(limit for low altitudes and high temperatures).} \qquad \text{(B-5)}$$

It is also interesting that the limiting low-altitude values of the effective sigma's are not very different from the average σ values that we determined from our data using the constant-σ assumption and Eq 5 (although the details of that analysis will not be shown).

In writing Eqs B-3 and B-4 we have not mentioned the fact that the meteors can be expected to melt before they vaporize to an appreciable extent. The present radar data seem to indicate that the meteors do not immediately disintegrate upon melting – i.e. the traces seem to be continuous when the temperatures are expected to exceed the melting points. Apparently the molten meteors are held together by surface tension. In writing Eq B-3 we have not specifically included the solid-liquid transition, and we have used the heat of sublimation ΔH_{sblm} as if the meteor evaporated directly from the solid phase.

The calculations in this section have been for hypothetical meteors composed of a single vaporizable material. Of course actual meteors are expected to be made of a mixture of materials, each of which would vaporize at its own rate. More detailed computations including mixtures of materials have been described in section IV . In the present section we have also ignored the effect of sputtering. It is to be expected that at the highest altitudes, where the meteor temperatures are relatively low, the meteor mass loss rate will be dominated by sputtering, so the effective σ should include an added constant term for the sputtering contribution. On the basis of laboratory experimental results and theoretical studies (Behrisch 1981, Bodhansky et al 1980, Lebedinets and Shushkova 1980, Ratcliff et al 1997, Rogers et al 2005, Tielens et al 1994), we would expect that the sputtering term should be of the order of 4×10^{-14} s^2/cm^2. However, our deceleration data suggest a much larger value, of order 2×10^{-12} . The laboratory sputtering measurements of course involved much lower collisional fluxes than those expected for an incoming meteor, and much lower temperatures, and solid rather than molten targets.

The physico-chemical parameters used in these calculation have been shown in Table 1 of Section IV.

References

Akopov, F.A., "Behavior of zirconium dioxide ceramic under the operating..." , (1999). (Google search on boiling point of FeO).

R. Behrisch, Ed., Topics in Applied Physics – Vol 47, "Sputtering by Particle Bombardment", 1981, Springer-Verlag, Berlin, Heidelberg, New York.

Bodhansky, J., J. Roth, and H.L. Bay, "An Analytical Formula and Important Parameters for Low-Energy Ion Sputtering", J. Appl. Phys., 51, (1980), 2861-2865.

Brewer, L., P.W. Giles, and F.R. Jenkins, "The Vapor Pressure and Heat of Sublimation of Graphite", J. Chem. Phys., 16, (1948), 797 .

Brewer, L., and R. F. Porter, "A Thermodynamic and Spectroscopic Study of Gaseous Magnesium Oxide", J. Chem. Phys., 22, (1954), 1867-1877.

Bronshten, V.A., "Physics of Meteor Phenomena", Reidel, Dordrecht, 1983.

Ceplecha, Z., J. Borovicka, W.G. Elford, D.O. Revelle, R.L. Hawkins, V. Porubcan, and M. Simek, "Meteor Phenomena and Bodies", Space Science Rev., 84, (1998).

CIRA 1961, "COSPAR International Reverence Ionosphere 1961".

Clarke, J.T. and B.R. Fox, "Rate and Heat of Vaporization of Graphite above 3000K", J. Chem. Phys.. 51, (1969), 3231-3240.

Close, S., M. Oppenheim, S. Hunt, and A. Coster, "A Technique for Calculating Meteor Plasma Density and Meteoroid Mass from Radar Head Echo Scattering", Icarus 168, (2004), 43-52.

Close, S., S.M. Hunt, M.J. Minardi, and F.M. McKeen, "Analysis of Perseid Meteor Head-Echo Data Collected using the Advanced Research Projects Agency Long-Range Tracking and Instrumentation Radar (ALTAIR)", Radio Science 35, (2000), 1233-1240.

Close, S., M. Oppenheim, S. Hunt, and L. Dyrud, "Scattering characteristics of high-resolution meteor head echoes detected and multiple frequencies," J. Geophys. Res., 107, (2002), A10, 1295, doi:10.1029/2002JA009253.

Close, S., S.M. Hunt, and F.M. McKeen, "Characterization of Leonid meteor head echo data collected using the VHF-UHF Advanced Research Projects Agency Long-Range Tracking and Instrumentation Radar (ALTAIR)", Radio Science, 37, (2002) 10.1029/2000RS002602.

Dyrud, L.P., L. Ray, M. Oppenheim, S. Close, and K. Denney, "Modelling high-power large-aperture meteor trails", J. Atm. And Solar-Terrestrial Phys, 67, (2005), 1171-1177.

Fabian, R., "Vacuum Technology: practical heat treating and brazing", (1993). 253 pages. (FeO vapor pressure). Google books.

Ferguson, F.R., J.A. Nuth III, and N. M. Johnson, "Thermogravimetric Measurement of the Vapor Pressure of Iron from 1573 K to 1973 K", J. Chem. Eng. Data, 49, (2004), 497-501.

Hunt, S., S. Close, M. Oppenheim, and L. Dyrud, "Two-frequency meteor observations using the Advanced Research Projects Agency Long-Range Tracking and Instrumentation Radar (ALTAIR)", (2000). In: (2001), pp 451-455. Abstract +References in Scopus (cited by Scopus).

Hunt, S.M., M. Oppenheim, S. Close, P.G. Brown, F. McKeen, and M. Minardi, "Determination of meteoroid velocity distribution at the Earth using high-gain radar", Icarus, 168, (2004), 34-42.

Janches, D., L..P. Dyrud, S.L. Broadley, and J.L.C. Plane, "First observations of micrometeoroid differential ablation in the atmosphere", Geophys, Res. Letters., 36, L06101, doi:1029/2009GL037389, (2009).

Jones, W., "Theoretical and observational determination of the ionization coefficient of meteors", Mon. Not. R. Astron. Soc., 288, (1997), 995-1003.

Korotev, R.L, "Chemical Composition of Meteorites", Washington University in St. Louis. (2006). HTTP://meteorites.wustl.edu/metcomp/index.htm

Lebedinets, V.N., and V.B. Shushkova, "Micrometeorite Sputtering in the Ionosphere", Planet Space Sci., 18, (1970), 1653-1659.

Lebedinets, V.N., "Evolutionary and physical properties of meteoroids," (1973). In: Proceedings of the IAU Colloc., 13th, Albany, NY. I4-17 June, 1971. NASA SP vol 319, (1973), p259.

Loveland, R., A. Macdonell, S. Close, M. Oppenheim, and P. Colestock, "Comparison of Methods of Determining Meteoroid Range Rates from LFM Chirped Pulses", submitted to Radio Science, 2010.

Opik, E.J., "Physics of Meteor Flight in the Atmosphere", Interscience, New York, 1958.

Pradyot Patnaik, "Handbook of Inorganic Chemicals", McGraw-Hill, (2002). (properties of FeO).

Pecina, P., and Z. Ceplecha, "New aspects in single-body meteor physics", Bull. Astron. Inst. Czechosl. 34, (1983), 102-121.

Ratcliff, P.R., M.J. Burchell, M.J. Cole, T.W. Murphy, and F. Allahdadi, "Experimental Measurements of Hypervelocity Impact Plasma Yield and Energetics", Int. J. Impact Engng, 20, (1997), 663-674.

Rogers, L.A., K.A. Hill, and R.L Hawkes, "Mass Loss due to Sputtering and Thermal Processes in Meteoroid Ablation", Planetary and Space Sci., 53, (2005), 1341-1354.

Taylor, J.B., and I. Langmuir, "The rates of evaporation of atoms, ions and electrons from caesium films on tungsten", Phys Rev. 44 (1933), 423. (Irving Langmuir Nobel lecture).

Taylor, A.D., "The Harvard radio meteor project velocity distribution reappraised", Icarus, 116 (1995), 154-158.

Tielens, A.G.G.M., C.F. McKee, C.G. Seab, and D.J. Hollenbach, "The Physics of Grain-grain Collisions and Gas-Grain Sputtering in Interstellar Shocks", The Astrophysical Journal, 431, (1994), 321-340.

Vondrak, T., J.M.C. Plane, S. Broadley, and D. Janches, "A chemical model of meteor ablation", Atmos. Chem. Phys. Discuss., 8, (2008), 14557-14606.

Wickramasinghe, N.C., and K.S. Krishna Swamy, "Comments on the Possibility of Interstellar Quartz Grains", The Astrophysical Journal, 154, (1968), 397-400. (SiO2 vapor pressure).

Meteoroid Fragmentation as Revealed in Head- and Trail-echoes Observed with the Arecibo UHF and VHF Radars

J. D. Mathews • A. Malhotra

Abstract We report recent 46.8/430 MHz (VHF/UHF) radar meteor observations at Arecibo Observatory (AO) that reveal many previously unreported features in the radar meteor return–including flare-trails at both UHF and VHF– that are consistent with meteoroid fragmentation. Signature features of fragmentation include strong intra-pulse and pulse-to-pulse fading as the result of interference between or among multiple meteor head-echo returns and between head-echo and impulsive flare or "point" trail-echoes. That strong interference fading occurs implies that these scatterers exhibit well defined phase centers and are thus small compared with the wavelength. These results are consistent with and offer advances beyond a long history of optical and radar meteoroid fragmentation studies. Further, at AO, fragmenting and flare events are found to be a large fraction of the total events even though these meteoroids are likely the smallest observed by the major radars. Fragmentation is found to be a major though not dominate component of the meteors observed at other HPLA radars that are sensitive to larger meteoroids.

Keywords meteor radar · meteoroid fragmentation · meteor flare

1 Introduction

Here we provide an update to Mathews et al. (2010) who present Arecibo Observatory (AO) radar meteor results that are consistent with meteoroid fragmentation. While this conclusion has proven to be controversial; the finding that fragmenting meteoroids are observed both optically and with radar has a long history. In reference to fragmentation, Verniani (1969) notes that "At present, the structure and composition of meteoroids is a matter of controversy, with contrasting views put forward by different investigators." In fact, Mathews (2004) notes evidence of meteoroid fragmentation and terminal flares dating to the first known radar meteor Range-Time-Intensity (RTI) image given in Hey et al. (1947) and in Hey & Stewart (1947). Additionally, there is much evidence and many papers on "gross fragmentation" in optical bolides – e.g., see Ceplecha et al. (1993) and references therein.

At the Meteoroids 2001 conference Elford & Campbell (2001) noted that "Radar reflections from meteor trails often differ from the predictions of simple models. There is a general consensus that these differences are probably the result of fragmentation of the meteoroid." Elford (2004) concluded that approximately 90% of all specular trail events are accessible to his Fresnel holography approach while only about 10% of these events can be analyzed via the classical approach and thus that fragmentation is a dominant process for 80% of the specular trail events.

"Terminal" radar meteor events–referred to here as terminal flares–are reported in the Arecibo UHF/VHF results by Mathews (2004). Kero et al. (2008) report smooth to complex meteor light-curves

J. D. Mathews (✉) • A. Malhotra
Radar Space Sciences Lab, The Pennsylvania State University, University Park, PA USA 16802. Email: JDMathews@psu.edu

(their Figures 1-3) that they interpret as simple ablation, two-fragment, & multiple-fragment events with interference of the various head-echo signals. Roy et al. (2009) use genetic algorithm techniques to explore details of fragmenting meteors observed at the Poker Flat Incoherent Scatter Radar (PFISR). They employ multiple model point scatterers as we will outline below and a genetic algorithm to find via the evolution of three multi-fragment meteor events in a piece-wise fashion over groups of five radar pulse voltage (as opposed to power and thus lost phase information) returns finding the speed, deceleration, and amplitude of each particle in the ensemble. Their fitting procedure yields relative speed resolutions of as little at 1 m/s. Briczinski et al. (2009) utilize statistical techniques to estimate the role of fragmentation and terminal flares in Arecibo UHF radar meteor data. They find that terminal flares constitute up to ~15% of all events and that low-SNR, short duration, and/or fragmentation explain the ~67% of all events for which deceleration cannot be determined.

2 Observational Technique

The observations reported here utilize both the AO 430 MHz and 46.8 MHz radar systems. It is important to emphasize that these radars are frequency and time coherent and thus phase coherent over very long periods – years – and so in principle offer the ability to resolve features or motions on the scale of a fraction of a wavelength and centimeters/sec, respectively. These properties are utilized in the observations reported here. These two radars employ co-axial feeds yielding an overlapping central illuminated volume thus yielding a sizeable fraction of events that are seen in both radars. Table 1 lists the relevant parameters of both radars while Figure 1 shows the 430 MHz linefeed that illuminates the spherical-cap surface along with the four Yagi feeds arranged co-axially around the linefeed. At 46.8 MHz the dish is effectively parabolic allow this "point" feed arrangement. In Table 1 the quality factor is transmitter power (MW) time the effective area of the antenna (m^2) divided by the system temperature (Kelvins). Clearly the UHF system is much more sensitive – by a factor of ~600 – than the VHF system.

Table 1. Arecibo V/UHF Radar Properties

Radar	Beamwidth	Gain (dBi)	Power	System Temp	Quality Factor
46.8 MHz	1.4°	40	~40 kW	3000 K	~3
430 MHz	0.17°	61	~2 MW	100 K	~1825

Figure 1. The VHF & UHF antenna feed layout of the AO carriage-house radars.

For the results given here a 20 μsec uncoded pulse was used at UHF while – due to system duty cycle limitation – a 10 μsec uncoded pulse delayed by 12 μsec relative to the UHF pulse start was employed at VHF. In both cases the receive system bandwidth was 1 MHz while in-phase and quadrature samples were taken at baseband with 1 μsec sampling intervals yielding 150 m range resolution. A 1 ms InterPulse Period (IPP) was utilized with the overall technique based on early AO D-region observations (Mathews 1984). The first dedicated AO meteor observations are reported by Mathews et al. (1997).

3 Observational Results

The results presented here were obtained during two ~12 hr observing sessions beginning at 2000 hr AST (Atlantic Standard Time) on 5 and 6 June 2008. Approximately 17,000 meteors were detected at UHF using automated detection software (Briczinski et al. 2009, Mathews et al. 2003, Wen et al. 2005, Wen et al. 2004). This approach was not separately applied at VHF due to the relatively short pulse and the high level of interference in the VHF band. The VHF results have been manually searched for large events that include some of the flare-trail results reported here.

Figure 2 displays Range-Time-Intensity (RTI) images of three meteor events that together characterize many of the ~17,000 UHF events we report here. Event 1, seen at 430 MHz, shows strong interference fading consistent with two slowly separating meteoroid fragments each of which has an individual head-echo. This interpretation builds on the results of Mathews et al. (2010) and Roy et al. (2009) and will be addressed further in the discussion section.

Event 2a, seen at VHF, shows several features including an underlying interference or fading pattern similar to event 1 but also an altitude-narrow trail that we attribute to an impulsive fragmentation "flare" occurring at about 40 ms and that results in a relatively small "blob" of plasma embedded in the background atmosphere. The term "flare" is adopted from optical meteor observations that often reveal impulsive brightening events. Event 2b shows the UHF return which defines the center of both beams. Note the strong intra-pulse fading that is due to the rapidly evolving particle distribution relative to the 69.7 cm wavelength. Figure 2, Event 3, shown only at VHF as the corresponding UHF event was very weak, shows mild fragmentation prior to 60 ms when a strong fragmentation flare occurs followed by a second flare at 110 ms. Of special interest in Event 3 is the strong interference fading – similar in effect to the Event 1 interference pattern – between the head-echo and trail-echoes. The implications of these various two-scatterer interference patterns will be explored in the discussion section and additional example events including those resembling differential ablation (Janches et al. 2009) and those opposite to differential ablation signature (intensity rises rapidly and falls slowly) are given in Mathews et al. (2010).

Next we consider some of the more subtle and perhaps surprising results from this dataset. Figure 3 displays a short UHF meteor head-echo that shows the beam pattern with a stronger central return and two side-lobe returns as the meteor moves across the beam. This event also displays a complex intra-pulse fading consistent with multiple, closely-spaced but rapidly dispersing, meteoroid "fragments" similar to those seen in Figure 2 event 2b. which is more slowly evolving (Roy et al. 2009). We are able to resolve these features of the meteoroid multi-head-echo evolution at the microsecond level due to the phase coherent nature of these radars. The Figure 3 VHF return is much longer than at UHF because – per Table 1 – the VHF beam is much wider. The combined U/VHF event is in common volume only over the span of the UHF return. The VHF meteor interference feature is simple like that of Figure 2, event 1 but fades more slowly than the Figure 3 UHF return as the wavelength is more than a factor of nine longer. The VHF return displays a clear terminal flare that is consistent with the LATE (Low-Altitude Trail-Echo) reported at Jicamarca (Malhotra & Mathews 2009). This type of event is relatively common.

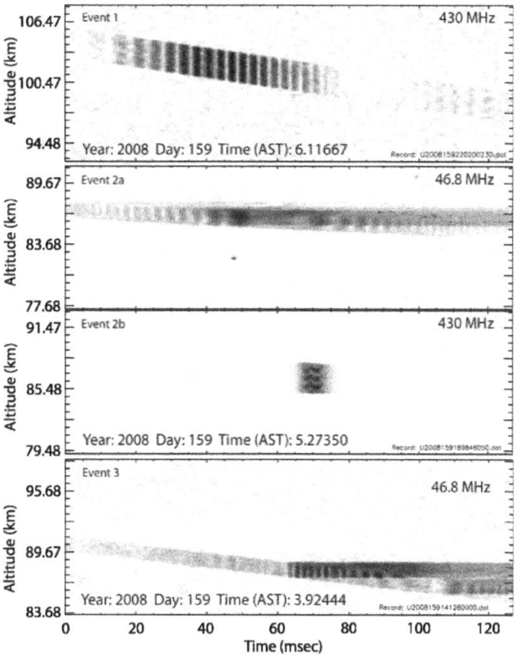

Figure 2. RTI images of three archetypal AO meteor events. Event 1, seen at UHF, shows a strong fading pattern consistent with two slowly separating meteoroid fragments each of which has an individual head-echo. The event 2 panels demonstrate the value of viewing the same event at two widely separated frequencies. Event 2a, seen at VHF, shows several features including an underlying interference pattern similar to event 1 but also an altitude-narrow trail that we attribute to a fragmentation "flare". Event 3 shows some Event 1 like fragmentation and two flare-trails. (Fig. 1 from Mathews et al. (2010))

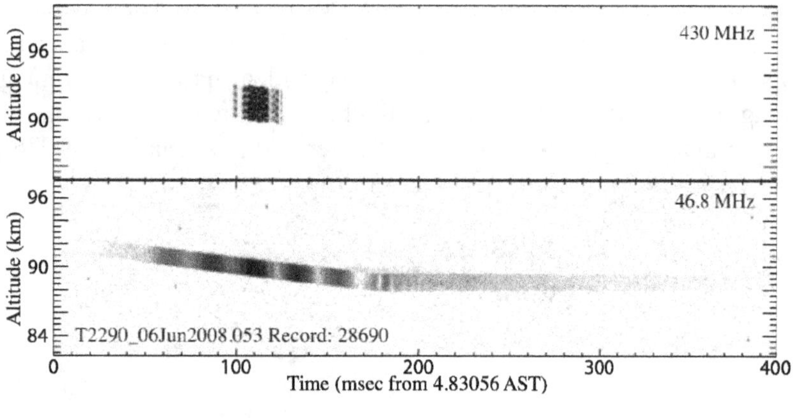

Figure 3. RTI images of a meteor event seen at both UHF and VHF. The UHF head-echo shows the beam-pattern as strong intra-pulse fading. The VHF echo has a terminal flare.

Figure 4 shows two UHF meteor events that are ambiguous and thus point to the wide range of knowledge potentially available via radar meteor studies. Event 1 is likely due to two or more particles that cause the strong intra-pulse interference fading visible in the ~60-70 IPP and ~82-95 IPP. It is unclear if the early event results in a trail and is then followed by a separate event that clearly results in the UHF trail. In any case, we might deem this total event a nano-shower in that almost certainly all particles were associated with a parent body at or just above atmospheric entry. Figure 4 event 2 is likely a terminal flare trail event similar to those reported at the 1280 MHz Sondrestrom Research Facility (SRF) terminal events (Mathews et al. 2008). It is also possible that this event is a "classical" trail event where the trajectory of the meteoroid is perpendicular to the zenith-pointing beam.

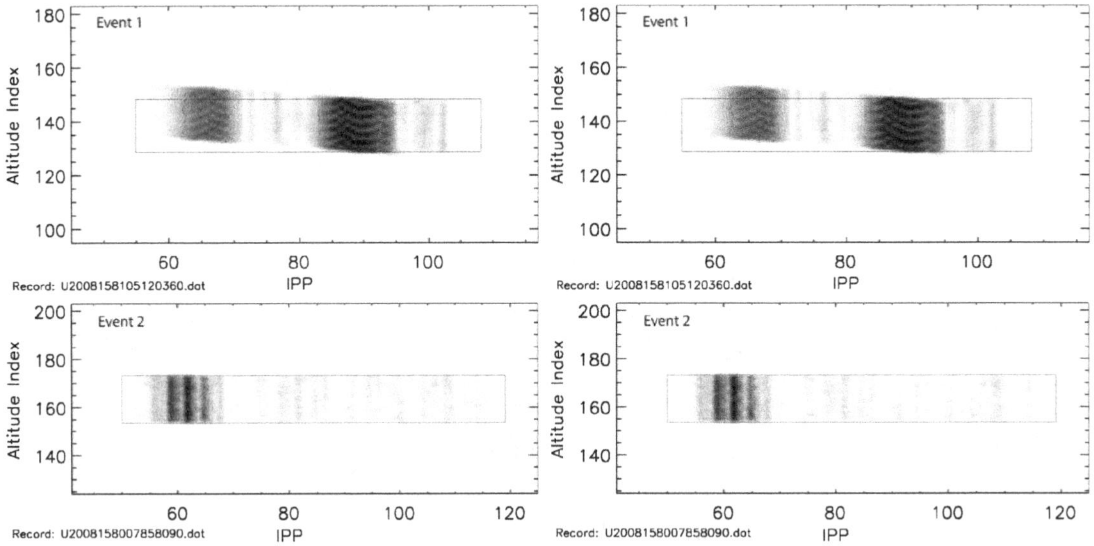

Figure 4. RTI images of two complex UHF head- and/or trail-echo events. Event 1 shows strong intra-pulse fading due to two or more individual head-echo producing meteoroids in close proximity. Event 1 displays a clear trail-echo after IPP 95. Event 2 is likely a terminal-flare trail but may be a classical trail-echo where the trajectory of the meteoroid is perpendicular to the zenith-pointing beam.

Figure 5 points to a new – previously unreported – class of radar meteor event. These long-lasting – for **k∠B** ≈ 45° – trail events appear to be the "fossil" remnants of a radar bolide event. That is, while the head-echo of the progenitor event is not always identifiable, the event generated sufficient (flare?) trail-producing plasma that the resultant trail-echo lasts a few seconds and may in fact drift into the VHF beam at the normal D-region wind speeds of order 100 m/s (Mathews 1976). Note the complex interference fading of the several regions of the trail-plasma.

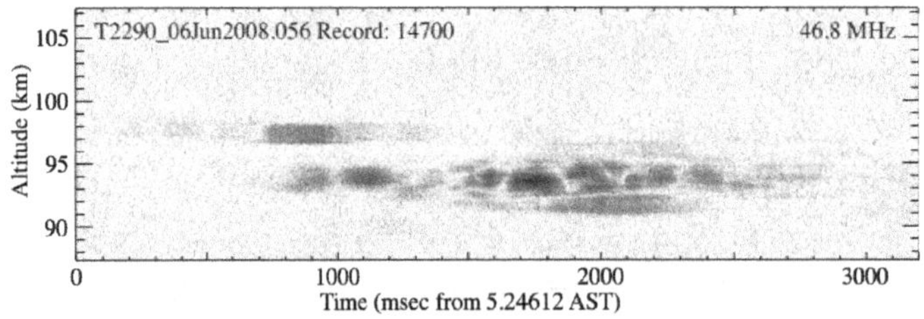

Figure 5. A likely "fossil" radar bolide event that has an ~3 sec lifetime. The progenitor event was not observed.

4 Discussion

We report several classes of V/UHF "common volume" radar meteor events that are, we argue, consistent with fragmenting meteoroids that produce multiple, interfering, head-echo events as well as "flare" and "terminal flare" trails that often display interference fading between/among the head- and trail-echo components. That fading occurs in a simple pattern (e.g., similar to the classic Young's point-source optics experiment) suggests a simple model of the meteor scattering process that, as we show below, appears both necessary and sufficient to explain what we observe. It is important to stress in introducing this model that it is successful in part due to the time and frequency and thus phase coherence of both the VHF and UHF radars that permit full use of the model we present. We also note that this capability has been intrinsic to most radars for many years but that full advantage of this "holographic" capability is just beginning for the modern geophysical radars.

In the model scenario we propose, each head- or trail-echo signal is consistent with a point target – i.e., each has a well-defined phase center – that is readily modeled at the receiver baseband as

$$x_n(t) = A_n \exp\left(\frac{j4\pi R_n(t)}{\lambda}\right) \tag{1}$$

where $R_n(t) = R_n(t_0) - v_n(t - t_0) + d_n(t - t_0)^2 / 2$ – the subscript n refers to the nth meteoroid fragment. In (1), $j = \sqrt{-1}$ and the multiple meteoroid fragments are taken to be traveling on the same trajectory at range $R_n(t)$ time t with t_0 the initial time and with constant speed and deceleration v_n, d_n, respectively. It is important to note that equation (1) is accurate only if all the meteor energy is contained in the received bandwidth at baseband – otherwise filter features such as ringing may occur. To this end we employ a 1 MHz bandwidth (actually 0.5 MHz at baseband for both the in-phase and quadrature channels thus satisfying the Nyquist sampling condition), 1 μs sample intervals, and a transmitter pulse of 10/20 μs at VHF/UHF, respectively. Thus the pulse spectrum is very narrow with respect to the sampled bandwidth so that the meteor Doppler shift (~22 kHz at VHF and ~200 kHz at UHF for a 72 km/s meteor) does not result in signal energy being lost outside the filter bandpass. Also note that eqn. (1) embodies the Doppler shift of the spectrum via the time rate of change of R(t) within a given pulse.

In an example of the successful use of equation (1), it can be seen that the signals from two slowly separating fragments alternately appear in- and out-of-phase as the net path from the two particles to the receiver varies over half a wavelength, $\lambda/2$. This results in a Young's experiment-like outcome as we demonstrate below. Use of (1) to successfully characterize meteor head-echo returns and extract Doppler information dates to the earliest meteor observations at Arecibo Observatory (Janches et al. 2003, Mathews et al. 2003, Mathews et al. 1997). Radio science implications are discussed by Mathews (2004).

Figure 6 bottom panel shows Figure 2, event 1 along with modeling results that employ eqn. (1) for two particles (head-echoes) at both AO radar frequencies. The model results include Gaussian distributed random noise in both the in-phase and quadrature channels. As noted in the caption, the two particles are taken to start together but then separate at speeds of 50.4 km/s and 50.3 km/s, respectively, with no deceleration. The particle head-echoes have equal scattering cross-sections. The model beam-pattern at UHF is modeled as a double Gaussian yielding main- and side-lobes that closely match the observed meteor return. The VHF beam-pattern is a single very wide Gaussian that causes slight intensity decrease at the model event edges.

The Figure 6 model results at 430 MHz are completely consistent with the observations. The matching could be "tuned" by adjusting the speeds, adding a slight deceleration, and adjusting the initial phase separations of the two particles. However, this complexity is unnecessary and will be left to actual multi-particle fitting algorithms (Briczinski et al. 2006) that are currently under development for the multi-particle case. Mathews et al. (2010) gives a similar modeling/data comparison but for head-echo fading with a stationary flare-trail while Roy et al. (2009) give details on using equation (1) fitting via genetic algorithms. The VHF model result in Figure 6 shows the slower fading rate at VHF relative to UHF. This model result is similar to the observational results given in Figure 3 where the UHF fading rate is very rapid while the VHF fading rate is quite similar to the Figure 6 VHF model result.

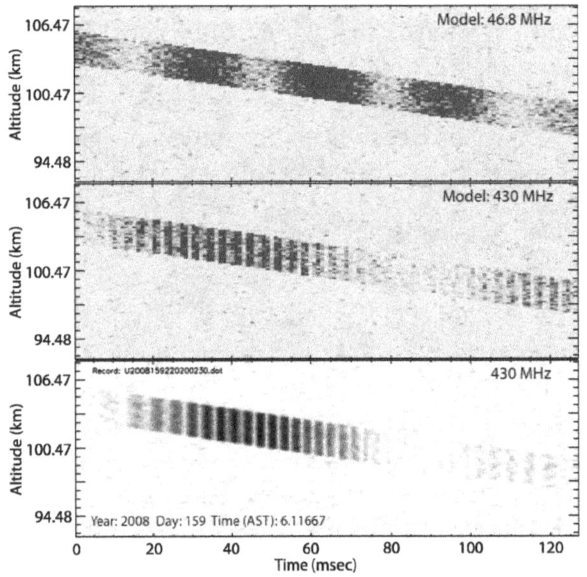

Figure 6. RTI images of an observed and modeled two-particle meteor event. The bottom panel event is just Figure 2, event 1. The model is eqn. (1) applied separately to two particles of equal scattering cross-section that start at the same location but separate as the speeds are taken to be 50.3 km/s and 50.4 km/s with no deceleration. The VHF and UHF fading rates are different due to the much longer wavelength (6.4 m vs. 0.697 m) at VHF. See text for details on the beam-patterns.

5 Conclusions

We have reported on common volume V/UHF radar meteor observations at Arecibo Observatory. These observations have revealed meteor head- and trail-echo features that are consistent with meteoroid fragmentation. Further, the VHF observations have revealed flare-related trail-echoes that, due to the interference fading between the head- and trail-echoes, are found to be "small" compared with a wavelength in that a well defined phase center exists. We additionally find that both a necessary and sufficient description of the head- and trail-echoes is given by eqn. (1) which simply models point-target scattering at receiver baseband with no Doppler spreading of the spectrum as this has not proven necessary. We give modeling results supporting this conclusion. These results go beyond those given by Mathews et al. (2010) and Roy et al. (2009) and provide necessary insight into the radio science aspects of radar meteor observations (Mathews 2004).

We additionally report observations of UHF trail-echoes and UHF meteor echoes that are consistent with meteoroid terminal "flare" events and/or "classical" meteor echoes from a meteor traveling perpendicular to the radar pointing direction that is at zenith for these results. Also, we report what may be a new class of radar meteor events that we term "radar bolides". Thus far, radar bolides appear only as large (i.e., intense, distributed in altitude, and long-lived) trail-events in that the progenitor meteoroid head-echo has not been convincingly identified as it apparently falls outside the radar beam. The radar bolides last 100's of milliseconds through, thus far, to about 10 seconds and consist of multiple scattering centers distributed over several kilometers in range. Apparently these "trails" drift into the radar beam due to the ~100 m/s winds in the upper mesosphere (Mathews 1976). It seems likely that this pattern of scatterers is formed when a large meteoroid breaks into a pattern of still large meteoroids with significant horizontal dispersion at 90 km altitude where we observe the "radar bolide" event. In any case, the "radar bolide" is quite intensive relative to the usual meteor events.

While Mathews et al. (2010) reports ~90% fragmentation signatures for this set of observations, a companion paper (Malhotra and Mathews, these proceedings), report a different distribution of meteor events from the Resolute Bay Incoherent Scatter Radar (RISR). At RISR they find an event type distribution of fragmentation (48%), simple ablation (32%), and differential ablation (20%). We suggest that this contrast is likely caused by AO "seeing" significantly smaller meteoroids than RISR – this due to the much higher sensitivity of AO relative to RISR.

Finally we note that the simultaneous presence of close meteoroid fragments renders a clear definition of dynamic mass (Fentzke et al. 2009, Janches & Chau 2005, Mathews et al. 2001), absolute scattering-cross section mass (Close et al. 2005), and meteoroid mass density (Briczinski et al. 2009, Novikov & Pecina 1990) difficult at best. Additionally, interpretation of details such as differential ablation (Janches et al. 2009) also becomes difficult as the ensemble of evolving particles appears to be capable of producing not only the lightcurves we expect for a differential ablation event but also the exact opposite (Mathews et al. 2010; Malhotra and Mathews, these proceedings). Put concisely, our results indicate that many meteoroids arrive at the top of the atmosphere as a "dustball" or an otherwise loosely-attached configuration of particles (Verniani 1969) that begin to separate immediately on encountering the atmosphere and/or as the system proceeds into the atmosphere and becomes visible as a radar meteor. These particles also undergo occasional instantaneous "flaring" whereby one of the ensemble of particles or a newly created particle is apparently terminally destroyed thus creating the plasma "blob" that we observe as the flare.

To paraphrase (Verniani 1969), the authors wish to conclude this section by quoting the thoughts of one of the historically-most-established leaders in meteor research: "I regard the process of fragmentation of meteor bodies as even more important than is recognized now. Therefore further studies of this process seem to be necessary. It is impossible to predict the course of fragmentation for an individual meteor particle but statistical regularities of the fragmentation process must exist and they should be studied. These statistical regularities are probably somewhat different for different meteor streams and also probably vary with the mass of the meteor particles." (Levin 1968)

Acknowledgements

The Arecibo Observatory is part of the National Astronomy and Ionosphere Center, which is operated by Cornell University under a cooperative agreement with the National Science Foundation. This effort was supported under NSF grant ATM 07-21613 to The Pennsylvania State University.

References

S.J. Briczinski, J.D. Mathews, & D.D. Meisel, J. Geophys. Res., **114** A04311 (2009)

S.J. Briczinski, C.-H. Wen, J.D. Mathews, J.F. Doherty, & Q.-N. Zhou, IEEE Trans. Geos. Remote Sens., **44** 3490 (2006)

Z. Ceplecha, P. Spurny, J. Borovicka, & J. Keclikova, Astron. Astrophys., **279** 615 (1993)

S. Close, M. Oppenheim, D. Durand, & L. Dyrud, J. Geophys. Res., **110** A09308 (2005)

W.G. Elford, Atmos. Chem. Physics, **4** 911 (2004)

W.G. Elford, & L. Campbell, Effects of meteoroid fragmentation on radar observations of meteor trails, Meteoroids 2001 Conference, ESA Publications, **SP-495**, ed. B. Warmbein (Swedish Institute of Space Physics, Kiruna, Sweden, 2001) pp. 419-423

J.T. Fentzke, D. Janches, & J.J. Sparks, J. Atmos. Solar-Terr. Phys., **71** (2009)

J.S. Hey, S.J. Parsons, & G.S. Stewart, Mon. Not. R. Astron. Soc., **107** 176 (1947)

J.S. Hey, & G.S. Stewart, Proc. Phys. Soc. Lond., **59** 858 (1947)

D. Janches, & J.L. Chau, J. Atmos. Solar-Terr. Phys., **67** (2005)

D. Janches, L.P. Dyrud, S.L. Broadley, & J.M.C. Plane, Geophys. Res. Lett., **36** L06101 (2009)

D. Janches, M.C. Nolan, D.D. Meisel, J.D. Mathews, Q.-H. Zhou, & D.E. Moser, J. Geophys. Res., **108** 1-1 (2003)

J. Kero, C. Szasz, A. Pellinen-Wannberg, G. Wannberg, A. Westman, & D.D. Meisel, Geophys. Res. Lett., **35** (2008)

B.Yu. Levin, Meteor Physics (Round-Table Discussion and Summary), Physics and Dynamics of Meteors, International Astronomical Union. Symposium no. 33, Dordrecht, D. Reidel, **33**, ed. L. Kresak, & P. M. Millman (Tatranska Lomnica, Czechoslovakia, 4-9 September, 1968) pp. 511-517

A. Malhotra, & J.D. Mathews, Geophys. Res. Lett., **36** L21106 (2009)

J.D. Mathews, J. Geophys. Res., **81** 4671 (1976)

J.D. Mathews, J. Atmos. Terr. Phys., **46** 975 (1984)

J.D. Mathews, J. Atmos. Solar-Terr. Phys., **66#3** 285 (2004)

J.D. Mathews, S.J. Briczinski, A. Malhotra, & J. Cross, Geophys. Res. Lett., **37** L04103 (2010)

J.D. Mathews, S.J. Briczinski, D.D. Meisel, & C.J. Heinselman, Earth, Moon, Plnts., **102** 365 (2008)

J.D. Mathews, J.F. Doherty, C.-H. Wen, S.J. Briczinski, D. Janches, & D.D. Meisel, J. Atmos. Solar-Terr. Phys., **65** 1139 (2003)

J.D. Mathews, D. Janches, D.D. Meisel, & Q.-H. Zhou, Geophys. Res. Lett., **28** (2001)

J.D. Mathews, D.D. Meisel, K.P. Hunter, V.S. Getman, & Q. Zhou, Icarus, **126** 157 (1997)

G.G. Novikov, & P. Pecina, Bul. Astron. Inst. Czechosl., **41** 387 (1990)

A. Roy, S.J. Briczinski, J.F. Doherty, & J.D. Mathews, IEEE Geosci. Remote Sens. Lett., **6** 363 (2009)

F. Verniani, Space Sci. Rev., **10** 230 (1969)

C.-H. Wen, J.F. Doherty, & J.D. Mathews, J. Atmos. Solar-Terr. Phys., **67** 1190 (2005)

C.-H. Wen, J.F. Doherty, J.D. Mathews, & D. Janches, IEEE Trans. Geos. Remote Sens., **42** 501 (2004)

A Study on Various Meteoroid Disintegration Mechanisms as Observed from the Resolute Bay Incoherent Scatter Radar (RISR)

A. Malhotra • J. D. Mathews

Abstract There has been much interest in the meteor physics community recently regarding the form that meteoroid mass flux arrives in the upper atmosphere. Of particular interest are the relative roles of simple ablation, differential ablation, and fragmentation in the meteoroid mass flux observed by the Incoherent Scatter Radars (ISR). We present here the first-ever statistical study showing the relative contribution of the above-mentioned three mechanisms. These are also one of the first meteor results from the newly-operational Resolute Bay ISR. These initial results emphasize that meteoroid disintegration into the upper atmosphere is a complex process in which all the three above-mentioned mechanisms play an important role though fragmentation seems to be the dominant mechanism. These results prove vital in studying how meteoroid mass is deposited in the upper atmosphere which has important implications to the aeronomy of the region and will also contribute in improving current meteoroid disintegration/ablation models.

Keywords meteor radar · meteoroid disintegration · meteoroid fragmentation · ablation

1 Introduction

Meteoroids are responsible for thousands of kilograms of mass flux into the earth's upper atmosphere annually (Mathews et al. 2001).These meteoroids are not only the only source of metallic ions in the upper atmosphere (Kelley 1989) but also, as a result of this very high mass flux, pose a threat to our space infrastructure (Caswell and McBride 1995) and are responsible for a variety of ionospheric phenomenon such as the Sporadic-E (Malhotra et al. 2008) and Polar Mesospheric Summer Echoes (Bellan 2008). This makes it imperative that we know and understand the form in which meteoroids disintegrate into the upper atmosphere in order to understand the aeronomy of the region.

As the meteoroid enters the earth's atmosphere, it collides with the air molecules and heats up. When the temperature reaches around 2000K – usually between 80-120km – surface particles start evaporating from the body. These particles quickly ionize, also ionizing the air molecules around them, forming a ball of plasma around the meteoroid. Radar scattering from this ball of plasma surrounding the meteoroid is called the head echo. Although meteor head echoes were first observed in the 1940s (Hey et al. 1947), their study gained momentum only in the 1990s when they were observed using the High Power Large Aperture [HPLA] radars (Mathews et al. 1997). Since then, these head echo observations have proved invaluable in determining meteoroid velocities (Janches et al. 2000), mass flux (Mathews et al. 2001) and radiants (Chau and Woodman 2004). More recently, these head echo observations are being studied to determine the form that meteoroid mass flux takes when it enters into

---

A. Malhotra • J.D. Mathews (✉)
Radar Space Sciences Lab, The Pennsylvania State University, University Park, PA USA 16802. Phone: 814-865-2354, Email: JDMathews@psu.edu

the earth's upper atmosphere. As outlined below, these initial studies on meteoroid disintegration using various HPLA radars have produced contrasting results, generating much interest and even controversy in the meteor community.

Kero et al. (2008), using the EISCAT 930 MHz UHF radar, provide "the first strong observational evidence of a submillimeter-sized meteoroid breaking apart into two distinct fragments" i.e. fragmentation. Fragmentation can take place either due to thermally induced stresses (Jones and Kaiser 1966) or due to the separation of a molten metal droplet from the lower density chondritic compounds of a heated meteoroid (Genge 2008). Kero et al. (2008) provide an example of a "beat pattern" light curve event [the light curve is defined as the pulse-integrated Signal-to-Noise Ratio (SNR)], their Figure 2, and interpret it as being due to interference from two distinct scattering centers. They show that the result is consistent with interference from two fragments of unequal cross-sectional area over mass ratio, separating from each other due to different deceleration along the trajectory of the parent meteoroid. They also provide examples of a "smooth" light curve where the measured SNR follows the antenna beam pattern (their Figure 1) i.e. simple ablation and a quasi-continuous disintegration event (their Figure 3).

Mathews at al. (2008), carrying out a similar analysis for the meteor echoes observed by the Sondrestorm Radar Facility (SRF) 1290 MHz radar, conclude that almost all the meteors observed by the SRF radar are fragmenting large meteoroids that are observed only in the terminal phase of their encounter with the upper atmosphere.

Roy et al. (2009) use a genetic-algorithm based search and fitting procedure to solve for the number of scatterers and their differential speeds in estimating the properties of complex light curves observed by the Poker Flat ISR radar. Based on the above-mentioned analysis, they conclude that fragmentation is the cause of complex light curves.

Dyrud and Janches (2008) determine meteoroid properties by comparing expected results from a theory based ablation model of the meteor head echo and observed meteor properties using the Arecibo 430 MHz UHF radar. They do not include the effects of fragmentation in their model as they "find no evidence that meteoroid fragmentation plays a role in the vast majority of head-echo observations at Arecibo". However, they also conclude that a simple ablation model cannot account for the non-smooth light curves observed by the radar.

Janches et al. (2009), using the Arecibo 430 MHz UHF radar, provide the first observations of differential observations in micrometeoroids. In the differential ablation process, the particle's more volatile components (Na and K) are released first when the temperature is still relatively low followed by the evaporation of less volatile components (Si, Fe and Mg) as the particle descends through the atmosphere, increasing its temperature. Events undergoing differential ablation are characterized by a sudden decrease or increase in the light curve. Though they observed features of the differential ablation process only in small percentage of the detected events, they still conclude that differential ablation is the main mechanism by which micron-sized particles deposit their mass in the upper atmosphere.

Mathews et al. (2010), using data collected from simultaneous observations using the same Arecibo 430 MHz UHF radar and the Arecibo 46.8 MHz common-volume VHF radar, present many unreported features in the radar meteor return that are consistent with meteoroid fragmentation. Based on modeling studies and statistical analysis, they conclude that fragmentation is the dominant process by which micrometeoroids deposit their mass in the upper atmosphere — a conclusion at direct odds with the one reached by Janches et al. (2009), though both the studies use the same radar.

It is clear from the above introduction that the process by which micrometeoroids deposit their mass in the upper atmosphere remains is a topic of much interest in the community and the relative roles of fragmentation, differential ablation and simple ablation is a subject of much debate and speculation.

However, to-date there has been no statistical study studying the relative contribution of the three mechanisms. In this paper, we present the results from first-ever such study. In Section 2, we present details of the observational set-up and radar parameters. The results are presented in Section 3, followed by the discussion in Section 4. We end with the conclusions of our study and the scope for future work in Section 5.

2 Observational Set Up

The results presented herein happen to be one of the first published results from the newly-operational 442.9 MHz Resolute Bay Incoherent Scatter Radar (RISR) located in Resolute Bay, Nunavut, Canada (74.72950° N, 94.90539° W). For these observations carried out on 24-25 and 26 August 2009 from 2140 to 0040 hours (UT) and 1120 to 1455 hours (UT) respectively (totally ~ 6.3 hours of data), the radar beam was pointed in a direction parallel to the Earth's rotation axis and the maximum power transmitted was ~ 1.7 MW. Transmission and reception was done using all the 128 panels of the radar. A pulse width of 90 μs with an IPP (Inter Pulse Period) of 2 ms was used for transmission.

3 Observational Results

We observe meteor signatures of all three micrometeoroid disintegration mechanisms, i.e. fragmentation, differential ablation and simple ablation using the Resolute Bay Incoherent Scatter Radar (RISR), enabling us to conduct a statistical analysis of the relative role of these mechanisms. We begin by presenting representative examples of all the three mechanisms as observed by RISR. These events will also serve to facilitate future similar studies using RISR and the other HPLA radars.

Figure 1a is a RTI (Range-Time-Intensity) plot of a typical fragmenting meteor event. Note the structure present in the meteor return. The beat pattern can be noticed even without the aid of the light curve. Figure 1b shows the light curve (pulse integrated SNR for each IPP) for the event shown in Figure 1a. As expected from the RTI plot, the beat pattern associated with fragmentation is observed. An explanation on the cause of this observed beat pattern is given in a companion paper by Mathews and Malhotra in this issue.

Figure 2a is a RTI plot of a typical meteor event undergoing differential ablation and the corresponding light curve is shown in Figure 2b. Notice the abrupt decrease in SNR received at ~ 32 ms; a possible signature of differential ablation [Figure 2 of Janches et al. (2009)]. Janches et al. (2009) attribute this sudden decrease in received power to the complete ablation of the main meteoroid constituents (Si, Fe and Mg). The reduced power received after the sudden decrease is due to the plasma created in ablation of the refractory metals (Ca, Al and Ti).

Figure 3a is a RTI plot of a typical meteor event undergoing simple ablation and Figure 3b is the corresponding light curve for this event. Notice the relatively smooth pattern (compare to the cases presented in Figure 1b and 2b) obtained for this event in Figure 3b. We assume simple ablation occurs due to the homogeneous composition of the meteoroids.

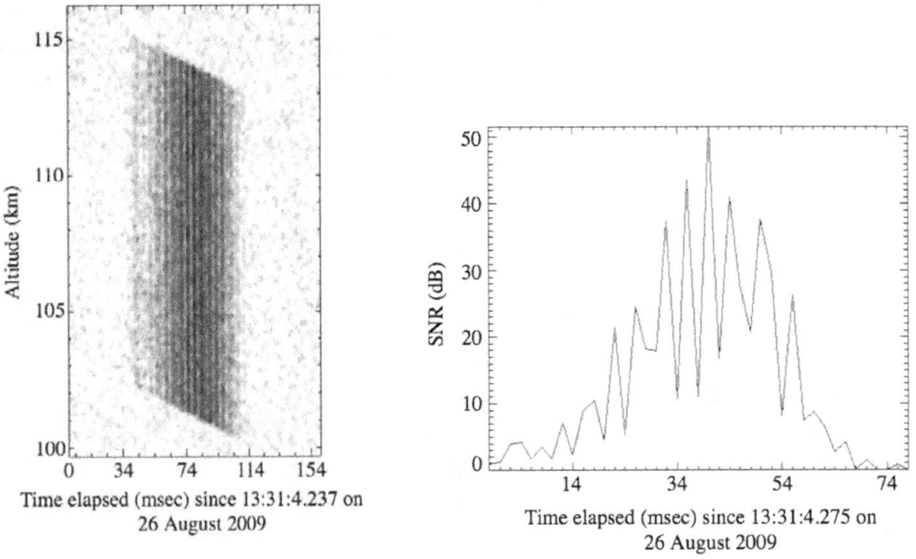

Figure 1. (a) Range Time Intensity (RTI) plot of a fragmenting meteor event. Notice the structure within the meteor return. (b) The light curve for this event. The "beat pattern" observed is obtained due to alternate in-phase and out-of-phase scattering due to change in separation between multiple particles.

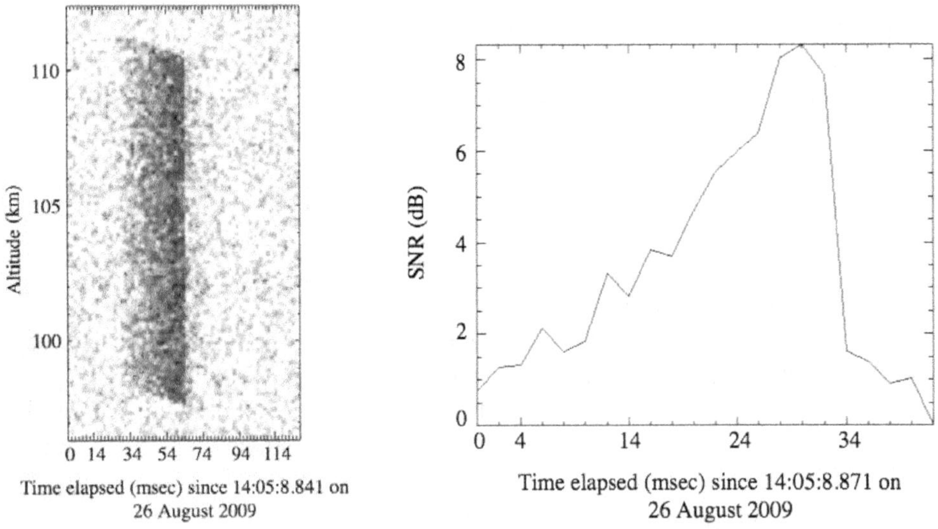

Figure 2. (a) RTI plot of a meteor event undergoing differential ablation. (b) The light curve for this event. The sudden drop in SNR is attributed to the complete ablation of the more volatile meteoroid constituents.

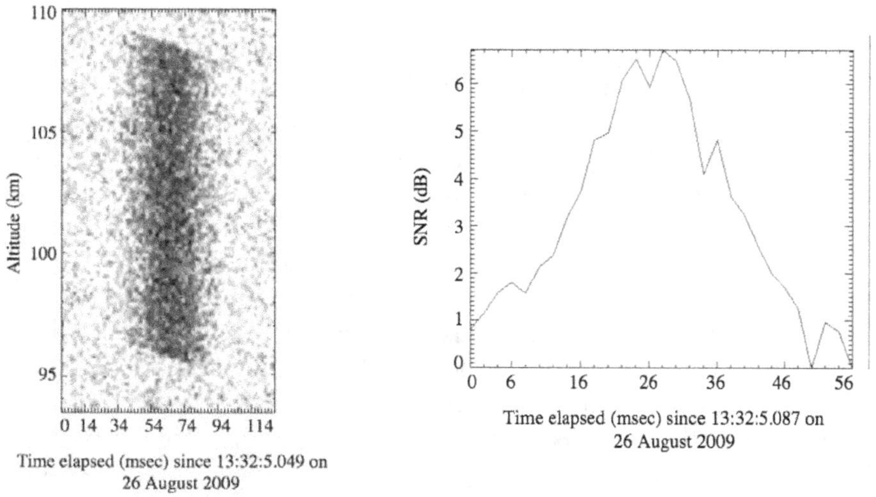

Figure 3. (a) RTI plot of a meteor event undergoing simple ablation. (b) The light curve for this event. Notice the relatively smooth profile compared to the events shown in Figure 1 and 2.

4 Discussion

Figures 1-3 show typical meteor events observed by RISR exhibiting fragmentation, differential ablation and simple ablation respectively. Note that we use Figures 1-3 to define what we interpret as these processes. We present the results from the statistical analysis determining the relative contributions of the three meteoroid disintegrating mechanisms. For the purpose of this analysis, we ignored low SNR events (SNR less than 2dB) as in these cases, even small changes in received power might result in giving an impression of a beat pattern, which might be wrongly interpreted as fragmentation. The events that exhibited two or more of the mechanisms were classified in all the relevant categories.

Following the above-mentioned criteria, we were able to classify 318 events in our data sets. 153 or ~48% of these events exhibited signatures of fragmentation, 62 or ~20% of the events exhibited signatures associated with differential ablation while 103 or ~32% of the events showed signatures of simple ablation. Fourteen events showed signatures of both fragmentation and differential ablation. Though we also observe events exhibiting both simple ablation and fragmentation, they are all low SNR cases and thus not included in the final count for the reasons mentioned above.

From these results, it is obvious that meteoroid disintegration in the upper atmosphere is a complex process in which all the three disintegration mechanisms play an important role, though from these results it seems that fragmentation is the dominant disintegration mechanism. This result has important implications on the aeronomy of the MLT (Mesosphere-Lower Thermosphere) region as it implies that majority of the mass flux from the micrometeoroids is deposited in form of dust rather than atomic metal form obtained due to ablation. The abundance of this meteoroic dust could also provide valuable insights into the formation of PMSEs. The fact that all the three mechanisms play a vital role in meteoroid disintegration is an equally important conclusion as it differs from the conclusions arrived at by Janches et al. (2009) and Mathews et al. (2010), which lay emphasis on differential ablation and fragmentation only, respectively. This result stresses the need for all the three disintegration mechanisms to be taken into account while coming up with any model for meteoroid ablation. The models currently in use to estimate radar meteor head echo properties consider only simple ablation and the above analysis shows that there clearly is a lot of scope for improvement in these models.

5 Conclusions

We have presented results from the first-ever study determining the relative importance of the three meteoroid disintegration mechanisms, namely fragmentation, differential ablation and simple ablation – a topic of much discussion and debate presently in the meteor community. We present "type specimen" meteor events that serve to define the presence of the three disintegration processes. Additionally, these results also constitute the one of the first reported observations from the newly-operational Resolute Bay Incoherent Scatter Radar. Our results suggest that meteoroid arrival and disintegration in the upper atmosphere observed by the UHF is a complex process in which all the three mechanisms play an important role though it seems that fragmentation is the dominant mechanism – an important result as it implies that majority of meteoroid mass flux is deposited in the upper atmosphere in dust rather than atomic form. The meteoroid disintegration process is further complicated by presence of events exhibiting signatures of more than one disintegration mechanism. Our results strongly suggest that any theoretical model explaining meteoroid disintegration should consider all the three disintegration mechanisms. Finally, we recommend that a similar classification study should be conducted not only at RISR with a larger data set but also at other radars such as the Arecibo, PFISR, SRF, ALTAIR and Jicamarca radars. Such a study would help in understanding the difference in the type of meteoroid flux observed by these radars at different locations operating at different frequencies and also lend further insights into the aeronomy of their respective MLT regions.

Acknowledgements

This effort was supported under NSF grants ATM 07-21613 and ITR/AP 04-27029 to The Pennsylvania State University.

References

P.M. Bellan, Journal of Geophysical Research (2008) doi:10.1029/2008JD009927.
R.D. Caswell and N. McBride, International Journal of Impact Engineering 17, 149 (1995).
J.L. Chau and R.F. Woodman, Atmos. Chem. Phys. Discuss. 4, 511 (2004).
L. Dyrud, D. Janches, Journal of Atmospheric and Solar-Terrestrial Physics 70, 1621 (2008).
M.J. Genge, Earth, Moon and Planets 102, 525 (2008).
J.S. Hey, S.J. Parsons, G.S. Stewart, Monthly Notices R. Astron. Soc. 107, 176 (1947).
D. Janches, L. Dyrud, S.L. Broadley, J.M.C. Plane, Geophysical Research Letters (2009) doi:10.1029/2009GL037389.
D. Janches, J.D. Mathews, D.D. Meisel, Q.H. Zhou, Icarus 145, 53 (2000).
J. Jones, T.R. Kaiser, Monthly Notices of the Royal Astronomical Society 133, 411 (1966).
M.C. Kelley. (Academic Press, New York, 1989) The Earth's Ionosphere: Plasma Physics and Electrodynamics.
J. Kero, C. Szasz, A. Pellinen-Wannberg, G. Wannberg, A. Westman, D.D. Meisel, Geophysical Research Letters (2008) doi:10.1029/2007GL032733.
A. Malhotra, J.D. Mathews, J. Urbina, Geophysical Research Letters (2008) doi:10.1029/2008GL034661.
J.D. Mathews, S.J. Briczinski, A. Malhotra, J. Cross, Geophysical Research Letters (2010) doi:10.1029/2009GL041967.
J.D. Mathews, S.J. Briczinski, D.D. Meisel, C.J. Heinselman, Earth, Moon, and Planets 102, 365 (2008).
J.D. Mathews, D. Janches, D.D. Meisel, Q.H. Zhou, Geophysical Research Letters 28, 1929 (2001).
J.D. Mathews, D.D. Meisel, K.P. Hunter, V.S. Getman, Q.H. Zhou, Icarus 126, 157 (1997).
A. Roy, S.J. Briczinski, J. Doherty, J.D. Mathews, IEEE Geoscience and Remote Sensing Letters 6, 27 (2009)

CHAPTER 9:

VIDEO AND OPTICAL OBSERVATIONS

Video Meteor Fluxes

M. D. Campbell-Brown • D. Braid

Keywords meteor · Eta Aquariid · sporadic · flux · video

1 Introduction

The flux of meteoroids, or number of meteoroids per unit area per unit time, is critical for calibrating models of meteoroid stream formation and for estimating the hazard to spacecraft from shower and sporadic meteors. Although observations of meteors in the millimetre to centimetre size range are common, flux measurements (particularly for sporadic meteors, which make up the majority of meteoroid flux) are less so. It is necessary to know the collecting area and collection time for a given set of observations, and to correct for observing biases and the sensitivity of the system.

Previous measurements of sporadic fluxes are summarized in Figure 1; the values are given as a total number of meteoroids striking the earth in one year to a given limiting mass. The Grün et al. (1985) flux model is included in the figure for reference. Fluxes for sporadic meteoroids impacting the Earth have been calculated for objects in the centimeter size range using Super-Schmidt observations (Hawkins & Upton, 1958); this study used about 300 meteors, and used only the physical area of overlap of the cameras at 90 km to calculate the flux, corrected for angular speed of meteors, since a large angular speed reduces the maximum brightness of the meteor on the film, and radiant elevation, which takes into account the geometric reduction in flux when the meteors are not perpendicular to the horizontal. They bring up corrections for both partial trails (which tends to increase the collecting area) and incomplete overlap at heights other than 90 km (which tends to decrease it) as effects that will affect the flux, but estimated that the two effects cancelled one another. Halliday et al. (1984) calculated the flux of meteorite-dropping fireballs with fragment masses greater than 50 g, over the physical area of sky accessible to the MORP fireball cameras, counting only observations in clear weather. In the micron size range, LDEF measurements of small craters on spacecraft have been used to estimate the flux (Love & Brownlee, 1993); here the physical area of the detector is well known, but the masses depend strongly on the unknown velocity distribution. In the same size range, Thomas & Netherway (1989) used the narrow-beam radar at Jindalee to calculate the flux of sporadics. In between these very large and very small sizes, a number of video and photographic observations were reduced by Ceplecha (2001). These fluxes were calculated (details are given in Ceplecha, 1988) taking the Halliday et al. (1984) MORP fireball fluxes, slightly corrected in mass, as a calibration, and adjusting the flux of small cameras to overlap with the number/mass relation from that work. Then faint video observations, which overlap with small cameras at their largest sizes, were similarly calibrated using the small camera data. The flux data from Ceplecha's study between 10^{-6} and 10^{-4} kg does not fit the slope between the LDEF and Super-Schmidt data (Figure 1), so uncertainty remains in this region. The flux in this size range is of particular importance, since much of the mass lost by comets is in particles of this size; also, the greatest danger to

M. D. Campbell-Brown (✉)
University of Western Ontario, London ON N6A 3K7 Canada. E-mail: margaret.campbell@uwo.ca

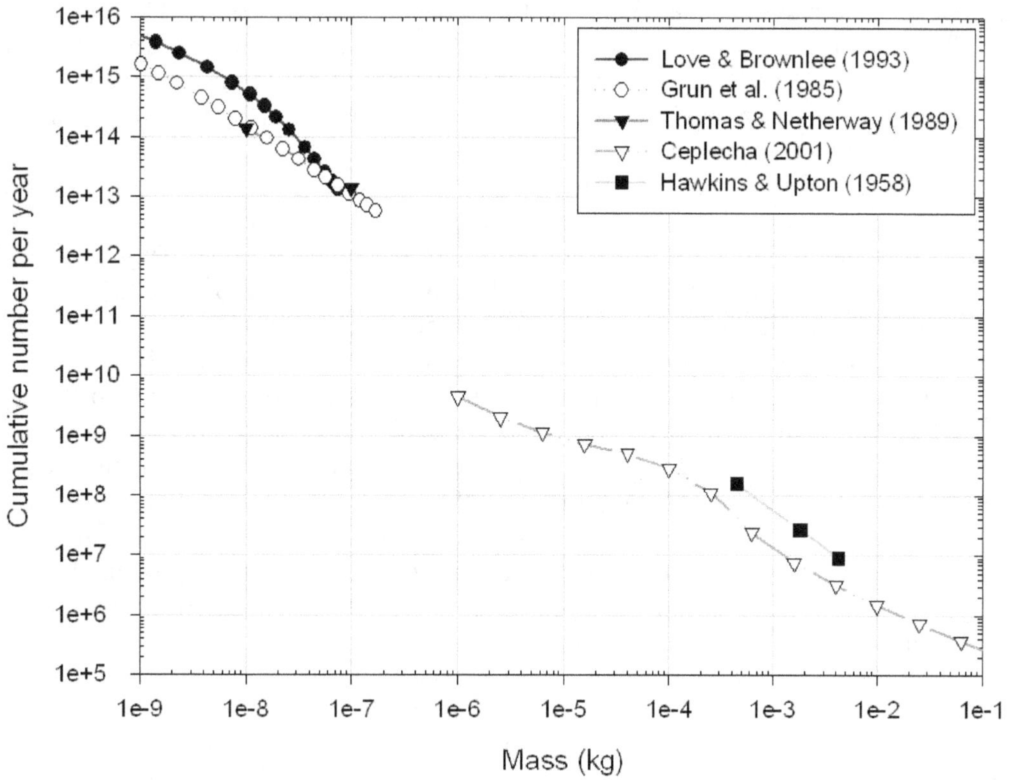

Figure. 1. Plot of meteoroid fluxes on the Earth from previous studies.

spacecraft comes from particles common enough to pose a real threat, and large enough to cause damage.

Shower fluxes have been estimated from visual observations (Brown & Rendtel, 1996), and from photographic and video observations. The usual method (employed in calculating Leonid fluxes by Koten et al. (2007), for example), uses the physical area observed by a pair of cameras at 100 km and applies a correction for radiant elevation. The most rigorous optical fluxes have been calculated for Leonids, Orionids and some minor showers (e.g. Gural et al., 2004; Trigo-Rodriguez et al., 2007, 2008) using a thorough simulation of the observing systems, including the camera sensitivity, range biases, and angular speed of the meteors on each camera. Details of the simulation are given in Molau et al. (2002).

In this work, we rigorously calculate the collecting area for a set of two intensified video cameras deployed in Arizona in 2006. The collecting area calculation was tested on the Eta Aquariid meteor shower and then applied to the antihelion, apex and north toroidal sporadic sources to obtain a sporadic flux.

2 Observations & Data Analysis

The data used in this study were taken from two sites in Arizona: the Fred Lawrence Whipple Observatory (31.675°N, 110.953°W) and Kitt Peak National Observatory (31.962°N, 111.60°W),using identical cameras, during a nine-night campaign in 2006. The baseline between the two sites was approximately 75 km. Both systems had 25 mm, f/0.85 objective lenses, Gen III ITT image intensifiers, and Cohu 4910 video cameras. Each system produces 30 interlaced frames per second, with standard video resolution of 640×480 pixels and 8 bits per pixel. The data were recorded on digital tapes for later analysis. Two nights of data were analyzed for this project: April 27 and May 6, 2006. The latter is the peak of the eta Aquariid meteor shower.

The MeteorScan software package (Gural, 1997) was used to identify meteors in the data. A total of 235 meteors simultaneously observed with both cameras were identified. The astrometry and photometry were measured using an in-house software package called PhotoM. Trajectories of the two-station meteors were calculated using MILIG, developed by J. Borovička (Borovička, 1990). Photometric masses were calculated for each of the meteors, and the distribution of these masses was used to find the sporadic mass index, $s = 2.02 \pm 0.02$, and the limiting mass, 2.06×10^{-6} kg.

In order to calculate the flux of meteoroids from a particular radiant, the number of meteoroids must be counted. Rather than calculate a partial trail correction, we accept only meteors for which the maximum of the light curve occurred in the common volume of the two cameras. There is some uncertainty even in this strict criterion: many meteor light curves are nearly flat at the peak, so judging whether the maximum was just inside or just outside the volume can be difficult. Some meteors were growing fainter when they entered the field of view of both cameras, and some growing brighter as they left both cameras: while the first or last observed frame might have been the maximum, these meteors were excluded. This left 121 meteors in the sample.

Figure 2 shows the radiant distribution in ecliptic coordinates. The apex of the Earth's way is in the centre of the plot, and the antihelion source to the right, near the antihelion point at 180° ecliptic longitude. The antihelion source is the clearest feature: the north apex source is also identifiable. Although there are meteors in the region of the north toroidal source, its borders are not clearly defined. The Eta Aquariids are visible as a small cluster of radiants to the left of the apex source, just above the ecliptic around longitude 295°. There are virtually no meteors in the region of the south apex source, and only one close to the helion source. There are a large number of meteors which are not within the 15 degree radius of any of the sporadic sources.

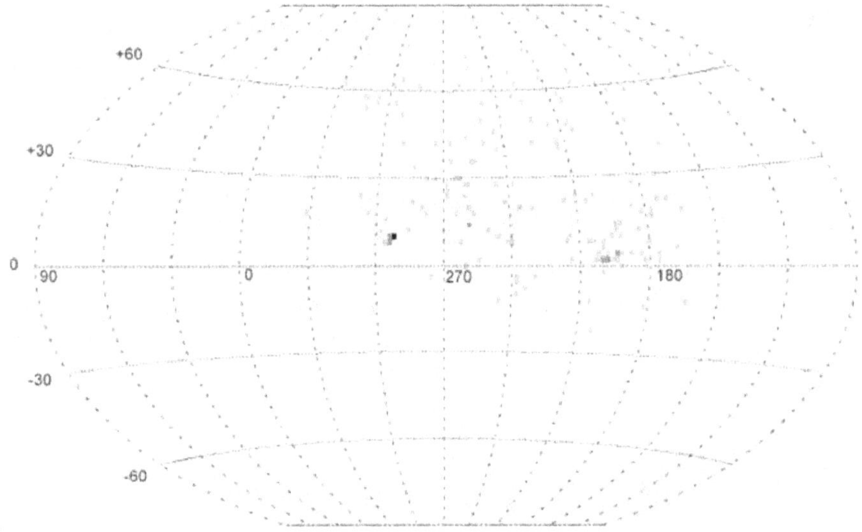

Figure 2. Radiants of meteors used in the flux study. The horizontal axis is the ecliptic plane; the apex of the Earth's way is in the centre (270° longitude) and the sun is at (0,0). The darkest dot represents six meteors with very close radiants; the lightest dots have only one meteor per 1 degree bin.

3 Collecting Area

If meteoroids all ablated at the same height, and detectors were uniformly sensitive, calculating the collecting area would be straightforward: the physical area covered by the detector at that height could be found quite simply. Even at a single height, the problem is more complicated: the sensitivity of a camera is generally a function of the position on the detector (with the most sensitivity generally occurring in the center of the field of view, and the least at the edges, mainly due to vignetting from the objective lens). The area in the sky is not at a uniform distance from the camera, so the limiting sensitivity will vary according to the range. Finally, the angular speed of the meteor as seen at the detector will influence whether or not it will be detected: a meteor coming straight at the camera may not be identified as a meteor at all, since it does not trace out a line, while one which is moving perpendicular to the line of sight will have its light in each frame spread over more pixels, which may reduce the signal until it is lost in the noise. All of these effects should properly be taken into account when calculating flux.

Even for shower meteors, the heights of meteors vary significantly from one to another, and meteors may not all cross one particular surface of constant height. In that case, the collecting area must be calculated at different heights, with a weighting for the probability of observing a meteor at that height.

The sensitivity of each camera was calculated from flatfields for each system. The optical centre of the image was found, using the highest pixel values in the flatfield to find the region of maximum sensitivity. The distance of each pixel in degrees from this optical centre was determined, and a fit performed to find the sensitivity as a fraction of the maximum as a function of angle from the centre.

For a particular radiant, the collecting area was calculated for half hour intervals throughout the night. For each time interval, slices from 80 to 120 km, with a spacing of 2 km, were taken; the corrected area of each slice was calculated, and a weighting factor was applied according to the height distribution of maximum luminosities of the meteors in the dataset. The weights, found using the distribution of maximum heights in the data set, were distributed as a Gaussian, with a maximum at 98 km and a standard deviation of 13 km; the final collecting area was normalized by dividing by the sum of the weights. Each slice was divided into squares 4 km × 4 km; the area of each square was weighted by the sensitivity of each camera, compared to the maximum sensitivity, the range to each camera squared, and the angular speed of a meteor from the given radiant at that position on each camera. If the trails at that point would be less than 3 pixels long, the area of that square was set to zero, assuming the meteor would not have been detected. The area was also weighted for the cosine of the zenith angle of the radiant, since the rate depends geometrically on the angle between the radiant and the surface. The total weighting factor was taken to the power of $s - 1$; if the mass index is large, there are many faint meteors, and more meteors will be missed in the less sensitive areas. If s is small, there are many bright meteors and fewer will be missed, so the collecting area is larger.

The integrated nightly collecting area for all heliocentric radiants is shown in Figure 3. It can be seen that the maximum collecting area occurs outside the sporadic sources, and partly explains the large number of meteors observed outside the sources. The collecting area for the north apex and antihelion sources are actually low compared to other parts of the sky. The region where the radiants pass through the fields of view of the cameras is also clearly visible as a cuved line of lower collecting areas in the middle of the maximum area.

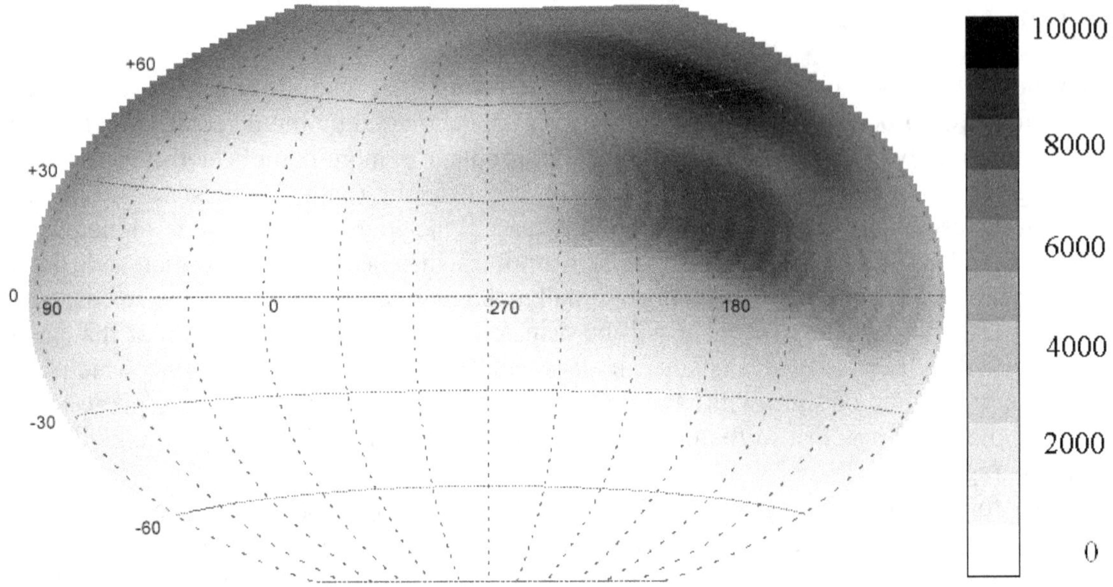

Figure 3. Integrated daily collecting area of the video system in heliocentric coordinates (as in Figure 2).

4 Eta Aquariid Fluxes

May 6, 2006 was the peak of the Eta Aquariid meteor shower. Although the radiant rose only about two hours before dawn at the observing site, and only 8 two-station Eta Aquariids had their light curve maximum in the common volume, we calculated the shower flux as a test of the method. The IMO value of the mass index, 1.95, was used (Dubietis, 2003), even though this is for larger visual meteoroids, since there were not enough Eta Aquariid meteors in our sample to calculate the mass index. The collecting area of the system for the Eta Aquariid radiant is shown in Figure 4.

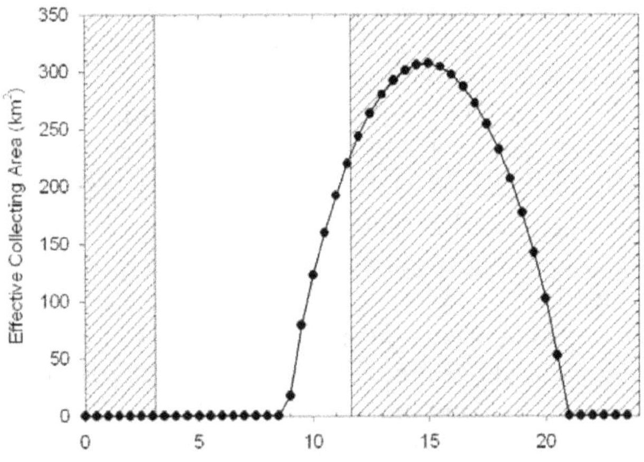

Figure 4. Collecting area for the Eta Aquariid radiant at half-hour intervals. The shaded regions indicate times when the sky was too bright to observe, starting and ending at nautical twilight.

The non-zero collecting areas were summed and the flux obtained for the two hour period was 0.0028 ± 0.0009 meteoroids km^{-2} hr^{-1}. This corresponds to a zenithal hourly rate of 65 (see Brown & Rendtel, 1996, for the formula to convert between ZHR and flux), which is very close to the value recorded for visual observations that year by the IMO (imo.net). This is certainly due partly to chance; since the number of meteors used in the flux calculation was so small, there are significant uncertainties in the estimate, but it gives us confidence that the collecting area calculation is correct.

5 Sporadic Fluxes

Sporadic fluxes are slightly more complicated than shower fluxes. The sources are diffuse, so the radius chosen will strongly affect the number of echoes included and therefore the flux. The collecting area also varies significantly across the source: the leading edge of the source can rise more than an hour before the trailing edge. When calculating the angular speed, there are uncertainties not only because of the large radiant area, but also because the speeds of the meteors have a broad distribution around the average, instead of being tightly confined as shower speeds are. For this study, we take a simple approach. Each source is divided into four quadrants, and the collecting area for each quadrant is calculated in half hour intervals. The average of these four values is used as the true collecting area. This approach is more efficient than the more rigorous version, which would involve calculating the collecting area for dozens of points around the source and then performing a weighted average reflecting the differing activity of each small point around the source, and it correctly reproduces the slow rise in collecting area as the radiant moves above the horizon. In calculating angular velocity, the average speed for each source (30 km/s for the antihelion, 35 km/s for the north toroidal, and 60 km/s for the north toroidal) was used rather than a distribution. The collecting area should be slightly lower for meteors moving faster than the average, and slightly higher for slower meteors, but the total collecting area should be the same if the velocity distributions are Gaussian.

Fluxes were calculated separately for the two nights of data, since the collecting areas for each source vary very slightly in that time period. Since the number of observed meteors was low, hourly fluxes were not calculated; the total number of meteors from each source was divided by the average collecting area. It was not possible to calculate a mass index for each source individually from the small numbers, so a mass index of 2.0 was assumed for each source, consistent with the s measured for all the sporadics observed in the dataset.

A total of 24 antihelion, 21 north apex, and 15 north toroidal meteors were recorded on the two nights. When divided by collecting area (pictured in Figures 5-7), this produced fluxes of 0.039 ± 0.006 meteoroids km^{-2} hr^{-1} for the antihelion source, 0.041 ± 0.006 meteoroids km^{-2} hr^{-1} for the north apex, and 0.012 ± 0.002 meteoroids km^{-2} hr^{-1} for the north toroidal. The errors were calculated using Poisson statistics for the small numbers, plus estimates of the error due to assuming a mass index and height distribution based on small numbers. The collecting area was varied to look at a reasonable range of mass indices for each source, and was found to vary by about 10%. The change in the weighted area of a slice from 90 km to 110 km was also found to be close to 10%.

To find the total sporadic flux, the flux from each of the three observed sources was doubled to account for its unobserved pair: the helion, south apex and south toroidal sources. This ignores the fact that the flux of the helion and antihelion sources vary through the year and the maxima and minima do not coincide (Campbell-Brown & Jones, 2006). It is believed that the pairs of sources have very close to symmetrical flux values when summed over the year, so this method should give a good annual value if there was more data. We proceed with this value, knowing that it is based on too little data, to see how it compares to previous studies.

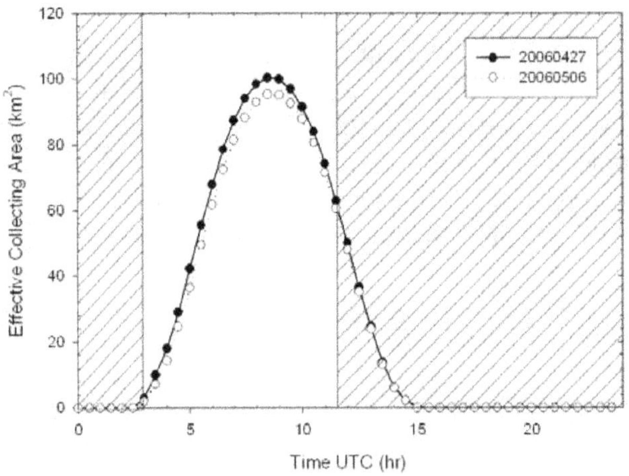

Figure 5. Collecting area for the antihelion source. The shaded regions indicate daytime until nautical twilight, when the sky was too bright to observe.

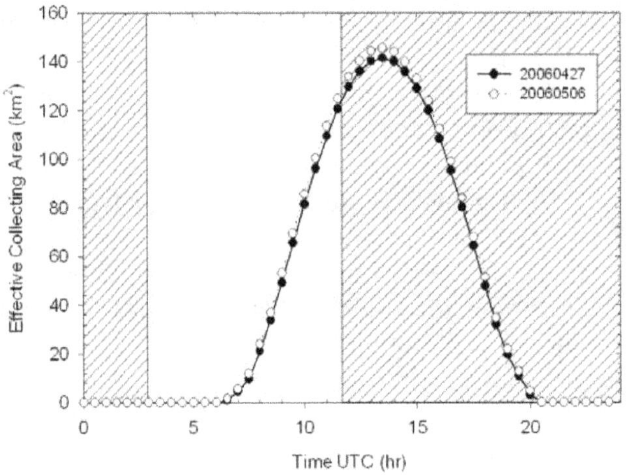

Figure 6. Collecting area for the north apex source.

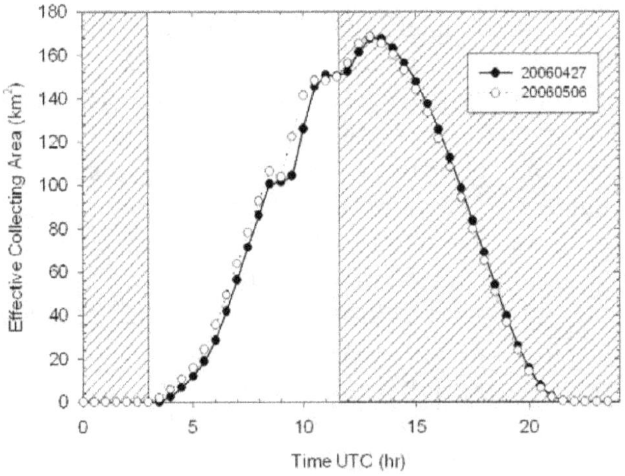

Figure 7. Collecting area for the north toroidal source.

The total sporadic flux from all the sources is 0.18 ± 0.04 meteoroids $km^{-2} hr^{-1}$. To compare this to the studies mentioned in the introduction, we convert this to a fluence over the whole Earth over a year, by multiplying by the cross-sectional area of the Earth and the number of hours in a year. The total is $(2.0 \pm 0.4) \times 10^{11}$ meteoroids.

The error bars include only errors in our measured value: they do not reflect the fact that the sporadic flux changes over the course of a year and that figures for part of two days are being used to estimate the flux over a full year. Figure 8 shows this result with previous studies. Note that the error bars are smaller than the symbol, because of the logarithmic scale.

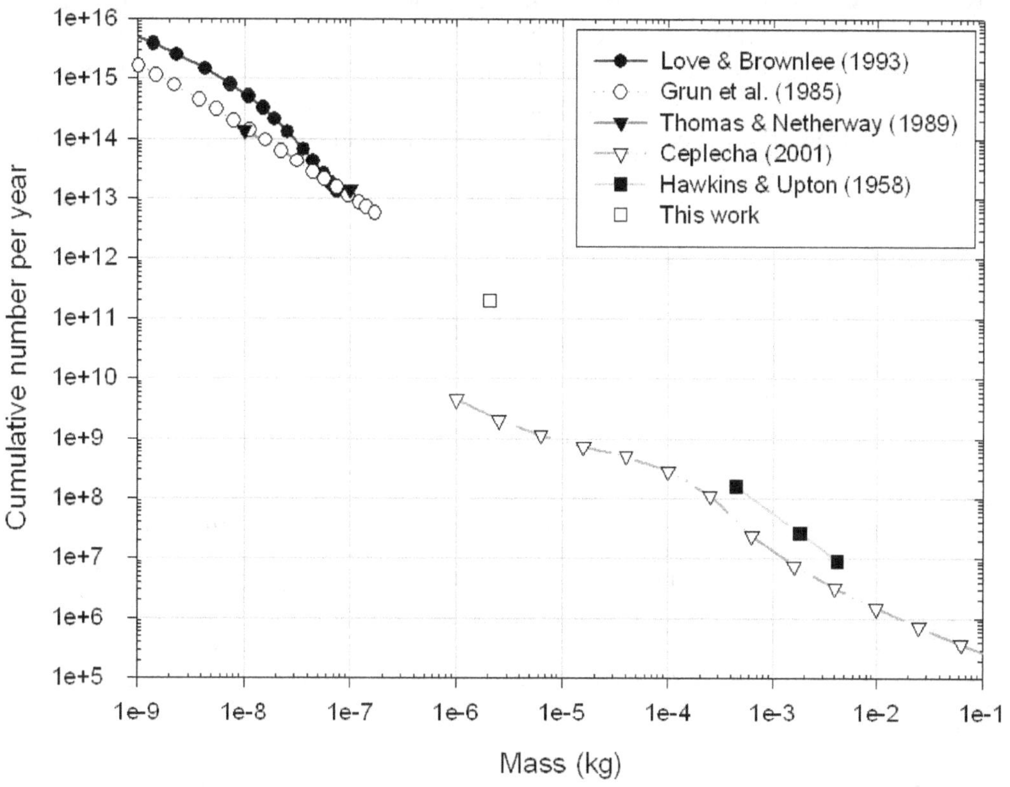

Figure 8. Plot of meteoroid fluxes on the Earth from previous studies, with the data point from the current study.

6 Discussion

The flux results for the Eta Aquariid meteor shower, though based on few meteors, are very promising, and give confidence that our method of calculating collecting area for particular radiants gives reasonable results. Shower fluxes are easier to calculate than sporadic fluxes, because of the higher numbers and narrow range of radiants and velocities, and more measurements with other systems are available for comparison, so future studies will examine more showers to further validate the method.

The total sporadic flux measured in this study fits surprisingly well on a line joining the fireball camera data to the Grün model, and is well above the flux from video studies by Ceplecha (2001). The fit is more surprising considering that it is based on only two nights of data from one part of the year, and a total of only 60 meteors.

The flux reported here reflects only meteoroids with radiants in one of three sporadic sources. An additional 61 meteors with maximum luminosity in the common volume were not included in the flux calculations because their radiants lay outside the sources. While this would seem to introduce a factor of two error in our measurement, we believe that the actual change in flux would be small if these other meteors were included. Inspection of Figures 2 and 3 shows that most of the meteors from radiants outside the sources occur in regions of the sky with very large collecting areas, meaning that the flux from those areas will be low.

For the past year, we have been running an automated two-station video system at the University of Western Ontario, and have collected over 1500 two-station meteor observations, mostly sporadic meteors. This dataset will be the subject of the next flux study, which will use a much larger dataset collected over a much more extensive range of solar longitudes to calculate the flux of sporadic meteors. In addition to the flux from the sporadic sources, this new study will calculate the fluxes from the whole visible sky, something which will be possible with much larger numbers. This new flux value, and the mass index which will accompany it, will better fill in the gap in our understanding of meteoroids in the millimetre to centimetre size range.

Acknowledgements

Thanks to Jean-Baptiste Kikwaya and Shannon Nudds, who collected the video data, and to the NASA Meteoroid Environment Office for funding.

References

Borovička J., "The Comparison of Two Methods of Determining Meteor Trajectories from Photographs", Astronomical Institutes of Czechoslovakia, Bulletin 41, 391-396 (1990)

Brown P., Rendtel J., "The Perseid Meteoroid Stream: Characterization of recent activity from visual observations", Icarus, 124, 414-428 (1996)

Campbell-Brown M., Jones J., "Annual Variation of Sporadic Radar Meteor Rates", MNRAS 367, 709-716 (2006)

Ceplecha Z., "Earth's Influx of Different Populations of Sporadic Meteoroids from Photographic and Television Data", BAICz 39, 221-236 (1988)

Ceplecha Z., "The Meteoroidal Influx to the Earth", in Collisional Processes in the Solar System, 35-50. Kluwer, Dordrecht (2001)

Dubietis A., "Long-term Activity of Meteor Showers from Comet 1P/Halley", JIMO 31, 43-48 (2003)

Halliday I., Blackewell A.T., Griffin A.A., "The frequency of meteorite falls on the earth", Science 223, 1405-1407 (1984)

Hawkins G., Upton E., "The Influx Rate of Meteors in the Earth's Atmosphere", ApJ, 128, 727-735 (1958)

Grün E., Zook H.A., Fechtig H., Geise R.H., "Collisional Balance of the Meteoritic Complex", Icarus 62, 244-272 (1985)

Gural P., "An Operational Autonomous Meteor Detector: Development Issues and Early Results", JIMO 25, 136-140 (1997)

Gural P., Jenniskens P., Koop M., Jones M., Houston-Jones J., Holman D., Richardson J. "The Relative Activity of the 2001 Leonid Storm Peaks and Implications for the 2002 Return", AdSpR 33, 1501-1506 (2004)

Koten P., Borovička J., Spurny P., Evans S., Štork R., Elliott A., "Video Observations of the 2006 Leonid Outburst", EM&P 102, 151-156 (2007)

Love S., Brownlee D., "A Direct Measurement of the Terrestrial Mass Accretion Rate of Cosmic Dust", Science, 262, 550-553 (1993)

Molau S., Gural P., Okamura O., "Comparison of the `American' and the `Asian' 2001 Leonid Meteor Storm", JIMO 30, 3-21 (2002)

Thomas R.M., Netherway D.J., "Observations of Meteors Using an Over-the-horizon Radar", PASAu 8, 88-93 (1989)

Trigo-Rodriguez J., Madiedo J., Llorca J., Gural P., Pujols P., Tezel T., "The 2006 Orionid Outburst Imaged by All-sky CCD Cameras from Spain: Meteoroid spatial fluxes and orbital elements", MNRAS 380, 126-132 (2007)

Trigo-Rodriguez J., Madiedo J., Gural P., Castro-Tirado A., Llorca J., Fabregat J., Vitek S., Pujols P., "Determination of Meteoroid Orbits and Spatial Fluxes by Using High-Resolution All-Sky CCD Cameras", EM&P 102, 231-240 (2008)

Searching for Serendipitous Meteoroid Images in Sky Surveys

D. L. Clark • P. Wiegert

Abstract The Fireball Retrieval on Survey Telescopic Image (FROSTI) project seeks to locate meteoroids on pre-existing sky survey images. Fireball detection systems, such as the University of Western Ontario's ASGARD system, provide fireball state vector information used to determine a pre-contact trajectory. This trajectory is utilized to search databases of sky survey image descriptions to identify serendipitous observations of the impactor within the hours prior to atmospheric contact. Commonly used analytic methods for meteoroid orbit determination proved insufficient in modeling meteoroid approach, so a RADAU based gravitational integrator was developed. Code was also written to represent the description of an arbitrary survey image in a survey independent fashion, with survey specific plug-ins periodically updating a centralized image description catalogue. Pre-processing of image descriptions supports an innovative image search strategy that easily accounts for arbitrary object and observer position and motion.

Keywords meteor · meteoroid · pre-detection · sky survey · frustum · image search

1 Introduction

The association of in-space and in-atmosphere images provides a unique opportunity to correlate results from differing observation and modelling techniques. In-space and in-atmosphere observations both directly and indirectly yield conclusions as to object size, composition and dynamics. With the two observations of the same object, one is able confirm consistency, or highlight discrepancies, in existing methods. One would hope as well that the discovery of a pre-fireball meteoroid (PFM) would add to the understanding of the visual properties of Earth-impacting objects. The discovery of a PFM in space would serve to confirm or suggest refinements to methods used to calculate heliocentric orbits from fireball observations.

When work began on the FROSTI project in the summer of 2007, there had not been a single fireball object which had both been recorded in space on its approach to Earth, and recorded in the atmosphere as a fireball. The goal of FROSTI is to discover such dual observations through a systematic search of historical sky survey images for objects detected in all-sky camera systems. The initial data image survey targeted was the Canada-France-Hawaii Telescope Legacy Survey (CFHTLS) image catalogue (CFHT, 2009). A lofty goal of FROSTI was to be the first to relate in-space and in-atmosphere observations of a common object. However, that accomplishment was met with the pre-contact discovery of object 2008 TC3 by the Catalina Sky Survey (Jenniskens, et al., 2009) prior to the object's atmospheric entry over Liberia, and its subsequent meteorite deposit. Regardless, the FROSTI project continues with the intent to systematically arrive at further like observations.

D. L. Clark (✉) • P. Wiegert
Department of Physics and Astronomy, The University of Western Ontario, London, Ontario, Canada. Phone: +1-519- 657-6825; E- mail: dclark56@uwo.ca

The software used in this project is a pre-existing astronomical simulation package (ClearSky) developed by the author. Figure 1 depicts the flow of processing involved in searching for serendipitous images of PFMs using this software. (1) The atmospheric contact position and velocity state of the object, with error bars, are made available to ClearSky. This may involve the simple keying of an individual event or the development of custom plug-ins for event collections. The contact state information required is contact longitude, latitude and elevation, apparent radiant right ascension and declination, and the contact velocity, all with error bars. The software handles a variety of coordinate systems and reference frames. (2) A probability cloud of positional probability members is sampled from the input data and error bars. Each of these members is gravitationally integrated back in time for 48 hours, resulting in an ephemeris over time for each member. An orbit at infinity is calculated at the end of the integration of each member. The entire cloud of probability members is used to report a statistical orbit at infinity estimate with error bars. This orbit may be used as verification against published orbit elements, typically arrived at by analytic methods. (3) In preparation for image searching, sky survey updates are periodically downloaded to maintain a local generic image description catalogue. (4) The image catalogue is searched for candidate images using the generated ephemerides, and a simulated image is created for each candidate. (5) Using the simulated image as a guide, the actual sky survey image is manually searched for the PFM.

2 Modelling PFM Visibility

2.1 Primitive Modelling

The initial goal of modelling PFM visibility was to determine whether these objects are in fact visible for any significant duration of time prior to contact. Frequency distributions were not initially considered. PFM characteristics affecting visibility are size, distance from Earth and the Sun, phase angle, and albedo. Wiegert et al. (2007), extending on Bowell et al. (1989), document a relationship of asteroid diameter D in kilometers to absolute magnitude H_k and albedo A_k for colour filter k as:

$$D = \frac{1347 \times 10^{-H_k/5}}{A_k^{1/2}} \quad (1)$$

Disregarding colour filters, rearranging and combining with (7) and (8), and assuming a constant approach speed v such $\Delta = vt$ that for a time t prior to contact, we derive a formula for apparent magnitude m: follows:

$$m = -5log_{10}(\frac{DA^{1/2}}{1347}) + 5log_{10}(rvt) - 2.5log_{10}((1-G)\phi_1 + G\phi_2) \quad (2)$$

where:

$$\phi_1 = e^{-3.33(\tan(\frac{\alpha}{2}))^{0.63}}, \qquad \phi_2 = e^{-1.87(\tan(\frac{\alpha}{2}))^{1.22}}$$

We now have an expression for apparent magnitude in terms of object diameter (D) in metres, albedo (A), velocity (v) expressed consistently in units such that vt is in AU, phase angle (α) and time (t), as well as solar distance (r) and slope parameter (G). Assuming $r \sim 1$ AU in the proximity of Earth, and $G = .15$ typical for low albedo asteroids, we are able to plot m against a sampling of reasonable D, A, v at α values, for various time periods.

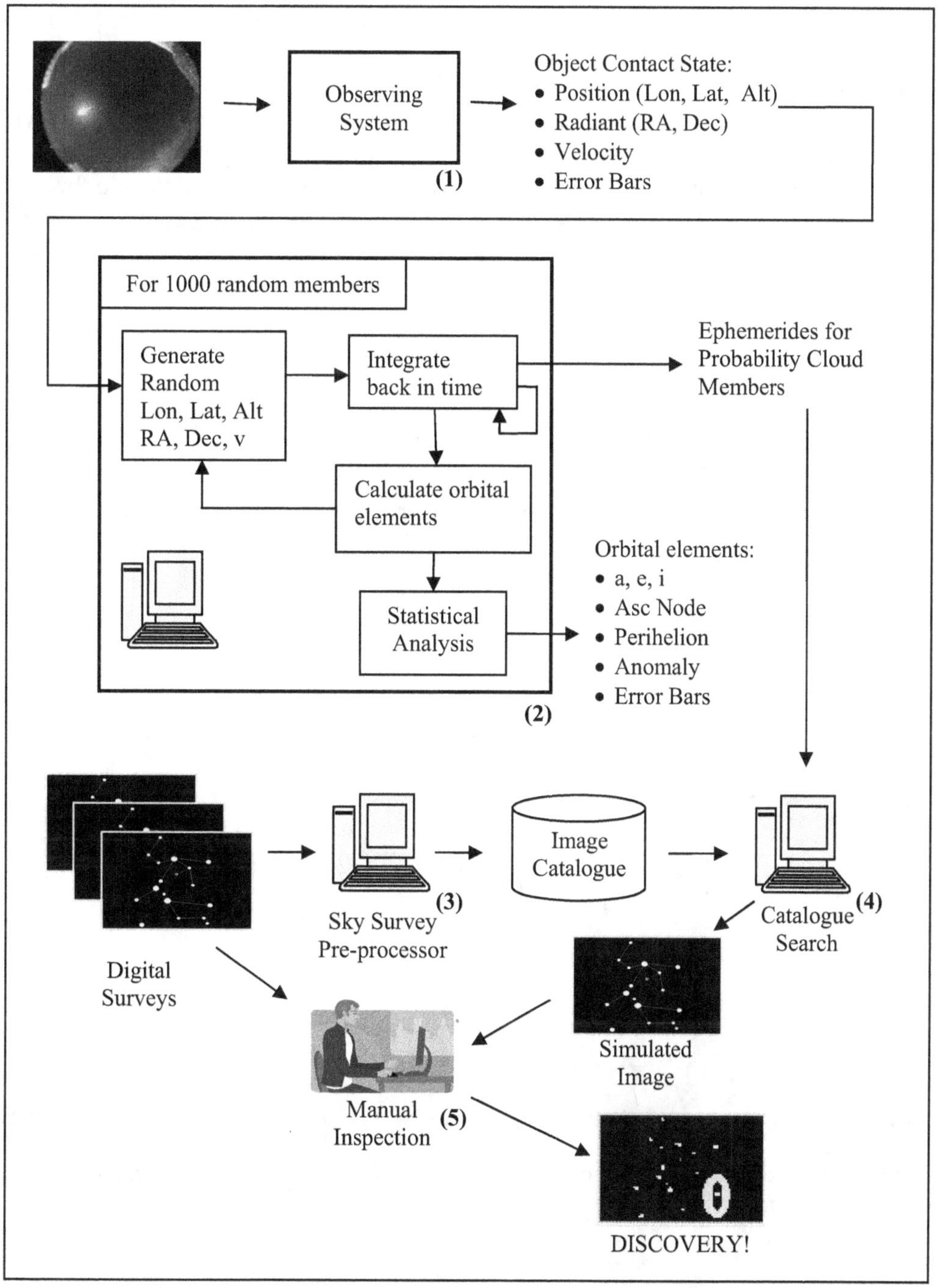

Figure 1. The process flow for PFM image searching, showing: (1) the transcription or importing of event contact state information and error bars, (2) the selection of a PFM probability cloud members, and the gravitational back integration of each member, resulting in ephemerides for each member, and a determination of orbit-at-infinity orbital elements, (3) the preprocessing of sky survey image descriptions into a generic image catalogue, (4) the searching for images based on PFM ephemerides, and (5) the manual inspection of candidate images for the PFM.

In Figure 2 visual magnitudes are plotted for objects with $A = 0.05$ and 0.25, $D = .25$ and 1.0 metres, $v = 20$, 30, 50 and 70 kms^{-1}, and $\alpha = 0°$, 30°, 60°, 90° and 120° at 3 hour intervals from 3 hours to 48 hours prior to contact. Symbols in the plot represent each time interval, with lines connecting points of like interval. The CFHTLS visibility limit of 24th magnitude is shown for comparison. One observes in the plot that there are indeed combinations of PFM physical and dynamical attributes which support predetections. In addition to the expected favouring of higher albedo, larger diameter, slower speed, and lower phase angle objects, this plot demonstrates that very few objects remain visible for time periods in the range of the original project target of 48 hours, and that visibility ranges of 6-12 hours are more representative.

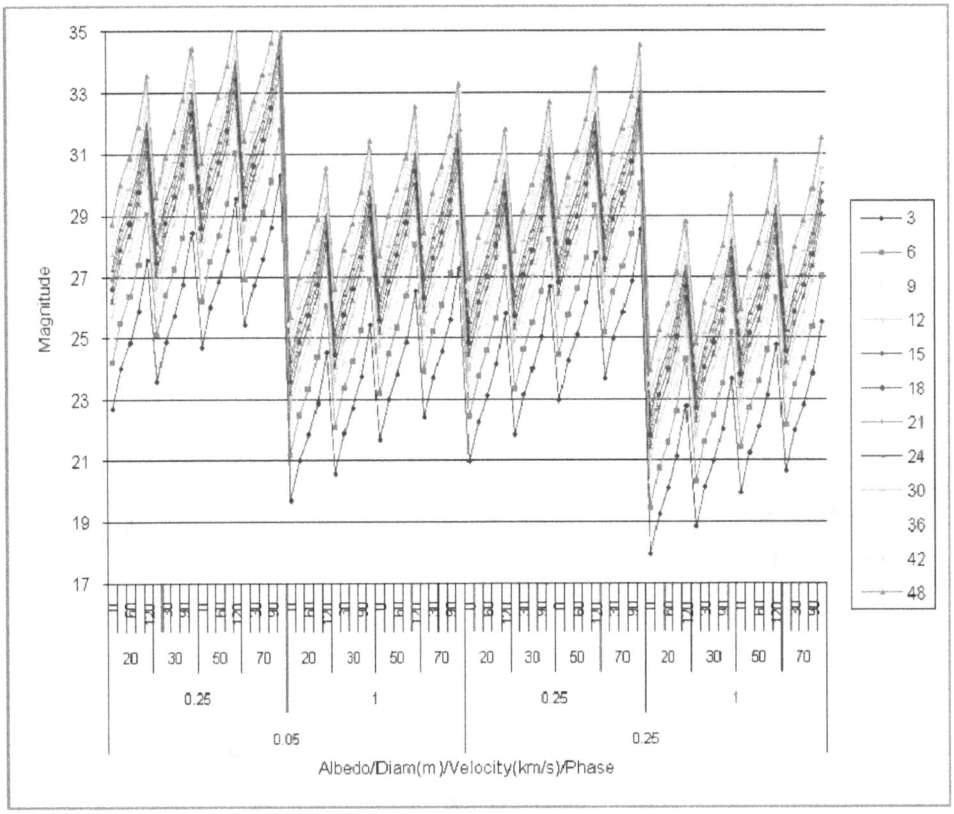

Figure 2. Plot of apparent magnitude over time of a variety of objects having albedo of 0.05 and 0.25, diameter of .25 and 1.0 metres, approach velocity of 20, 30, 50 and 70 kms-1, and phase angle 0°, 30°, 60°, 90° and 120°, assuming linear approach. The gray shaded area represents visibility within the CFHTLS images. Lines join points of equal visibility duration.

2.2 Bottke/Brown/Morbidelli Modelling

The simplistic modelling above, although reassuring that object prediction images could exist, does not provide insight into the frequencies of objects with attributes permitting successful predetections. For this we turn Near Earth Asteroid (NEA) dynamical models of Bottke et al. (2002a), fireball size frequency distribution and flux model of Brown et al. (2002), and the albedo model of Morbidelli et al. (2002a).

The Bottke 2002a NEA distribution is a 5-intermediate source model of NEA distribution binned over orbit semi-major axis (a), eccentricity (e), and inclination (i). In addition to a, e, and i, values for longitude of the ascending node (Ω), the argument of perihelion (ω) and true anomaly (f) are required. In the case of the general NEA population, the three angles Ω, ω, and f may be uniformly selected from the full 0-360° range, as there is no natural anti-symmetric bias to these elements. However, PFMs are characterized within the NEA population as objects which have the immediate potential to collide with the Earth. A standard equation for Keplerian motion is:

$$r = \frac{a(1-e^2)}{1+e\cos f} \tag{3}$$

where r is the object-Sun distance. Re-arranging, we have:

$$f = \pm cos^{-1}\frac{a(1-e^2)-r}{er} \tag{4}$$

Selecting a uniformly random time t in the time range of interest, we are able to determine r by assuming r very closely approximates the Earth-Sun distance. The Earth-Sun distance is readily available from published theories such as DE405 (NASA JPL planetary position ephemerides available as tables of Chebyshev coefficients and supporting code).

Since the argument of perihelion ω is defined as an angle from the ecliptic, the circumstance of Earth-object collision occurs on the ecliptic, and f is defined as an angle from ω, we are able to determine ω from f. There are four possible relationships among f, ω, and Ω characterized by the object being at the ascending node or descending node, and whether the object is inbound or outbound in its orbit in relation to the Sun. These four cases are selected uniformly:

1) Ascending node, outbound: $f = \cos^{-1}$ ..., $\omega = -f, \Omega = L$
2) Ascending node, inbound: $f = -\cos^{-1}$ ..., $\omega = -f, \Omega = L$
3) Descending node, outbound: $f = \cos^{-1}$ ..., $\omega = \pi - f, \Omega = \pi + L$
4) Descending node, inbound: $f = -\cos^{-1}$ ..., $\omega = \pi - f, \Omega = \pi + L$

With approach characteristics handled, we must now model the size and albedo distributions which will impact visibility. Brown et al. (2002) describe a power law for the cumulative number of objects (N) colliding with Earth per year with diameter $\geq D$ in metres as:

$$log\, N = c_0 - d_0\, log\, D \tag{5}$$

where $c_0 = 1.568 \pm 0.03$ and $d_0 = 2.70 \pm 0.08$. Assuming a diameter of at least .2 m is required for visibility in telescopic images, equation (5) yields a flux of 2800 objects/year. This is not a large sample size at all when we consider that the samples are distributed over more than 15,000 a, e, i bins in the Bottke distribution, and we still require a distribution over an albedo range. Therefore, the significantly larger sample is used, and resulting frequencies must be scaled back accordingly.

For albedo modelling we turn to Morbidelli et al. (2002a) who define 5 NEO albedo classes: Hig(h), Mod(erate), Int(ermediate), Low, and Com(etary) with a mean albedo for each, and albedo ranges for all but the Com class (for which we will assume the mean value for all samples). They then assign differing slope parameter values for each class to simulate a phase angle affect. Finally, they model a frequency distribution by class for the NEO population.

A sample of 10,000,000 objects was generated using the above NEA, bolide size and albedo models. This sample size is a compromise of reasonable required computation time against granularity of result binning. For the strict needs of visual magnitude analysis, a smaller sample size could be used. However, other analyses (below) were performed on the model which benefited from the increased sample size. Figure 3 shows the visual magnitude distribution of the sample objects plotted over various times from 5 minutes to 24 hours prior to Earth contact. As in the simple model of above, a significant portion of objects are potentially visible (magnitude <24) in sky surveys in the minutes prior to contact. However, this visible proportion trails off very quickly in the hours prior to contact, to the extent that almost none of the model population have visual magnitude less than 6 hours prior.

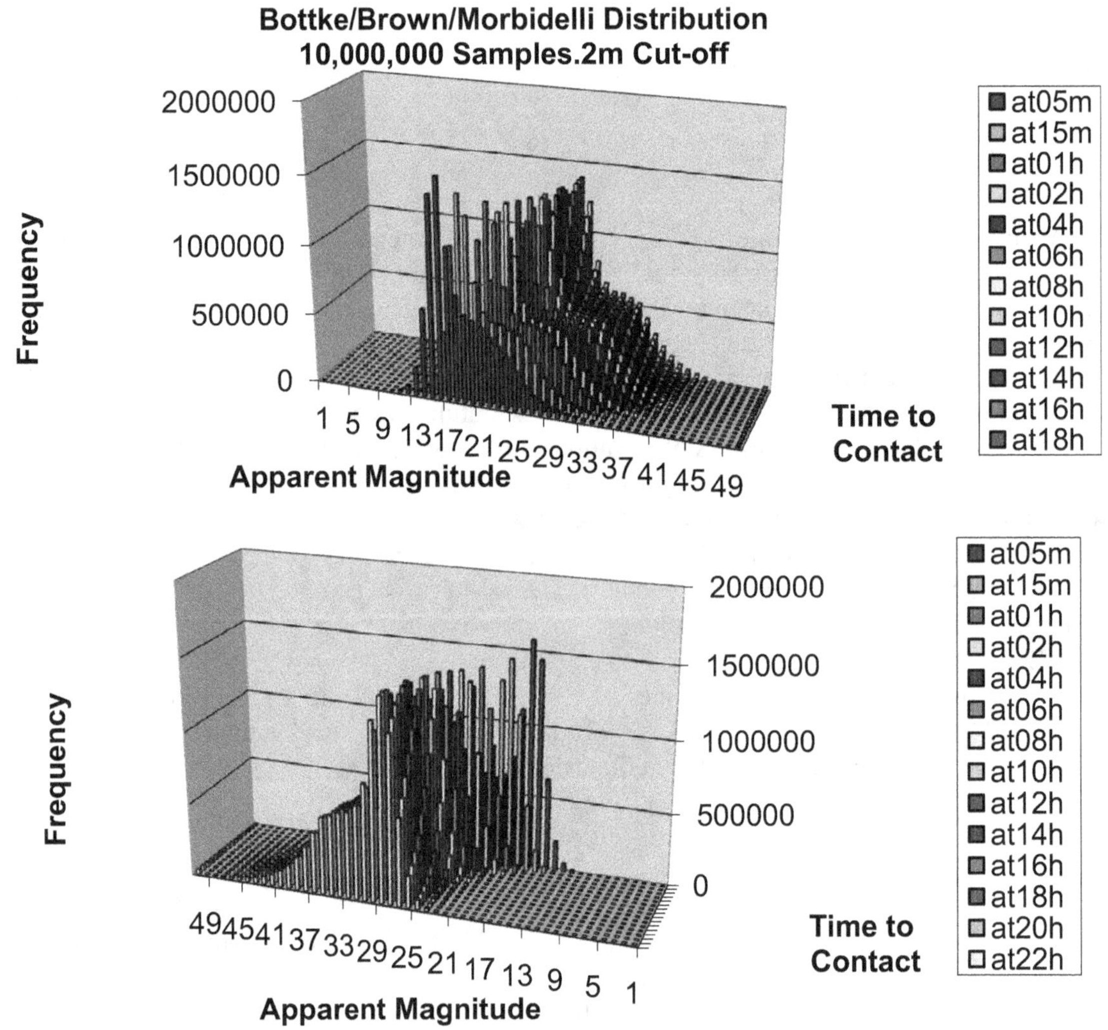

Figure 3. A visual magnitude plot of 10,000,000 simulated objects generated from the Bottke et al. (2002a) NEA distribution, the Brown et al. (2002) bolide size-frequency distribution, and the Morbidelli et al. (2002a) albedo distribution. Only objects above .2m in diameter are considered. Objects are selected by adjusting the argument of perihelion such that objects collide with Earth. The plot shows the number of objects falling into 1-magnitude wide bins over a series of time intervals prior to contact. Top: From foreground to background are the time intervals of 5 minutes, 15 minutes, 1 hour, and 2 hours to 24 hours in 2-hour increments. Bottom: The same plot with earlier times in the foreground.

3 Determination of PFM Trajectory

The search for a nearby object in sky surveys requires an accurate ephemeris for the object, with a good understanding of the errors in position over time. It is insufficient to use the published orbital elements of a PFM, elements that are typically derived using analytical means. The elements are expressed in far too little accuracy to be useful; for example a semi-major axis expressed to the precision of .001 AU yields errors rivalling the radius of the orbit of the Moon. As well, published orbits are orbits at infinity, not reflecting the impact of the Earth's gravity on the approach trajectory. Instead, an accurate translation of the PFM's contact state into heliocentric coordinates and a full gravitational integration are required.

3.1 Calculating the PFM Heliocentric Contact State

The heliocentric contact state of an object is represented as a cloud of probability members, each member having longitude λ_B, latitude φ_B, height h_B, radiant right ascension and declination α_R and δ_R, and velocity v_∞, all at an epoch t, where each of these values, including the epoch, are generated from a Gaussian distribution defined by the reported mean and standard deviations. Each member's contact state is converted to heliocentric coordinates, in preparation for the integration of each member, as follows:

1) The geocentric coordinates centred on Greenwich (x_G, y_G, z_G) are calculated using the WGS84 theory.
2) The mean rotation of the Earth θ is calculated using the methods of Meeus (1991) Chapter 11.
3) The apparent sidereal rotation of the Earth θ' is calculated from θ as described in Chapters 11 and 21 of Meeus (1991). This involves the calculation of the mean obliquity of the ecliptic ε_0, the nutation in longitude $\Delta\Psi$, and the nutation in obliquity $\Delta\varepsilon$. The calculations of nutation and obliquity require that the time of the event be expressed in Dynamical Time (*TD*), not universal time (UT). This difference in these timeframes is taken from a table of adjustments available on the US Naval Observatory web site (USNO, 2010).
4) We are then able to rotate (x_G, y_G, z_G) by θ' giving the Earth-centred equatorial coordinates with respect to the equinox of the date (x_E, y_E, z_E).
5) (x_E, y_E, z_E) are converted to the equinox J2000 (x_{EJ}, y_{EJ}, z_{EJ}) by converting to spherical coordinates, precessed to J2000 by the methods of Meeus (1991) Chapter 20, and converting back to rectangular coordinates. We retain the right J2000 right ascension α_E for later use.
6) The apparent contact velocity of the PFM equinox J2000 (v_{xoJ}, v_{yoJ}, v_{zoJ}) is calculated directly from α_R, δ_R, and $-v_\infty$.
7) The velocity due to the rotation of the Earth (v_{xRot}, v_{yRot}, v_{zRot}) is the tangent vector at the Earth-centred position expressed in equatorial coordinates for the epoch of the date. The magnitude v_{ROT} of the velocity is taken from a complete rotation of the earth at the object's distance and declination. Care must be taken when velocities are tracked in software with respect to solar time scales; we must make a sidereal adjustment. (v_{xRot}, v_{yRot}, v_{zRot}) is then calculated from v_{ROT} and α_E, v_{zRot} being 0.
8) (v_{xRot}, v_{yRot}, v_{zRot}) are converted to equinox J2000 (v_{xRotJ}, v_{yRotJ}, v_{zRotJ}) as in 5) above.
9) We arrive at an Earth-centred equatorial J2000 velocity (v_{xEJ}, v_{yEJ}, v_{zEJ}) by summing (v_{xoJ}, v_{yoJ}, v_{zoJ}) and (v_{xRotJ}, v_{yRotJ}, v_{zRotJ}).

10) The Earth-centred equatorial J2000 position (x_{EJ}, y_{EJ}, z_{EJ}) and velocity (v_{xEJ}, v_{yEJ}, v_{zEJ}) are converted to heliocentric coordinates (x_{EH}, y_{EH}, z_{EH}) and (v_{xEH}, v_{yEH}, v_{zEH}) by adding the Earth's position at the epoch using the JPL Horizons DE405 ephemeris. The epoch must be expressed in Terrestrial Time (*TT*), equivalent to *TD* as calculated in 3).
11) These equatorial coordinates are converted to heliocentric ecliptical coordinates (x_H, y_H, z_H) and (v_{xH}, v_{yH}, v_{zH}) by converting to spherical coordinates, converting to ecliptical coordinates as in Meeus (1991) Chapter 12, and converting back to rectangular coordinates. These calculations again require ε_0 and $\Delta\varepsilon$ as calculated in 3).

3.2 Integrating the PFM Trajectory

Not wanting to re-invent the wheel in the field of numerical integrators, and understanding that this application did not require sophisticated optimizations or approaches in performance, we decided on a quick rework of an existing C-language implementation of RADAU-15, a 15th-order differential equation integrator documented in Everhart (1985). The RADAU family of integrators is characterized by the use of Gauss-Radau spacings for sequence sub-steps. The work of porting and integrating the publically available C-code involved converting C code to C++, the language used in the remainder of the project coding, and abstracting the concepts of an integrator, force calculations, and physical objects into C++ interfaces and implementations to facilitate substitution of trial implementations. The initial implementation of the RADAU integrator was tested by integrating the major objects of the solar system over 100 years, and comparing the results to the JPL DE405/DE406 ephemerides. A full integrator-to-DE405/406 comparison was performed, with acceptable results (for example: .6" error in solar longitude for Mars, with an oscillating .000003 AU error in solar distance after 100 years). This test required the implementation of post-Newtonian adjustments, a refinement not required for the integration of meteoroid objects on Earth-approach. These post-Newtonian adjustments require knowledge of velocity state within the inter-object force calculations that is not required by PFM integrations.

The resulting integration back in time of a collection of probability points generated from contact state value and error bars yields a slowly expanding probability cloud representing the possible meteoroid paths. Figure 4 is a sample illustration of the Bunburra Rockhole event, generated from an initial contact sate provided by Pavel Spurný in a private correspondence (Spurný, 2009). The convergence of the probability cloud towards the eventual error bars in the original state is evident.

3.3 Comparison to Ceplecha Analytical Orbits

The analytical orbit-at-infinity calculation methods of Ceplecha (1987) provide the means to verify the resulting orbits from the back-integration technique. Two sets of fireball orbits derived using Ceplecha's calculations were used to perform this comparison: the ten largest mass European Network events from 1993-1996 documented in Spurný (1997), and 10 more recent unpublished European Network events provided by Spurný in a private correspondence (Spurný, 2010). For the purposes of orbit-at-infinity calculations, the back-integrations are stopped at 2-months prior to Earth contact. Figure 5 demonstrates the good correspondence between the methods for the Spurný (1997) events. Results on the later events await Spurný's publication of his results.

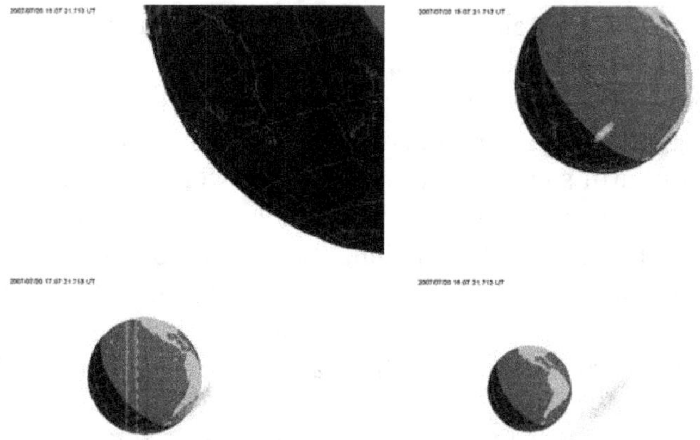

Figure 4. The RADAU-15 generated probability cloud for the Bunburra Rockhole effect. Meteoroid contact position, velocity and error bars were provide by Pavel Spurný in private correspondence (Spurný, 2009). The illustrations show the probability at the time of contact, and 1, 2 and 3 hours prior to contact. The viewer is a consistent 9100 km distance from the cloud's mean point.

Event	Source	a	a SD	e	e SD	I	I SD	node	node SD	peri	peri SD
EN070594(Leszno)	ClearSky	2.1070	0.0172	0.5328	0.0037	6.8953	0.0838	227.1320	0.0004	338.2490	0.2630
	Published	2.1000	0.0200	0.5320	0.0040	6.9100	0.0700	227.1100	0.0001	338.2000	0.2000
	Delta	0.0070	-0.0028	0.0008	-0.0003	-0.0147	0.0138	0.0220	0.0003	0.0490	0.0630
EN070893(Polnã)	ClearSky	2.0040	0.0283	0.5163	0.0070	18.8955	0.2539	135.4610	0.0004	209.5240	0.0727
	Published	2.0030	0.0060	0.5162	0.0013	18.9000	0.0300	135.4410	0.0002	209.5200	0.0700
	Delta	0.0010	0.0223	0.0001	0.0057	-0.0045	0.2239	0.0200	0.0002	0.0040	0.0027
EN150294(Dresden)	ClearSky	2.3386	0.0132	0.5784	0.0023	33.8443	0.0347	327.1350	0.0000	173.9130	0.1768
	Published	2.3380	0.0030	0.5783	0.0006	33.8410	0.0120	327.1300	0.0001	173.9000	0.0200
	Delta	0.0006	0.0102	0.0001	0.0017	0.0033	0.0227	0.0050	-0.0001	0.0130	0.1568
EN150396(Dobriš II)	ClearSky	7.5345	1.4256	0.8779	0.0204	8.3313	0.4868	355.5680	0.0007	141.1660	0.9380
	Published	7.2000	1.1000	0.8800	0.0200	8.3000	0.5000	355.5530	0.0001	141.2000	0.9000
	Delta	0.3344	0.3256	-0.0021	0.0004	0.0313	-0.0132	0.0150	0.0006	-0.0340	0.0380
EN220293(Meuse)	ClearSky	1.5069	0.0181	0.5682	0.0036	32.5929	0.1559	334.4100	0.0000	266.8560	0.8100
	Published	1.5000	0.0200	0.5670	0.0040	32.6000	0.2000	334.4070	0.0001	266.9000	0.8000
	Delta	0.0069	-0.0019	0.0012	-0.0004	-0.0071	-0.0441	0.0030	-0.0001	-0.0440	0.0100
EN220495A(Koutim)	ClearSky	2.3878	0.0105	0.7886	0.0008	4.1310	0.0462	32.4134	0.0002	277.3980	0.0788
	Published	2.3740	0.0040	0.7878	0.0003	4.1190	0.0120	32.3858	0.0001	277.5800	0.0500
	Delta	0.0138	0.0065	0.0008	0.0005	0.0120	0.0342	0.0276	0.0001	-0.1820	0.0288
EN231195(J. Hradec)	ClearSky	3.4436	0.1283	0.7813	0.0070	11.8921	0.5959	240.3480	0.0002	242.7970	1.9673
	Published	3.3900	0.0500	0.7790	0.0030	11.9900	0.0200	240.3360	0.0007	243.3000	0.3000
	Delta	0.0536	0.0783	0.0023	0.0040	-0.0979	0.5759	0.0120	-0.0005	-0.5030	1.6673
EN241095B(Odra)	ClearSky	1.3055	0.0859	0.5663	0.0193	52.7598	0.5480	211.0410	0.0006	281.4500	4.2692
	Published	1.3270	0.0110	0.5710	0.0020	52.8000	0.2000	211.0380	0.0007	280.2000	0.4000
	Delta	-0.0216	0.0749	-0.0047	0.0173	-0.0402	0.3480	0.0030	-0.0001	1.2500	3.8692
EN250594(Ulm)	ClearSky	2.0128	0.1134	0.5548	0.0214	2.5661	0.7510	244.4780	0.0293	312.0790	1.7704
	Published	2.0400	0.0200	0.5600	0.0030	2.5000	0.0400	244.5260	0.0007	313.1000	0.3000
	Delta	-0.0272	0.0934	-0.0052	0.0184	0.0661	0.7110	-0.0480	0.0286	-1.0210	1.4704
EN251095A(Tisza)	ClearSky	1.0780	0.0069	0.8068	0.0010	6.1339	0.1655	31.2538	0.0003	140.4080	0.3306
	Published	1.0770	0.0090	0.8067	0.0010	6.2000	0.2000	31.2595	0.0001	140.4000	0.4000
	Delta	0.0010	-0.0021	0.0001	0.0000	-0.0661	-0.0345	-0.0057	0.0002	0.0080	-0.0694

Figure 5. A list of the 10 highest mass events from Spurný (1997) showing the published orbital elements calculated using Ceplecha (1987) calculations compared to the orbit at infinity elements calculated using the project's software ClearSky's integration technique. Semimajor axis, eccentricity, inclination, longitude of the ascending node and argument of perihelion are listed. Standard deviations are listed beside each element. ClearSky elements are displayed in blue if outside the published error bars. Published elements are displayed in red if outside the ClearSky calculated error bars. Note that the ascending nodes are numerically close, but are consistently flagged as being out of the corresponding error bars.

The close correspondence of orbit elements from the Ceplecha and integration techniques serves as both a validation for the time-honoured analytical method, and as a confirmation the integration technique does accurately reproduce object approach trajectories. However, the small but systematic variance in longitude of ascending node garnered further attention. Section 11 of Ceplecha (1987) describes in detail the impact of Earth's gravity on calculating velocity and radiant direction of a meteor, this impact being removed prior to the calculation of orbital elements. However, in formula (48) of Section 11, Ceplecha makes the assumption that the longitude of the ascending node (Ω) of the orbit can be directly derived from the solar longitude of the Earth (L_{SUN}) at the time of impact. This is true of the instantaneous orbit of the meteoroid, but not its orbit at infinity. The instantaneous Ω is drawn towards the limiting value L_{SUN} as the meteoroid approaches the Earth. The magnitude of this shift in Ω depends on the approach characteristics and the length of time the meteoroid is influenced by Earth's gravity. The largest calculated variance in Ω is .15° for Spurný (2010) event EN231006. Figure 6 demonstrates the shift in ascending node of approximately .1° of the Bunburra Rockhole event. A consequence of this variance in Ω that has not been quantified is the dependency in the Ceplecha calculations of all other orbit elements except semimajor axis on Ω and L_{SUN}. Further quantification of the impact is noted as possible future work.

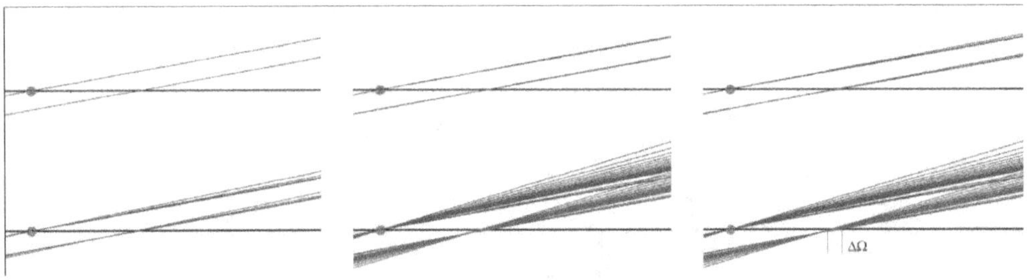

Figure 6. The shift $\Delta\Omega$ of the ascending node Ω of the instantaneous orbit of the Bunburra Rockhole meteoroid. From left to right, top to bottom, the progress of the shift is shown, 4, 3, 2, and 1 hours prior to contact, at contact, and at contact with the shift highlighted. The rightmost diagonal lines represent the instantaneous orbits at the ascending node on the near side of the Sun. The leftmost lines represent the orbits at the descending node on the far side of the Sun.

4 Searching Sky Surveys

4.1 Image Frustums

Astronomical images are typically thought of as two-dimensional rectangular projections onto the celestial sphere. Such images may be defined by the right ascension (α) and declination (δ) of the four corners of the image, or by the α and δ of the image centre, width and height of the image, and the rotation around the image centre. The computations involved in determining the location of a fast moving object in relation to a long image exposure involves several conversions of the object's position into observer centred α and δ as both the object and observer move over time. These conversions, although not complex, are computationally expensive as they involve trigonometric transformations. Since the position of an object is relative to the observer, there is little opportunity to optimize this heavy computation against multiple images, or against multiple image surveys. I therefore developed an image representation scheme which supports a front-end loaded one-time optimization of individual image representations, while reducing the object-image computation complexity.

A single survey image in reality is a projection of a three-dimensional volume of space. Assuming a rectangular image, this volume is a frustum as shown in Figure 7. A frustum is defined as the portion of a solid lying between two planes. An image frustum is the portion of a square pyramid lying between a front viewing plane and an arbitrary depth of field plane. This real-world frustum space can be transformed into a three dimensional rectangular 2x2x1 frustum space to which object position intersections are easily calculated. The determination of the image frustum and the calculation of the transformation into a rectangular frustum space are costly, but may be performed once per image description with the resulting transformation being stored and associated with the image description. Viewing frustums and the related transformations have been used for decades in rendering three dimensional world scenes onto a two-dimensional view port (screen). In particular, I have leveraged the unpublished lecture notes and course exercises by Beatty (1980).

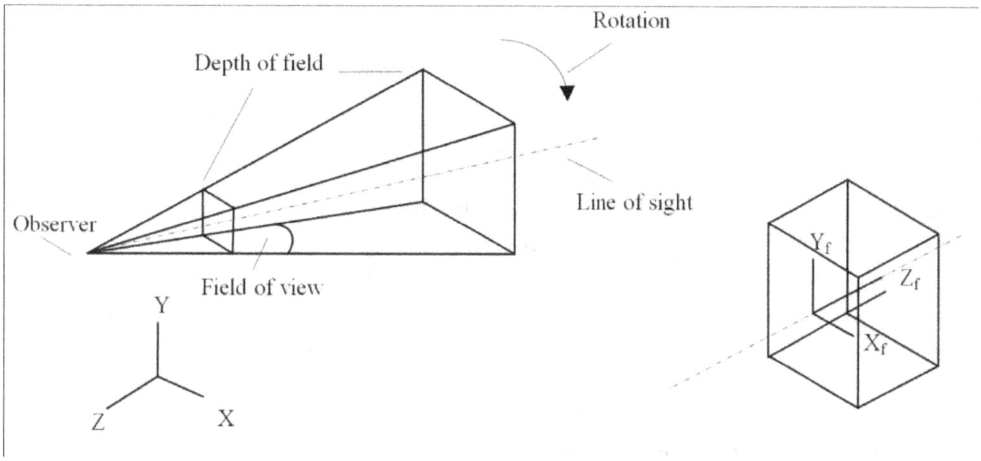

Figure 7. An image frustum in heliocentric space, and the rectangular frustum space. Any astronomical image is a representation of the three dimensional frustum volume defined by the observer location, the field and depth of view of the image, the line of sight, and the rotation of the image. Points in heliocentric space may be easily transformed into points in the frustum space, facilitating image-object intersection and object motion characterization.

An image frustum is defined by a set of parameters: observer position (d_x, d_y, d_z), a line of sight (s_x, s_y, s_z), horizontal and vertical fields of view (α_h, α_v), a near view distance (d, taken as a distance closer to the observer than the typical in-atmosphere meteor distance), and a depth of field (f, taken as infinity). It can be shown that the real-world frustum space may be transformed into a rectangular frustum space based on a number of constituent transformations using the above frustum parameters. A PFM position (or the position of any solar system object) in heliocentric coordinates (x, y, z) may be translated to the image frustum space (x_f, y_f, z_f) by the simple application of a 4x4 transformation matrix T_f to the homogenous coordinates (x, y, z, 1), where T_f is the matrix product of a series of constituent transformations:

$$[x, y, z, 1][T_F] = [x_f, y_f, z_f, w_f] \qquad (6)$$

An object's position with respect to the 2x2 image frustum front face is:

$$x_I = \frac{x_f}{w_f}, \; y_I = \frac{y_f}{w_f} \quad (7)$$

where $-1 \le x_I \le 1$ and $-1 \le y_I \le 1$ correspond to the object being on the image. Additional outputs of the above transformation are four values called edge coordinates, which allow for quick object-image checking, prior to performing the above divisions in calculating x_I and y_I:

$e_1 : w_f + x_f$: left edge
$e_2 : w_f - x_f$: right edge
$e_3 : w_f + y_f$: bottom edge
$e_4 : w_f - y_f$: top edge

In all cases, the edge coordinate e_i exhibits the properties:

$e_i > 0$ the position is inside the edge
$e_i < 0$ the position is outside the edge

As long as the 4x4 T_f transformation matrix is calculated before search time, the computation required to determine an object-image intersection at a single point in time is the 12 multiplications and 9 additions required to calculate the x_f, y_f, and w_f values, and the 4 additions and 4 comparisons required to perform the edge coordinate checks. The edge coordinates provide a means to quickly determine the characteristics of an object's motion on, off, or through the image by comparing beginning-of-exposure coordinates to end-of-exposure coordinates.

4.2 Automated Downloads and Pre-processing

The object-image intersection process described above assumes the opportunity to possess pre-calculated image frustum descriptions and transformations prior to executing object searches. I have developed a generic image catalogue data base which serves to: 1) maintain local copies of available image databases, 2) provide a generic representation of the images databases, 3) support the storage of the above pre-calculated data, and 4) provide indexing to support various object search use cases. The image catalogue developed for this project is implemented as a flat file of generic image data referring back to local copies of downloaded image databases. Indexes are maintained on the image catalogue to facilitate searching by survey name, time frame, or both. The image catalogue is recreated from scratch on every download cycle. This approach was taken to avoid the concerns of needing a sophisticated underlying database technology that could handle both efficient insertion and querying. The download of updates from each image survey can be implemented as either a periodic replacement of the survey data, or as a net-change update. If the downloaded survey catalogue is represented in an easily parsed fixed form, then the survey import may be implemented by simply defining the survey catalogue in meta-data. If the survey data is difficult to parse, a survey plug-in may be developed to implement the import. Figure 8 describes the contents of the generic image catalogue records.

4.3 General Use of the Image Search Algorithm

The image representation described above and the resulting search based on ephemerides is a generic method which is independent of the source of the image and the target being searched. The

Image Collection	The name of the image survey, database, or collection. E.g. "CFHT Catalogue", "Catalina Catalogue".
Image File	The local copy of each image survey is described by one or more file names, and a description of the format of these files. The image catalogue points to the file number and file offset corresponding to an image in this set of files.
Image Offset	The position in the image file where the source description of the image may be found (see Image File).
Right Ascension	The right ascension of the centre of the image.
Declination	The declination of the centre of the image.
Width	The width of the image in radians.
Height	The height of the image in radians.
Start Time	The start date and time of the image exposure.
Exposure	The length of the exposure in seconds.
Starting Frustum	The parameters used to describe the image frustum at the beginning of the exposure, and the contents of the 4x4 transformation matrix used to convert object positions to the frustum space. The frustum description includes: • Frustum rotations around each axis • Dimensions of the frustum front face • Observer distances to the front and back faces • 16 numbers corresponding to the T_f image frustum transformation.
Ending Frustum	The parameters used to describe the image frustum at the end of the exposure (See Starting Exposure)

Figure 8. Image catalogue record description.

technique is therefore easily adapted to any use where the image source and target object are in motion. Whereas the original need of this project was to provide a flexible system that would support a variety of telescopic sky surveys, the approach can be easily extended to cover other image sources, such as spacecraft image databases, and amateur astrophoto collections. In addition, support sky surveys may be searched for any solar system object. For example, in support of the article Gilbert and Wiegert (2009), ClearSky was used to search the CFHTLS for images of three main-belt comets: 133P/Elst-Pizzaro, P/2005 U1 (Read), and 176P/LINEAR. Images of 176P/LINEAR were located, including a set of three images dated 2007 January 15. Figure 9 shows the image search result file for that period. Figure 10 shows the three image simulations, and the corresponding close inspection images created by Wiegert from CFHTLS image downloads.

5 Results to Date and Future Work

At the time of writing, this project has not yielded a discovery of a PFM image, however searching will continue. The results of the project have been of a more indirect nature, with our gaining understanding of PFM visibility and the nature of PFM orbits. The project has also contributed to the science with a confirmation of the heavily relied-upon analytical methods of the past 30 years, and by providing a useful general image catalogue search technique.

5.1 Image Searches

CFHTLS image searches have been performed on two collections of European Network events as documented in Spurný (1997) and Spurný (2010), as well as three individual events: Grimsby (McCausland, et al., 2010), Bunburra Rockhole (Bland, et al., 2009) and (Spurný, 2009), and Buzzard Coulee (Hildebrand, et al., 2009). No images have yet been found. The implementation of Spacewatch

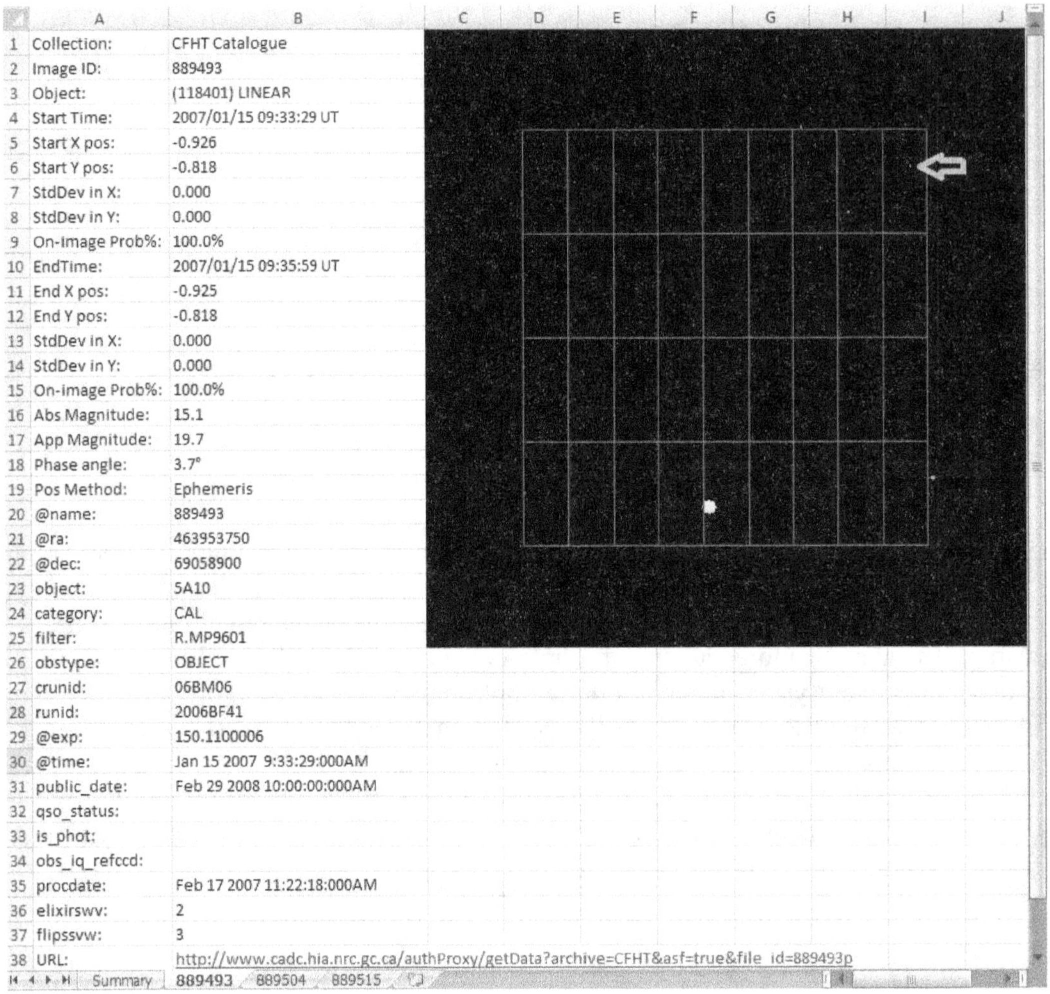

Figure 9. One image result from the CFHTLS catalogue search result file for 176P/LINEAR for the period of January 2007. Three images were located, the details of each circumstance represented in an Microsoft Excel spreadsheet tab. The result file text contains image description information (including the URL for downloading the CFHTLS image), object visibility, and the position of the object in the image. The X,Y coordinates of the object in the image (ranging from -1 to 1) are reversed due to the orientation of the image. The green arrow has been added to show the position of the object.

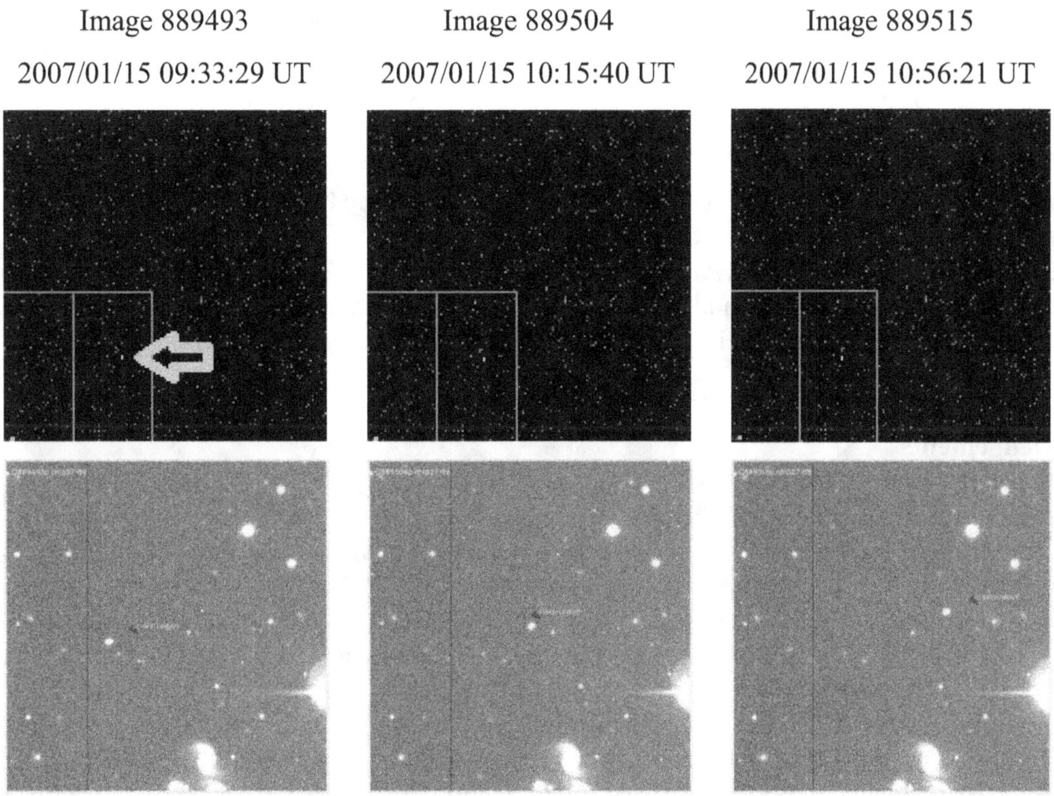

Figure 10. Top row: Detail from the three simulated CFHTLS images from 2007 January 15 showing object 176P/LINEAR. The green arrow has been added to highlight the object position. Note the slight movement in the object. Bottom row: Extractions from the actual CFHTLS images created by Dr. Paul Wiegert. The red arrow indicates the actual object; the blue arrow indicated the expected position based on the MPC orbit published at the time. Orientations between the simulations and the extractions are opposite.

(McMillan, 2010) and Catalina (Beshore, 2010) sky survey downloads are complete, and discussions are ongoing regarding access to the PanSTARRs (Jedicke, 2010) survey data. Contact state information on any additional significant fireball events is most welcome. As described above, the search techniques have shown useful in discovering serendipitous images of other solar system objects.

5.2 Modelling Results

As the project progressed, the realization developed that the chance of linking a fireball observation to a serendipitous PFM image is low. The modelling performed shows that a sufficiently large event in any one given meteor detection network is, at the optimistic end, a yearly event. Conservatively, in the case of ASGARD, it is a decadal event. Survey effectiveness analyses show that the chances for telescope detection can be severely reduced by object or solar geometry, as well as the quickly reducing apparent magnitude of the PFM. Finally, the chance of success is further reduced by the unmodelled but evidently small probability of a survey telescope being pointed at the correct field in relation to its overall sky coverage capability or preference. However the possibility of success remains enticing. Of particular note is the level of genuine interest shown for this project by the many people with whom I have discussed the work. Those involved in sky surveys seem eager to make their data available, and those analysing various fireball events are eager to provide and transform their data for use in this project.

5.3 Orbit Determination Results

A major accomplishment of this project was the confirmation of the Ceplecha (1987) orbit determination methods, and the offering of an adjustment of that work to account for a shift in the PFM's longitude of the ascending node. In verbal conversations with Dr. Pavel Spurný and others, it has been stated that this validation of Ceplecha's methods has not been done before, even though they are widely used. There can now be an increased level of confidence in the orbits derived from those methods.

Acknowledgements

I thank Dr. Paul Wiegert (UWO) for his support and technical assistance in this project, Dr. Peter Brown (UWO) and Dr. Pauline Barmy (UWO) for their support and advice, Rob Weryk (UWO) and Zbyszek Krzeminski (UWO) for their infinite patience in explaining the ASGARD system and for preparing data for my analysis, Dr. Pavel Spurný (Astronomical Institute, Academy of Sciences of the Czech Republic) for his explanations and data allowing me to compare methods to the Ceplecha work, and my wife Angela for her proof-reading.

This work is based in part on observations obtained with MegaPrime / MegaCam, a joint project of CFHT and CEA/DAPNIA, at the Canada-France-Hawaii Telescope (CFHT) which is operated by the National Research Council (NRC) of Canada, the Institut National des Science de l'Univers of the Centre National de la Recherche Scientifique (CNRS) of France, and the University of Hawaii. This work is based in part on data products produced at TERAPIX and the Canadian Astronomy Data Centre as part of the Canada-France-Hawaii Telescope Legacy Survey, a collaborative project of NRC and CNRS

References

Beatty, J. (1980). Computer Science 688. University of Waterloo, Ontario: Unpublished.

Beshore, E. (2010, June 02). Access to the CatalinaSky Survey Catalogue (Personal email).

Bland, P. A., Spurný, P., Towner, M. C., Bevan, A. W., Singleton, A. T., Chesley, S. R., et al. (2009). A Eucrite Delivered from an Aten-type Orbit: The Last Link in the Chain from 4 Vesta to Earth. *40th Lunar and Planetary Science Conference, (Lunar and Planetary Science XL), held March 23-27, 2009 in The Woodlands, Texas, id.1664* .

Bottke, W. F., Morbidelli, A., Jedicke, R., Petit, J.-M., Levison, H. F., Michel, P., et al. (2002a). Debiased Orbital and Absolute Magnitude Distribution of the Near-Earth Objects. *Icarus* , *156* (2), 399-433.

Bowell, E., Hapke, B., Domingue, D., Lumme, K., Peltoniemi, J., & Harris, A. W. (1989). Application of photometric models to asteroids. *Asteroids II; Proceedings of the Conference, Tucson, AZ, Mar. 8-11, 1988 (A90-27001 10-91)* , 524-556.

Brown, P., Spalding, R. E., ReVelle, D. O., Tagliaferri, E., & Worden, S. P. (2002). The flux of small near-Earth objects colliding with the Earth. *Nature* , *420*, 294-296.

Ceplecha, Z. (1987). Geometric, dynamic, orbital and photometric data on meteoroids from photographic fireball networks. *Astronomical Institutes of Czechoslovakia, Bulletin* , *38* (July 1987), 222-234.

CFHT. (2009, 11 12). *Canada-France-Hawaii Telescope Legacy Survey*. Retrieved 11 14, 2009, from Canada-France-Hawaii Telescope: http://www.cfht.hawaii.edu/Science/CFHTLS/

Gilbert, A. M., & Wiegert, P. A. (2009, June). Searching for main-belt comets using the Canada-France-Hawaii Telescope Legacy Survey. *Icarus* , *201* (2), pp. 714-718.

Hildebrand, A. R., Milley, E. P., Brown, P. G., McCausland, P. J., Edwards, W., Beech, M., et al. (2009). Characteristics of a Bright Fireball and Meteorite Fall at Buzzard Coulee, Saskatchewan, Canada, November 20, 2008. *40th Lunar and Planetary Science Conference, (Lunar and Planetary Science XL), held March 23-27, 2009 in The Woodlands, Texas, id.2505* .

Jedicke, R. (2010, May 27). Access to the PanSTARRS Image Catalogue (Private conversation). Breckenridge, Colorado.

Jenniskens, P., Shaddad, M. H., Numan, D., Elsir, S., Kudoda, A. M., Zolensky, M. E., et al. (2009). The impact and recovery of asteroid 2008 TC3. *Nature , 458* (7237), 485-488.

McCausland, P. J., Brown, P. G., Hildebrand, A. R., Flemming, R. L., Barker, I., Moser, D. E., et al. (2010). Fall of the Grimsby H5 Chondrite. *41st Lunar and Planetary Science Conference, held March 1-5, 2010 in The Woodlands, Texas.* , p. 2716.

McMillan, R. (2010, June 01). The Spacewatch image catalogue textual log (Personal email).

Meeus, J. (1991). *Astronomical Algorithms.* Richmond, Virginia: Willman-Bell, Inc.

Morbidelli, A., Jedicke, R., Bottke, W. F., Michel, P., & Tedesco, E. F. (2002a). From Magnitudes to Diameters: The Albedo Distribution of Near Earth Objects and the Earth Collision Hazard. *Icarus , 158* (2), 329-342.

Spurný, P. (2009, February 27). *State vector for the Bunburra Rockhole meteor (Personal email)* .

Spurný, P. (1997). Exceptional fireballs photographed in central Europe during the period 1993-1996. *Planetary and Space Science , 45*, 541-555.

Spurný, P. (2010, May 13). Recent European Network event tables (Personal email).

Wiegert, P., Balam, D., Moss, A., Veillet, C., Connors, M., & Shelton, I. (2007). Evidence for a Color Dependence in the Size Distribution of Main-Belt Asteroids. *The Astronomical Journal , 133* (4), 1609-1614.

Data Reduction and Control Software for Meteor Observing Stations Based on CCD Video Systems

J. M. Madiedo • J. M. Trigo-Rodríguez • E. Lyytinen

Abstract The SPanish Meteor Network (SPMN) is performing a continuous monitoring of meteor activity over Spain and neighbouring countries. The huge amount of data obtained by the 25 video observing stations that this network is currently operating made it necessary to develop new software packages to accomplish some tasks, such as data reduction and remote operation of autonomous systems based on high-sensitivity CCD video devices. The main characteristics of this software are described here.

Keywords meteor · meteoroid · fireball · software · meteor showers

1 Introduction

Since 2006 the SPanish Meteor Network (SPMN) has performed continuous monitoring of meteor and fireball activity over Spain and neighbouring countries. For this purpose, we mainly employ all-sky CCD cameras and high-sensitivity CCD video devices to monitor the night sky (Trigo-Rodríguez et al., 2006a, 2007a, 2007b; Madiedo, 2007). In addition, we have employed daytime CCD video cameras since 2007 in order to monitor fireball activity over 24 hours and increase the opportunities for meteorite recovery in Spain. As a result of this effort, a total of 25 observing stations are currently in operation. Several of them have been configured to work in a fully autonomous way. The two main cores of the Network are located in the regions of Catalonia and Andalusia. As these are separated by about 1000 km, there is a higher probability of clear skies and of meteor activity being recorded every night. The establishment of 25 meteor observing stations implies that a large amount of data needs to be reduced. This made it necessary to develop new software tools in order to perform a fast analysis of our data. The main features of these new packages are presented here.

2 Description of the Observing Stations and Procedures

Trigo-Rodríguez et al. (2004) previously reported the first steps in the development of the SPMN that employed low-scan-rate all-sky CCD cameras with +2/+3 meteor limiting magnitude. Since 2006 the

J. M. Madiedo (✉)
Facultad de Ciencias Experimentales, Universidad de Huelva. 21071 Huelva (Spain). E-mail: madiedo@uhu.es

J. M. Trigo-Rodríguez
Institut de Ciències de l'Espai–CSIC, Campus UAB, Facultat de Ciències, Torre C5-parell-2ª, 08193 Bellaterra, Barcelona, Spain; Institu d'Estudis Espacials de Catalunya (IEEC), Edif. Nexus, c/Gran Capità, 2-4, 08034 Barcelona, Spain

E. Lyytinen
Kehäkukantie 3B, 00720 Helsinki, Finland.

SPMN started to establish observing stations based on video systems to analyze meteor activity (Madiedo and Trigo-Rodríguez, 2007; Trigo-Rodríguez, 2007a, 2007b). These employ several high-sensitivity Watec CCD video cameras (models 902H and 902H Ultimate from Watec Corporation, Japan) to monitor the night sky. The cameras generate video imagery at 25 fps with a resolution of 720x576 pixels. These cameras are connected to PC computers via a video acquisition card. The computers use the UFOCapture software (Sonotaco, Japan) to automatically detect meteor trails and store the corresponding video sequences on hard disk. The cameras are arranged in such a way that the whole sky is monitored from every station and, so, this maximizes the common atmospheric volume recorded by the different systems. These devices are equipped with a 1/2" Sony interline transfer CCD image sensor with their minimum lux rating ranging from 0.01 to 0.0001 lux at f1.4. Aspherical fast lenses with focal length ranging from 3.8 to 6 mm and focal ratio between 1.2 and 0.8 are used for the imaging objective lens. In this way, different areas of the sky can be covered by every camera and point-like star images are obtained across the entire field of view.

Since 2007 we also started to employ CCD video systems to monitor the sky during the day (Madiedo and Trigo-Rodríguez, 2008). Daytime CCD video cameras work in the same way as nocturnal cameras do, but in this case lower-sensitivity devices are employed in order to avoid image saturation due to sunlight. These are endowed with slower optics (f1.4) and are arranged so that a part of the landscape falls within their field of view. In this way, identifiable structures or buildings appearing in the images can be used for image calibration to obtain the equatorial coordinates of fireballs.

3 Data Reduction Software

For data reduction we have developed a new software called Amalthea. This is a MS-Windows compatible package that has been programmed in C and C++ programming languages. The main characteristics of this software are described below.

3.1 Image and Video Processing

Amalthea was designed to analyze CCD images containing meteor trails and also video files recorded by our high-sensitivity CCD video devices. In many cases these images contain artefacts that may negatively interfere with data analysis. For this purpose, a wide number of image transformation filters have been implemented in the software. These include, for instance, light-pollution removal, brightness and contrast enhancement and video deinterlace filters. Some of these image transformation procedures allow enhancement of the video images before they are used for the astrometric analysis. For instance, our software automatically stacks the frames contained in video files in order to increase the number of stars available for the astrometric analysis described below.

3.2 Astrometry

Meteor and stars positions are obtained from static CCD images or from video sequences recorded by devices that monitor the night sky. The procedure we follow to obtain the equatorial coordinates of the meteor along its path have been described by Trigo-Rodriguez et. al (2007a). In a first stage, reference stars must be specified in order to apply a fitting method that allows conversion between plaque coordinates and equatorial coordinates. Then, by measuring the plaque coordinates of the meteor, these positions are automatically transformed by Amalthea into their corresponding equatorial counterparts.

It must be taken into account that in most cases we employ video devices that provide interlaced video sequences. These sequences must be deinterlaced by our software in order to remove some artefacts that could interfere in the astrometric reduction. Besides, depending on observing conditions in the area where the video stations are located, the number of reference stars in the corresponding video files can be very low, which is not enough to perform a good astrometric analysis. To solve this problem the software follows two different strategies. On one side, it stacks the frames contained in the video files in order to increase the signal to noise ratio of the resulting image. The number of reference stars in the resulting image is significantly higher. On the other side, as the cameras are pointed towards fixed altitude and azimuth coordinates, reference stars obtained at different times can be taken into account for a given measurement.

To perform the astrometric reduction the user manually clicks on the reference stars that must be taken into consideration for the corresponding calculations. Then, the user selects which fitting method must be used to convert from plaque coordinates to equatorial coordinates. Several options are available for this. Then the calculation is performed and the position of the reference stars is back-calculated by Amalthea in order to establish the error (standard deviation) of this calculation. In this way, the user, if necessary, can repeat the calculations by removing those stars which give rise to higher errors or include new ones.

Once we can convert between plaque and equatorial coordinates, plaque coordinates of meteors are specified by the user by clicking on the corresponding positions along the meteor trail. Their equatorial counterparts are then automatically provided by Amalthea. This can be done on static CCD images or on animated video sequences. In the latter case, time information necessary to calculate meteor velocities and decelerations is automatically obtained from the video file. In the former case, time information can be specified by the user if, for instance, a rotary shutter has been used.

3.3 Meteor Atmospheric Trajectory

Amalthea keeps a database with the geographic position of all the observing stations established by the SPMN. For meteors recorded simultaneously from at least two different observing stations the software can calculate its atmospheric trajectory and radiant once the above-described astrometric procedure has been performed. In order to do this the software uses the well-known planes intersection method (Ceplecha, 1987). If time information is available, velocities and decelerations are also calculated along the meteor trail. This allows us also to obtain the pre-atmospheric value of the meteor velocity, V_{inf}.

3.4 Orbital Parameters

The orbital and radiant parameters of the meteor are calculated according to the procedure described by Ceplecha et al. (1987). For this purpose, the values of the pre-atmospheric velocity, V_{inf}, radiant position and meteor apparition time are used, together with the average velocity corresponding to an averaged meteor position (latitude, longitude and altitude) along the meteor trail.

The procedure implemented in the Amalthea software has been tested with the Dutch Meteor Society (DMS) orbit calculation software (Langbroek, 2004) and the MORB software developed by the Ondrejov Observatory (Ceplecha et al., 2000). Although the DMS software does not provide any error parameters, we always found that the results provided by this package and Amalthea are very similar, with differences that are very small and within the error bars provided by Amalthea (Tables 1 and 2). However, significant discrepancies were found for the case of the MORB software. When this situation was analyzed in detail, we found that the origin of these is related to a bug in the calculation of the geocentric radiant in the MORB software.

Table 1. Comparison between orbital parameters calculated by Amalthea and the DMS software for different meteors recorded by the SPMN. Equinox (2000.00).

SPMN Code	Software	*q*(AU)	*a*(AU)	*e*	*i*(°)	*ω* (°)	*Ω* (°)
080806	Amalthea	0.9482±0.0003	9.65±1.5	0.902±0.016	113.58±0.15	149.82±0.32	139.1491±0.00003
	DMS	0.948	9.63	0.902	113.60	149.80	139.15
	MORB	9.558±0.001	21.36±7.76	0.955±0.016	111.79±0.15	152.87±0.35	139.1481±0.0003
210110	Amalthea	0.9543±0.0002	2.85±0.02	0.665±0.003	48.80±0.06	202.46±0.08	301.615±0.00002
	DMS	0.954	2.83	0.663	48.81	202.47	301.612
	MORB	0.9667±0.0006	15.47±5.11	0.937±0.020	44.92±0.28	195.55±0.38	301.6113±0.00003
071106	Amalthea	0.9774±0.0003	17.31±5.6	0.943±0.018	161.35±0.09	167.74±0.22	236.5034±0.00003
	DMS	0.977	16.96	0.942	161.36	167.68	236.505
	MORB	0.9864±0.0017	63.44±88.29	0.984±0.021	156.93±0.80	185.0±2.3	236.50565±0.00007

Table 2. Comparison between geocentric radiant position and pre-atmospheric velocities (geocentric, *Vg* and heliocentric, *Vh*) calculated by Amalthea and the DMS software for different meteors recorded by the SPMN. Equinox (2000.00).

SPMN Code	Software	R.A.(°)	DEC.(°)	*Vg*(km/s)	*Vh*(km/s)
080806	Amalthea	46.85±0.051	57.29±0.05	58.79±0.20	40.73±0.20
	DMS	46.85	57.29	58.79	40.73
	MORB	46.47±0.13	57.070±0.05	58.78±0.20	41.34±0.18
210110	Amalthea	230.03±0.1	66.57±0.03	29.80±0.20	38.61±0.21
	DMS	230.10	66.54	29.79	38.59
	MORB	227.23±0.10	67.41±0.03	29.79±0.05	41.77±0.22
071106	Amalthea	156.51±0.1	21.41±0.05	70.90±0.20	41.76±0.20
	DMS	156.52	21.40	70.80	41.65
	MORB	156.20±0.30	21.53±0.30	70.90±0.20	42.20±0.22

3.5 Meteorite Fall Analysis

Very bright fireballs can be the source of potential meteorite producing events. So, the analysis of these events is fundamental in order to locate, recover and study the corresponding fragments. For this purpose it is necessary to calculate the atmospheric trajectory of the fireball and also to model the so-called dark flight, which is the portion of the trajectory followed once the particle has been decelerated in such a way that no light is emitted. The atmospheric trajectory is determined by following the procedures described above. Then, our Amalthea software solves the aerodynamic equations that describe the dark flight of the meteoroid. To do this, information about the particle and its terminal point must be entered. A standard Runge-Kutta procedure is followed in order to integrate the position of the particle from the terminal height to the ground. Wind data are also taken into account by entering, as a function of height, latitude and longitude, the values of atmospheric pressure, temperature and wind velocity and direction. The resulting meteorite impact position is shown both numerically and drawn on a map.

In order to test the calculation procedure implemented in Amalthea, we have compared the results provided by our software to those provided by the software developed by Z. Ceplecha, P. Spurny and J. Borovicka for the case of the Villalbeto de La Peña meteorite fall (Trigo-Rodríguez et al., 2006b). The fall of this L6 chondrite occurred on January 4, 2004 in north-west Spain (Trigo-Rodríguez et al., 2006b; Llorca et al., 2005). Both software packages provide the same result.

3.6 Meteor Spectra Analysis

Some of the all-sky CCD cameras and high-sensitivity video devices employed by the SPMN are endowed with holographic diffraction gratings (600 to 1200 lines/mm) in order to obtain meteor spectra. Typically we can obtain these spectra for meteors as bright as mag. -4 or lower without using any image intensifier device. These spectra are very useful in obtaining chemical information about meteoroids (Trigo-Rodríguez et al., 2009).

The Amalthea software is able to analyze these spectra when they are recorded on AVI video files or on static all-sky CCD images. These raw spectra must be calibrated in order to take into account the response of the camera to different wavelengths. This information can be taken from the documentation provided by the manufacturer of the camera or can be experimentally obtained by comparing a known spectrum of an astronomical object with the spectrum obtained by the camera for the same source. Once this calibration is performed, the software identifies the main meteor emission lines. Figure 1 shows an example of the emission spectrum of a sporadic fireball recorded by the SPMN from Sevilla on May 27, 2010, at 3h19m40.1 ± 1s UTC. The two most prominent lines correspond to Na_I -1 (589.5 nm) and Ca_I-1 (422.7 nm).

4 Description of the SPMN Video Station Control Software

The number of SPMN video meteor stations has increased from 2 in 2006 to 25 in 2010. This has resulted in the necessity to address two main issues that have arisen from this situation: the establishment of a system to check the large volume of data provided by these observing stations in order to identify multiple-station events and potential meteorite dropping events, and the necessity to locate several autonomous meteor observing stations in remote locations or in places where no direct human intervention is always possible. Thus, during 2009 new software packages have been developed in order to achieve a fully robotic operation of several video meteor observing stations. These new packages also allow the systems to be remotely controlled though an Internet connection. A prototype robotic video station was setup by the University of Huelva in the environment of the Doñana Natural Park (south-west of Spain) in April 2009 (Madiedo et al, 2010). This station was fully operative till the robotic system was completely developed in August 2009. Nowadays, three robotic video meteor stations are operative in the south of Spain. Two of them are located in the western area of Andalusia, in the provinces of Huelva and Sevilla. The latest one has been setup in Sierra Nevada (Granada), in the eastern part of this region.

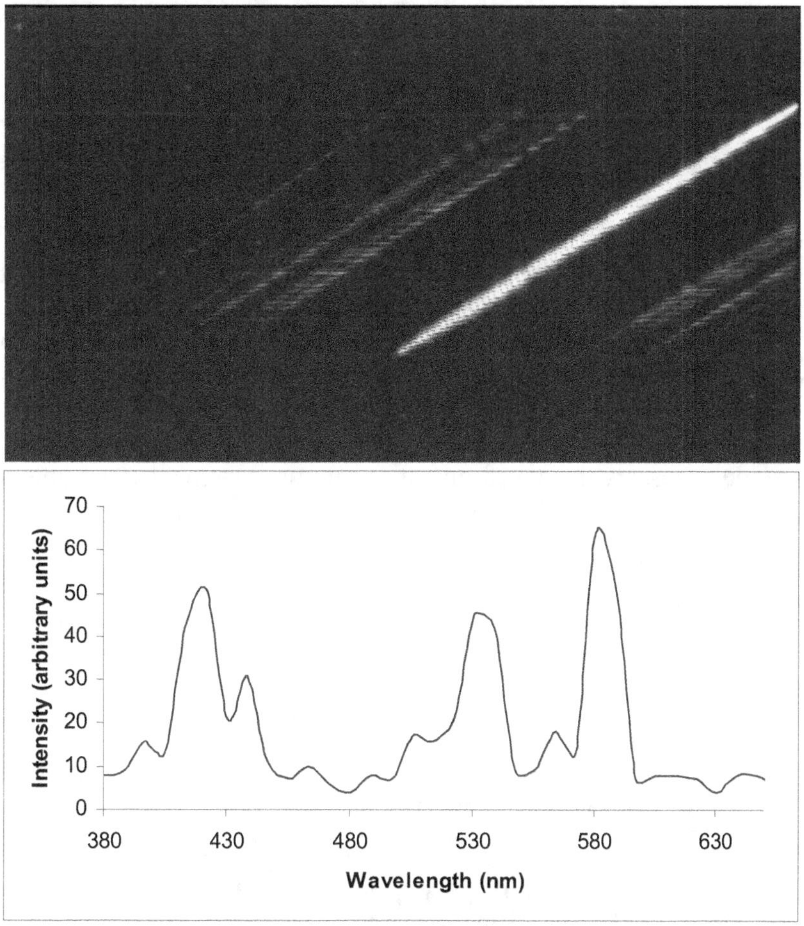

Figure 1. Sporadic fireball recorded from Sevilla on May 27, 2010, at 3h19m40.1 ± 1s UTC with a high-sensitivity CCD video camera endowed with a diffraction grating and its emission spectrum analyzed by the software Amalthea.

The PC computers that control the new robotic stations have been configured so that they are automatically switched on and off when data acquisition must start and finish, respectively. When the PCs are started, one of our recently developed software packages starts the data acquisition software (UFOCapture) and monitors that this application is properly working during the whole observing session. We have also developed another software package that automatically checks the meteor trails recorded by the system. Then, if a very bright fireball is detected (typically mag. -12 or brighter) which could give rise to a potential meteorite fall, an email is automatically sent to the operator together with the corresponding images.

These robotic stations provide a huge volume of data, as they are currently an average of over 1000 meteors per month. In order to handle this information, we have also developed a software package that is automatically started when the observing session is over. Then, this application checks every meteor trail recorded on hard disk. These data are compressed and sent to a FTP server for further processing. The application also checks other data stored by this FTP server, as data from different robotic stations are also placed there, and identifies which meteor trails have been simultaneously recorded from at least two different locations. The operator receives an automatic email with these data, which can be reduced to obtain the atmospheric trajectory of the corresponding meteoroid and also radiant and orbital information.

5 Conclusions

A continuous effort is being made by the Spanish Meteor Network in order to improve and expand our meteor observing stations based on high-sensitivity CCD video devices. Software engineering has been one of our priorities in the latest years and, as a result, a new software package for data reduction has been developed and successfully tested with other existing applications. However, significant discrepancies have been detected with the results provided by Ondrejov's orbits calculation software (MORB). This is due to the fact that MORB software does not calculate as a result of an incorrect calculation of the geocentric radiant. Besides, the possibility of installing high-sensitivity video systems in remote locations made necessary the development of robotic systems that are able to operate in a fully autonomous way and also to automatically notify the occurrence of remarkable events. Several software packages have been also developed in order to accomplish these tasks.

References

Ceplecha, Z. (1987) "Geometric, dynamic, orbital and photometric data on meteoroids from photographic meteor networks", Bull. Astron. Inst. Czechols. 38, 222-234.

Ceplecha Z., P. Spurný, J. Borovička (2000) Meteor Orbit (MORB) software. Ondrejov Observatory. Czech Republic.

Langbroek, M. (2004) Meteor Orbit Calculation software. Dutch Meteor Society.

Llorca, J., Trigo-Rodríguez, J. M., Ortiz, J. L., Docobo, J. A., García-Guinea, J., Castro-Tirado, A. J., Rubin, A. E., Eugster, O., Edwards, W., Laubenstein, M., Casanova, I. (2005) " The Villalbeto de la Peña meteorite fall: I. Fireball energy, meteorite recovery, strewn field, and petrography", Meteoritics and Planetary Science 40:6, 795-804.

Madiedo, J.M. and Trigo-Rodríguez, J.M. (2007) "Multi-Station Video Orbits of Minor Meteor Showers", *Earth, Moon, and Planets*, 102, 133-139.

Madiedo, J.M., Trigo-Rodríguez, J. M., Ortiz, J. L., Morales, N. (2010) "Robotic Systems for Meteor Observing and Moon Impact Flashes Detection in Spain", Advances in Astronomy, doi:10.1155/2010/167494.

Madiedo, J.M. and Trigo-Rodríguez, J.M. (2008) "On the development of new SPMN diurnal video systems for daylight fireball monitoring", *European Planetary Science Congress, A*bstract EPSC2008-A-00319.

Trigo-Rodríguez J.M., Llorca J., Castro-Tirado A.J., Ortiz J.L., Docobo J.A., and Fabregat J. (2006a) "The Spanish Fireball Network", Astron. & Geoph. 47:6, 26.

Trigo-Rodríguez, J.M., Borovička, J., Spurný, P., Ortiz, J.L., Docobo, J.A.., Castro-Tirado, A..J., Llorca, J. (2006b) " The Villalbeto de la Peña meteorite fall: II. Determination of atmospheric trajectory and orbit", Meteoritics and Planetary Science 41:4, 505-517.

Trigo-Rodríguez J.M., Madiedo J.M., Castro-Tirado A.J., Ortiz J.L., Llorca J., Fabregat J., Vítek S., Gural P.S., Troughton B., Pujols P., and Gálvez F. (2007a) "Spanish Meteor Network: 2006 continuous monitoring results", WGN J. of IMO 35, 13-22.

Trigo-Rodríguez, J.M., Madiedo, J.M., Llorca, J., Gural, P.S., Pujols, P., Tezel, T. (2007b) “The 2006 Orionid outburst imaged by all-sky CCD cameras from Spain: meteoroid spatial fluxes and orbital elements”, Mon. Not. R. Astron. Soc. 380, 126-132.

Trigo-Rodriguez, J.M., Madiedo, J.M. Williams, I. P. (2009) "The outburst of the κ Cygnids in 2007: clues about the catastrophic break up of a comet to produce an Earth-crossing meteoroid stream". Mon. Not. R. Astron. Soc. 392, 367–375.

The Updated IAU MDC Catalogue of Photographic Meteor Orbits

V. Porubcan • J. Svoren • L. Neslusan • E. Schunova

Abstract The database of photographic meteor orbits of the IAU Meteor Data Center at the Astronomical Institute SAS has gradually been updated. To the 2003 version of 4581 photographic orbits compiled from 17 different stations and obtained in the period 1936-1996, additional new 211 orbits compiled from 7 sources have been added. Thus, the updated version of the catalogue contains 4792 photographic orbits (equinox J2000.0) available either in two separate orbital and geophysical data files or a file with the merged data. All the updated files with relevant documentation are available at the web of the IAU Meteor Data Center.

Keywords astronomical databases · photographic meteor orbits

1 Introduction

Meteoroid orbits are a basic tool for investigation of distribution and spatial structure of the meteoroid population in the close surroundings of the Earth's orbit. However, information about them is usually widely scattered in literature and often in publications with limited circulation. Therefore, the IAU Comm. 22 during the 1976 IAU General Assembly proposed to establish a meteor data center for collection of meteor orbits recorded by photographic and radio techniques. The decision was confirmed by the next IAU GA in 1982 and the data center was established (Lindblad, 1987).

The purpose of the data center was to acquire, format, check and disseminate information on precise meteoroid orbits obtained by multi-station techniques and the database gradually extended as documented in previous reports on the activity of the Meteor Data Center by Lindblad (1987, 1995, 1999 and 2001) or Lindblad and Steel (1993).

Up to present, the database consists of 4581 photographic meteor orbits (Lindblad et al., 2005), 63.330 radar determined orbit: Harvard Meteor Project (1961-1965, 1968-1969), Adelaide (1960-1961, 1968-1969), Kharkov (1975), Obninsk (1967-1968), Mogadish (1969-1970) and 1425 video-recordings (Lindblad, 1999) to which additional 817 video meteors orbits published by Koten el al. (2003) were added.

V. Porubcan (✉)
Astronomical Institute of the Slovak Academy of Sciences, 059 60 Tatranská Lomnica, Slovakia; Faculty of Mathematics, Physics and Informatics, Comenius University, 84248 Bratislava, Slovakia. E-mail: porubcan@fmpb.uniba.sk

J. Svoren • L. Neslusan
Astronomical Institute of the Slovak Academy of Sciences, 059 60 Tatranská Lomnica, Slovakia

E. Schunova
Faculty of Mathematics, Physics and Informatics, Comenius University, 84248 Bratislava, Slovakia

2 Photographic Orbits

In 2001 the MDC was moved from the Lund Observatory to the Astronomical Institute of the Slovak Academy of Sciences in Bratislava. As it is operating for about twenty years and collecting data acquired already since 1936, it accumulated a large number of meteoroid orbits obtained from different sources providing them to researchers for various studies.

The last but one version of photographic data, the 1990 version, contained data on 3518 meteors with precisely reduced orbits, but only of about 90% of them had also at disposal also complete geophysical data (Lindblad, 2001). In many cases the orbital and geophysical data were not published at the same time and in some cases one set was not published and the data had to be obtained by correspondence with the investigator or are remain missing. This was also the case with the last 2003 version of the catalogue containing 4581 precisely reduced meteor orbits (Lindblad et al., 2005), where originally not all orbits were published with all the catalogued parameters.

The data for the 2003 version were compiled from 17 different stations or investigators listed in Table 1, where *Code* in the first column is the letter code assigned to the investigator and the catalogue comprises the orbits obtained over the period of sixty years (1936-1996).

Traditionally, the previous versions consisted of two independent files separately with the orbital file (*orb.dat*) and geophysical file (*geo.dat*). In the new version and for convenience of further treatment with the data both files were also merged, introduced in a new format and writing the data in a single file designated *all2003.dat* and sorted by the date of meteor detection.

The database can be downloaded from the IAU MDC site from the address: *http:/www.astro.sk/~ne/IAUMDC/Ph2003/database.html* together with relevant documentation.

Table 1. IAU MDC catalogue of photographic orbits. Code – investigator or station code letter, N – the total number of reduced orbits per station / investigator.

Code	N	Investigator/Station
W	166	Whipple (small camera)
J	413	Jacchia (Super-Schmidt)
H	313	Hawkins and Southworth (Super-Schmidt)
P	353	Posen and McCrosky (Super-Schmidt)
S	314	McCrosky and Shao (Super-Schmidt)
G	25	Gale Harvey, New Mexico State University
D	636	Babadzhanov et al., Dushanbe (small camera)
O	459	Shestaka et al., Odessa (small camera)
K	206	Kiev (small camera)
C	103	Ceplecha (small camera)
E	335	Ceplecha and Spurný, European Network
F	334	McCrosky, Prairie Network
I	259	Halliday et al., MORP Network
N	259	Koseki, Nippon Meteor Society
T	85	Ohtsuka, Tokyo Meteor Network
U	66	Ochai et al., Nippon Meteor Society
B	435	Betlem et al., Dutch Meteor Society
R	22	Trigo-Rodriguez et al., Spanish Meteor Network
Sum	4792	

3 New Version of the Catalogue

Since 1996 additional new photographic meteor orbits are acquired and can be catalogued. Some of them are older and were not yet included in the catalogue and the rest are new orbits recorded in the last few years. Though the present catalogue of 4581 orbits is covering the whole year observations rather uniformly, dominant in the number of orbits are the Perseids and Geminids. The recent exceptional returns of the Leonids greatly improved their orbital statistics, but more of them were not yet published and cannot be included in the catalogue.

At the present, there are 211 new meteoroid orbits which can be included in the new version of the catalogue and thus the number of the orbits of the MDC catalogue increased to 4792. The new orbits are compiled from seven sources and consists of 27 Perseids recorded during the exceptional shower return in 1993 (Spurny, 1995), 75 Leonids observed during the 1998 shower outburst (Betlem et al., 1999), 10 Leonids from the 2002 storm (Trigo-Rodriguez et al., 2004), 19 bright fireballs recorded within the European Fireball Network between 1990-2004, 22 fireballs recorded within the Spanish Meteor Network in 1991-2004, 52 exceptional bolides from the EN observed in 2006-2007 (Spurny, 2010) and 6 fireballs recorded in Japan in the beginning of nineties.

In order to resolve any potential inconsistencies in the published orbital and geophysical parameters, these have been checked by a standard checking procedure as applied to the sets of orbits in the previous version of the catalogue.

The check of consistency of the orbital and geophysical data is made in two steps: (a) Assuming that the published radiant and velocity at the time of detection were correct, the orbital elements are recalculated; (b) As sometimes errors appear also in the published geophysical (encounter) data, the radiant coordinates and the geocentric velocity are recalculated from the published orbital elements utilizing the optimal method of theoretical radiant prediction for a given orbital geometry (Neslušan et al., 1998).

In the new version of the catalogue also the error bars of the orbital and geophysical parameters derived by individual investigators, if available will be added and included to the both data files.

At the next step we plan to innovate the catalogue more regularly that is after any new set of meteoroid orbits is published or submitted for inclusion in the catalogue and information about any updating of the database will be announced on web page of the IAU MDC.

Acknowledgements

The authors acknowledge support from VEGA - the Slovak Grant Agency for Science (grants Nos. 1/0636, 2/0022 and 2/0011).

References

H. Betlem, P. Jenniskens, J. van't Leven, R. Ter Kuile, C. Johannink, H. Zhao, C. Lei, G. Li, J. Zhu, S. Evans, P. Spurny, Very precise of 1998 Leonid meteors. Meteoritics and Planet. Sci. **34**, 979 (1999)

P. Koten, P. Spurny, J. Borovicka, R. Stork, Catalogue of video meteor orbits 1. Publ. Astron. Inst. Sci. Czech Rep. **91**, 1–32 (2003)

B.A. Lindblad, The IAU Meteor Data Center in Lund. In: Z. Ceplecha, P. Pecina (eds.) Interplanetary Matter, Proc. 10th ERAM. Astron. Inst. Czechosl. Acad. Sci., Prague, 201–204 (1987)

B.A. Lindblad, D.I. Steel, In: Asteroids, Comets, Meteors, Milani et al. (eds.), 497 (1993)

B.A. Lindblad, The IAU Meteor Data Center in Lund. Earth, Moon and Planets, **68**, 405 (1995)

B.A. Lindblad, A survey of meteoroid orbits obtained by two station video observations. In: Meteoroids 1998, W.J. Baggaley and V. Porubcan (eds.), Bratislava, 274 (1999)
B.A. Linbdlad, IAU Meteor Data Center. In: Meteoroids 2001, B. Warmbein (ed.), ESA SP-495, 71 (2001)
B.A. Lindblad, L. Neslusan, V. Porubcan, J. Svoren, IAU Meteor Database of photographic orbits-version 2003. Earth Moon Planets, **93,** 249–260 (2005)
L. Neslusan, J. Svoren J, V. Porubcan, A computer program for calculation of a theoretical meteor-stream radiant. Astron. Astrophys., **331**, 411 (1998)
P. Spurny, EN Photographic Perseids. Earth, Moon and Planets, **68**, 529 (1995)
P. Spurny (2010) *Private comm.*
J.M. Trigo-Rodriguez, J. Llorca, E. Lyytinen, J.L. Ortiz, A.S. Caso, C. Pineda, S. Torrell, 2002 Leonid storm fluxes and related orbital elements. Icarus **171**, 219 (2004)

CHAPTER 10:

THE FUTURE OF OBSERVATIONAL TECHNIQUES AND METEOR DETECTION PROGRAMS

French Meteor Network for High Precision Orbits of Meteoroids

P. Atreya • J. Vaubaillon • F. Colas • S. Bouley • B. Gaillard • I. Sauli • M.-K. Kwon

Abstract There is a lack of precise meteoroids orbit from video observations as most of the meteor stations use off-the-shelf CCD cameras. Few meteoroids orbit with precise semi-major axis are available using film photographic method. Precise orbits are necessary to compute the dust flux in the Earth's vicinity, and to estimate the ejection time of the meteoroids accurately by comparing them with the theoretical evolution model. We investigate the use of large CCD sensors to observe multi-station meteors and to compute precise orbit of these meteoroids. An ideal spatial and temporal resolution to get an accuracy to those similar of photographic plates are discussed. Various problems faced due to the use of large CCD, such as increasing the spatial and the temporal resolution at the same time and computational problems in finding the meteor position are illustrated.

Keywords meteor · CCD · observation

1 Introduction

Meteor astronomy has implemented video techniques for all-night observations in addition to visual and photographic during the past decade. The IMO Video Meteor Network [1], Polish Fireball Network [2], Spanish Meteor Network [3], Dutch Meteor Society [4] and Czech Meteor Network [5] are few of the meteor networks actively operating in Europe. Detection and analysis software such as Metrec, Meteorscan and UFOcapture eases the tedious setup and encourages professional and amateur astronomers alike to set up meteor stations.

These networks have thrived from off-the-shelf video cameras and lenses. One of the major drawbacks from these configurations is the astrometric quality of data acquired. The typical cameras used (Watec, Mintron etc.) with 640 × 480 pixels, along with medium angled lens (~50°), have spatial resolution of 0.08°/pixel. This corresponds to a resolution of 140 m if the meteor is at 100 km distance from the camera. This causes a large uncertainty in velocity, and thereafter, in the semi-major axis of the meteoroids. For example, a typical meteor with velocity of 39.9 km/s corresponds to semi major axis of 5 AU, whereas, velocity of 40.0 km/s corresponds to 10 AU, a change of 0.1km/s in velocity corresponding to 5 AU in this particular case [6]. Semi-major axis is very sensitive to velocity component of the meteoroids, and thus we need 10 order of magnitude better spatial resolution than most of these off-the-shelf cameras for precise meteor velocity and semi-major axis.

The need for more precise orbital elements of meteoroids is imminent for modeler's. Figure 1 shows the node of different trails (1817-1913 AD) of Draconid stream for the year 2011. Only with high resolution observation and precise semi major axis, it is possible to identify the exact trail that will cause the outburst on 8th October, 2011. Similarly, the 2009 Leonids outbursts were predicted to occur due to

P. Atreya (✉) • J. Vaubaillon • F. Colas • S. Bouley • I. Sauli • M.-K. Kwon
IMCCE, 77 Avenue Denfert Rochereau, 75014, Paris, France. E-mail: atreya@imcce.fr

B. Gaillard
Lheritier, 10 avenue de l'Entreprise, 95862 Cergy, France

trails from 1533 AD and 1536 AD [7], whose radiants differ by 0.9°. But the observations made from current instruments were not able to identify the exact trail from which the outburst occurred. So there is need for better accuracy, which can be obtained by higher temporal and spatial resolutions.

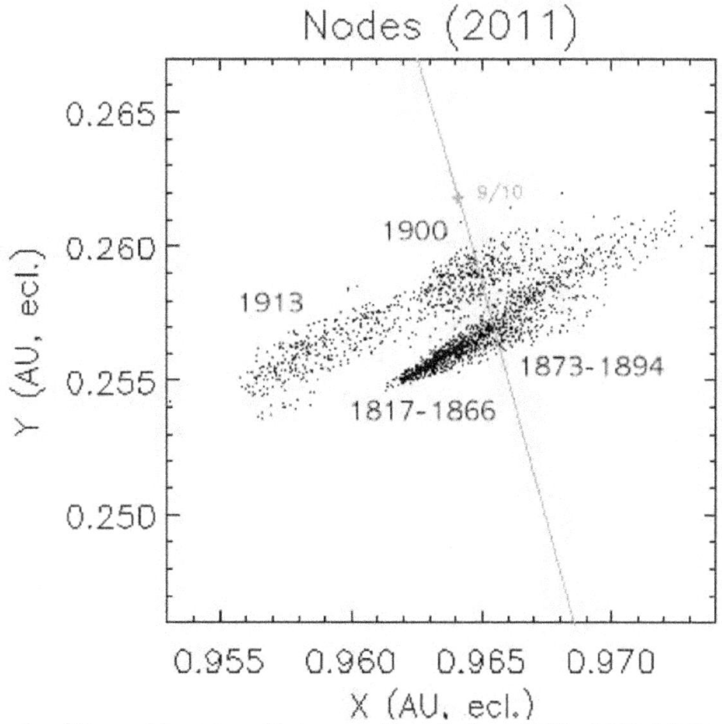

Figure 1. The node of Draconids meteoroid stream with trails from 1817 – 1913 AD for the 2011 outburst.

There are generally two ways to increase the spatial resolution of the camera system, i.e. to use a large CCD and to use a narrow lens. However, only increasing the spatial resolution is not sufficient for obtaining precise velocity. A simple simulation was performed to identify the spread of meteor in a single image for a fixed high resolution of 0.01°/pixel for various frame rates. Figure 2 shows the length of a meteor (in pixels) for different temporal resolution for slow (angular speed of 10°/s) and fast meteors (angular speed of 30°/s).

For the typical video frame rate of 25–30 fps the slow meteor will be spread out across 40 pixels, whereas the fast meteor will be spread more than 100 pixels. Thus it will be difficult to compute the position of meteor within 1 pixel accuracy as its spread across too many pixels. When the frame rate is increased to 100 fps, the meteor is spread across only 10–30 pixels. Thus increasing frame rate is essential if spatial resolution is increased to estimate the meteor position accurately in the images. Due to this reason, even if narrow lens are used with small CCDs with video frame rate, it will have computational disadvantages.

There are also several disadvantages of using large CCDs. The time to read a single frame in a CCD is inversely proportional to the size of the CCD. Table 1 shows the frame rate of different size of CCDs from JAI camera company. Only the smallest CCD have temporal resolution of 100 or better, which is preferred for good spatial resolution. An external rotating shutter can be used to provide good temporal resolution.

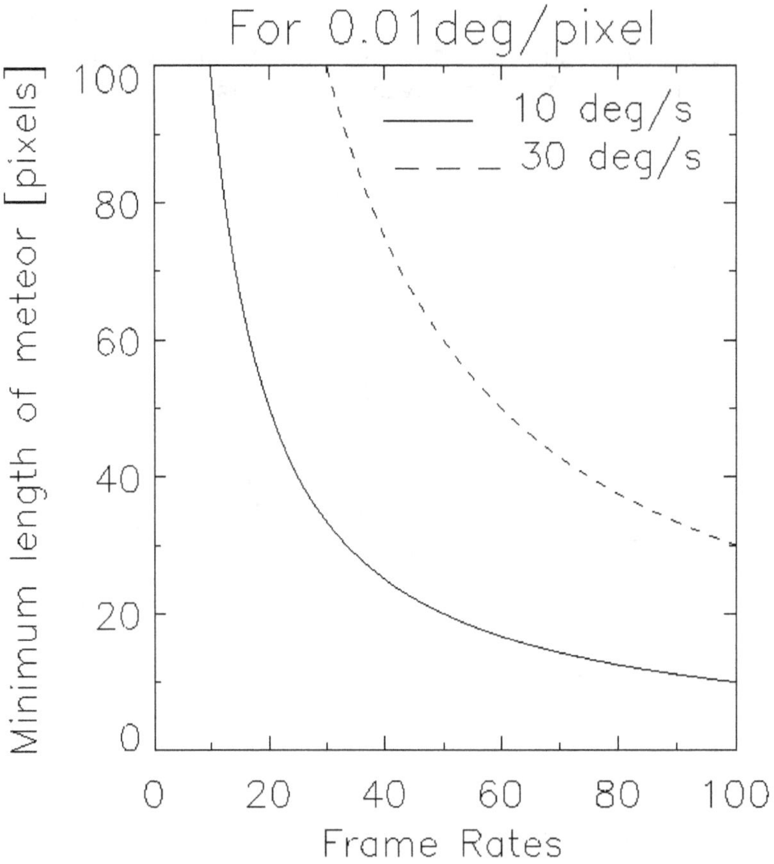

Figure 2. The length of a meteor for different temporal resolution for two typical meteor speed. The spatial resolution of set to 0.01 deg/pixel

Table 1. The frame rate of different sizes of CCDs.

Camera ID	Size of CCD [Pixels]	Max Frame Rate
RM-6740CL	648 × 484	200
CM-140GE	1392 × 1040	31
CM-200GE	1628 × 1236	25
BM-500GE	2456 × 2048	15
AM-1600GE-F	4872 × 3248	3

2 Lheritier Camera

Lheritier, a French camera and system vision company, has developed an interline progressive scan camera LH11000. It uses Kodak Kai 11002 sensor with 4032 × 2688 effective pixels of 9 μm size. The readout noise is ~30 e^-, with gain of 0–30 DB. The images are saved in 16 bit format. It takes 149 ms to read a single image, and the frame rate is 6.7 fps. The minimum and maximum shutter speeds are 0.8 ms and 52428 ms respectively. This camera was tested for military and aviation environment and can withstand −10°C to 50°C and humidity of 100% non condensing.

Figure 3 shows the modification of CCD readout method to mimic electronic shutter system. The top part of the figure shows the basic procedure of CCD system. The signal denotes the exposure duration (10 ms), and all the images are stored separately. This is not ideal because of the "dead time" of 149 ms after every image. A modification to a CCD system was made by introducing a "break" period (10 ms) which means that during this period the CCD is closed as shown in the bottom section of the figure. The images are stacked onto the CCD (N = 50 times), and read out only once. This method not only decreases the dead time of the camera, but due to stacking of the images, causes an effect very similar to the external rotating mechanical shutter. The "signal", "break" and the total exposure duration can be modified to suit fast and slow meteors. This method was possible as Lheritier company developed a way to "close" the CCD completely, which is used as "break" duration. This also implies that there is no upper limit to the size of CCD that can be used for the detection of meteors due to low temporal resolution. The dead time of reading a single image can be reduced by making longer exposures.

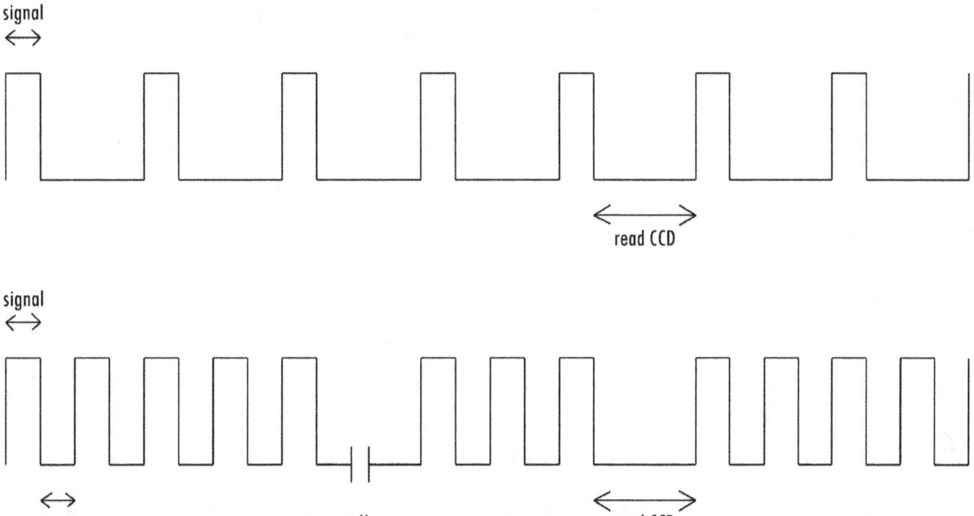

Figure 3. Modification of basic CCD readout method to mimic electronic shutter system.

Figure 4 shows an example of meteor detected (raw image) with LH11000 camera, modified shutter system and Nikon 85 mm F1.4 lens (FOV 28° × 18°). The meteor was captured during Lyrid observation campaign on 22nd April, 2010 at 00:56:50 UT from Observatory of Haute Provence (OHP), France. The spatial resolution is ~0.007° or ~25". The signal and break duration was set to 10 ms and the total exposure of the image was 1 second (N = 50 loops).

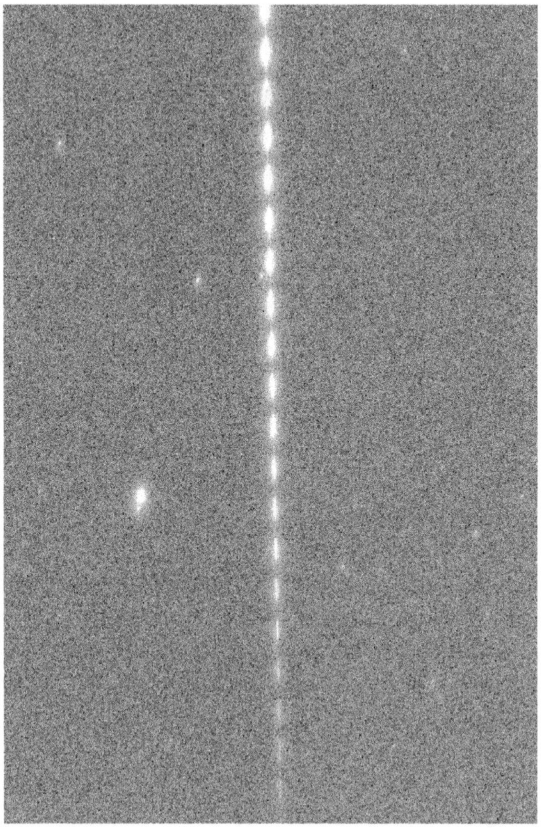

Figure 4. Example of a meteor (raw image) with 10 ms signal, 10 ms break and total exposure of 1 seconds.

The intensity plot for few breaks is shown in Figure 5. The distance between the peaks are ~40 pixels. With the spatial and temporal resolution of 0.007°/pixel and 10 ms respectively, the angular velocity is ~28°/s, which agrees with those of general meteors. Precise position and velocity can be computed by fitting different types of curves in the reduced data. Figure 4 and smoothness of the intensity curve of Figure 5 proves that this new method of electronic shutter works.

3 French Meteor Network

The first video meteor network will be started in France under PoDeT-MET project. The primary aim of this network is to get high precision orbits of meteoroids. The Lh11000 camera is equipped with Cannon 50mm F1.2 lens (FOV 40° x 27°) and have spatial resolution of 0.01° which is ~8 times higher than the numerous off-the-shelf systems. This corresponds to 17.4m resolution if the observed meteor is at a distance of 100 km. The first dedicated triple station network will be set up in south of France during 2010–2011. The first station will be set up at Pic du midi Observatory, at the height of 3000m. The other two stations will be set ~100 km further away. An all sky camera will also be installed to compliment the high precision cameras. The network will be expanded to other areas of France in future.

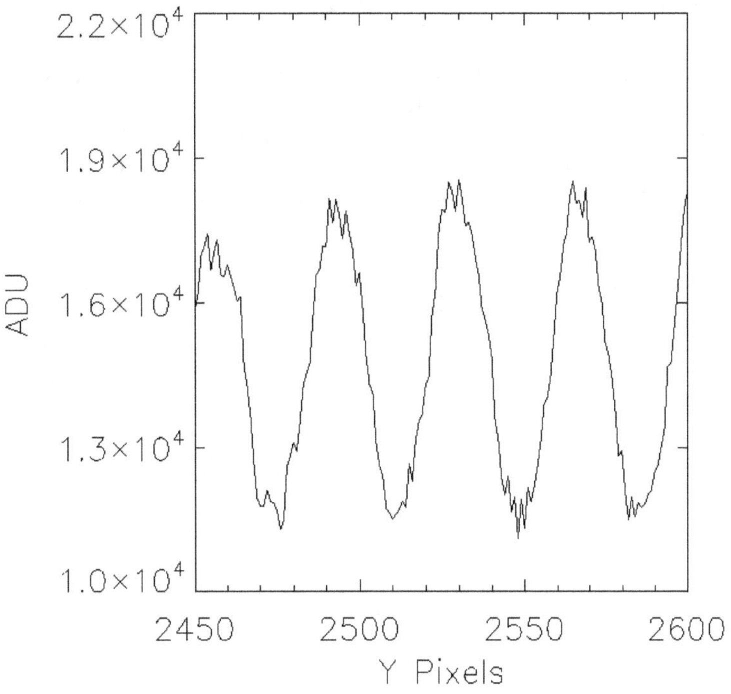

Figure 5. The intensity of meteor breaks along Y axis.

Starting a remote and automated camera network has many challenges. A single image from the LH11000 camera is 20 MB in size. This results in 400+ GB of raw data every night from a single camera. One of the big challenge is to develop a very fast and yet efficient and automated method to detect meteor. The false detections have to be limited, as the number of images that can be transferred are limited by low bandwidth internet. One of the most important factors in computing precise orbits of meteoroids is to estimate the meteor position in the image with less than 1 pixel uncertainty. Most of the software for processing meteor images were written for off-the-shelf cameras, and many modifications would have to be made so that the required accuracy can be achieved. Other challenges include developing a cooling system for the camera to reduce CCD noise while heating up the lens area to prevent condensation.

Acknowledgements

The authors would like to thank Lheritier company for all the support. This project is funded primarily by Paris city grants

References

1. Molau, S.: 2003, The AKM video meteor network,Proceedings of the Meteoroids 2001 Conference, pp. 315 -318
2. Olech, A.; Zoladek, P.; Wisniewski, M.; Krasnowski M.; Kwinta, M.; Fajfer, T.; Fietkiewicz, K.; Dorosz, D.; Kowalski, L.; Olejnik, J.; Mularczyk, K.; Zloczewski, K.:2006, Polish Fireball Network, Proceedings of the IMC 2005, pp.53-62
3. Trigo-Rodriguez, J. M.; Madiedo, J. M.; Castro-Tirado, A. J.; Ortiz, J. L.; Llorca, J.; Fabregat, J.; Vitek, S.; Gural, P. S.;

Troughton, B.; Pujols, P.; Galvez, F.: 2007, Spanish Meteor Network: 2006 continuous monitoring results, WGN, 35, 1, pp. 13-22

4. Miskotte, K.; Johannink, C.: 2006, Taurids 2005: results of the Dutch Meteor Society, WGN, 34, 1, pp. 11-14
5. Koten, P.; Borovicka.J.; Spurn, P.; Evans,S.; Stork,R.; Elliott, A.: 2006, Double station and spectroscopic observations of the Quadrantid meteor shower and the implications for its parent body, MNRAS, 366, 4, pp. 1367-1372.
6. Barentsen, G., 2009, Meteor Astrometry: What accuracy do we need, Proceedings of the IMC 2009, in press.
7. Jenniskens, P., 2006, Meteor Showers and their Parent Comets, Cambridge University Press, UK

BRAMS: the Belgian RAdio Meteor Stations

H. Lamy • S. Ranvier • J. De Keyser • S. Calders • E. Gamby • C. Verbeeck

Abstract In the last months, the Belgian Institute for Space Aeronomy has been developing a Belgian network for observing radio meteors using forward scattering technique. This network is called BRAMS for Belgian RAdio Meteor Stations. Two beacons emitting a circularly polarized pure sine wave toward the zenith act as the transmitters at frequencies of 49.97 and 49.99 MHz. The first one located in Dourbes (Southern Belgium) emits a constant power of 150 Watts while the one located in Ieper (Western Belgium) emits a constant power of 50 Watts. The receiving network consists of about 20 stations hosted mainly by radio amateurs. Two stations have crossed-Yagi antennas measuring horizontal and vertical polarizations of the waves reflected off meteor trails. This will enable a detailed analysis of the meteor power profiles from which physical parameters of the meteoroids can be obtained. An interferometer consisting of 5 Yagi-antennas will be installed at the site of Humain in order to determine the angular detection of one reflection point, allowing us to determine meteoroid trajectories. We describe this new meteor observing facility and present the goals we expect to achieve with the network.

Keywords radio meteors · forward scattering

1 Introduction

The Earth's atmosphere is constantly hit by thousands of meteoroids with sizes ranging from submillimeters to several meters. Their estimated cumulative mass is in the range [40-100] tons per day. They play a crucial role in a number of astronomical and aeronomical studies and, given their intercept velocities in excess of 11 km/s, they pose a significant threat to spacecraft. Traditionally, they have been detected by visual means or with radars during their interaction with the atmosphere. Here we propose to study the meteoroid population with the BRAMS (Belgian RAdio Meteor Stations) network, a set of radio receiving stations using forward scattering techniques and two dedicated beacons as transmitters. The BRAMS network will be finished by the end of 2010. Its current state is described in section 2 while section 3 is devoted to the objectives of the project. In the conclusion, the advantages of a forward scattering system over traditional radar systems will be shortly discussed.

2 The BRAMS Network

In 2009 the Belgian Institute for Space Aeronomy (BISA) initiated the development of BRAMS, a Belgian network of radio receiving stations using forward scattering techniques to detect meteors. This

H. Lamy (✉) • S. Ranvier • J. De Keyser • S. Calders • E. Gamby
Belgian Institute for Space Aeronomy, Avenue Circulaire 3, 1180 Brussels, Belgium. E-mail: herve.lamy@aeronomie.be

C. Verbeeck
Royal Observatory of Belgium, Avenue Circulaire 3, 1180 Brussels, Belgium

project is carried out in collaboration with about 20 Belgian radio amateurs or groups of amateur astronomers which will host several stations throughout the country (see black squares on Figure 1).

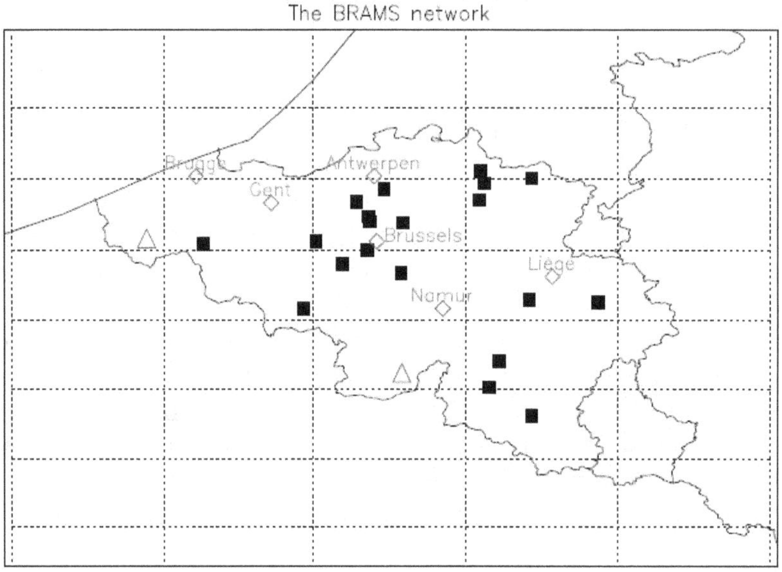

Figure 1. Geographical distribution of the stations of the BRAMS network. The black squares represent the receiving stations while the two triangles represent the two beacons (Ieper in the West and Dourbes in the South-East). The biggest cities of Belgium, represented by rhombs, have been added to facilitate identification of the stations.

The transmitters are two dedicated beacons located in Ieper and Dourbes (the two triangles on Figure 1). They emit a pure sinusoidal wave with a constant power of respectively 50 and 150 watts, at frequencies respectively of 49.99 and 49.97 MHz. These frequencies are protected avoiding ambiguity about the origin of detected meteor echoes. The beacon in Ieper is a crossed 2-element Yagi antenna which has been constructed in 2008 by radio amateurs from the VVS (Vereniging Voor Sterrenkunde - Flemish association of amateur astronomers). It emits toward the zenith with a HPBW (Half-Power Beam Width) of ~100° (see Figure 2) which covers a large surface of the sky at altitudes between 90 and 120 km. The theoretical total gain of this antenna is 5.9 dBi. The beacon in Dourbes is under construction and will be active in September 2010. It will be similar to the one in Ieper but will use a 8m× 8m grid as the reflector to increase the power emitted upward (HPBW ~64° -see Figure 2). With this grid, the total gain is increased by 3.6 dB as compared to the Ieper beacon. The stability of the frequency of the Dourbes beacon is secured with a GPS disciplinated OCXO (Oven-Controlled Crystal Oscillator) 10 MHz reference.

A typical receiving station is made of a 3-element Yagi antenna linked to a ICOM ICR75 receiver by a coaxial cable. The received signal is sampled and stored on a local PC. With 20 receiving stations spread over Belgium, we will increase both the number of meteor detections and the chances of having the same meteor detected simultaneously by several stations. This last point is essential for the determination of meteoroid trajectories since each detection gives information only about one point of the trajectory. This requires a very good synchronization of the stations. The stations will be synchronized by using the NTP (Network Time Protocol) application and a local GPS as reference clock. Accuracy of 1 ms is expected. Each station will generate approximately 1 GB of data per day. First these data will be temporarily stored on local hard disks then later archived with the IT facilities of

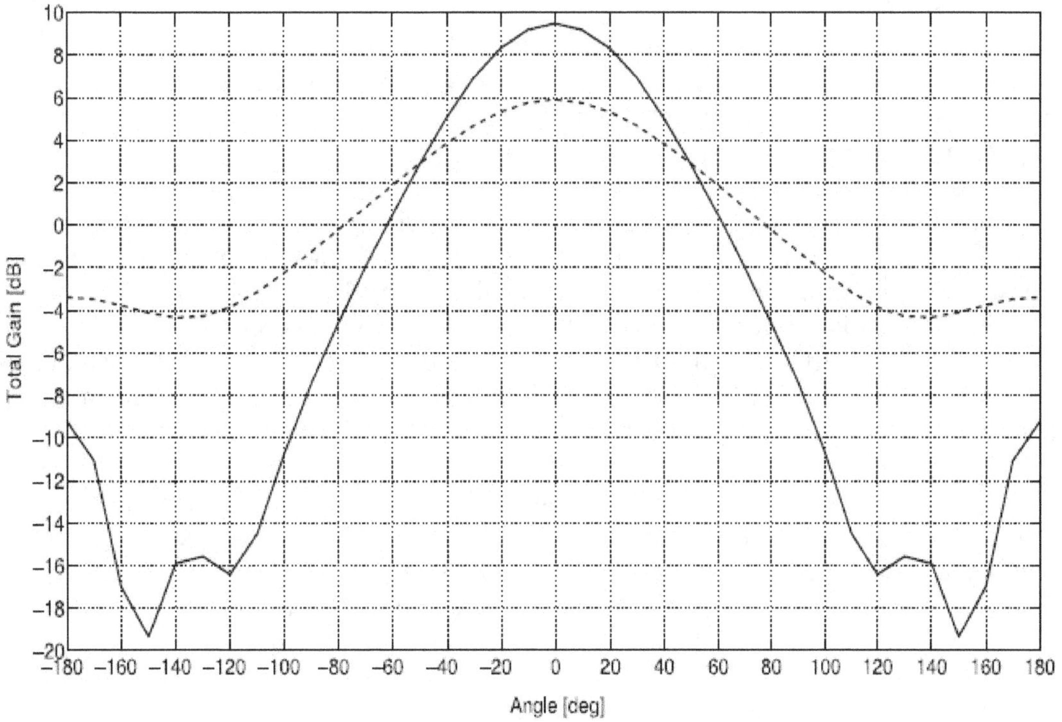

Figure 2. Theoretical radiation pattern (in dB) of the two beacon antennas (circularly polarized). Solid line: Dourbes beacon (with a 8m × 8m grid). Dashed line: Ieper beacon.

BISA, where all data will be analyzed and become available to the scientific community. In these data there will be a large number of meteor echoes but also a large number of "spurious" echoes such as reflection of radio waves on planes, sporadic Es, thunderstorms, etc. as well as local broad-band interferences. A software is currently developed to automatically select meteor echoes only.

The station in BISA (located in Uccle in the South of Brussels) is also equipped with an additional antenna, a crossed 3-element Yagi, connected to a low noise amplifier (MITEQ AU-1138) and a software defined radio receiver (USRP2). With this setup, we will measure both polarizations (horizontal and vertical) of the meteor echoes and combine them to obtain the total power reflected by the meteor. This will allow us to perform quantitative analyses of the signal in order to retrieve physical parameters of the meteoroids.

An interferometer made of 5 Yagi antennas will be installed in the radio astronomical site of Humain in order to determine the angular direction of one reflection point with an accuracy ≤ 1-2°. Humain is located at approximately 60 km north-east of Dourbes. We will use the 5 antenna configuration of [8] consisting of two orthogonal linear arrays of 3 antennas with the central one common to both arrays. The central antenna will be a crossed 3-element Yagi antenna used in order to obtain the total power of the incoming waves like in the BISA station. To get information on the angular direction of the reflection point, the phase of the signals from the 5 antennas are compared. For this purpose, the analog-to digital converters of the five USRP2 receivers are synchronized. In addition, the frequency stability of the receivers is ensured by a 10 MHz reference similar to the one used for the beacon in Dourbes.

3 Objectives of the Project

By using the BRAMS network, we plan to reach the objectives described below. They are separated according to whether they can be achieved with a single station, with several stations or with several stations and the interferometer.

1. By using data from a single receiving station, the activity profiles of meteor showers (number of echoes versus time) can be obtained by subtracting sporadic meteor echoes from all echo counts. These activity profiles must be corrected for the observability function of the station, which describes its sensitivity for detecting meteoroid echoes from a specific direction in the sky ([7], [15], [13]). The activity profiles of the main meteor showers will be studied and compared to results in the literature. Mass indexes and flux densities will be calculated for meteor showers and sporadic background using the method developed by [1] and [11], [12].

2. Simultaneous detection of a meteor by several stations allows in principle to retrieve some information about the meteoroid trajectory. We will test a method proposed by [9] exploiting the fact that a meteor trail is tangential to a family of ellipsoids with the receivers as one focal point and the transmitter as the other one (specularity condition). If the meteor is also detected by the interferometric station in Humain, the problem is simplified since then we know the angular direction of the reflection point for this particular geometry. The height of this reflection point can be obtained for shower meteors but is still unknown for sporadic meteors. However, for underdense meteors, it can be estimated from the exponential decay of the power profile (with the use of a good atmospheric model such as MSISE-00) and for overdense meteors, it can be estimated with the method proposed by [5]. With the 3-D position of one specular point and with detections of the meteor by other stations, we can retrieve the trajectory of the meteoroid even for sporadic meteors ([16]). Eventually the goal is to produce a map in Sun-centered ecliptical coordinates of the distribution of meteor radiants and to compare it to similar ones obtained by optical or radar means (e.g. [4]).

3. For a given meteor, when the geometry is fully resolved, the study of the power profiles give access to several important physical parameters if the technical characteristics of the receiving station (antenna gains, polarization of the reflected wave, calibration of the acquisition card, etc.) are perfectly known [16]. These characteristics will only be known for the stations in BISA and in Humain. For the others, we will only have reasonable estimates. The electron line density at the reflection point can be computed from the maximum of the power profile. If the electron line density is obtained in several points of the meteoroid trail (from multi-station observations), the initial mass of the meteoroid can be estimated with models of meteoroid ablation in the Earth's atmosphere (e.g. [3]). The ambipolar diffusion coefficient can be obtained from the exponential decay of underdense meteors and yields information on the meteor's height. The speed of the meteoroid can be determined by several methods: a) from Fresnel oscillations if these are present and if the signal-to-noise ratio of the data is large enough, b) from initial rise times of the power profiles if the meteor is observed by several stations (requiring a good synchronization from the various stations) and c) from the Doppler effect for head echoes (see below).

4. Head echoes are associated with the ionized region in front of the meteoroid. Therefore, these echoes in spectrograms show a large Doppler effect due to the high velocity component of the

object along the line-of-sight. If such an echo is detected by at least 3 stations, we can in principle retrieve the total velocity of the meteoroid [10], [14]. In combination with an ablation model, the meteoroid mass and density can be estimated. Comparison with the results on head echoes obtained with High-Power Large-Aperture radars will be considered.

5. Once we have the radiant of individual meteors and also an estimate of the relative intercept velocity of the incoming particle, the next step is to calculate the orbital parameters of the detected meteoroids. A correction must be applied for the acceleration produced by the gravitational focusing of Earth as well as for the Earth's heliocentric velocity. Knowing the orbital parameters contributes to a better understanding of the distribution and evolution of material in the Solar System.

6. Another advantage of having stations with dual polarized antennas is the ability to investigate the polarization of the received waves. Since most of receiving stations can detect only one polarization, it is very important to determine the depolarization coefficient of the reflections on the meteor trail. The rotation of the polarization plane of the incoming wave can be due to ionospheric Faraday rotation or as a result of the scattering from meteor trails. We will investigate whether the depolarization depends on the type, the trajectory, or the size of the meteor trail. To our knowledge, very few similar measurements have been done with a forward-scatter system (see however [2]).

4 Conclusion

The BRAMS network will be a very useful tool to better characterize the distribution of meteoroids in the Solar System. Most researchers employ a backscatter set-up rather than a forward-scatter system, as the latter has a much more complicated geometry. However, backscatter systems suffer from the echo ceiling selection effect, which limits their views on faint and fast meteors. Forward-scatter setups are much less vulnerable to this selection effect, hence yielding a less biased meteoroid population. To our knowledge, there are only two other forward-scatter systems run by professional astronomers: the Bologna-Lecce-Modra system [6] and the HRO system at the Kochi University in Japan [17]. The former has only 2 receiving stations and no interferometric capabilities while the latter has only 6 stations. BRAMS will feature at least 20 stations that will allow improved multi-station analysis and will also have the additional advantage of an interferometric system.

Acknowledgements

BRAMS is a project of the Belgian Institute for Space Aeronomy which is funded by the Belgian Solar-Terrestrial Center of Excellence (STCE -http://www.stce.be). This project is carried out in collaboration with many radio amateurs from Belgium, in particular with several members from the VVS (Vereniging Voor Sterrenkunde). We would like to thank them for their participation in this project.

References

1. Belkovich O. et al., Processing of radar observations, Proceedings of the Radio Meteor School 2005, Oostmalle, Eds. C. Verbeeck and J.-M. Wislez, (2006).
2. Billam E.R. & Browne I.C., Characteristics of Radio Echoes from Meteor Trails IV: Polarization Effects, Proceedings of the Physical Society, section B, Volume 69, Number 1, (1955).
3. Campbell-Brown M.D., Koschny D., Model of the ablation of faint meteors, A&A, 418, 751-758, (2004).
4. Campbell-Brown M.D., High Resolution radiant distribution and orbits of sporadic radar meteoroids, Icarus 196, 144-163, (2008).
5. Carbognani A., De Meyere M., Foschini L., Steyaert C., On the meteor height from forward scatter radio observations, A&A,361, 293-297, (2000).
6. Cevolani G. et al, Baseline effect on the forward scatter radar reflections from meteor trains, Il Nuovo Cimento C, 19, 447-450, (1996)
7. Hines C.O., Diurnal variations in the number of shower meteors detected by the forward-scattering of radio waves -Part III. Ellipsoidal theory, Can. J. Phys., 36, 117-126 (1958).
8. Jones J., Webster A.R., Hocking W.K., An improved interferometer design for use with meteor radars, Radio Sci., 33, 55-65 (1998).
9. Nedeljkovic S., Meteor forward scattering at multiple frequencies, Proceedings of the Radio Meteor School 2005, Oostmalle, Eds. C. Verbeeck and J.-M. Wislez, (2006).
10. Richardson J. & Kuneth W., Revisiting the radio Doppler effect from forward-scatter meteor head echoes, WGN, Journal of the International Meteor Organization, 26, 117-130, (1998).
11. Ryabova G.O., Calculation of the incident flux density of meteors by numerical integration I, WGN, Journal of the International Meteor Organization, 36, 120-123, (2008).
12. Ryabova G.O., Calculation of the incident flux density of meteors by numerical integration II, WGN, Journal of the International Meteor Organization, 37, 63-67, (2009).
13. Steyaert C., Brower J., Verbelen F., A numerical method to aid in the combined determination of stream activity and Observability Function, WGN, Journal of the International Meteor Organization, 34, 87-93, (2006).
14. Steyaert C., Verbelen F. et al, Meteor Trajectory from Multiple Station Head Echo Doppler Observations, submitted to WGN.
15. Verbeeck C., Calculating the sensitivity of a forward scatter setup for underdense shower meteors, Proceedings of the International Meteor Conference 1996 (Eds. A. Knöfel and P. Roggemans), 122-132 (1997).
16. Wislez J-M., Meteor astronomy using a forward scatter set-up, Proceedings of the Radio Meteor School 2005, Oostmalle, Eds. C. Verbeeck and J.-M. Wislez, (2006).
17. Yamamoto M.-Y. et al, Development of HRO interferometer at Kochi University of Technology, Proceedings of the International Meteor Conference 2006, Eds. F. Bettonvil & J. Kac, 117-125, (2007).

The New Meteor Radar at Penn State: Design and First Observations

J. Urbina • R. Seal • L. Dyrud

Abstract In an effort to provide new and improved meteor radar sensing capabilities, Penn State has been developing advanced instruments and technologies for future meteor radars, with primary objectives of making such instruments more capable and more cost effective in order to study the basic properties of the global meteor flux, such as average mass, velocity, and chemical composition. Using low-cost field programmable gate arrays (FPGAs), combined with open source software tools, we describe a design methodology enabling one to develop state-of-the art radar instrumentation, by developing a generalized instrumentation core that can be customized using specialized output stage hardware. Furthermore, using object-oriented programming (OOP) techniques and open-source tools, we illustrate a technique to provide a cost-effective, generalized software framework to uniquely define an instrument's functionality through a customizable interface, implemented by the designer. The new instrument is intended to provide instantaneous profiles of atmospheric parameters and climatology on a daily basis throughout the year. An overview of the instrument design concepts and some of the emerging technologies developed for this meteor radar are presented.

Keywords meteor radar · FPGAs · software radar · open source

1 Introduction

Meteoroids impact and disintegrate in the Earth's atmosphere daily. Current estimates for this global meteor flux vary from 2,000-200,000 tons per year, and estimates for the average velocity range between 10 km/s and 70 km/s [Taylor, 1995; Ceplecha et al al., 1998; Janches et al., 2000b; Cziczo et al., 2001; Mathews et al., 2001]. The understanding of the properties of the meteor flux is important for several fields of study which range from solar system evolution to imaging of gravity waves in the mesosphere. For example, meteoric metals are one of the sources of metal and ion layers in the mesosphere/lower thermosphere (MLT) region. They are also the source of condensation nuclei which is needed for the formation of noctilucent clouds (NLC) [Kelly and Gelinas, 2000; Smith et al., 2000; Liu et al., 2002; Rapp et al., 2003]. Yet, the basic properties of this global meteor flux, such as average mass, velocity, and chemical composition remain poorly understood [Mathews et al., 2001].

It is still unknown how the changes in the meteor flux will influence these phenomena, because current modeling efforts of the physics and chemistry of meteor atmospheric entry and ablation require better observational constraints [McNeil et al., 2002; Pellinen-Wannberg et al., 2004; Plane, 2004]. We

J. Urbina (✉)
315 EE East, University Park, PA 16802. Phone: 814-863-5326; Fax: 814-863-8457; E-mail: jvu1@psu.edu

R. Seal
University Park, PA 16802

L. Dyrud
Applied Physics Laboratory, John Hopkins University, Columbia, MD, 20723

believe much of the mystery surrounding the basic parameters of the meteor input exists for two reasons: 1) The unknown sampling biases of different meteor observation techniques, and 2) The lack of continuous and routine measurements of radar meteors using advanced techniques. In an effort to provide new and improved meteor radar sensing capabilities, Penn State has been developing advanced instruments and technologies for future meteor radars, with primary objectives of making such instruments more capable and more cost effective in order to study the basic properties of the global meteor flux. With the rapid emergence of new standards and protocols in wireless communication, many functions of traditional radio receivers are being implemented in software [Mitola, 2000; Reed, 2002]. These new radio receivers are called software radios since their implementation relies heavily on digital signal processing techniques and require fewer radio frequency components than classic analog radios.

We describe in the this paper the current implementation of an open source VHF software radar system as a first step towards developing a new generation of radar systems for meteor and aeronomical science. In section 2, we describe the analysis and design of the system. Section 3 is devoted to a discussion of the software architecture and system configuration. We present first meteor observations in section 4. Finally, in section 5, a summary of the paper is presented.

2 System Analysis and Design

Precise definition of the problem domain is the first stage of design known as requirements analysis [Fowler, 2004]. This section discusses the techniques used to analyze, model, and implement the design of the data acquisition presented in this paper. Detailed discussion of both hardware and software are combined to better communicate their interdependence.

2.1 Requirements Analysis

Definition of the system's capabilities and features are best defined by users (domain experts) of the system. A preliminary list of requirements were created as a first step in the design process:

1. System users are primarily scientists.
2. Users need the ability to configure the system to meet their own specifications.
3. System configuration should be stored for later re-use.
4. Multiple configurations, cycled at predetermined intervals, are sometimes necessary for a single experiment.
5. The system will use the Linux Operating System.
6. Some experiments require large bandwidths and high storage rates.
7. Minimal real-time processing and plotting tools are necessary to ensure proper setup and equipment function.
8. Data headers are needed for data storage and retrieval. A standard format will be required.
9. Analysis software will be required for post processing.
10. Software should be able to accommodate newer hardware revisions with minimal effort.
11. Documentation is a critical component.

This list served as the framework for the design and each item was categorized according to a function. Primary tasks were identified and further refinement produced smaller, well-defined subtasks.

From this requirement analysis, 5 primary functions and 4 supporting subfunctions were defined. These tasks, having well-defined boundaries, partitioned the system into 5 primary programs of operation: 1) System Configuration Program, 2) Data Collection Program, 3) Real-Time plotting Program, 4) Post Processing Software, and 5) Data Formatting Routines. Next, the process of hardware selection followed. Hardware was categorized as follows: 1) Wide-Band digital receiver, 2) High speed general purpose computer, 3) High speed, large capacity data storage, 4) Radar pulse controller, and 5) Antenna Configuration and Control System. Due to space limitation and relevance of the acquisition system, only the wide-band receiver implementation is described below. For a complete discussion of the integration of these four components, please see [Seal, 2010].

2.2 Software Radar System

The software-defined radio for radar systems takes advantage of existing open source radio software created by the GNU Software Radio for AM, FM, and HDTV signal detection [http://www.gnu.org/software/gnuradio/]. The system is built around a PC with a 2.6 GHz AMD Phenom X4 Quad-core processor, 4 GB DDR2 RAM, with 16 1 TB SATA hard drives, and Gentoo Linux operating system. Commercially available PC boards will be used for radar controller and digitization/processing purposes. The functional diagram of the transmitting and receiving modules of the VHF radar system is shown in Figure 1.

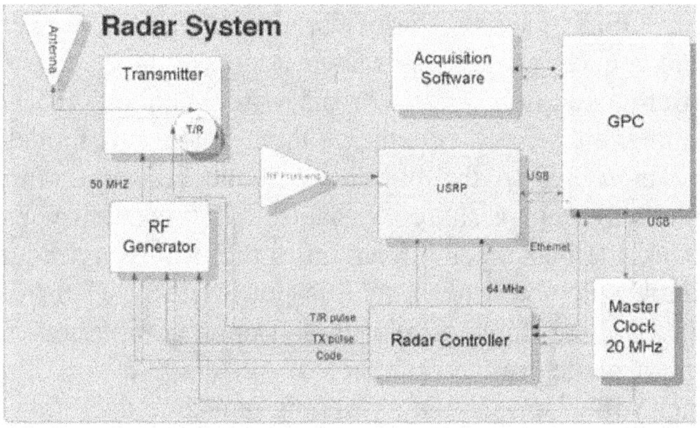

Figure 1. Functional diagram of the transmitting and receiving module of the 50 MHz Defined Radar for Meteor and Aeronomical Science.

The received signal from the antenna is passed through an analog band-pass filter tuned to the desired operating frequency. The next stage is a low-noise amplifier with programmable gain that boosts the signal level further. The output of this amplifier is passed through a protection circuit before it is sent to the digital receiver units where the carrier signal is sampled at 64 MHz (50 MHz carrier signal is translated to 14 MHz into the first Nyquist zone of 32 MHz) and then is digitally down converted and decimated in software to produce quadrature and in-phase baseband signals. The dynamic range of the system is about 90 dB. To ensure coherence of the transmitter/receiver, a 10 MHz oven controlled local oscillator is used to produce the clocks required by different parts of the system including the radio frequency gated signal for the transmitter. The desired frequency of operation is selected by software and set to 50 MHz (the system can be tuned to any value between 1 and 100 MHz with adequate RF front-end circuitry). Each A/D inside the digital receiver card can operate at a maximum speed of 80 Msamp/s per unit channel. With one digital receiver board the system will support a total of 4-complex channels to serve a 4-receiver interferometric radar. If more channels are needed, the system can easily be extended to 8, 16, 32, 64, or 128 complex channels with additional receiver boards.

The system operates as follows: data acquisition software is used to load and initiate the frequency synthesizer board (that provides all the required clock for the system), load the radar controller card, configure the digital receiver boards, initiate data collection, and route the digital samples to a disk file as they become available. The radar controller will provide four control signals: a sample start trigger to the receiver board, blanking, T/R switching, and RF pulse. Additional control pulses, e.g., coding, etc., can also be provided by the counter/timer card if needed. The frequency synthesizer card will have three (more can be provided if needed) additional frequency outputs, which may be useful in frequency domain interferometry measurements. Four 5kW (20kW total peak power) transmitters with two pairs of T/R switches are used to excite the interferometric antenna. The transmitter has about 1 MHz bandwidth and a duty cycle of 10%.

Radar data collection begins when a trigger signal is applied to the external gate of the universal software radio peripheral (USRP). Next, the user application requests data from the USRP, via the driver's interface. When data is available, the user application selects a segment of data and copies it into a secondary, user allocated buffer system. This secondary system allows data sharing among multiple processes by utilizing the POSIX shared memory library and the tmpfs [Robbins, Online] file system. Tmpfs transparently allows large regions of PC RAM to be used as a standard storage device. This filesystem is, by default, dynamically resizable through the use of swap space. For high speed operation, a fixed size tmpfs is required; preventing interaction with swap space which drastically degrades performance. The POSIX shared memory library uses the tmpfs filesystem to allocate regions of memory included in the requesting process's own address space; providing data sharing among processes. Efficient buffer operation in a read/write system is accomplished using the producer/consumer (P/C) threading model [Binstock, Online]. The P/C model requires two threads: the producer thread handles data writes to the buffers, and the consumer thread manages data reads. Synchronization is controlled through a shared variable that tracks dirty buffers (buffers containing pending data). In this model, the consumer thread starts the sequence, requesting data from the first buffer. If data is not available, the consumer thread is put to sleep. Then, the producer begins filling buffers at a continuous rate; waking the consumer thread upon completion of a buffer. This operation continues indefinitely using a predetermined number of buffers. Non-real-time operating systems can impose unpredictable latencies [Seelam, Online]; violating real-time operation. Applying the P/C model, latencies can be masked through buffering; bypassing the need for a real-time operating system. Additionally, use of this model, combined with shared memory regions, allow for multiple levels of real-time processing to occur simultaneously. This approach preserves storage device bandwidth which is critical for high speed, real-time data writing. This provides a major advantage over older systems in which the storage device spent a large amount time seeking to satisfy system reads and writes; further limiting bandwidth.

3 First Radar Observations

We have conducted first radar observations with the software radar system in conjunction with four 5-element Yagi antennas for transmission and a 50-MHz transmitter with peak power of 20 kW. On reception, we used five 5-element Yagi antennas in a cross configuration. The experiment was carried out with an inter-pulse period (IPP) of 1 ms and pulse width of 1 km range resolution. After a quick analysis of the meteor trails of the received data, the first results of the new system look promising.

Figure 2 shows In-Phase and Quadrature raw voltages of an underdense meteor trail. Clearly present in the signal is the attenuation or classical exponential decay of these type of reflections.

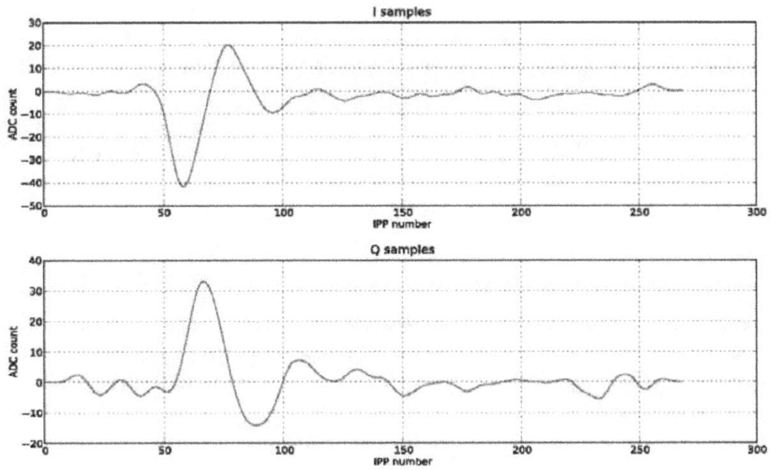

Figure 2. In-Phase and Quadrature raw voltages an underdense meteor trail detected on May 5, 2010.

4 Summary

We have presented an overview of the implementation of a new meteor system based on open source hardware and software tools. This system will be used by Communication and Space Sciences Laboratory at Penn State University to conduct meteor research. The acquisition system enables the operation of the radar with bandwidths approaching 10 MHz and data throughput greater than 30 MB/s. The system is flexible and is easily reconfigurable, allowing the user to implement newer ionospheric experiments. We will make our software radar control programs available freely through the Open Source software development web site of SourceForge at: [http://sourceforge.net].

Acknowledgements

This work is supported by the National Science Foundation under grants: ATM-0638624 and ATM-0457156 to Penn State University.

References

Binstock, A.,The producer/consumer threading model. [http://www.intel.com/cd/ids/ developer/asmo-na/eng/columns/performance/52523.htm].

Ceplecha, Z., J. Borovicka,W. G. Elford, D. O. Revelle, R. L. Hawkes, V. Porubcan, and M. Simek, Meteor phenomena and bodies, Space Science Reviews, 84, 327471, 1998.

Cziczo, D. J., D. S. Thomson, and D. M. Murphy, Ablation, flux, and atmospheric implications of meteors inferred from stratospheric aerosol, Science, 291, 17721775, 2001.

Fowler, M. UML Distilled Third Edition. Pearson Education, Inc., 2004.

Janches, D., J. D. Mathews, D. D. Meisel, and Q. H. Zhou, Micrometeor Observa¬tions Using the Arecibo 430 MHz Radar, Icarus, 145, 5363, 2000b.

Kelly, M. C., and L. J. Gelinas, Gradient drift instabililty in midlatitude sporadic E layers: localization of physical and wavenumber space, Geophysical Research Letters, 27, 457, 2000.
Liu, A. Z., W. K. Hocking, S. J. Franke, and T. Thayaparan, Comparison of Na lidar and meteor radar wind measurements at Starfire Optical Range, NM, USA, Journal of Atmospheric and Terrestrial Physics, 64, 3140, 2002.
Mathews, J. D., D. Janches, D. D. Meisel, and Q.-H. Zhou, The micrometeoroid mass flux into the upper atmosphere: Arecibo results and a comparison with prior estimates, Geophys. Res. Lett., 28, 1929, 2001.
Mitola, J. III Software Radio Architecture: Object Oriented Approaches to Wireless Systems Engineering, John Wiley and Sons Inc., 2000.
McNeil,W. J., E. Murad, and A. J. M. C. Plane, Models of Meteoric Metals in the Atmosphere, pp.265, Meteors in the Earths atmosphere. Edited by Edmond Murad and Iwan P. Williams. Publisher: Cambridge, UK: Cambridge University Press, 2002., p.265, 2002.
Pellinen-Wannberg, A., E.Murad, B. Gustavsson, U. Brandstrom, C. Enell, C. Roth,
I. P. Williams, and A. Steen, Optical observations of water in Leonid meteor trails, Geophys. Res. Lett., 31, 3812, 2004.
Plane, J. M. C., A time-resolved model of the mesospheric Na layer: constraints on the meteor input function, Atmospheric Chemistry and Physics, 4, 627638, 2004.
Rapp, M., F. Lubken, P. Hoffmann, R. Latteck, G. Baumgarten, and T. A. Blix, PMSE dependence on aerosol charge number density and aerosol size, Journal of Geophysical Research (Atmospheres), 108,81, 2003.
Reed, J. H., Software Radio A Modern Approach to Radio Engineering, Prentice Hall Communications Engineering and Emerging Technologies Series, 2002.
Robbins, D., Common threads: Advanced filesystem implementors guide, part 3.
[http:// www-128.ibm.com/developerworks/library/l-fs3.html].
Seal, R., J. Urbina, M. Sulzer, S. Gonzalez, N. Aponte, Design of an FPGA-based radar controller, National Radio Science Meeting, Boulder, CO, January 3-6, 2008.
Seal, R., A new generation of meteor radar systems, MS thesis, Penn State University, University Park, PA, 16802, 2010.
Seelam, S., J. S. Babu, and P. Teller, Automatic i/o scheduler selection for latency and bandwidth optimization, 4th International Conference on Parallel Architectures and Compilation Techniques, 2005. [http://pact05.ce.ucsc.edu/].
Smith, S. M., M. Mendillo, J. Baumgardner, and R. R. Clark, Mesospheric gravity wave imaging at a subauroral site: First results from Millstone Hill, Journal of Geo-physical Research, 105, 27,11927,130, 2000.
Taylor, A.D., The Harvard Radio Meteor Project meteor velocity distribution reap-praised, Icarus, 116, 205-209,1995.

Maximizing the Performance of Automated Low Cost All-sky Cameras

F. Bettonvil

Abstract Thanks to the wide spread of digital camera technology in the consumer market, a steady increase in the number of active All-sky camera has be noticed European wide. In this paper I look into the details of such All-sky systems and try to optimize the performance in terms of accuracy of the astrometry, the velocity determination and photometry. Having autonomous operation in mind, suggestions are done for the optimal low cost All-sky camera.

Keywords all-sky · meteor camera · performance

1 Introduction

Since the 1960s and 1970s automated networks of meteor cameras have been in use to collect data on fireballs and recover meteorites. Well known are the Prairie Network (United States), the Meteorite Observation and Recovery Program MORP (Canada) and the European Network (former Eastern Europe), being examples of professional projects from that time. The European Network is still in operation nowadays [Spurný 2010, Flohrer 2006] and also other networks arose both on professional level (e.g. ASGARD in Canada [Brown 2010]; DFN in Australia [Bland 2008]) as well as amateur networks like the Polish Fireball Network [Olech 2006], and many others. Building and operating fireball patrol stations have always been well in reach for amateurs. Nowadays, with digital recording methods being used everywhere, this is true even more. Much digital imaging is done with sensitive (intensified) video cameras; in this paper on the contrary I will look into DSLR cameras with fisheye lens, because of their much higher resolution (10 Mpixel and more, compared to ~600x800 for standard video techniques) and which could be purchased for just under 1000EUR. I aim in this paper at an autonomously working station, which is easy to built and easy in use. Details of the setup are given in Table 1.

With the above as baseline, the goal we try to achieve is: (1) Accurate astrometry (error in semi major axis Δa < 0.01AU) [Vaubaillon 2007], (2) Accurate velocity determination (idem), (3) Proper photometry (for mass estimates and trail density distributions, but no requirement defined).

Table 1. Evaluated hardware setup

Camera	Canon EOS 350D (6 Mpxl)
Lens	Full frame Sigma 4.5mm F/2.8 fisheye
Exposure control	Canon TC80N3 timer controller; twilight switch. No PC
Timing	GPS/DCF clock for time reference marks in star trails
Chopper	LC-TEC optical shutter (modulation freq. 10-100 Hz)

F. Bettonvil (✉)
Astronomical Institute, Utrecht University, Princetonplein 5, NL-3584 CC Utrecht, The Netherlands. E-mail: F.C.M.Bettonvil@astro.uu.nl

2 Astrometry

One camera pixel equals on average ~5' and the plate reduction error of the combination camera-lens turns out to be of the same order [Bettonvil 2006]. When assuming that the error in radiant position caused by astrometry is of the same amount, we can calculate the effect on the orbital elements, which is illustrated in Table 2 for an asteroidal fireball (being an example of the type of fireballs that is of most interest for meteorite recovery). It shows that the error in the semi major axis is just within our goal[2].

Table 2. Effect of errors (respectively in radiant position (error A) and velocity (error B)) on the orbital elements for an ι-Aquarid [Bettonvil 2006].

		Radiant			
Radiant	Observed	Geocentr.	Heliocentric	Error A	Error B
R.A. [°]	342°,959	343°,201		±0,100	-
Decl [°]	-05°,281	-07°,367		±0,100	-
Heliocn. Longitude [°]			288°,647	-	-
Heliocent. Latitude [°]			-0°,179	-	-
Velocity [km/s]	32,292	30,183	34,967	-	±0,096
		Orbital elements			
Longitude of ascending node [°]		(Ω)	322°,528	±0,339	±0,024
Inclination [°]		(i)	0°,322	±0,161	±0.019
Argument of perihelion [°]		(ω)	131°,029	±0,432	±0,057
Semi major axis [AU]		(a)	1,6758	±0,0095	±0,0135
Perihelion distance [AU]		(q)	0,2415	±0,0012	±0,0011
Aphelion distance [AU]		(Q)	3,1102	±0,0178	±0,0282
Eccentricity [AU]		(e)	0,8559	±0.0001	±0,0018

For automated operation a window cover is required however, which either could be a hemispherical acrylic dome or watchmaker's glass (Figure 1 left). Although the first seems preferred due to the (insensitive) perpendicular penetration of the light beam, performance was measured too be bad due to local irregularities (Figure 1 right). A watchmaker's glass appears to be fine.

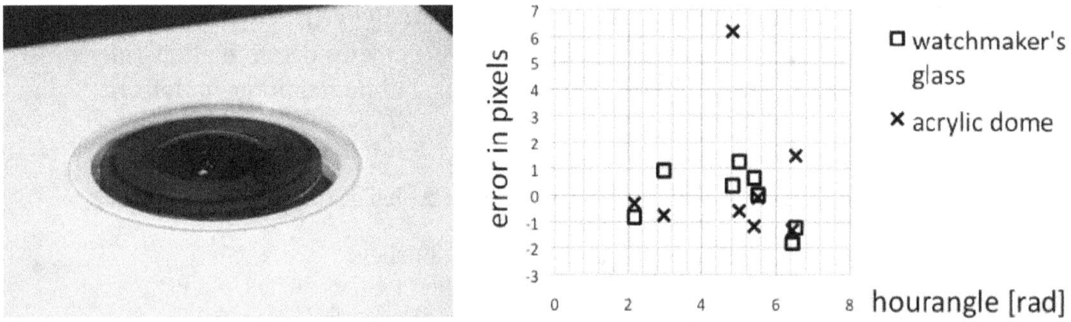

Figure 1. (left) Watchmaker's glass; (right) measured difference in star position between an acrylic dome and watchmaker's glass for several stars in azimuthal direction. The measurement resolution was 1 camera pixel. The single high outlier for the acrylic dome is real.

[2]Also the other orbital elements are affected by a change in radiant position, and even Ω, which is due to the ecliptical origin of the investigated meteor.

3 Velocity Determination

For measurement of the velocity, I chose not to use a conventional rotating chopper in front of the lens, but instead a liquid crystal shutter (LC-TEC 2010), mounted between lens and camera [Bettonvil 2007b, 2010]. Advantage is that the chopper frequency can be more accurate (crystal operation, no rotating parts, no wind influence). Optical ray tracing and measurements showed that no significant aberrations occur [Bettonvil 2010].

Instead of direct measurement of the chopper breaks, frequency analysis with FFT is used, which, after doing simulations and tests, gave velocity errors of 0.2-0.5%, 3-5 times better than conventional choppers [Bettonvil 2008]. A 0.3% velocity error (Table 2, error B) is required to stay within our requirements for the error in semi major axis ($\Delta a < 0.01$AU).

4 Photometry

DSLR camera's, when read out in RAW mode, allow up to 10-15 bit dynamical range, much more than video camera's (7-8 bit). Nevertheless, because in photography exposures are being integrated over a (much) longer time, the noise and dark level easily go up to unacceptable levels, reducing the effective dynamical range. Dark level and noise is mitigated by: (a) subtraction of 2 successive images; (b) the use of low camera sensitivity (i.e. ISO setting); (c) short exposure times. Sensor cooling with Peltier elements is a very effective method [Bettonvil 2007a], but disregarded here, due to its complexity.

Saturation of fireballs is also to be avoided, which is the second reason for choosing low sensitivity. Figure 2 (left) shows the relation between ISO setting and brightness for fireballs and stars. It confirms that ISO 100 (is lowest value, resulting in noise counts of ~100) is preferred for allocating sufficient dynamical range for bright fireballs ($\sim m_v = -12$). The sensitivity for stars is of course low then (+2), which affects the number of reference stars for astrometry. It is solved by combining multiple exposures. Figure 2 (right) shows the measured linearity of the camera as derived from stars. It shows non-linearity in the order of 0.1mag.

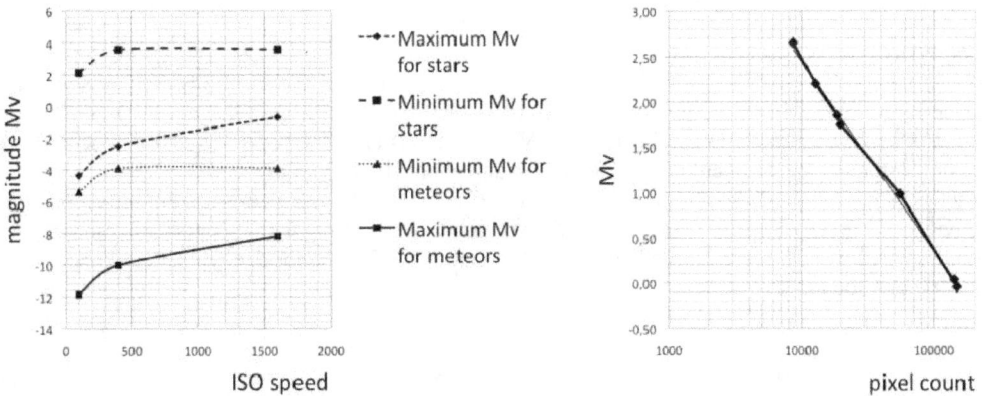

Figure 2. (left) Faintest detectable stars and meteors, as derived from the noise level, and brightest detectable stars and meteors before saturation occurs, as function of ISO speed; (right) linearity of the camera response for stars.

For cometary fireballs [Jenniskens 2006], with $v = 25$ km/s and at 100 km distance, we can write for the relation between mass M and brightness m_v:

$$\frac{dM}{dm_v^{abs}} = -0.92 \cdot 10^{0.933-0.4\,m_v} \qquad (1)$$

A photometry error of 0.1 mag results then in a ~10% error in the estimate of the mass of the meteoroid, which seems acceptable.

5 Conclusions

It seems that useful astrometry, velocity determination and photometry, of bright fireballs can be done with a DSLR camera with full frame fisheye lens. Part of the camera is an optical Liquid Crystal shutter. Operation with very low sensitivity (ISO100) as well as short exposure times is recommended to maximize dynamic range and avoid saturation.

References

Bettonvil F (2010) Digital All-sky cameras V: Liquid Crystal Optical Shutters. In: Proceedings of the International Meteor Conference 2009. International Meteor Organization, Hove. in print

Bettonvil F (2008) Determination of the velocity of meteors based on sinodial modulation and frequency analysis. Earth, Moon and Planets 102, pp 205-208.

Bettonvil F (2007b) Digital All-sky cameras III: A new method for velocity determination. In: Bettonvil F, Kac J (eds) Proceedings of the International Meteor Conference 2006. International Meteor Organization, Hove, pp 138-141

Bettonvil F (2007a) Digital All-sky cameras II: Effect of Peltier cooling on fixed pattern noise. In: Bettonvil F, Kac J (eds) Proceedings of the International Meteor Conference 2006. International Meteor Organization, Hove, pp 134-137

Bettonvil F (2006) Orbit Calculation of the August 15, 2002 Fireball over the Netherlands. In: Bastiaens L, Verbert J, Wislez J, Verbeeck C (eds) Proceedings of the International Meteor Conference 2005. International Meteor Organization, Hove, pp 171-178

Brown P, Weryk R, Kohut S, Edwards W, Krzeminski Z (2010). WGN. pp. 25-30.

Bland P, Spurný P, Shrbený L, Borovička J, Bevan A, Towner M, McClafferty T, Vaughan D, Deacon G (2008). Asteroids, Comets, Meteors 2008. LPI Contribution No. 1405 paper id. 8246

Flohrer J, Oberst J, Heinlein D, Grau T, Spurný P (2006) European Planetary Science Congress 2006. Berlin. pp 518

LC-TEC (2010) http://www.lctecdisplays.com, LC-TEC Displays AB, Tunvgen 281, 781 73 Borlänge, Sweden

Olech A, Zoladek P, Wisniewski M, Krasnowski M, Kwinta M, Fajfer T, Fietkiewicz K, Dorosz D, Kowalski L, Olejnik J, Mularczyk K, Zloczewski K. (2006) In: Bastiaens L, Verbert J, Wislez J, Verbeeck C (eds) Proceedings of the International Meteor Conference 2005. International Meteor Organization, Hove, pp 53-62

Jenniskens P (2006) Meteor Showers and Their Parent Comets. Cambridge University Press, Cambridge. pp 802

Spurný P, Borovička J, Shrbený L (2007) Near Earth Objects, our Celestial Neighbors: Opportunity and Risk. In: Valsecchi G, Vokrouhlický D, Milani A (eds) Proceedings of IAU Symposium 236. Cambridge University Press, pp 121-130

Vaubaillon J (2007) priv. comm.

www.ingramcontent.com/pod-product-compliance
Lightning Source LLC
Chambersburg PA
CBHW081716170526
45167CB00009B/3603
* 9 7 8 1 5 0 3 2 3 5 3 6 6 *